International Conference of Young Scientists and Students "Topical Problems of Mechanical Engineering" (ToPME 2020)

IOP Conference Series: Materials Science and Engineering
Volume 1129

Moscow, Russia
2 - 4 December 2020

ISBN: 978-1-7138-4612-3
ISSN: 1757-8981

Printed from e-media with permission by:

Curran Associates, Inc.
57 Morehouse Lane
Red Hook, NY 12571

Some format issues inherent in the e-media version may also appear in this print version.

Institute of Physics
Dirac House, Temple Back
Bristol BS1 6BE UK

Phone: 44 1 17 929 7481
Fax: 44 1 17 920 0979

techtracking@iop.org

Additional copies of this publication are available from:

Curran Associates, Inc.
57 Morehouse Lane
Red Hook, NY 12571 USA
Phone: 845-758-0400
Fax: 845-758-2633
Email: curran@proceedings.com
Web: www.proceedings.com

TABLE OF CONTENTS

IOP Conf. Series: Materials Science and Engineering 1129 (2021) 011001 doi:10.1088/1757-899X/1129/1/011001

32th International Conference of Young Scientists and Students "Topical Problems of Mechanical Engineering 2020" (TopME 2020), 2-4 December 2020, Moscow, Russia.

We are pleased to introduce the Proceedings of the 32th International Conference of Young Scientists and Students "Topical Problems of Mechanical Engineering 2020" (TopME 2020). The Conference was held from December 2th to December 4th, 2020 in Moscow, Russia. It was organized by the Blagonravov Mechanical Engineering Research Institute of the Russian Academy of Sciences (IMASH RAN). The scientific sections of the Conference covered a wide range of topics on materials science, strength, survivability and safety of machines, tribology, vibration processes and systems, machines mechanics and biomechanics.

In 2020, the conference was held virtually due to restrictions related to COVID19 using the ZOOM platform. The organizers were located at the Blagonravov Mechanical Engineering Research Institute of the Russian Academy of Sciences, Moscow, Russia.

The meetings were held in the format of a speech with an oral presentation. For each participant was allotted 15 minutes including questions. Plenary sessions were held on the first day of the conference.

150 scientists from various cities of Russia, England and Sweden participated in the conference in 2020. The organizing committee selected 72 articles of the participants for publication in the Proceeding of conference.
The conference passed without major problems. Organizing committee connected many participants virtually. The exhibition part of the conference was impossible due to the remote presentation.

List of Conference Committee

Ganiev R. F., Honorary Conference Chairman, academician of RAN, Scientific Director of IMASH RAN, Mechanical Engineering Research Institute of the Russian Academy of Sciences, Moscow, Russia

Glazunov V. A., Chairman of Program Committee, Ph.D., Director of IMASH RAN, Mechanical Engineering Research Institute of the Russian Academy of Sciences, Moscow, Russia

Prof. A. N. Polilov, Chairman of Organizing Committee

A. Yu. Albagachiev, IMASH RAN, Moscow, Russia

V. K. Astashev, IMASH RAN, Moscow, Russia

Yu. I. Bobrovnitskii, IMASH RAN, Moscow, Russia

S. M. Kaplunov, IMASH RAN, Moscow, Russia

O. I. Kosarev, IMASH RAN, Moscow, Russia

Yu. G. Matvienko, IMASH RAN, Moscow, Russia

N. A. Makhutov, IMASH RAN, Moscow, Russia

G. Ya. Panovko, IMASH RAN, Moscow, Russia

I. A. Razumovsky, IMASH RAN, Moscow, Russia

A. N. Romanov, IMASH RAN, Moscow, Russia

V. O. Solov'ev, IMASH RAN, Moscow, Russia

I. V. Stepanyan, IMASH RAN, Moscow, Russia

A. V. Frolova, IMASH RAN, Moscow, Russia

A. N. Nikiforov, IMASH RAN, Moscow, Russia

M. V. Prozhega, IMASH RAN, Moscow, Russia

M. S. Pugachev, IMASH RAN, Moscow, Russia

E.O. Reschikov, IMASH RAN, Moscow, Russia

K. B. Salamandra, IMASH RAN, Moscow, Russia

V. I. Svirin, IMASH RAN, Moscow, Russia

A. E. Shokhin, IMASH RAN, Moscow, Russia

N. A. Tatous', IMASH RAN, Moscow, Russia

I. P. Lyan, IMASH RAN, Moscow, Russia

O.I. Kulakov, IMASH RAN, Moscow, Russia

O. Yu. Sklemina, IMASH RAN, Moscow, Russia

V. N. Samusenko, IMASH RAN, Moscow, Russia

N. G. Sharonov, VolGTU, Volgograd, Russia

Y.V. Kalinin, VolGTU, Volgograd, Russia

IOP Conf. Series: Materials Science and Engineering 1129 (2021) 011002 doi:10.1088/1757-899X/1129/1/011002

Peer review declaration

All papers published in this volume of *IOP Conference Series: Materials Science and Engineering* have been peer reviewed through processes administered by the Editors. Reviews were conducted by expert referees to the professional and scientific standards expected of a proceedings journal published by IOP Publishing.

- **Type of peer review: Open**

For Peer review of submissions was used open system. Based on the results of the speeches, the editors of the Conference selected the most interesting scientific reports and proposed to publish the article in the Proceedings. All Papers were checked for plagiarism. At the next stage, the articles were checked for scientific novelty and practical significance. Reviews were compiled for each Paper. The authors had the opportunity to make changes to the article after the Editor's comments. The Papers of the participants who made an oral report at the conference were allowed to be published.

- **Conference submission management system:**

The organizing committee used the Conference Management Toolkit (CMT) for applying submissions.

- **Number of submissions received:**

The organizing committee received 164 submissions.

- **Number of submissions sent for review:**

150 submissions were sent for review.

- **Number of submissions accepted:**

76 submissions were accepted by Editors.

- **Acceptance Rate (Number of Submissions Accepted / Number of Submissions Received X 100):**

Accepted Rate is 46.

- **Average number of reviews per paper:**

One review for paper was prepared.

- **Total number of reviewers involved:**

15 reviewers worked with applications.

- **Any additional info on review process:**

- **Contact person for queries:**
  Ph.D. Maksim Prozhega
  Laboratory for the Research of wear at boundary lubrication
  Mechanical Engineering Research Institute of the Russian Academy of Sciences
  (IMASH RAN)
  101990, 4 Maly Kharitonyevsky Pereulok, Moscow, RUSSIA
  prmaksim@gmail.com

Influence of parameters of gas-dynamic spraying on substructure of "brass" coating type

V E Arkhipov, G V Moskvitin and M S Pugachev

Mechanical Engineering Research Institute of the Russian Academy of Sciences, Moscow, 101990, Russia

GVMoskvitin@yandex.ru

Abstract. Investigations of the influence of the parameters of applying a gas-dynamic coating of copper-zinc on its substructure have been carried out. It was found that the coating has a gradient hardness and different phase composition. Increasing the temperature of spraying and coating with 3mm overlap leads to a decrease in the size of the substructures. The structure of the coating, which is gradient in phase composition and hardness, should provide high tribotechnical characteristics of the friction surface in contact.

1. Introduction

In the field of mechanical engineering, powder coatings of various application methods are widely used. The main areas of application of such coatings are friction pairs, sealing of joints, restoration of geometry, and anti-corrosion protection. One of the modern methods of forming a coating from a powder mixture is gas-dynamic spraying. It allows the application of coatings of ductile metals with a relative elongation of more than 40%, zinc, nickel, aluminum and copper on a steel supporting structure. In this case, the supporting structure does not heat up more than 140°C. The coatings have high values of adhesion strength and cohesive strength of the coating metal.

The quality of such coatings depends on the technological modes of spraying: composition and fraction of the sprayed mechanical mixture, number of passes, spraying temperature, powder consumption, distance from the spray nozzle exit to the spraying surface, etc.

One of the factors that have a significant effect on the mechanical properties of the considered coatings is their substructure.

The aim of the work was to assess the influence of the parameters of gas-dynamic spraying on the substructure of the coating of the “brass” type.

2. Equipment and research method

The coating was sprayed on the backing of 5140H steel using gas-dynamic device DIMET-404 at air temperature 270°C, 360°C and 450°C, the rate of replaceable nozzle displacement in relation to the sample surface 10 $mm \cdot s^{-1}$, the distance from the nozzle cut to the surface equal to 10 mm and nozzle displacement of 2 mm (contact ratio 94%) and 3 mm (contact ratio 55%) (table 1). To apply the coating, the authors used the mechanical mixture of copper and zinc particles and aluminum oxide (Al_2O_3) of as received grade C-01-11 with ingredients weight ratio $Cu{:}Zn{:}Al_2O_3$ = 35%:35%:30%, manufactured at Obninsk Powder Spray Center [1].

Copper-zinc (brass) coatings were applied to a 5140H steel supporting structure using (a gas-dynamic device manufactured by DIMET-404 at air flow temperatures from 270°C to 450°C in 90°C increments, at a nozzle velocity along the sample of 10 $mm \cdot s^{-1}$ and the distance from the end of the nozzle to the spray surface is 10 mm. In this case, to obtain a continuous coating of a flat surface, the nozzle moved relative to the previously deposited layer by 2 mm (with an overlap ratio of 64%) and 3 mm (with an overlap ratio of 55%) (table 1). The initial powder mixture of grade S-01-11 consisted of particles of aluminum oxide (Al_2O_3), zinc and copper with a powder mass ratio of Al_2O_3: Zn: Cu = 30%:35%:35%, produced in Obninsk [1].

Table 1. Results of the analysis of surface treatment by particle flow.

Nozzle displacement, mm	Surface exposed to particles, times						Ratio of displacement, %
	1		2		3		
	time, seconds	area, mm^2	time, seconds	area, mm^2	time, seconds	area, mm^2	
2	14	20	28	90	42	113	64
3	12	74	24	149	-	-	55

The phase composition of the coatings was investigated on a Rigaku Ultima IV multifunctional X-ray diffractometer using CuKα radiation and a parallel beam.

The analysis of diffraction patterns is carried out on the basis of the PDXL software product (Rigaku) using the PDF-2 database. The quantitative analysis and analysis of the broadening of diffraction lines in order to determine the parameters of the fine crystal structure (block size, microstrains) is carried out using the Rietveld method implemented in the PDXL software package (Rigaku).

The hardness of the coating was measured by the Vickers hardness test in accordance with GOST 2999-75 at a load of 245 mN and a holding time of 10 seconds using a SHIMADZU HMV-2 hardness tester.

3. Results of the research

The change in the technological parameters of the coating deposition based on a mixture of copper and zinc particles is accompanied by the formation of a metal layer, in which copper, zinc and two solid solutions of electronic type based on $CuZn_3$ (ε - phase) and Cu_5Zn_8 (γ - phase) are revealed by X-ray diffraction phase analysis (table 2).

An increase in the spraying temperature is accompanied by a decrease in the copper content and an increase in the content of zinc and electronic compounds.

The research results show the presence of the process of diffusion of copper into zinc, which takes place more intensively when the nozzle is displaced by 2 mm. When spraying at temperatures of 360°C and 450°C, the γ-phase is revealed in the coating structure, the mass fraction of which is 17.0% and 33.0%, respectively. The hardness of zinc does not exceed 50 HV, while the hardness of copper and electronic compounds reaches 120 HV and 180 HV [2].

The coating received does not belong to copper and zinc alloys, since X-ray diffraction analysis shows the presence of copper and zinc, which does not correspond to the composition of brass, which is obtained by melting copper and zinc [3].

This structure, which is gradient in phase composition and hardness, should provide high tribotechnical properties of the friction surface [4]. In addition, the properties of the coating surface as a whole are influenced by the state of the substructure, namely, the magnitude of microstrains and the size of blocks of coherent scattering regions (CSR) [5].

Table 2. Phase analysis of coverage.

Nozzle displacement, mm	Mass fraction of metals, phases and compounds,%			
	Cu	Zn	ε - phase	γ - phase
		T=270°C		
2	87.9±1.0	1.44±0.17	5.3±0.9	-
3	83.8±0.8	1.3±0.1	8.5±0.9	-
		T=360°C		
2	62.0±0.9	7.9±0.6	7.9±0.8	17.0±2.0
3	69.0±0.7	12.6±0.6	10.4±0.5	-
		T=450°C		
2	40.5±0.8	9.0±0.6	11.4±0.7	33.0±0.2
3	56.0±0.8	24.0±0.6	12.0±0.7	-

Measurements of the lattice parameters of copper and zinc show that a change in the technological modes of spraying does not affect the lattice period of metals. A decrease in the displacement distance of the nozzle from 3 mm to 2 mm leads to a significant increase from 0.3870 nm to 0.4303 nm in the size of one of the faces of the crystal lattice of the electronic compound of the ε-phase (table 3). This result can be interpreted as the beginning of the rearrangement of the structure of the electronic compound (ε-phase), which has a hexagonal close-packed lattice, into another crystallographic system with the formation of a new compound.

Table 3. Parameters of metal lattice, phases and compounds.

Nozzle displacement, mm	Lattice period, nm			
	Cu	Zn	ε	γ
		T=270°C		
2	0.3619	0.2667/0.4953	0.2760/0.4303	-
3	0.3617	0.2666/0.4945	0.2759/0.3870	-
		T=360°C		
2	0.3620	0.2666/0.4953	0.2757/0.430	0.8888
3	0.3615	0.2666/0.4944	0.2759/0.4505	-
		T=450°C		
2	0.3617	0.2666/0.4946	0.2754/0.4296	0.8900
3	0.3620	0.2665/0.4943	0.2753/0.4298	-

Measurements of the lattice parameters of copper, zinc and intermetallic compounds (ε and γ-phases) revealed that an increase in the deposition temperature from 270°C to 360°C with a nozzle displacement by 3 mm does not lead to changes in the lattice size for copper and zinc (table 3). In the electronic compound of the ε-phase, the change in the temperature of the air flow is accompanied by a significant increase in the size of one of the lattice faces from 0.387 nm to 0.4305 nm. The change in the parameters of the crystal lattice of the ε-phase is associated with additional enrichment with copper atoms, its rearrangement in accordance with the copper-zinc state diagram and subsequent transformation into the gamma phase.

When using a nozzle offset of 2 mm, an increase in the spraying temperature has no significant effect on the lattice parameters of all structural components (table 3).

Changing the technological parameters of spraying (L = 3 mm → 2 mm) does not significantly affect the lattice parameters of copper, zinc and corundum. In the electronic compound of the ε-phase, one of the lattice faces decreases in size and its parameters become close to those given in the technical literature. For the newly formed compound γ-phase, which has a body-centered cubic lattice, the parameters also correspond to those given in the literature (table 3) [3].

An increase in the spraying temperature from 360°C to 450°C with a nozzle displacement of 2 mm does not significantly affect the lattice parameters of all structural components (table 3). The use of a technological displacement of the nozzle relative to the previously deposited metal layer of 3 mm also does not significantly affect the lattice parameters of all structural components except for the ε-phase. For this electronic compound based on $CuZn_3$, with an increase in the deposition temperature, one of the lattice periods significantly changes its size (0.4505 nm → 0.4298 nm) and the lattice itself has parameters that are given in scientific papers (table 3).

The distance by which the nozzle is displaced at the same temperature of coating spraying, taking into account the measurement and calculation error, has practically no effect on the magnitude of microstrains of copper and electronic compound.

An increase in the deposition temperature leads to a significant decrease in the magnitude of microstrain in copper when the nozzle is displaced by 2 mm (0.187% → 0.149%) and 3 mm (0.187% → 0.149%) (table 4). An increase in the temperature of the air flow from 270°C to 360°C is accompanied, though not significant, but an increase in the energy of the particle flow, which should lead to an increase in deformation [6]. Therefore, a decrease in the amount of deformation is due to the heating of the coating with air and the return (rest) processes in the structure. As a rule, the more significant the deformation, the lower the temperature at which the processes of migration of atoms and rearrangement of boundaries take place. An increase in the deposition temperature does not significantly affect the magnitude of microdeformations of zinc and an electronic compound (ε-phase). Zinc, as a brittle metal, collapses upon impact due to the displacement of the metal volumes relative to each other along certain crystallographic planes (extraplanes), and an electronic compound is formed on its basis, which should also not affect the magnitude of microstrains.

With a decrease in the displacement of the nozzle (3 mm → 2 mm) and the use of one air flow temperature of 360°C, the value of copper microstrains increases from 0.129% to 0.149%, which is associated with the additional effect of the particle flow on the coating metals (table 4). The change in the deposition parameters does not affect the magnitude of microstrains of the remaining structural components.

Table 4. Deformation of metals, phases and compounds.

Nozzle displacement, mm	The magnitude of microdeformations (microstrains), %			
	Cu	Zn	ε	γ
	T=270°C			
2	0.187±0.04	0.10±0.11	0.11±0.08	-
3	0.189±0.011	0.00±0.016	0.10±0.04	-
	T=360°C			
2	0.149±0.04	0.03±0.12	0.13±0.04	0.307±0.016
3	0.129±0.003	0.00±0.12	0.18±0.03	-
	T=450°C			
2	0.119±0.010	0.004±0.008	0.21±0.03	0.17±0.02
3	0.13±0.002	0.005±0.10	0.20±0.03	-

An increase in the spraying temperature with a displacement of the nozzle by 3 mm does not affect the magnitude of microstrains in all structural components of the coating based on a mixture of copper, zinc and corundum particles. A decrease in the displacement distance of the nozzle to 2 mm is accompanied by a significant decrease in the magnitude of microstrains in copper (0.149% → 0.119%) and an electronic compound based on Cu_5Zn_8 (γ-phase) from 0.307% to 0.17%.

With a decrease in the displacement of the nozzle (3mm → 2mm) and the use of one air flow temperature of 450°C, no significant changes in the magnitude of microdeformations are observed.

Calculation of the size of the substructure of metals and electronic compounds shows a significant effect of the technological parameters of spraying on the value of the CSR of the structural components of the coating based on copper, zinc and corundum. An increase in the mixing of the nozzle from 2 mm to 3 mm leads to a significant refinement of the substructure from > 200 nm to 21.4 nm for copper and from >100 nm to 22.4 for zinc (table 5). The size of the substructure of the electronic compound with an increase in the displacement of the nozzle decreases almost twofold from 70.5 nm to 36.2 nm and differs slightly from the size of the zinc substructure.

Table 5. Results of calculating the sizes of structural components.

Nozzle displacement, mm	Size of CSR, nm			
	Cu	Zn	ε	γ
		T=270°C		
2	>200	>100	70.5±4.7	-
3	21.4±0.4	22.4±1.6	36.2±3.6	-
		T=360°C		
2	>200	>200	82.8±41.5	>200
3	61.9±3.6	19.9±0.7	30.4±3.0	-
		T=450°C		
2	89.8±19.5	63.7±6.2	86.5±76.7	62.0±15.3
3	20.8±0.6	23.5±0.7	21.2±1.2	-

An increase in the gas flow temperature leads to an almost threefold increase in the size of the copper substructure from 21.4 nm to 61.9 nm with a nozzle displacement of 3 mm, which is caused by an increase in the deposition temperature and a change in the mechanical properties of copper (table 5). The size of the substructure of other structural components, namely zinc and ε-phase, is not affected by an increase in the deposition temperature. When using the technological mode of spraying, in which the nozzle is displaced by 2 mm, an increase in the spraying temperature from 270°C to 360°C has practically no effect on the size of the substructure of the main structural components, with the exception of zinc, where it increases from >100 nm to >200 nm (table 5). The size of the substructure of the new electronic compound of the γ - phase does not differ from the size of the substructure of zinc, on the basis of which it is formed.

An increase in the deposition temperature from 360°C to 450°C with a nozzle displacement of 3 mm is accompanied by a significant decrease in the size of the CSR of copper and electronic compound ε-phase from 61.9 nm to 20.8 nm and from 30.4 nm up to 21.2 nm, respectively (table 5). The size of the zinc substructure does not change significantly with an increase in the deposition temperature: 19.9 nm and 23.5 nm.

A decrease in the distance to 2 mm is accompanied by a more significant change in the size of the substructure of the main phase components of the structure. The size of the CSR of copper, zinc, and intermetallic compound (γ-phase) sharply decreases from > 200 nm to 89.8 nm, > 200 nm to 63.7 nm and >200 nm to 62.0 nm, respectively. An increase in the deposition temperature from 360°C to 450°C is accompanied by a more significant increase in the energy of the particle flux than an increase from 270°C to 360°C, which has a significant effect on the refinement of the structural components (table 4).

Spraying the coating at an air flow temperature of 450°C with a nozzle displacement of 2 mm and 3 mm is accompanied by the formation of a coating in which, with a smaller displacement of the nozzle relative to the surface, the size of the copper, zinc and ε-phase substructures is significantly larger than with a nozzle displacement of 3 mm (table 5). This difference may be due to the additional direct effect of the heated air flow on the coating.

4. Conclusions

Gas-dynamic spraying of a mechanical mixture of particles of copper, zinc and aluminum oxide (corundum is accompanied by the formation of a "brass" -type coating with a gradient in hardness and phase composition).

An increase in the deposition temperature significantly reduces the size of the substructure of the structural components: for copper from 200 nm to 89.8 nm, zinc from 200 nm to 63.7 nm, γ - phases from > 200 nm to 62.0 nm; the size of the ε-phase substructure changes insignificantly and is in the range of 75 nm - 86.5 nm. In this case, the deposition of the coating with a nozzle displacement of 3 mm leads to a greater refinement of the substructure of the phase components.

An increase in the deposition temperature leads to a significant decrease in the amount of deformation in copper from 0.187% to 0.119% and in an electronic compound based on Cu_5Zn_8 (γ-phase), from 0.307% to 0.17%. The value of microdeformations of the ε-phase increases slightly from 0.11% to 0.21% with an increase in the spraying temperature.

References

[1] Dimet Application of technology and equipment 2020 Available at: http://www.dimet-r.narod.ru/

[2] Arkhipov V E, Landarski A F, Moskvitin G V and Pugachev M S 2020 *J. Bulletin of mechanical engineering* chapter 3 pp 45-9

[3] Efremov B N 2016 *BRASSES from phase structure to structure and properties* (Moscow: INFRA-M) p 314

[4] Chichinadze A V, Brown E D and Boucher N A 2001 *Fundamentals of tribology (friction, wear, lubrication)* (Moscow: Engineering) p 312

[5] Lakhtin Yu M and Leontieva V P 1990 *Material science: Textbook for higher technical educational institutions* (Moscow: Engineering) p 528

[6] Arkhipov V E, Landarski A F, Moskvitin G V and Pugachev M S 2017 *Gas-dynamic coating: structure and properties of coatings* (Moscow: KRASAND) p 240

Virtual modelling of the crash cushion operation with projected destruction

I V Demyanushko, I A Karpov, P S Mikheev and A A Mukhametova

Moscow Automobile and Road Construction State Technical University (MADI), 64, Leningradsky Prospect, Moscow, 125319, Russia

ilya.karpov@outlook.com

Abstract. The article discusses a technique for virtual modeling of the process of deformation and fracture of the structure of a crash cushion, which is used to protect the places of road works, which occurs when a car hits the device during a road accident. The peculiarity of the approach consists in taking into account and modeling the processes of fracture of special energy-absorbing elements used in devices for absorbing impact energy due to the directional destruction of these elements.

1. Introduction

Pull and mounted type crash cushions (often called a truck-mounted attenuator (TMA)) are the effective devices to protect the places of road works and prevent severe consequences in case of a road traffic accident, which occur when a car hits these places (figure 1) [1]. The kinetic energy of an uncontrolled car hitting at high speed is absorbed by means of special energy-absorbing elements integrated in the crash cushion structures, including crash cartridges, often with a cell structure, and/or structural elements with directional projected destruction.

Figure 1. Protection car with attached crash cushion.

In case of a road traffic accident the car hits the crash cushion directly (frontal impact) or at a certain angle, and deformation of the crash cushion energy-absorbing elements made as cell cartridges is described in [2]. The behavior mechanics of crash cushions and cell energy-absorbing elements was analyzed by a virtual testing approach with the use of virtual modelling in Ls-Dyna Software.

In some structures of crash cushions special devices are used, either along with cell cartridges or independently, which absorb energy through destruction losses, provided that the mechanics of such destruction shall certainly be predictable.

Modelling of crash cushion deformation and destruction, as well as the impact energy absorbing based on the used destruction forecast for special elements requires consideration of the fracturing process in case of a dynamic impact, which complicates the modelling task. Let us consider an example of such device (figure 2 (a), (b)), where the structure of energy-absorbing elements is comprised of rectangular tubes of various thickness with stress raisers (1) and special knives (2, 4) with most of them located inside the tubes.

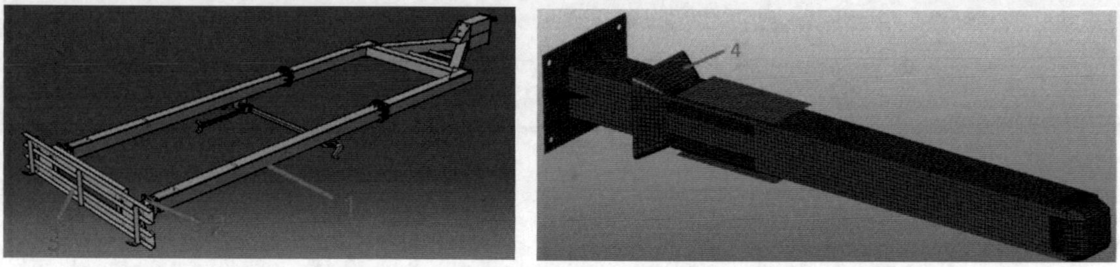

(a) (b)

Figure 2. Main cushion elements: (a) assembled cushion, (b) knives.

The device operates under the following principle: the coming car hits the front frame of the crash cushion (3 on figure 2), which is rigidly connected with the cushion knives (2). The frame starts pushing the knives forward, and they react on the tube (1) along the cutting surface (4). Since the cutting surface is much wider than the tube, the fracture starts growing on the edges with the growth direction determined by the grooves and initial cutting in the tube as the stress raisers (figure 3).

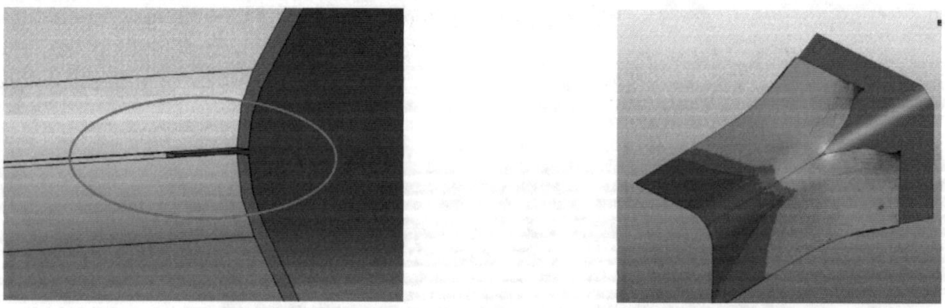

(a) (b)

Figure 3. Stress raisers at the tube edges: (a) tube initial geometry, (b) operating principle.

In this structure the kinetic energy produced when the car hits the frame is further used for the generation of free surfaces during the increase in fractures, which develop from stress raisers and directionally grow further, as well as for the knife friction with the tube surface, the plastic deformation of the tube sheet bending, and the plastic deformation of the front frame and the cushion bearing part. Since the fracture growth represents one of the main processes, let us consider it in detail.

2. Fracture Growth Modelling

A cohesive element with the linear softening law and bilinear fracture boundary stress law were used for the fracture growth modelling. These were related to the free energy velocity through the following expressions:

$$G_I = T * \frac{u_n}{2}, \tag{1}$$

$$G_{II} = S * \frac{u_t}{2}, \tag{2}$$

G_I, G_{II} is the free energy velocity by the fracture opening type I and II; S, T are the standard and tangential force vectors at the fracture boundary; and u_n, u_t are the fracture opening distances.

In order to set the model, the free energy velocity at the fracture opening, force vectors, as well as maximum material movements before opening were to be determined. These parameters were determined by the element-free Galerkin method (EFG) [3], as it provides for a high-precision destruction process modelling.

A part of the tube with symmetric boundary conditions for movements was considered for the tube operation modelling with the use of EFG methods. 1/4 part of the tube start was taken for the investigation. The model operating results are shown on figure 4.

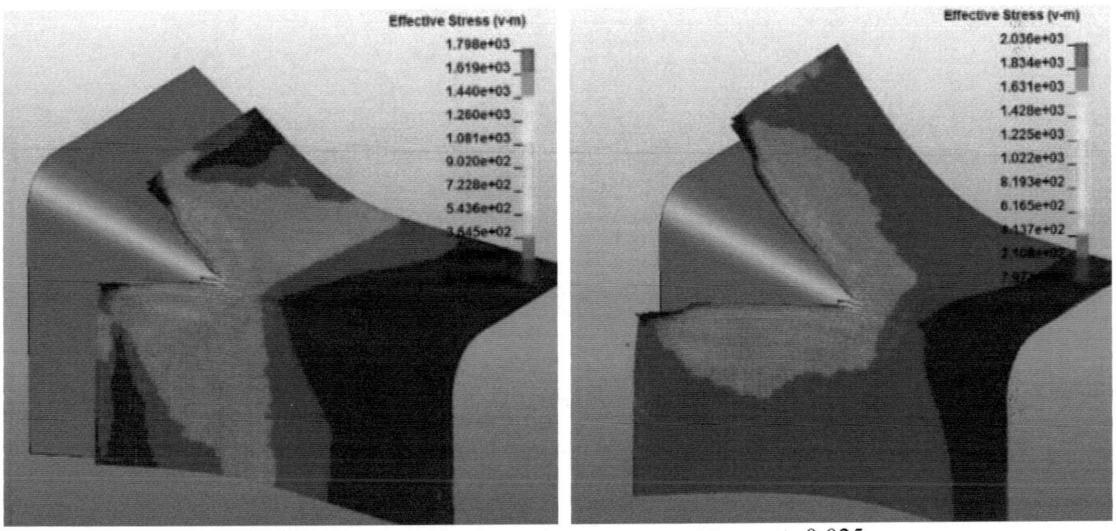

t=0.015 s t=0.025 s

Figure 4. Fracture opening. EFG Model.

The model provided for the calculation of force vectors at the fracture boundary, a J-integral [4] (figure 5) describing the free energy velocity by both types, as well as for the measurement of the fracture opening distances.

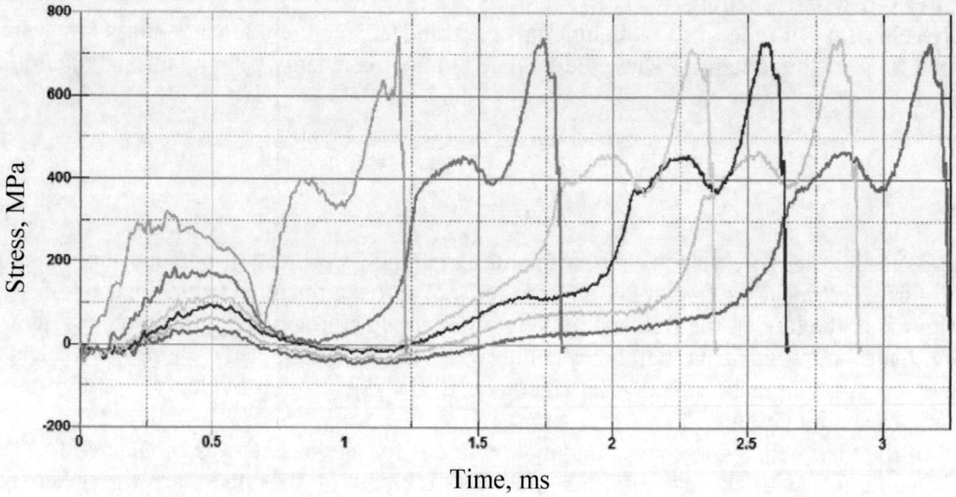

(a)

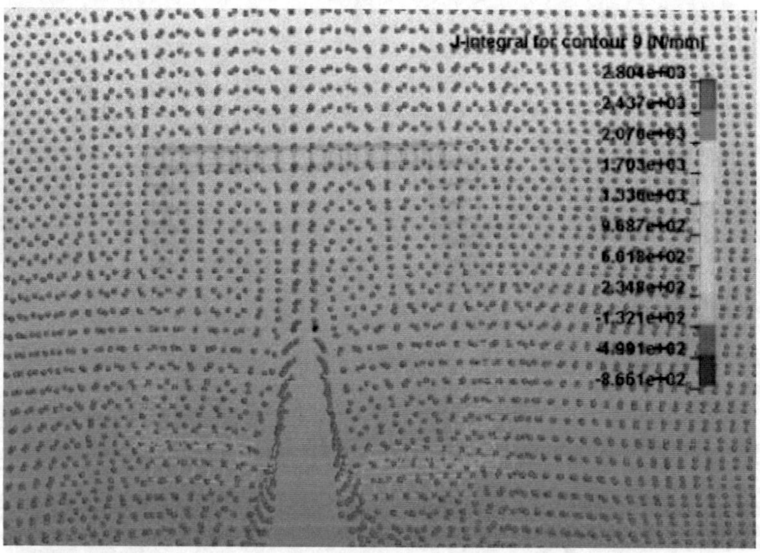

(b)

Figure 5. Fracture opening. EFG model: (a) surface forces, (b) J-integral contour.

Once the data required to set the operation of cohesive elements were received, they were uploaded to the model. The model operating results are shown on figure 6.

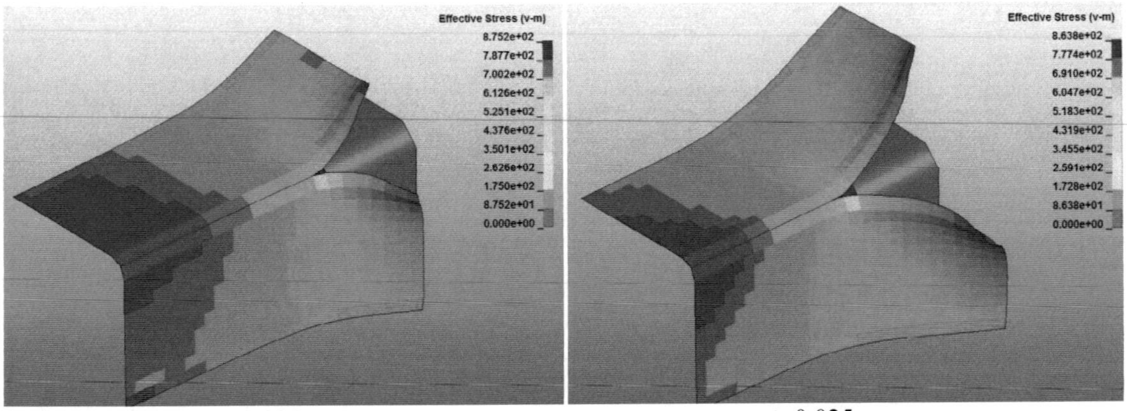

t=0.005 s t=0.025 s

Figure 6. Operation of the FEM model with cohesive elements with 5-mm grid.

In order to estimate the solution accuracy, the operating results of the EFG model were compared to the FEM models with the elements of 1 mm and 5 mm based on the efforts imposed on the tube by the knife. These results are shown on figure 6.

When analyzing the charts, it is apparent that the EFG model creates greater effort on the moving element than the FEM, especially on the stable fracture opening section. Maximum model difference is observed between EFG and 1 mm FEM and constitutes 15 percent. The 5 mm FEM model creates slightly greater efforts upon stabilization and has a difference of 6%. In general, such model accuracy can be considered as acceptable (figure 7).

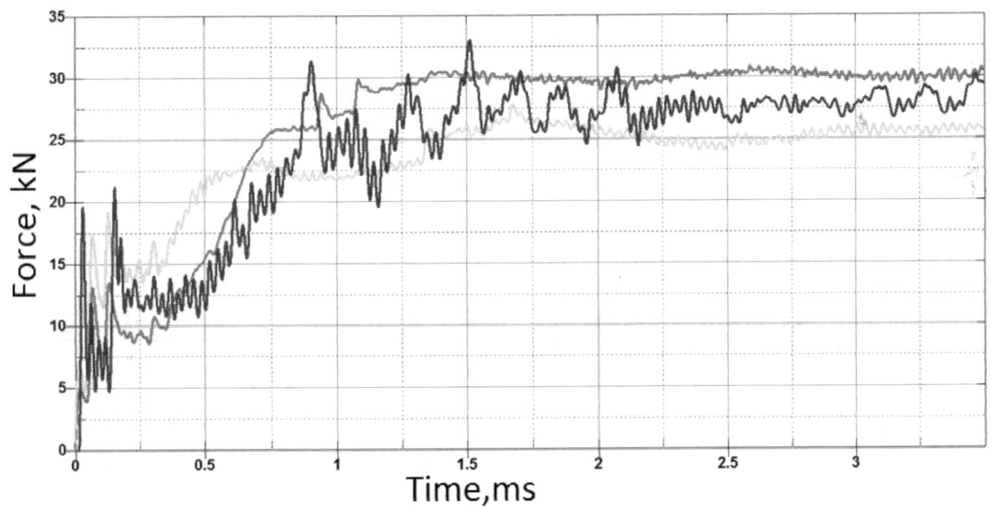

Figure 7. Forces: EFG (red), 1 mm FEM (green), 5 mm FEM (blue).

3. Entire cushion model

All elements described above were united into a general cushion model (figure 8). A basis was set for the model, through which no cushion or car parts can penetrate.

Figure 8. Assembled cushion model before impact.

The modelling and testing results are compared on figure 9. Good visual similarity of models can be observed.

t=0.038 s

t=0.09 s

t=0.21s

t=0.50 s t=0.822 s

Figure 9. Visual comparison of the model operation and experiment results.

IOP Conf. Series: Materials Science and Engineering 1129 (2021) 012002 doi:10.1088/1757-899X/1129/1/012002

To compare the model and the experiment, 3 parameters were used: the length of the barrier after deformation, amplitude values of accelerations in the center of gravity, and moving the front surface of the car. These parameters were chosen, as it was they who were specified in the report of the American crash test center. In the course of the quantitative comparison, the maximum movement of the car bumper under the impact was 3.97 m. Based on the testing results, the cushion moved by 3.89 m. The difference in the results was 2.1%.

When comparing the car accelerations in the center of gravity averaged for 10 ms, longitudinal and lateral accelerations were equal to 17.69 g and 4.11 g, respectively. Their values during the acceleration modelling were 17.5 g and 5.81 g. The difference in the results was 1.1% and 41%. The last difference is 1.8 g which is small in absolute terms when compared with other accelerations.

4. Conclusions

Virtual modelling of the crash cushion operation under the car impact has shown satisfactory results when compared to the field tests. Good visual consistence is observed between the model and tests. The quantitative comparison criteria demonstrated the difference of maximum 10% for all parameters, except for the lateral acceleration. The latter differed by 41% in the model against the field tests (4.11 g when tested and 5.89 g in the modelling), which was probably due to the different position of the wheel mounting mechanism compared to the tests.

References

[1] Manual for Assessing Safety Hardware 2009 p 277 ISBN Number: 978-1-56051-665-1

[2] Demyanushko I V, Karpov I A, Zharikova S V and Mukhametova A A 2018 *Engineering Science and Innovations. Int. Conf. of Young Scientists and Students* pp 379-82

[3] Wu C T 2009 *7th European LS-DYNA Conf.* p 15

[4] Pestrikov V M and Morozov E M 2012 *Fracture Mechanics. Course of lectures* (St. Petersburg: Professiya Education Programs Center) p 522 ISBN: 978-5-91884-034-4

Features of TRIP steel deformation at low and moderate temperatures

A V Frolova[1], **V V Stolyarov**[2], **J V Tilak Kumar**[3] **and J Sudha**[3]

[1] Advanced Forming Research Centre, University of Strathclyde, 85 Inchinnan Drive, Renfrewshire, PA4 9LJ, UK

[2] Mechanical Engineering Research Institute of the Russian Academy of Sciences, 4,M. Kharitonyevskiy Lane, Moscow, 101990, Russia

[3] Department of Mechanical Engineering, College of Engineering Guindy, Anna University, Chennai, 600025, India

anna.frolova@strath.ac.uk

Abstract. The TRIP effect and tensile behaviour of a TRIP steel were studied at temperatures ranging from -120 to 200°C. The initial microstructure comprises of austenite and martensite (50:50). Excellent combination of UTS (~2395 MPa) and elongation (28 %) was observed at -100°C and is attributed to the TRIP effect, that found absent above 100°C and below -120°C. Transformation temperatures were studied by DSC and DMA. Analysis of the tensile behaviour, phase changes and fracture surface suggest possible changes in the deformation mechanisms at different temperatures.

1. Introduction

The phenomenon through which martensitic transformation in metastable austenitic steels takes place during plastic deformation is called the TRIP effect (TRansformation-Induced Plasticity) [1]. Despite the limited understanding of its origins, scientific and practical interest in this phenomenon has remained high. The practical interest stems from the use of TRIP steels as structural materials in general engineering (high-strength, corrosion-resistant fasteners, wires, cables), aircraft parts (helicopter rotor torsions), instrumentation (springs and elastic elements of the thinnest sections), biomedical applications etc. [2], [3]. Such steels possess not only high corrosion resistance and weldability, but also a winning combination of high strength (about 2000 MPa) and ductility (up to 35%).

Scientific interest in the TRIP effect implies an understanding of the transformation kinetics and the features of the mechanism(s) causing the TRIP effect, as well as its control by internal (structure refinement, chemical and phase composition) and external (temperature, speed and degree of deformation during thermomechanical processing; cyclic loads, external fields) variables. Many of these factors have been studied quite thoroughly in previous works [4-11]. Some general ideas of developments in metastable austenitic steels are given in a review [4] and a book [5]. In addition to chemical composition, the important parameters controlling the phase transformation are grain size, degree and temperature of deformation [6-8]. It is demonstrated that the shear stresses and high pressures present during the refinement of the microstructure through intense plastic deformation by torsion in AISI 304L steel can lead to the appearance of an intermediate phase ε (hcp), so that the sequence of transformations γ (bcc) $\rightarrow \varepsilon \rightarrow \alpha'$ (hcp) is confirmed [6].

In places of deformation localization, hardening occurs due to the transformation of austenite (A) into martensite (M), and further deformation is concentrated in neighbouring volumes of austenite [7]. The completion of the phase transformation depends on the deformation temperature with respect to the critical temperatures, M_s (the start temperature of martensitic transformation upon cooling) and M_d (the maximum deformation temperature above which no transformation occurs) [7]. In accordance with previous studies [4], [7], [8], a distinctive feature of most TRIP steels is that $M_s < 0°C$, and $M_d >$ RT. Thus, even after quenching, the austenitic steel at ambient temperature remains as austenite. An increase in the degree of plastic deformation and refinement of the initial structure in AISI 304L metastable steel to the nano- and ultrafine- grain states contributes to an increase in the amount of martensite formed and additional hardening. However, ductility before fracture decreases [8]. In the metastable AISI 304 stainless steel, an increase in the tensile strain rate in the range of 10^{-4} s^{-1} - 10^{-1} s^{-1} leads to a noticeable increase in Ultimate Tensile Strength (UTS) and yield strength (YS) [9]. In contrast, a decrease in the same properties with respect to an increase in the tensile strain rate approximately in the same range from $4.1 \cdot 10^{-5}$ to $2.5 \cdot 10^{-2}$ s^{-1} in the austenitic – martensitic steel VNS9-Sh is also reported in literature [10]. With regard to external factors influencing the TRIP effect, it has been shown that the application of pulsed current during tensile deformation of the same steel led to a suppression of the TRIP effect [11].

Among the various factors that significantly affect the kinetics of martensitic transformation and, consequently, the TRIP effect, insufficient attention has been paid to the tensile test temperature [13-15]. Studies focusing on that, typically involve tests above ambient temperature. However, it is essential to include both, low and moderate, temperatures in order to determine the boundaries of the TRIP effect. An analysis of a few publications on similar studies in a wider temperature range, including very low and moderate temperatures, demonstrates a noticeable discrepancy in the results obtained [16-18]. For example, in [16] for TRIP Fe-Mn steels with addition of both, aluminium and silicon, the temperature dependence of increasing tensile elongation correlates with the energy of stacking faults and twinning, and the best plasticity is recorded at room temperature. However, in [17] for lightly alloyed TRIP steel 600 studied in the range from -60 to 120°C, it was shown that maximum plasticity is observed at 70°C. Note that in both those studies, the chemical composition of TRIP steels differs sharply from steels with high content of chromium and nickel. Finally, in a steel close in chemical composition to the Cr-Mn-Ni TRIP steel studied in this article, the authors found the maximum value of elongation at 100°C [18]. TRIP steel, developed at VIAM (Russia), with UTS up to 2000 MPa and ductility of up to 35%, is used in the production of high-strength fasteners, corrosion-resistant high-strength wires and cables, critical parts of aircraft (in particular, helicopter rotor torsions). The products from this steel display excellent characteristics of mechanical properties after cold plastic deformation, when the phase composition of the steel consists of approximately equal amounts of work-hardened austenite and strain-induced martensite (SIM). Therefore, TRIP steel is mainly used in the form of thin rolled sheets or wires.

The aim of this work is investigation and critical examination of the tensile deformation behaviour at low and moderate temperatures of a TRIP steel.

2. Material and methods

A 0.3 mm thick sheet made of VNS9-Sh steel was obtained by cold rolling of hot-rolled steel in several steps with intermediate heat treatment. The chemical composition is shown in table 1. The steel structure in the initial state was a mixture of SIM and work-hardened austenite (~ 50% martensite and ~ 50% austenite).

Quasistatic tensile tests of standard flat samples (gauge size 0.3x2.5x15 mm) were performed at temperatures in the range of - 120 to 200°C on a horizontal tensile testing machine IR 5081-20 with a strain rate of $\acute{\varepsilon} = 8.3 \cdot 10^{-4}$ s^{-1}. The desired temperature in the range of 20 - 200°C was achieved with an industrial dryer [15]. Tests at low temperatures were performed as follows: before the tensile test, a chamber in the form of a tube with a hole cut out was mounted to the grips with the sample fixed between them (figure 1). Then the sample was cooled by immersing the assembly in liquid nitrogen for 3 minutes to stabilise the temperature. The required test temperature was achieved by varying the distance between the chamber and the sample surface that is exposed to the liquid nitrogen vapour in the range of 1 - 5 cm. Liquid nitrogen was topped up as and when required in an appropriate manner to compensate for losses. The temperature was monitored by a UT321 contact temperature meter with a chromel-alumel thermocouple attached to the centre of the gauge length and with an accuracy of ± 5°C.

Table 1. TRIP steel composition.

Chemical element	C	Cr	Ni	Mo	Mn	Si	N	S	P	Fe
Weight, %	0.25	14.5	4.8-5.8	2.7-3.2	≤ 1	≤ 0.6	0.03-0.07	≤0.01	≤0.015	Bal.

Figure 1. Test chamber with liquid nitrogen and a sample.

The determination of phase transition temperatures was carried out by DSC using a calorimeter Sensys Evo (Setaram). The measurements were performed in the temperature range from -90 to 300°C at a speed of 3 deg·min^{-1}. The phase composition was determined by XRD analysis in CuKα radiation using a Rigaku Ultima IV X-ray diffractometer. XRD analysis was performed far from the place of final specimen fracture and the region between the grips section and the gauge part of the sample. The temperature dependence of internal friction was studied by Dynamic Mechanical Analysis (DMA) on a Q800 analyser using a three-point bending scheme in the temperature range from -150°C to 400°C with 1 mm amplitude and 1 Hz frequency. Driving Force was 3.8 - 3.1 N. The fracture surface of the tensile tested samples was studied by JSM-6610LV Electron microscope with accelerating voltage of 25 kV.

3. Experimental results

3.1. X-Ray analysis

Previously performed X-Ray analysis [19] clearly indicated that when the test temperature is reduced from 20 to -120°C, the relative intensity of the diffraction peak of austenite (220) drastically decreases and other two austenitic peaks (200) and (311) disappear. Simultaneously, the relative intensity of all the martensite diffraction peaks increases, including the strongest peak (211) (figure 2). The latter feature indicates the presence of martensite grains preferred orientation with crystallographic planes (110) parallel to the sample surface. Thus, when the test temperature drops below 20°C, the direct transformation of residual austenite to martensite increases, the volume fraction of which can reach 95% or more.

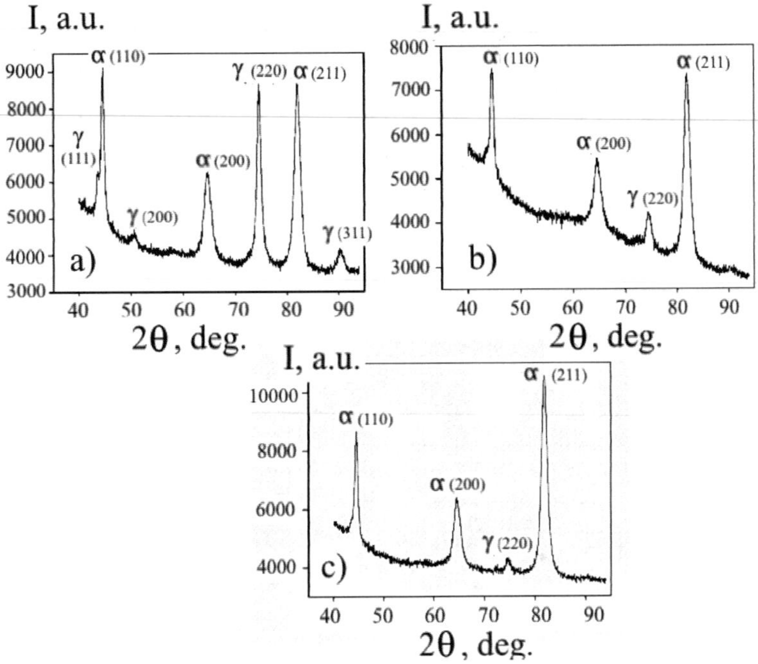

Figure 2. X-Ray diffraction spectra of samples: (a) 20°C; (b) 0°C; (c) 100°C [19].

3.2. Mechanical properties

Figure 3 shows the stress-strain curves at temperatures ranging from -120°C to 200°C. A yield stress and a strain hardening (SH) stage with a very low SH exponent can be observed at 23°C, which corresponds to a typical tensile curve when the TRIP effect is present [20]. Raising the test temperature to 100°C and above dramatically changes the shape of the deformation curves.

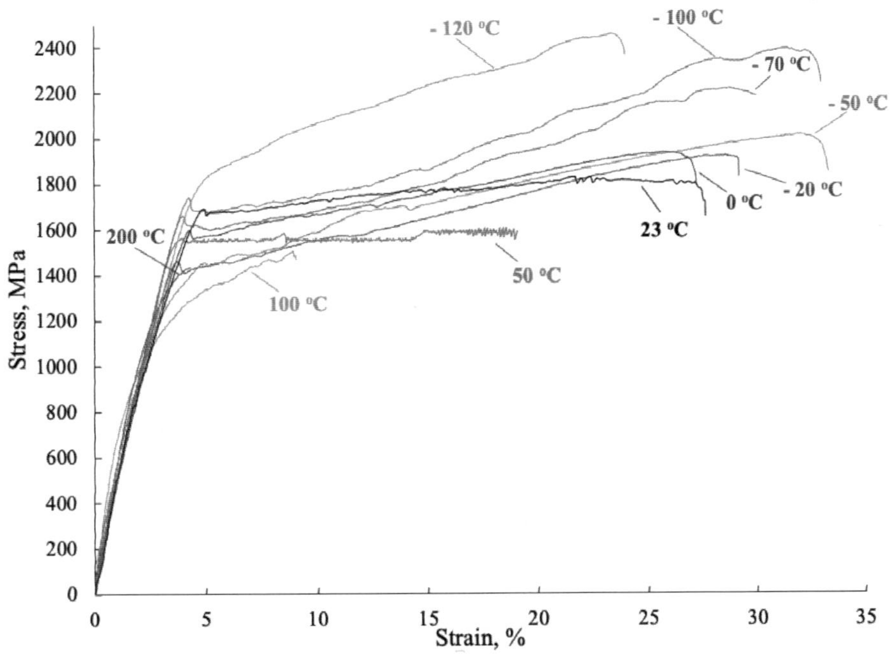

Figure 3. Stress-strain curves in the temperature range -120°C to 200°C.

There is no yield point, and the stage of SH has a reasonably high exponent. A significant decrease in the elongation at fracture (ε), as well as in the offset YS with an increase in temperature can be seen. The shape of the tensile curves at sub-zero temperatures indicates a significant difference in the SH exponent of the samples. In the range of 0 to -100°C, there are upper and lower yield points on the tensile curves, after which the stage of SH is observed. At a test temperature of –120°C, the upper and lower yield points are absent. Tensile properties at the different test temperatures are presented in table 2.

Table 2. Mechanical properties at different temperatures.

Test temperature, °C	UTS, MPa	YS, MPa	ε, %
200	1440	800	2
100	1505	810	7
50	1610	900	16
23	1830	1150	22
0	1940	1200	23
-20	1925	1100	25
-50	2015	1170	29
-70	2220	1450	26
-100	2395	1560	28
-120	2460	1730	18

Figure 4 shows the dependence of the UTS, YS, ε, and the α-phase content on the test temperature. A rise in the test temperature from ambient to 200°C leads to a rather sharp drop in UTS and ductility, as well as a decrease in the SIM formed (from 84% to 55%, respectively) (figure 4). A noticeable growth in UTS and YS and a slight increase in elongation were observed as temperature decreased. The maximum UTS (2460 MPa) was observed at -120°C (figure 4) with a sharp drop in ductility. The amount of SIM after deformation also grows with decreasing test temperature. At -100°C the volume fraction of martensite is 97% (comparing to initial ~50%). Maximum ε (29%) was observed at -50°C, and then it continuously reduces with increasing temperature up to 200°C (figure 4).

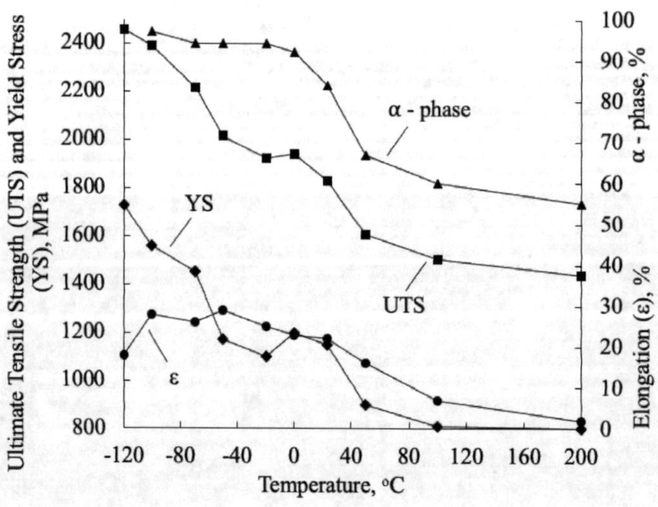

Figure 4. Temperature dependence of mechanical properties and α-phase content.

3.3. *DSC analysis in the initial state*

DSC analysis was used to determine the phase transformation temperatures in the initial dual-phase microstructure in the range from -90 to 300°C. During heating, two peaks (upward and downward) are seen in curve 1 (figure 5), following one after another: at 4.5°C (exothermic) and 11°C (endothermic). That could be associated with the start of the martensite decomposition into a ferrite-carbide mixture (upward peak) and the thermal transformation of residual austenite into martensite (downward peak). A repetition of the experiment with the same sample led to its reproducibility and a small shift of the peaks to a region of lower temperatures (figure 5, curve 2). No peaks were observed during cooling.

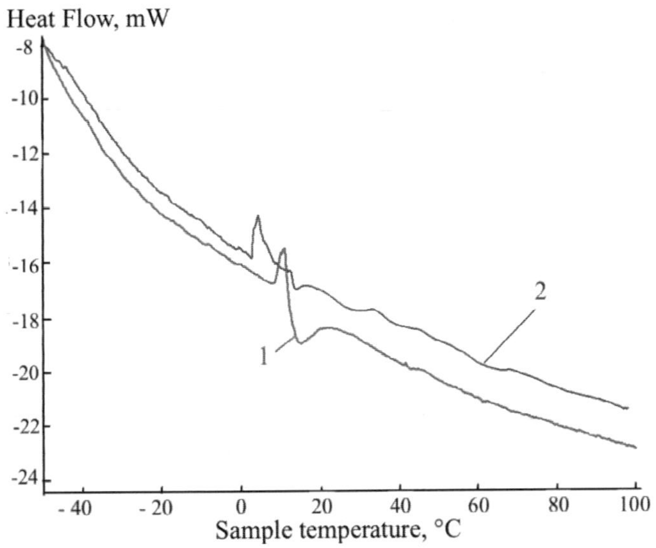

Figure 5. The heat flow dependence on temperature during heating.

3.4. Fracture analysis

Examination of the fracture surface (figure 6) showed a typical ductile dimple fracture across the test temperature range, however their size and presence of characteristic areas differ. The dimple size decreases with the test temperature, the dispersion becomes finer (figure 6 (a), (d)), the fracture surfaces appear rougher, and localised ductile shear regions are seen (figure 6 (d), (e)). At temperatures above 100°C, the fracture pattern begins to resemble the fracture surface of elevated temperature specimens.

4. Discussion

The reason for the slight deviation of the elastic modulus (figure 3), may be due to a difference in the rigidity of the "grips-sample" system, as well as the effect of nitrogen vapour or hot air on the material of the grips. Pronounced TRIP effect is observed at ambient temperature together with a low SH exponent and high ductility due to the phase transformation A → M. In samples tested at reduced temperatures, the SH exponent increases with deformation, as does the elongation at fracture. This may indicate a partial appearance of the TRIP effect at the early stages of deformation. The tensile curve at -120°C exhibits the maximum SH exponent as well as a slight decrease in ductility, which looks typical for a stable austenite phase. The shape of the stress-strain curves below -100°C indicates TRIP effect absence. Upper and lower yield points preceding the yield plateau can be seen in the range of 0 to -100°C. The high SH exponent at low temperatures can indicate a mechanism of additional SIM formation different from the necking mechanism at ambient temperature (compare curves at 23°C and 0°C in figure 3). In [20], the authors explained the presence of upper and lower yield points and the yield plateau in metastable austenitic steels by three features of the plastic deformation mechanism, viz. slip, twinning and martensitic transformation. In [21], it was noted, that TRIP steel VNS9-Sh does not exhibit the physical YS when it is in purely austenitic and purely martensitic states. That is consistent with disappearance of the upper and lower yield points in present study when the test temperature is decreased to -120°C (figure 3), as well as the increasing amount of α-phase (martensite) with temperature reduction (~ 95% at -120°C) (figure 4). The absence of TRIP effect features in the tensile curves at low temperatures (yield points, plateau and a low degree of SH) indicates the TRIP effect disappearance.

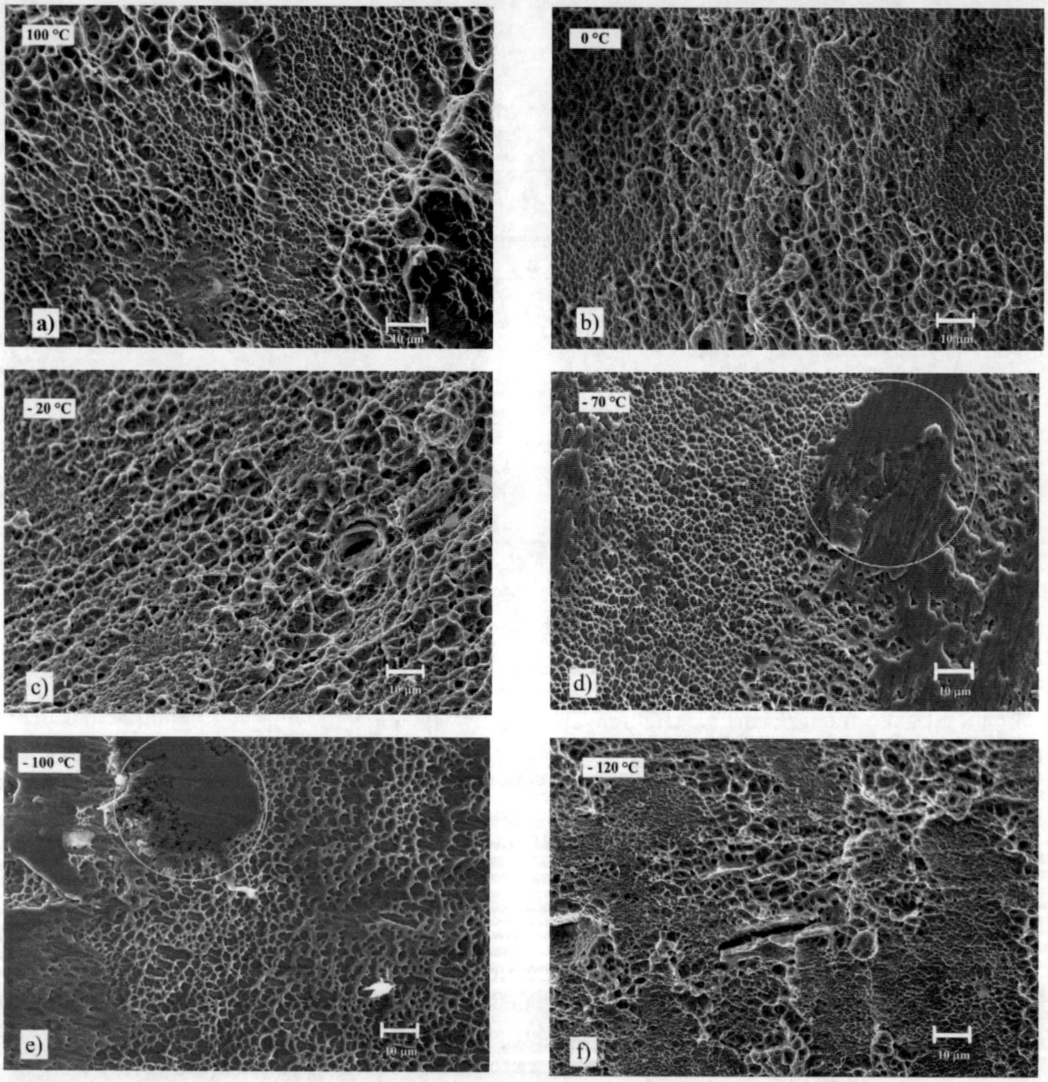

Figure 6. The sample fracture surfaces after tensile tests at different temperatures: red arrows indicate micro-cracks; white circles indicate ductile shear regions.

At a test temperature of 50°C, the amount of SIM is more than in the non-deformed initial material, which indicates that at this temperature the TRIP effect is present (formation of additional SIM). In the temperature range of 100 – 200°C, the deformation mechanism changes due to a partial reverse transformation of martensite to austenite, which is consistent with previous studies [15]. According to [22], the complete reverse M→A transformation in VNS-9Sh steel occurs after a tempering treatment at 600°C.

DSC analysis made it possible to experimentally establish that the temperature of the thermal stability of SIM is noticeably below the ambient temperature. Considering the possible temperature hysteresis, it can be assumed that the M_s temperature of the direct austenite to martensite transformation during cooling is close to 0°C or lower, which is consistent with published data [4], [7], [8]. Figure 4 confirms this assumption, since the increase in the volume fraction of martensite is practically exhausted for $T \leq 0$°C. In addition, especially for temperatures equal to and below 0°C, the degree of SH sharply increases (figure 3).

The microstructure of the fracture surfaces demonstrates the ductile nature of fracture, mainly by dimple fracture, which ensures good ductility, despite an increase in the martensite content. However, the size of dimples decreased with a lowering of test temperature and may be responsible for the increase in the tensile strength. Another feature of fracture is the appearance of ductile shear regions (figure 6 (d), (e)) and formation of micro-cracks (figure 6 (c), (f)).

5. Conclusions

The features present in the stress-strain curves (yield points, SH) and the mechanical properties of metastable austenitic TRIP steel in the temperature range of -120 to 200°C are reported.

The tensile behaviour at low and moderate temperatures, a change in the austenite/martensite phase ratio, and fracture analysis indicate drastic changes in the operating deformation mechanisms at different temperatures and, accordingly, the contribution to the TRIP effect.

An increase in the test temperature above the ambient leads to a decay in SIM formation. A decrease in temperature assists the formation of additional SIM. Changes in the phase composition lead to a decrease in the TRIP effect contribution at elevated (> 50°C) and reduced (<0°C) test temperatures and its disappearance at temperatures above 100°C and below -120°C.

Acknowledgments

This work was supported by the RFBR grant 16-58-48001 and the DST grant No. DST/INT/RFBR/IDIR/P-04/2016. The authors thank Prof. K. A. Padmanabhan for clarifying discussions, Prof. V. F. Terent'ev for the samples and XRD analysis and Dr. S. I. Sinyova for DSC results.

References

[1] Zackay V *et al* 1967 *Trans. of the ASM* **60(2)** 252-9
[2] Goldstein M I, Grachev S V and Vecksler Yu G 1999 *Special steels: Textbook 2nd ed* (Moscow: MISIS) p 408
[3] Chawla S L and Gupta R K 1997 *Materials Selection for Corrosion Control* (Materials Park (OH): ASM International) p 489
[4] Lo K H, Shek C H and Lai J K L 2009 *Mater. Sci. and Eng. R* **65** pp 39-104
[5] Fonstein N 2015 *Advanced High Strength Sheet Steels* (Basel: Springer) p 396
[6] Tavares S S M *et al* 2003 *Mater. Sci. and Eng. A* **358** pp 32-6
[7] Lakhtin Yu M and Leontieva V P 1990 *Materials Science* (Moscow: Mashinostroenie) p 528
[8] Shirdel M, Mirzadeh H and Parsa M H 2015 *Mater. Charact.* **103** pp 150-61
[9] Lu J Z *et al* 2013 *Surf. Coat. and Technol.* **221** pp 88-93
[10] Terentyev V F, Slizov A K and Prosvirnin D V 2016 *Deformatsiya i Razrushenie Materialov* **1** pp 14-8
[11] Stolyarov V V, Klyatskina E A and Terentyev V F 2016 *Lett. on Mater.* **6(4)** pp 355-9
[12] Martin S, Wolf S and Martin U 2011 *Solid State Phenom.* **172-174** pp 172-7
[13] Muёller A *et al* 2016 *Metall. and Mater. Trans A* **47** pp 59-74
[14] Tilak Kumar J V *et al* 2020 *Mater. Sci. and Eng. A* **777** 139046
[15] Stolyarov V V, Padmanabhan K A and Terentyev V F 2019 *Lett. on Mater* **9(1)** pp 113-7
[16] Grässel O, Frommeyer G, Derder C and Hofmann H 1997 *J. de Phys. IV France* **07(C5)** pp 383–8

[17] Berrahmoune M R, Berveiller S, Inal K, Moulin A and Patoor E 2004 *Mater. Sci. and Eng. A* **378** pp 304–7
[18] Tsuchida N, Yamaguchi Y, Morimoto Y, Tonan T, Takagi Y and Ueji R 2013 *ISIJ Int.* **53(10)** 1881–7
[19] Terentyev V F *et al* 2019 *Deformatsiya i Razrushenie Materialov* **11** pp 131-6
[20] Virakhovsky Yu G *et al* 1971 *The Phys. of Met. and Metallogr.* **32(2)** pp 348-63
[21] Terentyev V F *et al* 2017 *Deformatsiya i Razrushenie Materialov* **8** pp 39-47
[22] Terentyev V F *et al* 2018 *Deformatsiya i Razrushenie Materialov* **6** pp 20-5

IOP Conf. Series: Materials Science and Engineering 1129 (2021) 012004 doi:10.1088/1757-899X/1129/1/012004

Research of the occurrence and development of metal defects in sulfide stress cracking inducing environment

N I Gabelchenko, N A Kidalov, A A Belov, N V Grigoreva, Yu V Grebnev and N P Zhiltsov

Volgograd State Technical University, 28 Lenin Avenue, Volgograd, 400005, Russia

mitlp@vstu.ru

Abstract. Comparative studies of the metal of equipment after long-term operation in high hydrogen sulfide environment are carried out. To compare their resistance to sulfide stress cracking, the samples were hydrogenated under a stress of 0.6-0.9 of yield stress in a hydrogen sulfide environment. Characteristic of surface defects (blisters and bubbles) were identified, their size and time of occurrence were determined. To study the further development of the process, the samples were subjected to intense hydrogenation for 6 hours at maximum tensile stresses of 0.95 of the yield stress. Then they were cut across for conducting a metallographic study in order to identify the initiation and development of cracks in the metal. It is shown that steel containing cerium, which promoted the formation of relatively small globular sulfides, has the highest resistance to sulfide stress cracking. Such inclusions, in contrast to sulfide stringers of the FeS and MnS type, were not sources of crack initiation. Another source of crack initiation was the boundaries of pearlite and ferrite grains contaminated with microimpurities.

1. Introduction

Reliable operation of petrochemical equipment is a complex multifactorial task that requires solving a set of issues for its implementation (choice of material, operating regimes, manufacturing technology, control regulations, preventive measures, etc.). This problem has acquired particular importance in connection with an increase in the share of raw materials containing in their composition an increased amount of hydrogen sulfide in the total volume of extracted and processed oil products. Under such conditions, it is possible to saturate the metal of the equipment with diffusion-mobile hydrogen, which can lead to its embrittlement and destruction. The change in the initial properties of steel during operation can also be associated with temperature parameters, or with the simultaneous effect of the environment and temperature. Therefore, for reliable operation it is necessary to use metal that is resistant to stress corrosion cracking and achieve a homogeneous structure during manufacturing [1].

Currently, the most applicable method for controlling metal operating in hydrogen sulfide environments is the NACE TM0177 2005 method [2], which is effective for comparative testing of materials. For research purposes, other methods are also used, but all of them are based on the NACE method, with modified parameters of either the hydrogen-rich environment, or the loading scheme of the samples.

This paper considers the main changes in the metal surface and its microstructure after hydrogenation in a hydrogen sulfide environment under the tensile stresses in laboratory conditions.

2. Materials and methods

On the basis of the NACE stress corrosion cracking test data, flat samples of studied steels were prepared. Then, tensile stresses were created in the samples on the working side of the metal. A sample placed in four-point bending frame is shown in figure 1.

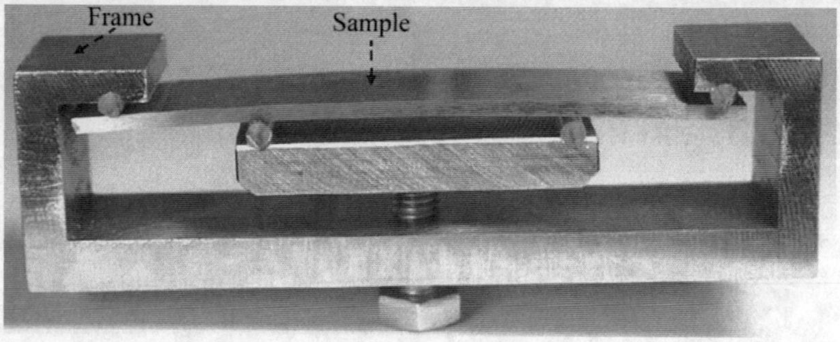

Figure 1. Frame with the fixed sample for immersion in the installation.

Then the frame with the sample was placed in a vessel with hydrogen sulfide solution. Hydrogenation was carried out in a standard NACE solution under a stress of 0.6-0.95 σ_y, while determining the time until the appearance of the first defects t_0, the number of defects during the basic hydrogenation time of 2 hours and calculating the average diameter of the formed bubbles D_{av}.

For research, a metal of approximately equal strength used for extraction and processing of petroleum products containing sulfur compounds in their composition was chosen. Samples were cut from the body and lid of a US-made apparatus and similar metal of Russian production – steel 20 and resistant to corrosion cracking steel grade 20YuCh (analog of AISI 1020 with additional doping with aluminum and rare earth metals), which are currently widely used in industry. The chemical composition of steels and their yield strength are given in table 1.

Table 1. Chemical composition and yield strength of the investigated steels.

Steel grade	Elements composition, %									Yield strength, σ_y, MPa
	C	Mn	Si	Cr	Cu	Al	Ce	S	P	
USA steel	0.19	0.55	0.09	0.04	0.06			0.033	0.011	230
Steel 20	0.16	0.49	0.18	0.04	0.12			0.031	0.019	230
Steel 20YuCh	0.20	0.66	0.24	0.04	0.10	0.06	0.01	0.017	0.019	240

3. Results and discussion

By analyzing the chemical composition of the three studied steels, taking into account our own research and research by other authors, it is possible to predict its effect on the resistance of steels to sulfide stress cracking [3 – 8].

The carbon content of all tested samples was within the permissible limits, namely 0.15-0.25%, which made it possible to obtain the desired structure.

The manganese content was 0.35-0.66%, which within these limits has a positive effect on the corrosion properties of the steel surface.

In the two investigated samples of steel 20 and steel 20YuCh, the silicon content in the permissible limits of 0.4-0.7%. In the USA steel sample, the silicon content exceeded the permissible limits, which may have led to a deterioration in resistance to corrosion destruction.

Copper in all samples has a positive effect on the strength characteristics of the samples, forming protective films on the metal surface.

Aluminum adversely affects the corrosion properties of the test samples.

The presence of rare earth metals in the composition of alloys used in a hydrogen sulfide environment increases the anticorrosion properties of the samples.

The results obtained after hydrogenation of the samples are shown in table 2 and figures 2-4.

Table 2. Results of hydrogenation of the studied steels.

Steel grade	Time before first surface defects appearance, min	Test time, min	Amount of surface defects, mm^2	Average diameter of defects, mm
Steel 20	10	120	2.16	0.078
Steel 20 YuCh	15	120	1.7	0.120
USA steel	5	60	13.9	0.152

These tests show that, in comparison with steel 20 and steel 20YuCh, the time for the appearance of the first surface defects on USA steel is shorter, and the number of bubbles formed is about 10 times is greater, although the chemical composition of steels is approximately the same. Obviously, the deterioration in resistance to stress corrosion cracking of USA metal is associated with its long-term operation. The surface of the samples is shown in figure 2.

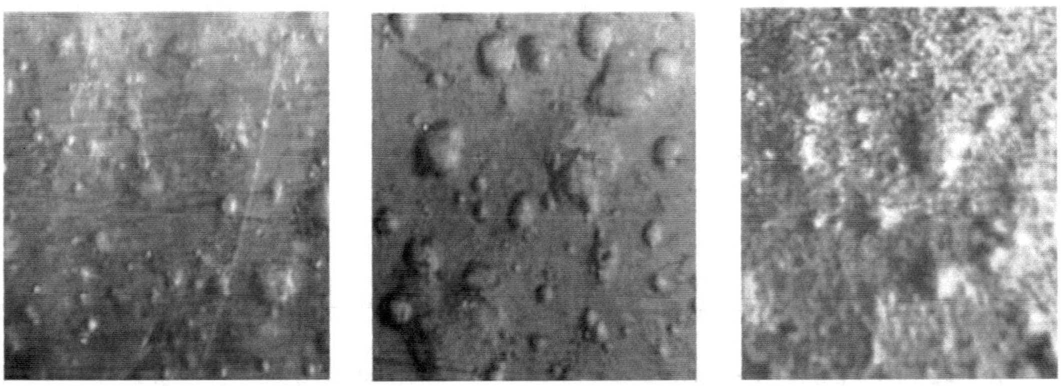

(a) (b) (c)

Figure 2. Bubbles on the surface of hydrogenated samples of steel 20 (a), steel 20YuCh (b), USA steel (c), ×10.

The best result for resistance to sulfide stress cracking was shown by steel 20YuCh (figure 3).

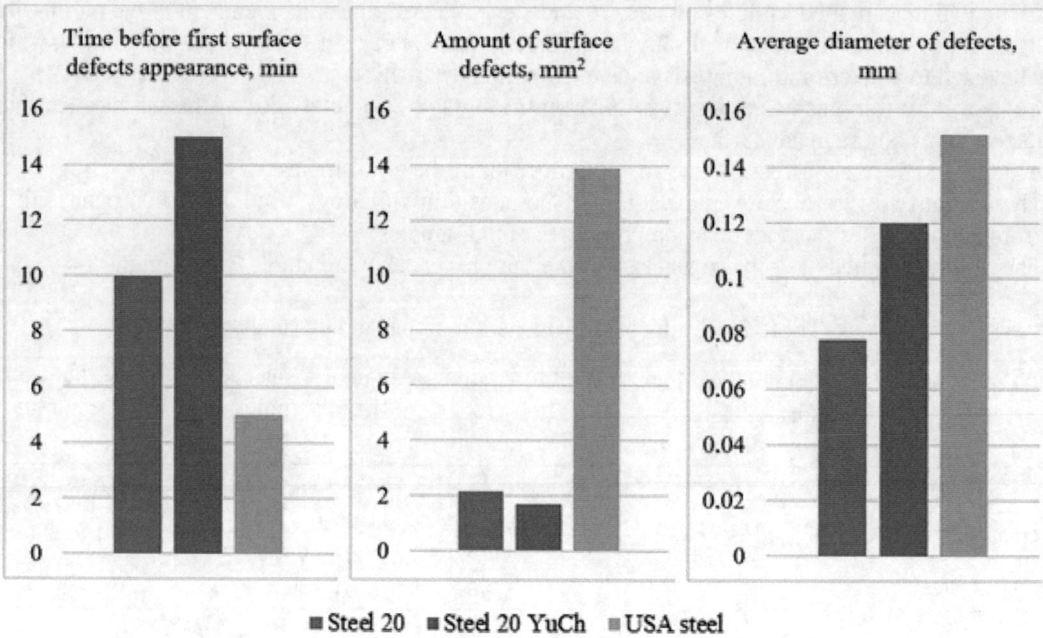

Figure 3. Analysis of the defects in the studied samples.

To study the development of stress corrosion cracking, the samples were subjected to intense hydrogenation for 6 hours. Then they were cut across and metallographic studies were carried out in order to identify the initiation and development of cracks in the metal.

The study of the microstructure and process of defects appearance was carried out on a microsections of studied steels. It is shown that in steel 20 cracks originate at the interface of pearlite and ferrite grains (figure 4 (a)) and along the sulfide stringers phase of the FeS MnS type (figure 4 (b)). Cracking of pearlite and ferrite grains boundary is due to their surface contamination with microimpurities.

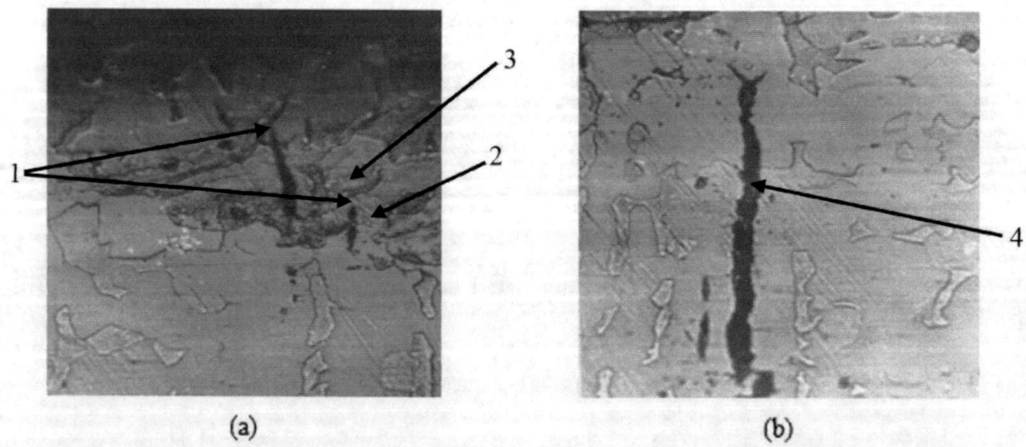

Figure 4. The initiation of microcracks (point 1) in steel 20 at the boundary of pearlite (point 2) and ferrite (point 3) grains (a), along the sulfide stringers phase (point 4) of the FeS·MnS type (b), ×100.

In the USA steel, microcracks (figure 5 (a)) spread along rolling direction and along boundaries of ferrite and pearlite grains, similar to steel 20. The opening of bubbles in the microstructure of USA steel is shown in figure 5 (b).

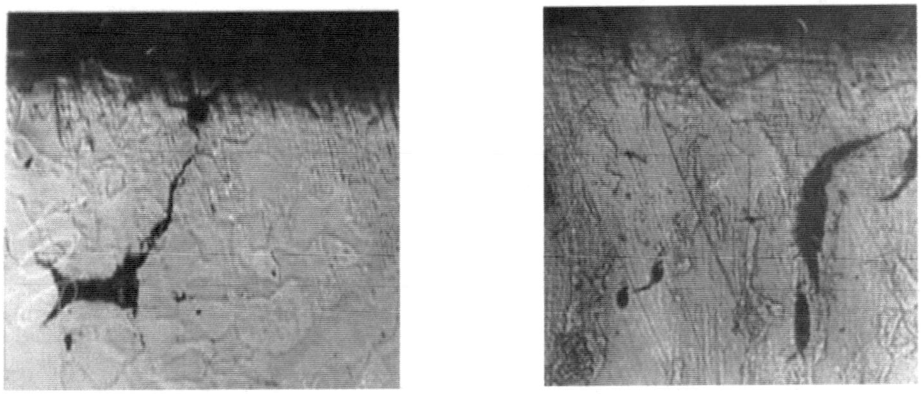

(a) (b)

Figure 5. Crack initiation (a) and bubble opening (b) in USA steel, ×100.

Since the resistance of steel to stress corrosion cracking significantly depends on the quantity and nature of non-metallic inclusions, especially sulfides and oxysulfides [5], the extraction of non-metallic inclusions and it's analysis was made for steel 20 and USA steel.

Isolation of nonmetallic inclusions was carried out by electrochemical method in Lukashevich-Duvanova electrolyte: 3% aqueous solution of $FeSO_4$ ($FeSO_4$ $7H_2O$) + 1% NaCl + 0.25% Rochelle salt ($KNaC_4H_4O_6$). Electrolysis mode: i ~ 0.025 N cm^2, t = 24 h, T = 20°C.

The resulting precipitate in addition to slag inclusions contained carbides. In the process of chemical dissolution of carbides, some types of inclusions could be destroyed or changed. Therefore, microscopic examination of inclusions was carried out before dissolving carbides. The inclusions were separated from the carbides and examined under a microscope. The precipitate was subjected to X-ray analysis (URS–70 X-ray machine with "Co" emitter, camera RKV, exposition 10h, U = 30 kV J = 10 mA). The result of X-ray diffraction analysis is: the crystal structure corresponds to compounds of the FeS type, cerium sulfide for steel 20YuCh and Al_2O_3, SiO_2, FeS (quartz) for USA steel. In the study of stable nonmetallic inclusions, the anodic precipitate was subjected to acid-oxidative treatment to dissolve carbides, consisting in the oxidation of carbides with ammonium persulfate and potassium permanganate with simultaneous heating. The released manganese dioxide was treated with a 50% citric acid solution. The precipitate of slag inclusions after the dissolving of carbides was studied again.

According to the results of X-ray structural analysis, the crystal structure corresponds to the following compounds: for steel 20YuCh – Al_2O_3, SiO_2, FeS, cerium sulfide; for USA steel – Al_2O_3, SiO_2, FeS.

4. Conclusion

1) The results of X-ray diffraction patterns of non-metallic inclusions of USA steel and steel 20YuCh can be made as follows:

- before the dissolution of carbides, sulfide inclusions in steel 20YuCh are elongated, dark, thin grains up to 0.21×0.017 mm in size. Round matte silicate with size up to 0.0595 mm and irregularly shaped matt translucent silicate with a red-yellow tint with size up to 0.034 mm;
- in USA steel a complex oxysulfide and aluminosilicate inclusions of elongated shape with intracrystalline transparent inclusions up to 0.12×0.051 mm and are dark rounded silicate

inclusions up to 0.085 mm are observed before the dissolution of carbides. Elongated shattered inclusions are also present;

- after dissolution of carbides in steel 20YuCh the aluminosilicate inclusions are transparent or translucent crystals of irregular shape up to 0.0595 mm and some amount of rounded matte and translucent silicate inclusions with rough edges with size up to 0.034 mm. In USA steel after dissolution of carbides the aluminosilicate inclusions are translucent with size up to 0.017 mm for irregular shape inclusions and up to 0.025 mm for rounded one. Also, there was observed a small amount of transparent needle-shaped inclusions (presumably silicon nitride);
- the results of inclusions analysis show that steel 20YuCh contain much fewer inclusions than USA steel, sulfides in it more dispersed and coagulated and significant part of sulfur tied with cerium. This explains the best resistance of 20YuCh steel against sulfide stress cracking.

2) As a result of hydrogenation of three considered steels: steel 20, steel 20YuCh and USA steel the steel 20YuCh showed the best resistance to stress corrosion cracking.

3) The high resistance of 20YuCh steel against sulfide stress cracking is explained by the purity of this steel in terms of inclusions, a much smaller amount and smaller size of sulfide stringers, and the fact that a significant part of the sulfur is tied with cerium.

References

[1] Romaniv O N and Nikiforchin G N 1986 *Mechanics of Corrosion Cracking of Structural Alloys* (Moscow: Metallurgy) p 294

[2] NACE Standard TM0177-2005 Standard Test Method Laboratory Testing of Metals for Resistance to Sulfide Stress Cracking and Stress Corrosion Cracking in H_2S Environments (Houston, Texas: NACE International)

[3] Rubenchik Yu I, Afanasenko E A and Legkostup N L 1980 *Improving the reliability of welded petrochemical equipment in environments that cause hydrogenation* (Moscow: TsINTIkhimneftemash) p 64

[4] Shparber I S, Schrader A V and Zhuk N P 1967 Corrosion and hydrogenation of steel in hydrogen sulfide solutions *Protection of metal* **5** pp 545-51

[5] Zikeev V N 1982 *Alloying and structure of hydrogen cracking resistant steels MiTOM* **5** pp 18–20

[6] Gabelchenko N I, Shevchuk G P and Sviridenko V Yu 1988 Corrosion-mechanical properties of steel 15MnCuNiMo *Physical and chemical mechanics of materials* **5** pp 107-8

[7] Rubenchik Yu I, Gabelchenko N I and Shevkun G P 1989 Improving the reliability of thick-walled castings from steel 20CrMnSiCuAl4L *Foundry Technologies and Equipment* **1** p 26

[8] Gabelchenko N I, Rubenchik Yu I and Zyuban N A 1997 Low-alloyed hydrogen-sulphide-resistant steels *Foundry Technologies and Equipment* **7** pp 18-9

Effect of the sol temperature on the structure and thickness of the SiO_2 coating of carbon fibers

S N Galyshev[1], **A I Gomzin**[2] **and N G Zaripov**[2]

[1] Institute of Solid State Physics of RAS, Chernogolovka, Moscow district, Russia
[2] Ufa State Aviation Technical University, Ufa, Russia

galyshew@gmail.com

Abstract. This paper deals with the development of carbon fiber coatings that act as diffusion barrier, and the role of the weak boundaries in the metal matrix composites. The developed coating will be a cost effective alternative to CVD-coatings. The paper considers the temperature dependence of the structure of SiO_2-sol and the coating thickness on the surface of the carbon fibers formed by sol-gel method. Detected sol temperature range optimum for coating. Possible reasons for the observed phenomena are proposed.

1. Introduction

Active research of directionally reinforced composites with a metal matrix and carbon fiber dates back to the 60s of the last century. However, compared with carbon fiber reinforced polymers, similar composites with a metal matrix have not found wide application in industry. As a first approximation, this is due to several natural reasons. Firstly, the chemical interaction between the matrix and the fiber introduces rough defects on the fiber surface, which significantly reduces its effective strength, and hence the strength of the composite [1]. Secondly, this interaction leads to the formation of a “strong” matrix/fiber bond, which in most cases is not desirable due to the need for uniform distribution of fibers in the matrix and limitation of the fiber volume fraction (no more than 30%) [2].

A possible solution to these problems is to create a barrier layer at the matrix/fiber interface. The most common method of applying barrier coatings that plays the role of “weak” boundaries is the CVD method. For example, in [3] the authors using a CVD coating of pyrolytic carbon, managed to obtain a composite with carbon fiber and an aluminum matrix (Al/Cf composite), the strength of which reaches 1360 MPa - the highest strength of an Al/Cf composite in open print today. Another example is work [4], in which the authors obtained a composite with an effective fiber strength of 95% of the initial one using gradient CVD coatings C/SiC/Si.

However, despite the successful elimination of the natural causes mentioned earlier and the creation of composites with high strength, such composites are still not widely used. This is primarily due to the complexity and low cost-effectiveness of the CVD method for coating carbon fiber.

A more accessible method for obtaining coatings from a practical point of view is the sol-gel technology, which does not require extreme conditions. Sol-gel technology makes it possible to obtain oxide coatings practically in “room” conditions. Such coatings serve as a reliable diffusion barrier [5, 6], and can also act as “weak” boundaries [7].

Today, there are several dozen works devoted to the deposition of oxide sol-gel coatings of various compositions on carbon fiber [8-10]. However, systematic research in this area is currently lacking.

In this work, the dependence of the structure and thickness of SiO_2 coatings on the carbon fiber surface formed by the sol-gel method on the sol temperature is considered. The optimum temperature range for the sol is determined for coating.

2. Materials and methods

UMT40-3K carbon fiber (manufactured by UMATEX Group, Russia) was used as a substrate for the deposition of SiO_2 coatings.

An oxide coating was applied to the carbon fiber by the method based on the sol-gel technology, similar to the method described in detail in [11, 12]. Carbon fiber is preliminarily cleaned from a sizing agent in acetone. Further, to create carboxyl groups on the fiber surface participating in the formation of sol-gel particles, the fiber is treated in nitric acid. After that, the fiber is immersed in a sol-gel solution of the following composition: tetraethoxysilane, ethanol, water and hydrochloric acid. In this work was studied the effect of the sol temperature on the structure of the coating formed on the surface of the carbon fiber. For this, five series of experiments were carried out at temperatures from 20 to 60°C.

The obtained sol-gel coatings on the carbon fiber surface were studied using a VERSA 3D scanning electron microscope. The images were obtained in the mode of secondary electrons with an accelerating voltage of 10 kV at magnifications up to 50000 times. It is known from preliminary experiments that the coating on the filaments on the carbon fiber tow surface may differ from the coating on the inside. Since the majority of filaments are concentrated precisely in the inner part of the fiber, for the study of coatings by scanning electron microscopy was used a technique for preparing samples, which allows obtaining more representative data in comparison with direct research of the bundle outer layers. After coating, a fragment of 10 mm was cut from a continuous bundle. The fragment was divided into two parts, after which one of the parts was placed on the surface of a carbon conductive tape fixed on the scanning electron microscope standard stage. After that, using tweezers the filaments of the selected fragment were distributed evenly over the conductive tape surface.

3. Results and discussion

Figure 1 shows the microstructure of the obtained coatings. At a sol temperature of 20°C, a coating is formed on the carbon fiber surface from a multitude of sol particles with average size is 10 ± 4 nm. When the temperature rises to 40°C, the thickness of the coating increases, and its structure becomes looser. At a temperature of 50°C there is an abrupt change in the thickness and structure of the coating, which becomes much thinner and more homogeneous in thickness. A further increase in temperature leads to a further decrease in the coating thickness.

The coating has a grained structure. Probably, the grains forming the coating are sol particles, and the coating is formed by the mechanism of their aggregation. It should be noted that an increase in temperature leads not only to a decrease in the coating thickness, but also a decrease in the size of the coating particles.

The observed effect is probably associated with the influence of temperature on the coagulation processes and the characteristics of the electric double layer on the surface of the sol particles. In the works of the classic of sol-gel processes Ralph K. Iler, it was noted that an increase in the temperature of some aqueous SiO_2 sols to 80°C, contrary to expectations, leads to an increase in the gelation time or, in other words, inhibits the aggregation process. This fully explains the effect observed in this work.

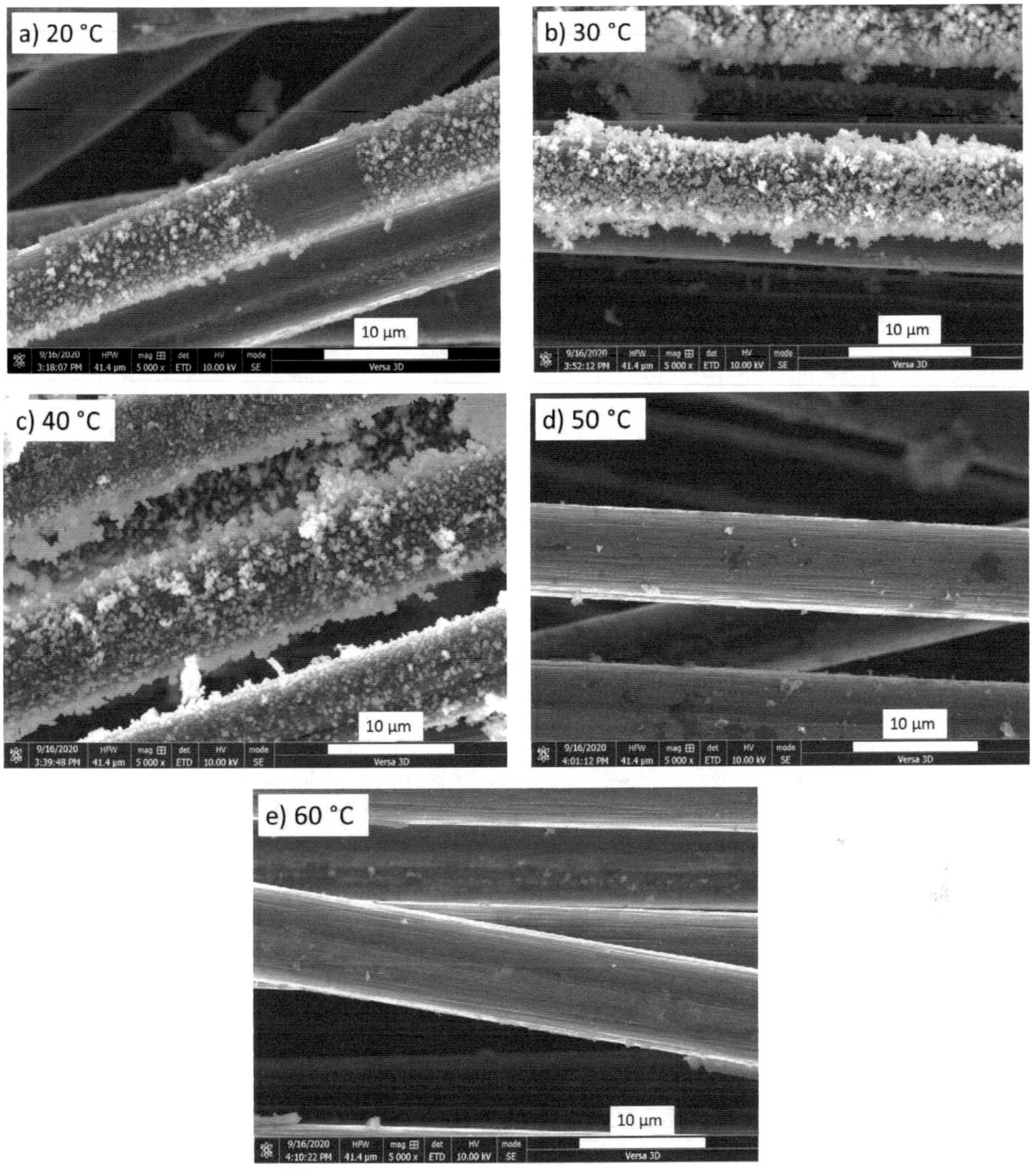

Figure 1. Microstructure of SiO_2 coatings of carbon fiber obtained at various sol temperatures: (a) 20°C, (b) 30°C, (c) 40°C, (d) 50°C, (e) 60°C.

Figure 2 shows the dependence of the average coating thickness on the sol temperature.

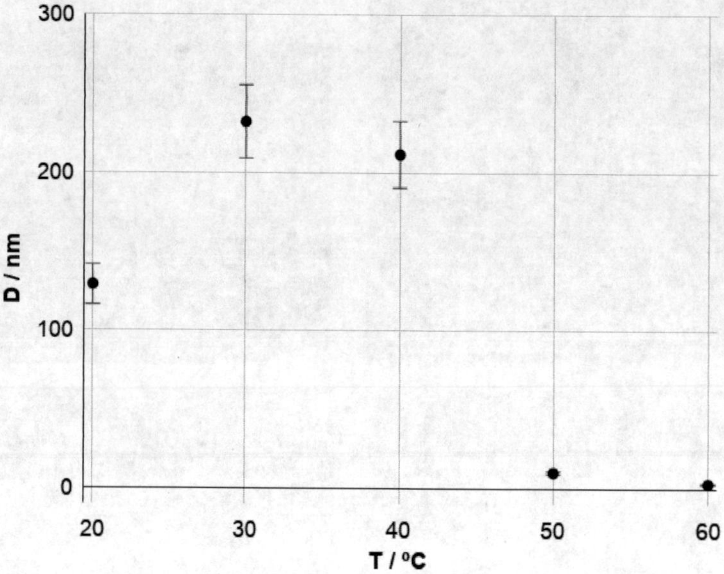

Figure 2. Dependence of the average coating thickness on the sol temperature.

The explanation of the above dependence undoubtedly requires further investigation of the process. Most likely, the sol temperature has such an effect on the structure and thickness of the coating due to the effect of temperature on the processes of polymerization, condensation, and dissolution in the colloidal system. Thus, it can be assumed that an increase in temperature up to 40°C promotes the condensation process. With a further increase in temperature, active dissolution of sol particles occurs, which leads to a significant decrease in the coating thickness.

4. Conclusion

The dependence on the temperature of the sol of the structure and thickness of SiO_2-coatings on the carbon fiber surface, formed by the sol-gel method, was determined. It is shown that the optimum temperature for coating is the temperature of the sol in the range from 20 to 40°C. The largest thickness of the obtained coatings was about 230 nm.

Acknowledgments

The authors are grateful to S.T. Mileiko, O.F. Shakhlevich, A.A. Kolchin and V.M. Masalov. The research part of this work was carried out in the Distributed Center for Collective Use of the ISSP RAS. The study was supported by a grant from the Russian Science Foundation (project No. 20-79-00004).

References

[1] Galyshev S, Gomzin A, Gallyamova R, Nazarov Yu, Vardanyan L, Shayakhmetov U and Musin F 2019 *IOP Conf. Ser.: Mater. Sci. Eng.* **489** 012004
[2] Mileiko S T 1997 *Metal and Ceramic Based Composite* (Amsterdam: Elsevier) p 690
[3] Vidal-Se´tif M H, Lancin M, Marhic C, Valle R, Raviart J-L, Daux J-C and Rabinovitch M 1999 *Mater. Sci. Eng.: A* **272** p 321-33
[4] Yu J, Li H and Shanga B 1994 *J. Mater. Sci.* **29** 2641
[5] Huang Yu, Ouyang Q, Zhang D, Zhu J, Li R and Yu H 2014 *Acta Metall. Sin.* **27 (5)** 775
[6] Tang Y, Deng Y, Zhang K, Liu L, Wu Y and Hu W 2008 *Ceramics Int.* **34** 1787

[7] Zeng Q 1998 *J. Appl. Polymer Sci.* **70** 177
[8] Rajan T, Pillai R and Pai B 1998 *J. Mater. Sci*. **33** 3491
[9] Clement J, Rack H, Wu K and Spencer H 1990 *Mater. Manuf. Process.* **5** 17
[10] Geiculescu C, Garth Spencer H, Rack H. and Sullivan B 1999 *Mater. Manuf. Process* **14** 489
[11] Vix-Gutesrl C and Ehrburger P 1997 *Carbon* **35** 1587
[12] Hao S, Lin T, Ning S, Qi Y, Deng Z and Wang Y 2019 *Mater. Sci. Semicond Process* **91** 181

Investigation of the influence of softening modifiers on the knockout of investment castings

N A Kidalov, N P Zhiltsov, N I Gabelchenko, A A Belov, N V Grigoreva and V F Zharkova

Volgograd State Technical University, 28 Lenin Avenue, Volgograd 400005, Russia

mitlp@vstu.ru

Abstract. The paper considers the problem of laboriousness of cleaning and knocking out investment casting castings from ceramic molds material. The influence of softening modifiers on the structure and residual strength of ceramic samples was analyzed. A magnesium sulfide and copper chloride were selected for the study because of their temperature of destruction, which is in the range of the calcination temperature of the ceramic shells and the temperature of pouring the mold with liquid metal. For comparative studies of particle size distribution after exposure to shock loading the samples were formed from ceramic suspension with introduction of 1.5 and 3 wt.% of softening modifiers. Fractures of ceramics were examined using a two-beam electron scanning microscope "Versa 3D". It was shown that the use of copper chloride as a softening modifier in the composition of ceramics is facilitate the process of knockout the mold of investment casting.

1. Introduction

Currently, one of the casting methods that provide the ability to manufacture high-quality castings of complex configuration, thin-walled castings, castings with low surface roughness, high accuracy class, minimum machining allowances, etc. is a promising method of investment casting in one-time multilayer ceramic forms. However, the materials used in this type of casting are often expensive, scarce and not environment friendly, and the shaping method is characterized by high labor intensity and duration. One of the reasons that increase the production time is the operation of knockout the casting from the shell and cleaning the internal cavities [1-3]. Despite the large number of binders that have already been developed and successfully used in investment casting, there is still no universal composition that makes it possible to obtain high-quality castings from a wide range of alloys with the necessary knockout ability, while meeting the production requirements and world environmental safety standards [4-7].

Thus, improving the technological properties of investment casting in order to reduce the duration and labor intensity of the process (the knockout ability and cleanability of the castings from the remains of ceramic mold) is an urgent problem of the foundry industry.

In the investment casting method, several cleaning methods are known, but each of them has its positive and negative aspects associated with inefficiency, high time and energy costs for cleaning, creating harmful production waste, as well as the high cost of the cleaning process itself [6-8].

The method proposed in this work for improving the process of cleaning castings from ceramic molds is free from these disadvantages.

2. Materials and methods

For the formation of easily knockout and easy-to-clean ceramic molds a suspension of the following composition (wt.%) was used: ethyl silicate (ETS32) - 51.0%, acetone - 40.0%, distilled water - 8.5 %, hydrochloric acid – 0.5 % and pulverized quartz refractory filler – the rest. Copper chloride ($CuCl_2$) and magnesium sulfate ($MgSO_4$) in amounts of 1.5 and 2.5 % (wt.) were added as technological softening modifiers [9]. The choice of these technological additives for introduction into the composition of the suspension is based on the principle – the temperature of calcination of ceramic shells (usually 900°C) < the additive temperature of destruction < temperature of pouring liquid metal [2, 9, 10].

In order to determine the residual strength of ceramic samples from the proposed suspension, control (without modifiers) and experimental bar-shaped samples were made with a cross section of 8×25 mm and a length of 120 mm, containing as modifiers – copper chloride or magnesium sulfate in amount of 1.5 or 3 % above weight (table 1). The amount of additive introduced into the suspension was determined based on preliminary studies [9, 10]. Before testing, the samples were heat treated up to a temperature of 1250°C in a "SNOL 80/12" chamber electric furnace, which corresponds to the temperature of pouring a ceramic mold with liquid iron.

The studied samples of ceramics were tested on a laboratory impact pendulum machine with a falling weight of 6.35 kg (± 0.015 kg) and a drop height of 50 (± 0.25 mm) [11]. The hammer was reequipped so that during the impact it was in full contact with the surface of the sample mounted on the plate. Control weighing of samples was carried out before testing. After three tests of the laboratory impact pendulum machine [11], the samples were sent for further study to the device for determining the particle size distribution, which includes a set of sieves with different mesh sizes [12].

The knockout of the ceramic mold was judged by the granulometric composition of the destroyed ceramic. For the study a set of sieves with a mesh size of 2.5-0.63 mm, as well as a tray – a pallet of a laboratory device was used. The ceramics were placed on the top sieve of the set, in which the sieves are arranged in descending order depending on the mesh size of the sieve. Sieving was carried out for 5 min, after which the remains of ceramics were transferred separately from each sieve onto glossy paper, weighed and the mass fraction of the remainder of ceramics on the sieves was calculated.

To estimate the effect of softening modifiers on the structure of ceramics, fractures of destroyed samples were investigated on a two-beam scanning electron microscope "Versa 3D". The dual beam system with integrated electron and ion beams (FIB/SEM) provides high quality images in compositional and topographic contrast.

3. Results and discussion

The test results and calculations of the mass fractions of ceramic residues on sieves are summarized in table 1.

Table 1. The test results.

Sample number	Modifier	Modifier content, in mass parts per 100 mass parts of suspension	Sample weight before testing m, g	After sieving	
				Sieve mesh size, mm	Sieve residue weight m_1, g
a	b	c	d	e	f
1	—	—	46.1	2.50	28.2
				1.60	10.4
				1.00	3.4
				0.63	2.5
				Tray	1.6

a	b	c	d	e	f
2	Copper chloride	1.5	44.5	2.50	18.1
				1.60	7.3
				1.00	6.3
				0.63	6.1
				Tray	6.7
3	Copper chloride	3	46.3	2.50	16.0
				1.60	6.7
				1.00	8.1
				0.63	8.1
				Tray	7.4
4	Magnesium sulfate	1.5	45.2	2.50	22.1
				1.60	8.8
				1.00	7.1
				0.63	4.0
				Tray	3.2
5	Magnesium sulfate	3	44.7	2.50	19.4
				1.60	8.1
				1.00	7.0
				0.63	5.4
				Tray	4.8

Comparative analysis of the granulometric composition of the control sample (without a modifier) and samples containing softening modifiers in the amount of 1.5 and 3 % after destruction on a laboratory impact pendulum is shown in figure 1.

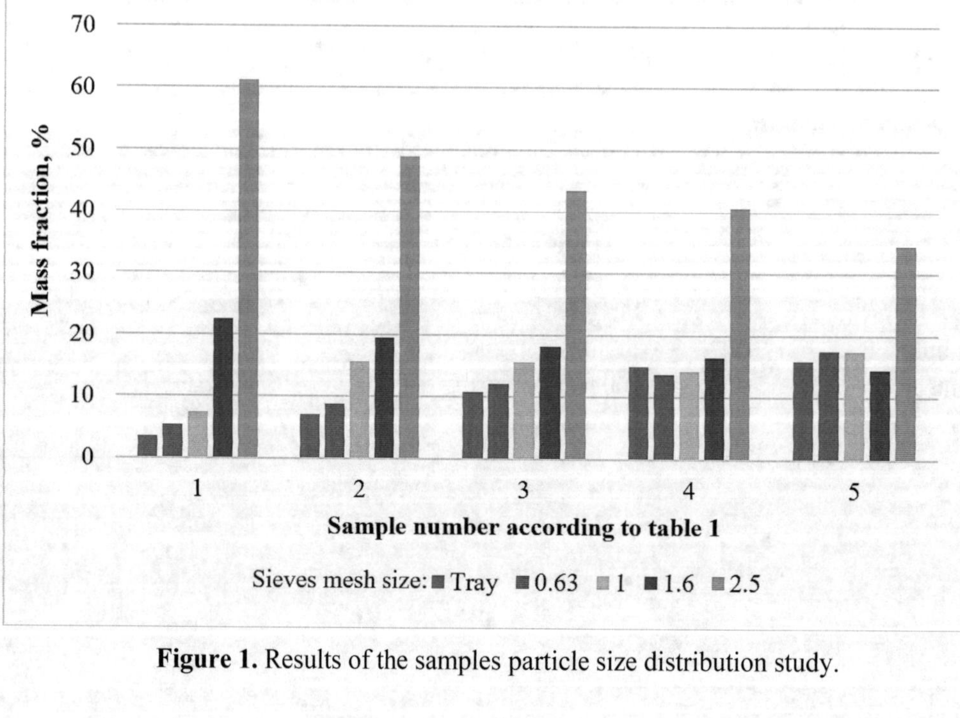

Figure 1. Results of the samples particle size distribution study.

Studies have shown that the control samples after testing on a laboratory impact pendulum collapsed partially – the mass fraction of large particles with a size of 2.5 mm and more was 61.17 %, which significantly exceeds the proportion of large particles in the samples with modifiers. Samples with modifiers softened more – the mass fraction of fine particles with a size of 0.63 mm and less in samples with copper chloride and magnesium sulfate in the composition exceeded the fraction of particles in control specimens without additives. The distribution of the remains of the experimental ceramics sample with the addition of copper chloride in an amount of 3 % (wt.) is mainly dispersed on sieves with mesh sizes from 1 mm to the tray. This indicates an easier knockout process for investment castings. Therefore, for comparative studies, fractures of ceramics with the addition of 3 % (wt.) copper chloride and ceramics without additives were selected.

The microstructures of the studied samples are shown in figure 2.

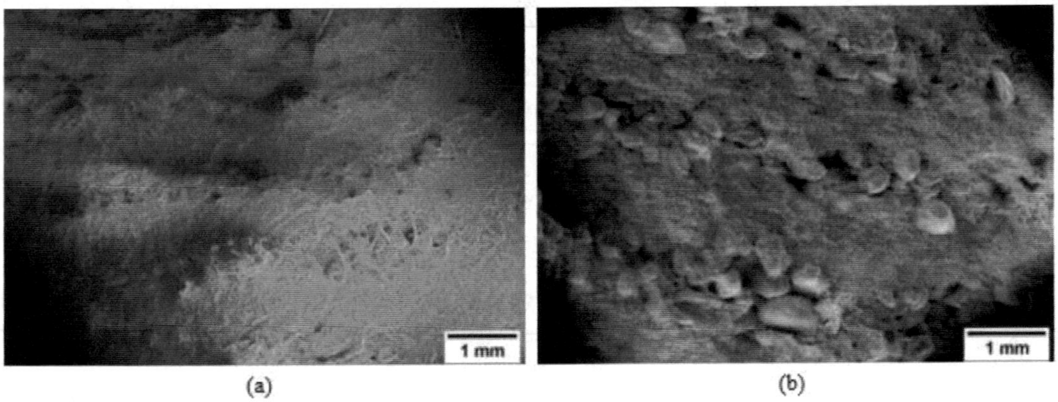

Figure 2. Microstructure of fractures of ceramic samples: (a) – control sample (without modifier), (b) – experimental sample with 3 % (wt.) of copper chloride.

It is shown that the microstructure of the experimental sample with the addition of 3% (wt.) copper chloride (figure 2 (b)), has microcracks in the binder film and the separation of the binder film from the surface of sand grains, in contrast to the continuous film of the binder in the control sample (figure 2 (a)). This is due to the fact that during pouring the liquid metal, the additive disintegrated with the release of gas, which, under its own pressure, softened the ceramics, creating internal stresses that contribute to the separation of the binder film from the sand grains. This improves the process of cleaning the castings from ceramic molds. The presence of such modifiers in the composition of the ceramic suspension makes it possible to facilitate the operation of knockout and cleaning the castings manufactured by the method of investment casting.

4. Conclusion

During the research, the following results were obtained:

- The granulometric composition of ceramics with the introduction of 3% (wt.) copper chloride is mainly concentrated on sieves with mesh sizes from 1 mm to the tray, which indicates that ceramic molds were knocked out with less effort.
- In the microstructure of ceramics modified with copper chloride, microcracks were revealed, which are serve as concentrators of internal stresses in the mold, as well as separation of the binder film from the surface of sand grains. This indicates a positive effect of the modifier on the softening of the ceramic mold during finished castings knockout.

References

[1] Gini E Ch, Zarubin A M and Rybkin V A 2005 *Foundry technology. Special types of casting* (Moscow: Academy) p 352

[2] Zhiltsov N P, Kidalov N A and Grebnev Yu V 2017 Research of knockout of ceramic moulds for investment casting *Procuring production in engineering (press-forging, foundry and other industries)* **7** pp 291-3
[3] Vertsyukh V A 2014 Resource-saving technology of shaping using aluminoborophosphate concentrate in investment casting *Dissertation of the candidate of technical sciences* (Chelyabinsk: Nosov Magnitogorsk State Technical University) p 126
[4] Vasil'ev V A 1996 Ceramic Mold Manufacture. Pattern Removing. Mold Roasting and Pouring. Finishing *Metallurg* **12** pp 35-7
[5] Znamenskii L G, Ivochkina O V and Varlamov A S 2019 Acceleration of the formation on the sodium silicate binder in investment casting *Mater. Sci. Forum* pp 673-7
[6] Dydak A, Kurdziel P, Zaba K, Puchlerska S and Szwachta G 2017 Influence of ceramic molds parameters on the quality of small-scale casts produced by the investment casting *METAL 2017 - 26th Int. Conf. on Metallurgy and Materials* vol 2017 (Czech Republic/Voronez: IBrno) pp 151-55
[7] Beckett B 1983 Cleaning investment castings the chemical way *Foundry Trade J.* **155** p 3270
[8] Wiese-Nielson K 1977 High pressure water cleaning of precision castings *Fonderie* **373** pp 397-403
[9] Kidalov N A and Zhiltsov N P 2018 Suspension for the manufacture of easy-to-clean ceramic casting molds *Patent for invention* **2673873**
[10] Kidalov N A and Zhiltsov N P 2018 A method of manufacturing easy-to-clean ceramic casting molds for investment casting *Patent for invention* **2673872**
[11] GOST 23409.7-78 1986 Moulding sands, moulding and core sand mixtures. Methods for determination of compressive, tensile, bending and shearing strength (Moscow: Standardinform)
[12] GOST 29234.3-91 2008 Moulding sands. Methods for determination of medium size of seed and coefficient of homogeneity (Moscow: Standardinform)

Strength analysis of composite plates with auxetic honeycomb at static bending by the finite element method

A V Mazaev

Voronezh State Technical University, Voronezh 304006, Russia

alexeymazaev@bk.ru

Abstract. In this paper, three-layer composite plates with continuous outer layers and an auxetic honeycomb interlayer of the chiral type based on rotating circles with tangentially attached rods are modeled. The physical and mechanical properties of the D16 aluminum alloy were chosen as the material properties of the layers of the composite plates. In the course of modeling, the discretization and volume of honeycomb structures in composites were varied at a constant thickness of the layers. Under the conditions of static bending of composite plates, load values were determined, at which maximum stresses in composites were equated to the conditional yield strength. The problem was solved using the finite element method in the framework of the theory of elasticity. The aim of this work is to study the influence of the discretization parameter and volume of the auxetic honeycomb interlayer on the magnitudes of load at yield strength stresses for the proposed composite plates.

1. Introduction

Sandwich composites based on honeycomb structures have a number of advantages over panels of continuous cross-section. In terms of structural applications, sandwich composites demonstrate primarily a reduced mass and sufficient strength, the value of which can be pre-determined by a rational choice of the cell geometry and the relative density of the honeycomb structure [1]. The geometric shape of the cells can form the auxetic behavior of the honeycomb structure, i.e. form a negative Poisson's ratio in the plane, the presence of which entails a number of useful advantages [2, 3]. It is known that auxetic materials have improved indentation resistance, high shear resistance and fracture toughness [3-5], synclastic curvature in bending [6], increased energy absorption [7], and others [3]. When developing sandwich composites with auxetic honeycomb, two variants of orientation of the plane with auxetic behavior are most often used: parallel to the plane [8, 9] or perpendicular to the plane [10, 11] of the composite panel, which directly affects the mechanical properties. The most popular method of manufacturing auxetic honeycomb structures is 3D printing. In this paper, we consider models of three-layer composite plates (figure 1), which include two continuous layers and a honeycomb auxetic interlayer of the chiral type based on rotating circles with tangentially attached rods. The plane, wherein the auxetic behavior of the honeycomb structure is manifested, is parallel to the plane of the composite plate.

Figure 1. A three-layer composite plate with an auxetic honeycomb interlayer.

To carry out numerical experiments, five chiral structures with different discretization were constructed, where the diameter of the circles d varied from 0.7 to 1.9 mm in increments of 0.3 mm (figure 2), and the side of the square grid cells a, in the nodes of which the centers of the circles are located, was determined by the coefficient $a = 1.6d$. The volume of honeycomb structures varied by changing the thickness of their walls t. As general values, the magnitudes of the chiral structure volume with $d = 0.7$ mm were established with changing the wall thickness t from 0.05 to 0.4 mm by the increments of 0.0875 mm. The obtained volume values do not lead to the loss of the geometric shape of chiral structures. The total thickness of the composite plate is 2 mm, the thickness of the outer layers and the honeycomb structure is 0.5 mm and 1 mm, respectively. Thus, 25 models of composite plates with five variants of the scale of cells of the honeycomb structure and five variants of the volume for each structure with a certain discretization were constructed. In so doing, 11 models of continuous plates were constructed for numerical experiments by varying the plate thickness from 1 to 2 mm by 0.1 mm increments.

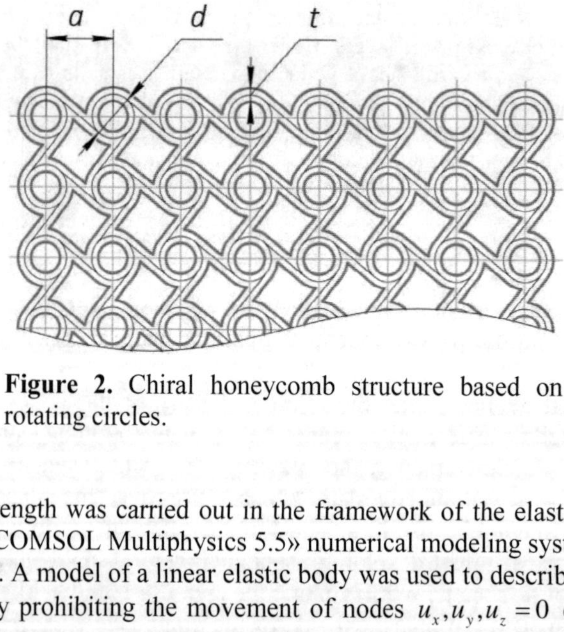

Figure 2. Chiral honeycomb structure based on rotating circles.

2. Calculation model

The analysis of plate strength was carried out in the framework of the elasticity theory by the finite element method in the «COMSOL Multiphysics 5.5» numerical modeling system using the «Structural mechanics» module [12]. A model of a linear elastic body was used to describe the material properties. The plates were fixed by prohibiting the movement of nodes $u_x, u_y, u_z = 0$ of finite element models which are located on shaded areas (figure 3). $u_{x,y,z}\left(0 \le x \le x_1, y = 0, 0 \le z \le b\right) = 0$, $u_{x,y,z}\left(x_2 \le x \le a, y = 0, 0 \le z \le b\right) = 0$, $u_{x,y,z}\left(0 \le x \le x_1, y = t, 0 \le z \le b\right) = 0$, $u_{x,y,z}\left(x_2 \le x \le a, y = t, 0 \le z \le b\right) = 0$, $a = 54$ mm, $b = 13$ mm, $t = 2$ mm, $l = a/2$, $x_1 = 12$, $x_2 = 42$, $z_1 = 1$, $z_2 = b - 1$. In case of calculation of composite plates, the external load was uniformly distributed over the nodes on a rectangular area $\left(l - 0.5 \le x \le l + 0.5, y = t, z_1 \le z \le z_2\right)$ with the width of 1 mm and the length of $\left(b - 2\right)$ mm. This condition allows one to exclude the false stress concentration at the lateral boundaries of the composite with a smaller width of the interlayer.

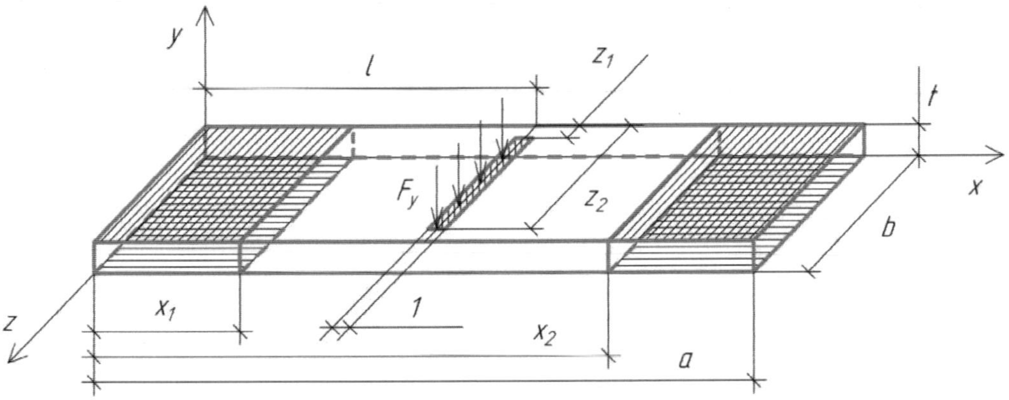

Figure 3. The boundary conditions of the plate for three-dimensional problem of theory of elasticity.

In the case of calculating continuous plates, the external load F_y was uniformly distributed on a straight-line segment $(x = l, y = t, 0 \le z \le b)$. When constructing a finite element mesh for a continuous plate, quadrilateral prisms were used. Discretization into quadrilateral prisms was performed by creating a two-dimensional mesh of quadrilaterals on the edge of the plate with subsequent broaching (figure 4 (a)). Discretization of composite plates into finite elements was carried out separately for each layer in a similar way when using triangular prisms (figure 4 (b)). The condition of continuity of field variables is established at the layer interfaces of composite plates.

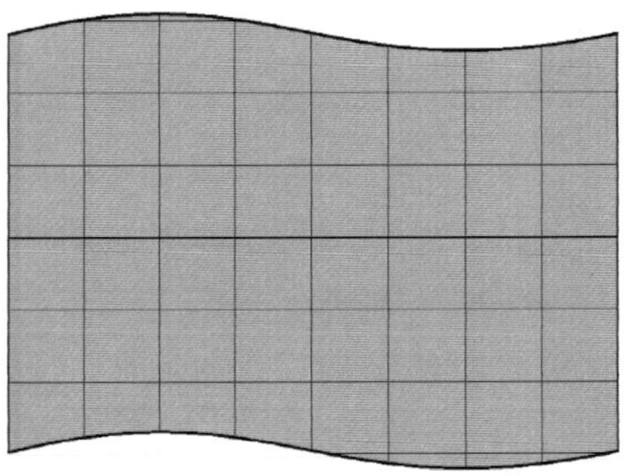

(a)

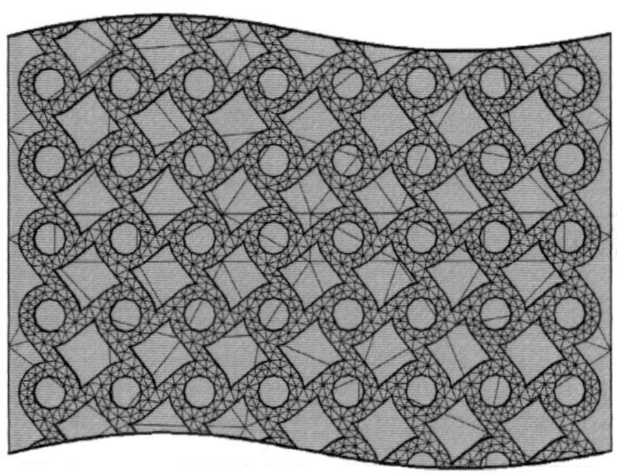

(b)

Figure 4. Finite element mesh of three-dimensional models: (a) elements of a continuous plate, (b) elements of a composite plate.

The physical and mechanical properties of aluminum alloy D16 [13] were used as the material properties of composite and continuous plates: elastic modulus $E = 72$ GPA, Poisson's ratio $\mu = 0.33$, density $\rho = 2780$ kg·m^{-3}, and conditional yield strength $\sigma_{0.2} = 290$ MPa. Under the static bending of composite and continuous plates, the load values F_y (N) were determined, at which the maximum stresses according to the Mises criterion were equated to the conditional yield strength $\sigma_{max} = \sigma_{0.2}$.

In order to verify the results of calculations of the «COMSOL Multiphysics 5.5» system, additional calculations of continuous plates were performed by the finite element method in displacements using the algorithm for solving a plane problem based on the known equations of elasticity theory [14, 15].

The algorithm of the method was adapted to the calculation of plates, where the stiffness matrix of the finite element k^e is determined by the expression

$$k^e_{r,s} = k^E_{r,s} + k^G_{r,s} \tag{1}$$

where k^E is the submatrix of normal deformations (2), k^G is the submatrix of shear deformations (3), *r* and *s* are numbers of the blocks of the matrices ($r = 1,2\ldots4$, $s = 1,2\ldots4$),

$$k^E_{r,s} = \frac{Eh}{4(1-\mu^2)} \begin{bmatrix} \gamma\xi_r\xi_s\left(1+\dfrac{\eta_r\eta_s}{3}\right) & \mu\xi_r\eta_s \\ \mu\eta_s\xi_r & \dfrac{\eta_r\eta_s}{\gamma}\left(1+\dfrac{\xi_r\xi_s}{3}\right) \end{bmatrix} \tag{2}$$

$$k^G_{r,s} = \frac{Gh}{4} \begin{bmatrix} \dfrac{\eta_r\eta_s}{\gamma} & \eta_r\xi_s \\ \xi_r\eta_s & \gamma\xi_r\xi_s \end{bmatrix} \tag{3}$$

E is the longitudinal elastic modulus, $G = E/2(1+\mu)$ is the shear modulus, μ is the Poisson's ratio, $\gamma = b/a$, *a* and *b* are dimensions of the sides of a rectangular (finite) element along the *x*- and *y*-axes, respectively, *h* is the size of the finite element on the *z*-axis, ξ and η are dimensionless coordinates of a rectangular element, $\xi_1 = -1$, $\xi_2 = 1$, $\xi_3 = 1$, $\xi_4 = -1$, $\eta_1 = -1$, $\eta_2 = -1$, $\eta_3 = 1$, $\eta_4 = 1$.

The matrix of matching the global numbers of nodes to the local numbers *N* is constructed according to the principle $N_{m,i} = r$, $r \in 1,2\ldots4$, where *m* is the global node number (figure 5 (a)), *i* is the number of the finite element (figure 5 (b)), and *r* is the local node number of the *i*-th finite element (figure 6), if $N_{m,i} \notin r$, then $N_{m,i} = 0$.

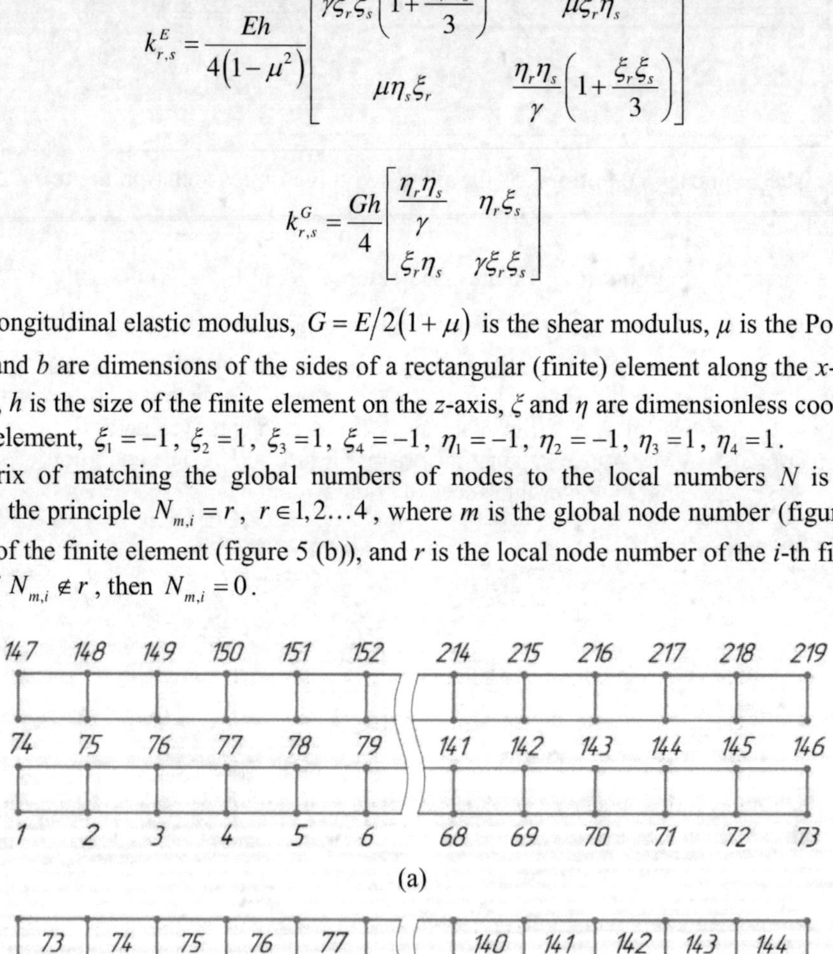

(b)

Figure 5. The finite element scheme of the plate: (a) global node numbering, (b) finite element numbering.

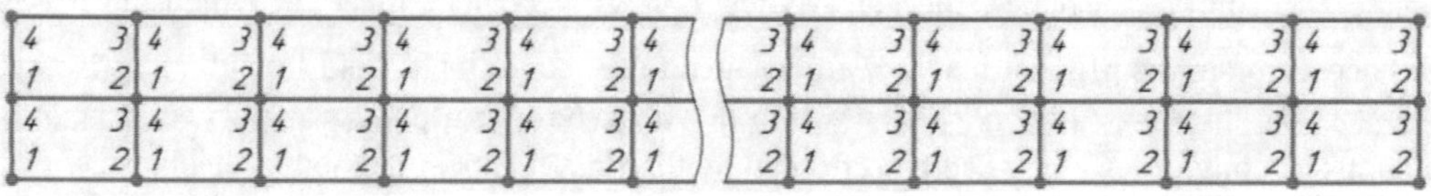

Figure 6. The scheme of the local node numbering of the finite elements.

The extended stiffness matrix k^{exp} is constructed using the $k^{exp}_{m,n}(i)=k^{e}_{r,s}$ principle, where $r=N_{m,i}$, $s=N_{n,i}$, $m,n\in 1,2\ldots c$, i is the number of the finite element, and c is the number of matrix rows N, if $r\vee s=0$ then $k^{e}_{r,s}=\begin{pmatrix}0&0\\0&0\end{pmatrix}$.

The stiffness matrix of the finite element model K is determined by summing the extended stiffness matrices $K=\sum_i k^{exp}(i)$. To account for the external fixation of a node in a finite element model, it is necessary to delete the rows $i_1=2m-1$, $i_2=2m$ and columns $j_1=2m-1$, $j_2=2m$ of the stiffness matrix K, where m is the global node number.

Node displacements are defined by the expression

$$u_a=K_a^{-1}P_a \tag{4}$$

where K_a^{-1} is the inverse matrix with respect to the stiffness matrix of the model considering the fixed nodes, P_a is the vector of nodal forces, $P_a=\left\{P^x_{a_1}\quad P^y_{a_2}\quad\cdots\quad P^x_{a_{n-1}}\quad P^y_{a_n}\right\}$ (here and further matrix-row in curly braces means the matrix-column), $n=2(c-p)$, $P^x_{a_{n-1}}$ and $P^y_{a_n}$ are the nodal forces along the x- and y-axes, respectively, c is the number of matrix rows N, and p is the number of fixed nodes.

The full displacement vector $u=\left\{u^x_1\quad u^y_2\quad\cdots\quad u^x_{n-1}\quad u^y_n\right\}$ ($n=2c$) includes zero displacements $u_{n_q}=0$, $q\in 1,2$, where $n_1=2m-1$, $n_2=2m$, m is the number of the fixed node, the $u_a=\left\{u^x_{a_1}\quad u^y_{a_2}\quad\cdots\quad u^x_{a_{b-1}}\quad u^y_{a_b}\right\}$ matrix is the submatrix of u, where $u_{a_b}\neq 0$, $b=2(c-p)$, c is the number of matrix rows N, and p is the number of fixed nodes. The displacement vectors along the x- and y-axes are defined by the expressions $x_i=u_m$ and $y_i=u_n$, respectively, where $m=2i-1$, $n=2i$, $i=1,2\ldots s/2$, u is the full vector of displacements, and s is the number of matrix rows in u.

The displacement vector of v nodes of the i-th finite element is constructed according to the $v_i=\left\{x_{m^1}\quad y_{m^1}\quad x_{m^2}\quad y_{m^2}\quad x_{m^3}\quad y_{m^3}\quad x_{m^4}\quad y_{m^4}\right\}$ principle, where $N_{m^1,i}=1$, $N_{m^2,i}=2$, $N_{m^3,i}=3$, $N_{m^4,i}=4$, x_{m^n} and y_{m^n} are the nodal displacements along the x- and y-axes, respectively.

The strain vector of ε of the i-th finite element is defined by the expression

$$\varepsilon(i,\xi,\eta)=\beta(\xi,\eta)\cdot v_i \tag{5}$$

where $\beta(\xi,\eta)$ is the matrix of the relationship between nodal displacements and strains,

$$\beta(\xi,\eta)=\frac{1}{2}\begin{pmatrix}b(1,\eta)&0&b(2,\eta)&0&b(3,\eta)&0&b(4,\eta)&0\\0&a(1,\xi)&0&a(2,\xi)&0&a(3,\xi)&0&a(4,\xi)\\a(1,\xi)&b(1,\eta)&a(2,\xi)&b(2,\eta)&a(3,\xi)&b(3,\eta)&a(4,\xi)&b(4,\eta)\end{pmatrix} \tag{6}$$

with $a(r,\xi)=\eta_r(1+\xi_r\xi)/b$, $b(r,\eta)=\xi_r(1+\eta_r\eta)/a$, $r=1,2\ldots 4$, a and b are dimensions of the sides of a rectangular element.

The nodal stress vector σ of the i-th finite element is defined by the expression

$$\sigma(i,\xi,\eta)=\chi\varepsilon(i,\xi,\eta) \tag{7}$$

wherein χ is the matrix of elastic constants, $\xi=\xi_r$, $\eta=\eta_r$, $r=1,2\ldots 4$, and

$$\chi = \frac{E}{1-\mu^2}\begin{pmatrix} 1 & \mu & 0 \\ \mu & 1 & 0 \\ 0 & 0 & 0 \end{pmatrix} \tag{8}$$

For the plane problem, the plates were fixed by prohibiting the displacements of nodes $u_x, u_y, u_z = 0$ of finite element models. Fixed nodes are located on shaded areas (figure 7).

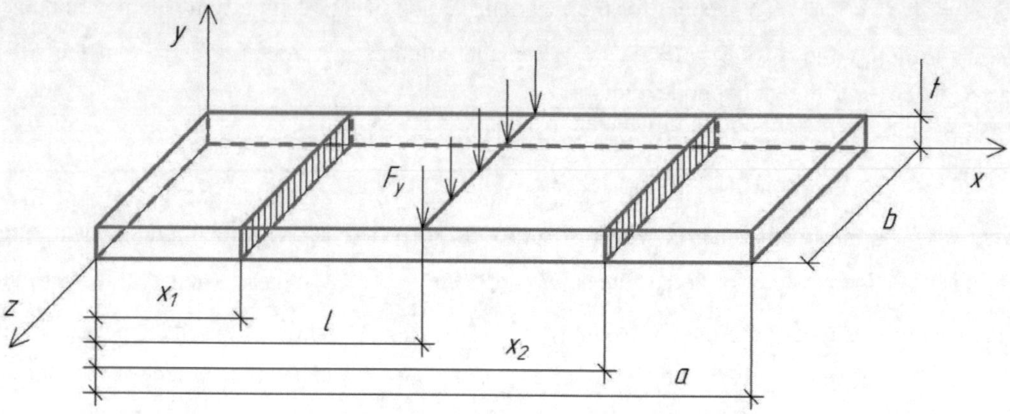

Figure 7. Boundary conditions of continuous plates for the plane problem of elasticity theory.

$u_{x,y,z}(x_1, 0 \le y \le t, 0 \le z \le b) = 0$, $u_{x,y,z}(x_2, 0 \le y \le t, 0 \le z \le b) = 0$, $a = 54$ mm, $b = 13$ mm, $t = 2$ mm, $l = a/2$, $x_1 = 12$, $x_2 = 42$. The external load F_y was uniformly distributed on a straight-line segment $(x = l, y = t, 0 \le z \le b)$.

3. Discussion of results and conclusion

For continuous plates with the thickness of $t = 1.5$ and $t = 2$ mm, the results of calculations via the «COMSOL Multiphysics 5.5» system were verified using the algorithm for solving a plane problem based on the finite element method. In the case of the plane problem for the plate at $t = 1.5$ mm, the finite element scheme shown in figure 5 was used, for the plate with $t = 2$ mm, a similar scheme was used for the number of rectangular elements equal to $i = 54$. As a result of calculations using the algorithm for solving the plane problem, matrices with scalar values of stresses in local nodes of finite elements are obtained. Using the contour graph for the stress matrix of the plate with the thickness $t = 1.5$ mm at $\sigma_{max} = \sigma_{0.2}$, the stress distribution diagram was constructed (figure 8 (a)). In the system «COMSOL Multiphysics 5.5» a similar stress distribution diagram for a plate at $t = 1.5$ mm was obtained (figure 8 (b)). Load value F_y at $\sigma_{max} = \sigma_{0.2}$ as a result of calculation using the algorithm for solving a plane problem for a plate with $t = 1.5$ mm and $t = 2$ mm is equal to 355.7 N and 627 N, respectively, while those obtained via «COMSOL Multiphysics 5.5» are 366.4 N and 660.5 N, respectively. Thus, the difference between the calculation results obtained using the «COMSOL Multiphysics 5.5» system and the algorithm for solving the plane problem for a plate with $t = 1.5$ mm and $t = 2$ mm is 3 % and 5.3 %, respectively. Based on the results of the analysis of the strength of composite and continuous plates, the plate's solid body volume V (mm^3) dependence of the load magnitude F_y at $\sigma_{max} = \sigma_{0.2}$ has been obtained and graphically presented in figure 9.

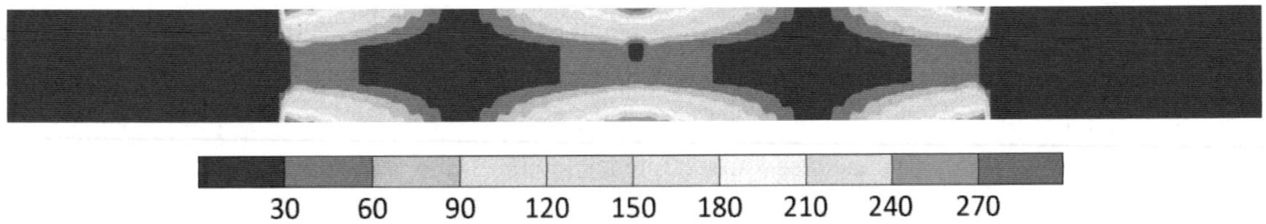

(a)

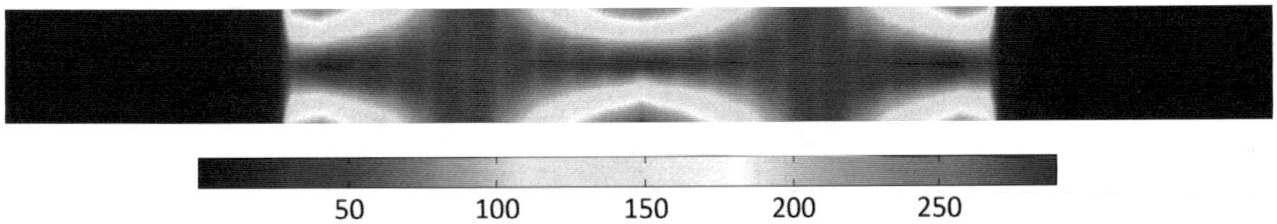

(b)

Figure 8. Stress distribution diagrams for a continuous plate with the thickness of $t = 1.5$ mm at $\sigma_{max} = \sigma_{0.2}$ obtained by the finite element method using two approaches: (a) the algorithm for solving a plane problem, (b) via «COMSOL Multiphysics 5.5».

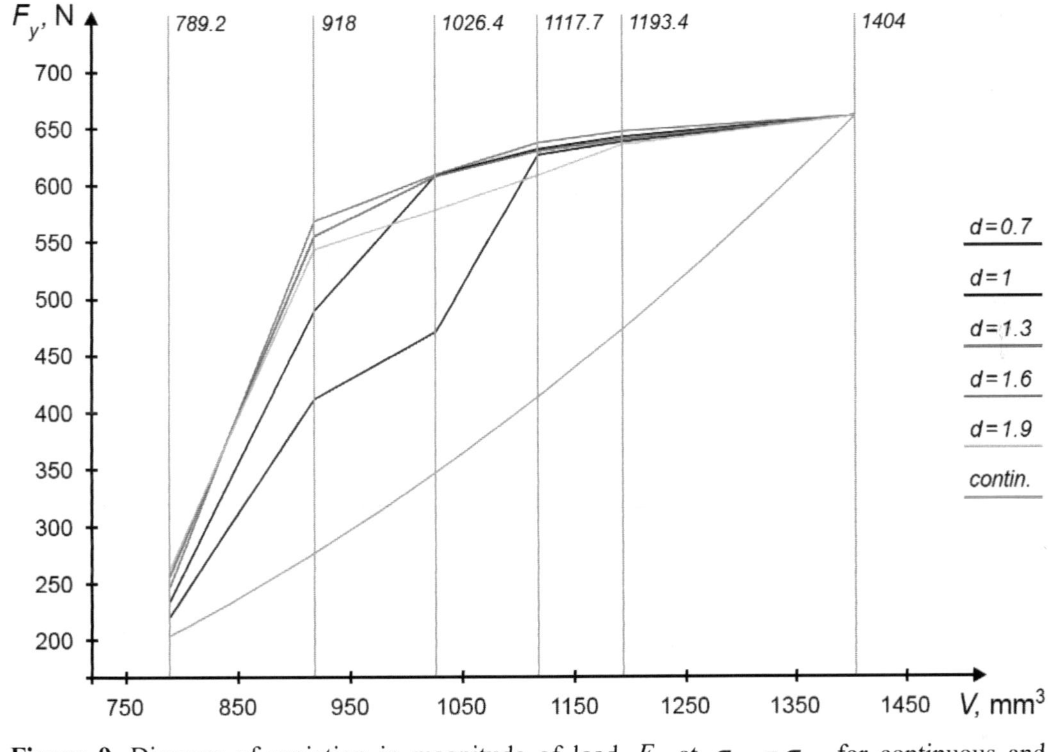

Figure 9. Diagram of variation in magnitude of load F_y at $\sigma_{max} = \sigma_{0.2}$ for continuous and composite plates as function of the solid body volume of the plate.

Vertical line-markers correspond to the volume values for composite plates, with the exception of the straight line $V = 1404$ mm^3, which is in line with the volume value for the continuous plate with $t = 2$ mm. The graphs show that there is a region with a significant difference in load values F_y at

$\sigma_{max} = \sigma_{0.2}$ for composite plates relative to a continuous plate with the same volume of solid body V. The magnitudes of F_y for composite plates during the discretization of chiral honeycomb with $d = 1.3$, 1.6 and 1.9 at the same volume V have an insignificant difference within the error range of the finite element method. The use of an auxetic honeycomb interlayer of the chiral type (figure 2) in the manufacture of three-layer composite plates (figure 1) could significantly reduce the consumption of the material used, and therefore decrease the weight of composite plates with an insignificant reduce in strength (figure 9) under static bending conditions (figure 3). With a decrease in the volume of the solid body of honeycomb interlayers, an advantage in strength is observed for composite plates with a lower discretization of chiral structures (figure 9). The considered composite plates can be manufactured using additive technologies.

Acknowledgement
The reported study was funded by RFBR, project number 20-38-90025.

References
[1] Zhang Q, Yang X, Li P, Huang G, Feng S, Shen C, Han B, Zhang X, Jin F, Xu F and Lu T J 2015 *Progress in Materials Sci.* **74** pp 332-400
[2] Saxena K K, Das R and Calius E P 2016 Three decades of auxetics research − materials with negative Poisson's ratio: a review *Advanced Engineering Materials* **18** pp 1847-70
[3] Mazaev A V, Ajeneza O and Shitikova M V 2020 Auxetics materials: classification, mechanical properties and applications *IOP Conf. Series: Materials Sci. and Engineering* **747**
[4] Underhill R S 2014 Defence applications of auxetic materials *Defense Systems Information Analysis Center J.* **1** pp 7-13
[5] Donoghue J P, Alderson K L and Evans K E 2009 The fracture toughness of composite laminates with a negative Poisson's ratio *Physica status solidi B* **246** pp 2011-17
[6] Naboni R and Mirante L 2015 Metamaterial computation and fabrication of auxetic patterns for architecture *Blucher Design Proc.* **2** 129-36
[7] Chekkal I, Bianchi M, Remillat C, Becot F X, Jaouen L and Scarpa F 2010 Vibro-acoustic properties of auxetic open cell foam: model and experimental results *Acta Acustica united with Acustica* **96** pp 266-74
[8] Klepka A, Staszewski W J, Uhl T, di Maio D, Scarpa F and Tee K F 2012 Impact damage detection in composite chiral sandwich panels *Key Engineering Materials* **518** pp 160-7
[9] Smardzewski J and Majewski A 2014 *Int. Sci. Conf.: New materials and technologies in the function of wooden products* pp 103-12
[10] Li T and Wang L 2017 Bending behavior of sandwich composite structures with tunable 3D-printed core materials *Composite Structures* **175** pp 46-57
[11] Chen G, Cheng Y, Zhang P, Liu J, Chen C and Cai S 2020 Design and modelling of auxetic double arrowhead honeycomb core sandwich panels for performance improvement under air blast loading *J. of Sandwich Structures & Materials*
[12] COMSOL AB 2019 *Structural Mechanics Module User's Guide* (Stockholm: COMSOL AB)
[13] Shalin R E 1982 *Aviation materials: reference book. Aluminum and beryllium alloys. Deformable aluminum alloys and beryllium-based alloys* (Moscow: VIAM)
[14] Obraztsov I F, Savel'ev L M and Khazanov K S 1985 *Finite element method in problems of structural mechanics of aircraft* (Moscow: Vysshaya Shkola)
[15] Zienkiewicz O C and Morice P B 1971 *The finite element method in engineering science* (London: McGraw-Hill)

Carbon-graphite preparation for impregnation with aluminum alloy

N Yu Miroshkin, V A Gulevskii and N A Kidalov

Volgograd State Technical University, 28 Lenin Avenue, Volgograd, 400005, Russia

nikolays34rus@gmail.com

Abstract. The possibility of increasing the efficiency of gasostat-free impregnation of porous carbon-graphite materials with aluminum alloys by applying thin high-purity metal alloying coatings by galvanic method to the surface of the pores of a carbon-graphite specimen is studied. It is shown that the preliminary application of such coatings by an electroplating method allows obtaining the effect of alloying with highly pure elements, increases the filling of pores with matrix melt and decreases the impregnation temperature (750°C). This technology makes it possible to obtain high-quality composite materials using simple technological equipment.

1. Introduction

Modern industry and manufacturing shows great interest in functional composite materials based on porous carbon materials impregnated with various metal alloys such as copper, antimony, aluminum. Metal-impregnated carbon composite materials occupy a significant share of the product range of top manufacturing companies of antifriction materials such as Ringsdorf (Germany), Schunk (Germany), Morgan (Great Britain), Mersen (France), Toyo (Japan) [1-4]. Interest in these composites is caused by characteristics that combine the advantages of a carbon material (high antifriction characteristics, self-lubrication, chemical resistance, ability to work successfully at high temperatures) with the properties of metals (high hardness, strength, electrical conductivity and thermal conductivity) [5, 6]. Composite materials with a copper matrix are most successfully used for the runners of the current collectors on electrified railway and urban transport under conditions of high currents. However, the disadvantage of a copper matrix is instability in some chemically active environments. Composite materials based on antimony matrix have high performance at high speeds and loads under dry friction conditions. In addition, these composites have high corrosion resistance, but they have low electrical conductivity, which limits their use [7, 8].

The aluminum matrix improves the strength characteristics of the composite material, such composite materials have high performance compared to materials based on copper and antimony. In addition, the aluminum matrix has high corrosion resistance in many corrosive environments [9, 10]. The combination of these properties makes aluminum alloys very promising as matrix alloys for obtaining composite materials. The difficulty of synthesizing such composites by liquid-state methods consists in the formation of an oxide film on the surface of aluminum, which prevents the chemical interaction of molten aluminum and the surface of carbon-graphite, impregnation is possible under conditions of very high temperatures (1200°C) [11]. Another way is to use high temperatures in combination with high pressure. However, there is a need to use complex and expensive autoclave equipment. An alternative solution is alloying the matrix alloy with surface active elements, which are elements with a low surface tension or metals with a low melting point. At the Department of Machines and Technology of Foundry of Volgograd State Technical University developed an experimental method for autoclave-free impregnation, and inexpensive standard furnace equipment and relatively simple technological equipment were used to implement this method [12]. The method consists of utilizing the pressure in a hermetic container made of a material with a low value of thermal expansion during heating and a metal (alloy) inside a device with a higher value of the coefficient of thermal expansion, for example, an aluminum matrix alloy and a device for impregnation made of a titanium alloy. As a result, the pressure inside the impregnating device is up to 5 MPa. However, the specified pressure is insufficient for the required filling of the pores of the carbon-graphite. To increase the filling of the pores, this method was improved by applying thin coatings of alloying high-purity metals on the inner surface of open pores and the perimeter of carbon frame by galvanic method.

The aim of this work was to improve the method of gasostat-free impregnation of porous carbon frame with aluminum alloys and to increase the quality and efficiency of impregnation.

2. Materials and methods

To increase the quality of impregnated carbon-graphite, the gas-free technology was divided into separate stages. At the first stage, preliminary preparation of carbon-graphite was carried out, during which galvanic coatings were applied to the walls of open pores. From a technological point of view, copper and nickel plating are acceptable. They can simply applied to carbon-graphite, differ in acceptable room requirements, available chemical reagents, do not require a powerful current source and are easy to prepare electrolyte. The disadvantage of these metal coatings is their recrystallization temperature, which for both metals is at the level of the melting point of aluminum. Moreover, the application of another alloying metal, such as zinc or chromium, without one of these coatings is already difficult or impossible. For example, to apply a zinc coating on carbon-graphite, it is necessary to use high current strength (20-40 A). And for those already covered with copper or nickel, these indicators decrease to 0.5-2 A. The initial solution was to apply the thinnest coatings, 5-10 μm thick, so that they did not affect the impregnation parameters, but reduced the contact angle and intensified the process. The solution was to apply the thinnest coatings, 5-10 μm thick, so that they did not affect the impregnation parameters, but reduced the contact angle and intensified the process.

Zinc was chosen as the main alloying element. Zinc has a high coefficient of thermal expansion, maximum fluidity and low melting point (419.5°C). Nevertheless, in the presence of oxygen and high temperature, zinc is rapidly oxidized, which leads to a significant decrease in the assimilation of the alloying element and environmental safety. This can be solved by applying an additional thin "protective" coating over the zinc, such as a layer of copper. In the impregnation process, zinc, having a low melting point, melts under the copper layer, increasing the initial infiltration of the matrix alloy into the pores. The copper coating will temporarily limit contact of the matrix alloy with zinc until oxygen is removed from the device.

Before coating, the carbon-graphite frame was degassed in a zinc electrolyte, and then a zinc coating was applied. This made it possible to fill open pores with zinc electrolyte and, upon further galvanization, to obtain a metal layer on the surface of the pores and to alloy the interface between the carbon-graphite and the matrix alloy with pure elements [13]. After applying a zinc layer, a copper coating was applied to the sample. After the completion of the formation of the copper coating, the sample was washed in water and dried. Then the carbon-graphite frame with galvanic coatings was placed in a small crucible with a volume of 100 ml, filled with an aluminum matrix alloy at a temperature (750°C), fixed the carbon-graphite with a special device from floating up and installed on a vibrating table. After that, it was covered with a vacuum hood, then the unit was connected to a vacuum pump, and the carbon-graphite frame was degassed in the matrix alloy, together with vibration treatment for 180 s at a temperature of 700°C. This treatment allows to dissolve a thin copper and zinc coating, and partially fill the pores of the carbon-graphite framework. After this treatment, the contents of the crucible were removed, fixed with a device that excludes the floating of carbon-graphite, and placed in a gasostat-free impregnation device, which was pre-filled by 2/3 with a crystallized matrix aluminum alloy. This makes it possible to achieve an additional increase in the volume of aluminum by 6% due to the fact that the transition of the metal to the liquid state is accompanied by an increase in volume and a decrease in density.

Then the impregnation device was covered with a lid heated to 800°C and the required amount of aluminum matrix alloy was added through the filling hole, and then hermetased with a plug heated to 900°C. At the next stage, the stopper was fixed with a cold wedge and the device for impregnation was placed in an oven heated to 750°C with isothermal holding for 20 minutes.

At each of these stages, the change in the weight of the carbon-graphite sample was recorded, which is shown in table 1.

Table 1. Change in the weight of carbon-graphite at the stages of its production.

Initial weight, g	After degassing in electrolyte, g	After drying, g	After impregnation, g
4.4	4.7	4.45	5.7
	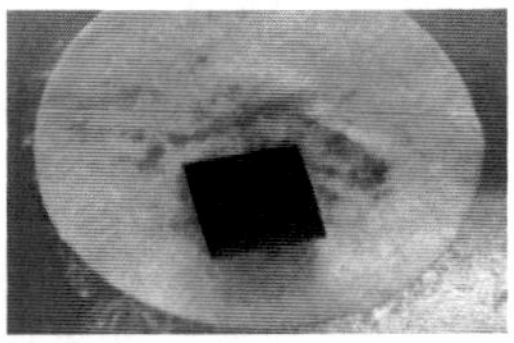	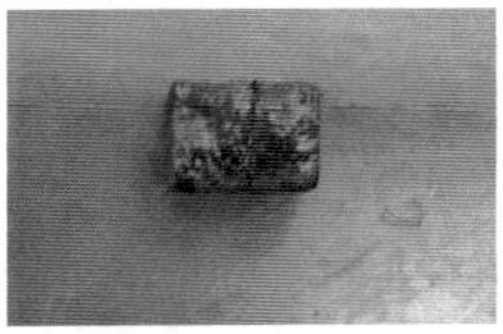	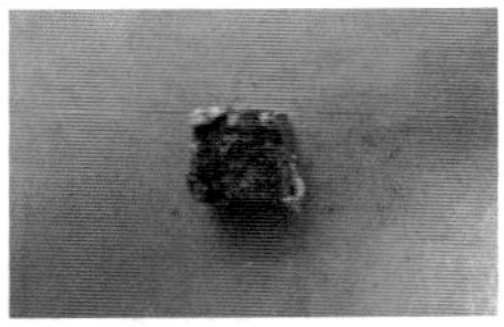

The final stage of fabricating composite material was its additional compaction with a lead alloy in a pressure form mold at a temperature of 500°C for 60 s [14]. Such thermomechanical treatment allows to compact the aluminum alloy in the pores of carbon-graphite, and modify the surface of the matrix alloy with lead, which has a positive effect on the antifriction characteristics of the composite material.

A sample for impregnation was made of AG-1500 carbon-graphite (open porosity of 15%) in the form of a cube with a side size of 30 mm. The sample volume was 900 mm^3 and the pore volume in it was 135 mm^3. The microstructure and chemical composition of the matrix metal in the composite material were investigated using a Versa 3D double beam electron scanning microscope in case of increase from 7000 to 22000 x, and acceleration voltage 15 kV.

3. Results and discussion

Figure 1 shows the microstructure of carbon-graphite after impregnation and compaction.

Figure 1. Microstructure of a carbon sample impregnation and compaction.

Studies of the chemical composition of the matrix alloy in the filled pore (figure 2) show that the matrix alloy in the initial state consisted of an alloy of aluminum and silicon, which is part of the A04130 alloy, and additionally the alloy was alloyed with magnesium (5 wt.%). Zinc and copper were applied to the walls of open pores of carbon-graphite in the form of thin galvanic coatings and their minimum amount confirms the effectiveness of their interaction with the matrix aluminum alloy, when they are completely dissolved.

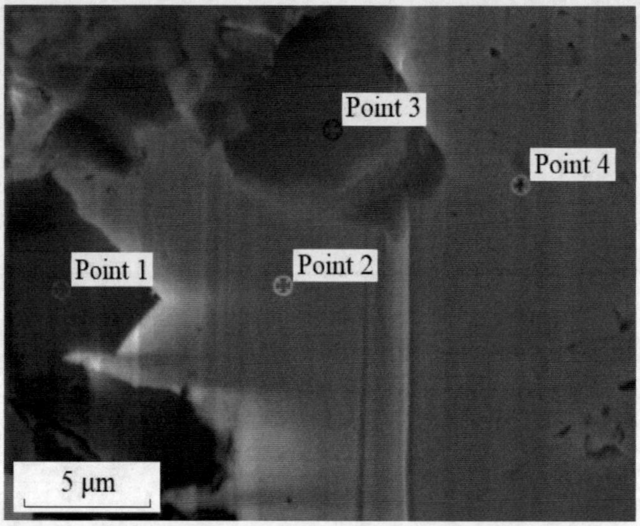

Figure 2. Studies of the chemical composition of the matrix alloy in the filled pore.

The presence of lead no more than 5% indicates (table 2) that in a short time (60 s) under the simultaneous action of temperature (500°C) and pressure, the lead alloy interacts with the surface of the matrix alloy, modifying it, which increases the antifriction characteristics of the composite material. It is worth noting the presence of titanium in the impregnated sample, titanium "came from the tooling material" into the alloy composition in the pore, since it was not introduced into the initial alloy for impregnation.

Table 2. The chemical composition of the aluminum alloy in the filled pore of carbon graphite.

Point	Element, at%							
	C	Al	Pb	Zn	Si	Cu	Mg	Ti
1	10.23	84.17	1.13	0.98	0.12	0.65	1.78	0.94
2	8.96	83.72	2.45	0.16	0.26	0.31	2.56	1.58
3	9.16	76.96	4.55	0.13	0.51	0.25	2.75	5.69
4	4.64	85.74	1.83	0.62	1.68	0.54	2.14	2.81

This is consistent with the presence of magnesium in the alloy, which, according to the electronegativity of Pauling's elements (figure 3), is located to the left of the other elements and has a higher selectivity with respect to carbon in the selected aluminum-based impregnation alloy.

22 Ti	13 Al	30 Zn	24 Cr	26 Fe	29 Cu	14 Si	28 Ni	51 Sb	15 P	82 Pb
1.54	1.61	1.65	1.66	1.83	1.90	1.90	1.91	2.05	2.19	2.33

a

Decrease in the activity of metal atoms →

K, Ca, Na, Mg, Al, Mn, Zn, Fe, Ni, Sn, Pb, H_2, Cu, Hg, Ag, Au

← Increasing the reducing properties of metal atoms

b

Figure 3. Electrochemical series of metal activity (a) - relative electronegativity, (b) - metal activity series.

This is consistent with the presence of magnesium in the alloy, which, according to the electronegativity of Pauling's elements (figure 3), is located to the left of the other elements and has a higher selectivity with respect to carbon in the selected aluminum-based impregnation alloy.

Molten aluminum is a very active material and its interaction with carbon should be further studied in conjunction with such not less active elements as titanium and magnesium.

4. Conclusion

Improving quality of impregnated graphite-carbon consists in the preparation and application of high-purity alloying metals (zinc, copper) to the inner surface of the pores by galvanic method. The combination of simultaneous exposure to temperature and pressure made it possible to obtain composite materials carbon-graphite aluminum alloy, economically using surface active elements in the form of galvanized layers (zinc, copper), reducing the impregnation temperature. Additional thermomechanical treatment of the aluminum alloy in the composition allows for the technological operation of compaction with the simultaneous modification of the surface of the matrix alloy with lead.

Acknowledgments

The reported study was funded by RFBR, project number 20-33-90234.

References

[1] Available at: https://www.schunkcarbontechnology.com Website of Public company “Schunk Carbon Technology”

[2] Available at: http://www.morgancarbon.com/ru Website of Public company “Morgan Advanced Materials”

[3] Website of Public company “Toyo Tanso” Available at: https://www.toyotanso.com/Products

[4] Website of Public company “Mersen” Available at: https://www.mersen.com

[5] Tuchinskii L I 1986 *Kompozitsionnyye materialy, poluchayemyye metodom propitki* (Moscow: Metallurgy) p 187

[6] Berent V Ya 2017 Perspektivy ispol'zovaniya metallouglerodnykh kontaktnykh vstavok dlya tokos’yema na zheleznykh dorogakh Rossii *Vestnik VNIIZHT* **76 (3)** pp 174–80

[7] Gulevsky V A, Antipov V I, Vinogradov L V, Kidalov N and Kolmakov A G 2017 *Propitka uglegrafita metallicheskimi splavami* (Volgograd: Volgograd State Technical University) p 271

[8] Galyshev S N, Gomzin A I, Gallyamova R F, NazarovA Yu, Vardanyan E L and Musin F F 2018 On the strength of carbon-fiber/aluminum-matrix composite wire *Composites & nanostructures* **10** pp 129-33

[9] Luts A R and Galochkina I A 2013 *Alyuminiyevyye Kompozitsionnyye Splavy - Splavy Budushchego* (Samara: SSTU) p 82
[10] Naidich J V 1981 *The Wettability of Solids by Liquid Metals* (New York: Academic Press) p 418
[11] Eustathopoulos N, Drevet N B and Nicholas M G 1999 *Wettability at High Temperatures* (Oxford: Pergamon) p 176
[12] Gulevsky V A, Vinogradov L V, Antipov V I, Miroshkin N Yu, Kolmakov A G, Kostikov V I, Mukhina Yu E and Baranov E E 2018 Razrabotka sposoba bezavtoklavnoy propitki poristogo uglegrafitovogo materiala liteynymi alyuminievymi splavami *Perspektivnye Materialy* **10** pp 73–9
[13] Gulevsky V A, Antipov V I, Vinogradov L V, Miroshkin N Yu, Gulevsky V V, Kolmakov A G, Mukhina Yu E and Baranov E E 2020 Use of electroplating for increasing the efficiency and quality of impregnation of a porous graphitized carbon material with copper alloys *Russian Metallurgy (Metally)* **7** pp 746-51
[14] Gulevsky V A, Miroshkin N Yu, Tsurikhin S N and Gundrov O Yu 2020 Podrobneye o propitke uglegrafita alyuminiyem *Izvestiya Volgogradskogo Gosudarstvennogo Tekhnicheskogo Universiteta* **7 (242)** pp 77-82

IOP Conf. Series: Materials Science and Engineering 1129 (2021) 012009 doi:10.1088/1757-899X/1129/1/012009

A simple model of additive process mechanics for controlling the evolution of stresses in filament-wound cylindrical composites during the manufacturing process

D A Parshin

Ishlinsky Institute for Problems in Mechanics of the Russian Academy of Sciences (IPMech RAS), Vernadskogo Avenue, 101 Bldg 1, Moscow 119526, Russia
Department of Applied Mathematics, Bauman Moscow State Technical University (BMSTU), 2nd Baumanskaya Street, Bldg 5, Moscow 105005, Russia

parshin@ipmnet.ru

Abstract. A relatively simple mathematical model is proposed and developed that allows using additive process mechanics approaches to trace the evolution of technological stress fields in cylindrical composites during their manufacture by slow multi-layer winding of fibre-reinforced material with an arbitrary variation in the pre-tension force and fibre packing density. The constructed closed analytical dependences make it possible to state problems on controlling the distributions of these stresses.

1. Introduction

Composites are widely used in creating modern materials and manufacturing products in various sectors of industry. This is largely due to the fact that their use allows to achieve the required strength and stiffness indicators of the resulting materials and products with a significant gain in weight and geometric parameters. At the same time, we can point out another remarkable feature of composites: many of the processes that are already used in practice or only potentially conceivable for their manufacture open up fundamental capabilities for controlled or even targeted formation of the required characteristics of the technological stress-strain state in the resulting materials and products. Such processes, of course, include all possible processes of additive — in the most general sense of the term — manufacturing of composites. It is obvious that in order to successfully use the mentioned capabilities, it is necessary to understand the mechanisms that cause the occurrence and development of mechanical stresses in specific composite manufacturing processes, and, moreover, to develop a mathematical description of these mechanisms.

It should be noted that traditional approaches to the study of mechanical problems associated with additive manufacturing processes operate with standard categories of mechanics of deformable solids, augmenting them with certain corrections designed to take into account the shape and properties of the solid that change during the manufacturing process. However, such approaches cannot be fully adequate due to the very special kinematics of the processes of deformation of solids subjected to unceasing supplementation with new material elements. Such solids, according to the already established terminology, are called built-up or growing solids. For their mechanical description, it is necessary to develop fundamentally new approaches, which is what the scientific works of the corresponding research school established in Russia under the leadership of Professor A.V. Manzhirov are devoted to (some of the latest results in this area can be found, for example, in [1–14]). The research presented in this paper is based on these approaches.

One of the long-used methods for additive manufacturing of composite materials and products is the method of multi-layer winding of the being reinforced material on a certain form. Pre-tension of the wound layers causes the emergence and further evolution of technological stress fields in the entire solid formed by winding. The nature and result of evolution will obviously depend in a decisive way on many specific parameters of the technological process under consideration. However, qualitative and, especially, quantitative forecasting of this dependence is impossible without adequate mathematical modelling of this process.

This work is devoted to the construction and analysis of one of the simplest mathematical models of additive process mechanics for the technological process of composite manufacturing by multi-layer power winding of a material being reinforced with parallel fibres onto a slowly rotating circular cylindrical form.

2. Calculation model

The simulated process is sequential applying material layers of relatively small thickness onto the outer surface of the manufactured solid. Each such layer contains one round of reinforcing fibres in the radial direction. The rounds of the fibres are oriented perpendicular to the generatrix of the current cylindrical application surface and are located along this surface without a fixed step and without any reference to the location of the rounds in the layer of material having been previously applied onto the solid. The fibre position in the binder stratum of each layer, in the radial direction, is also not clearly defined and can change even within a single layer. In the described case, we can assume that the equivalent mechanical properties of the solid obtained as a result of winding will be transversally isotropic with the local axis of isotropy coinciding with the circumferential direction [15]. Indeed, under the assumptions made, the neighborhood of an arbitrary point in any axial section of this solid is normally permeated by a large number of reinforcing fibres rather randomly distributed over this neighborhood.

We believe that it is acceptable to change the thickness of the elementary layers of material applied during the winding process. However, this change occurs independently of the cycles of wrapping these layers around the formed cylindrical solid. In this case, the total thickness of a large number of sufficiently thin elementary layers will, on average, remain the same in all axial sections throughout the entire process. This means that this process can be modelled as axisymmetric.

The problem on deforming the being formed solid will be considered in plane and quasi-static formulation. The first assumes a sufficient length of this solid along the axis of its rotation, the second assumes a smooth change in the rotation speed during the winding of the layers of additional material onto the solid, as well as the insignificance of dynamic effects caused by bringing these layers into contact with the being formed solid. We consider the rotation itself sufficiently slow in order to be able to not take into account the impact on the material of centrifugal forces of inertia (an example of solving the problem on layer-by-layer formation of a rotating isotropic viscoelastic cylinder taking into account centrifugal forces is presented in paper [16]). In addition, due to the relative smallness of the thickness of the applied individual layers, the following two simplifications will also be justified. On the one hand, it is correct to model the studied process of solid growth as a continuous one, in which an infinitely thin layer of new material is added to this solid for each infinitesimal time interval. On the other hand, we can refuse to take into account possible retarded deformation of the material, focusing only on the analysis of pure elastic deformation, since, as shown in [17], the influence of the deformation aftereffect on the process of stress development in a growing solid is leveled at a very low rate of growth of the latter.

We will investigate small strain in the being manufactured cylindrical solid. Then we can assume as given the function $R(t)$ describing the law of change over time t of the radius R of the outer surface of the growing solid under consideration, and this function will be strictly increasing. Indeed, this radius changes over time due to two factors: adding new material to the solid and the deformation of this solid. With a small strain, the latter factor can be ignored in comparison with the former.

In the proposed model, we will also not take into account the pliability of the form used, considering it to be essentially more rigid than the material wound on it. Thus, the internal radius R_0 of the being formed cylindrical composite will be considered constant. Note that this assumption is not of principle in the constructed model: it can be easily abandoned at the expense of complexity of all the dependencies obtained as a result of solving the corresponding mechanical problem.

Let φ hereinafter be the circumferential, z — the longitudinal, and ρ — the radial coordinate counted in a system that rotates along with the rigid cylindrical form on which the material is wound. The time in relation to which the growth process under study is considered is denoted by t.

As indicated above, the ort that defines the local isotropy axis indicates the direction of change of the curvilinear coordinate φ at each point. So orts that indicate the direction of change of coordinates z and ρ set a local orthogonal basis in the isotropy plane passing through this point.

We denote by v_φ, v_z, v_ρ the components of the velocity field associated with the deformation displacements of the points inside the growing solid under consideration, in the introduced moving frame of reference (φ, z, ρ). The plane strain state and the axial symmetry of the process give

$$v_z \equiv 0, \quad \frac{\partial v_\rho}{\partial z} \equiv \frac{\partial v_\varphi}{\partial z} \equiv 0; \quad \frac{\partial v_\rho}{\partial \varphi} \equiv \frac{\partial v_\varphi}{\partial \varphi} \equiv 0 .$$

Ignoring those deformation displacements of the points of the being formed solid, which twirl it in the considered coordinate system around the axis of symmetry, we will also have a mirror symmetry of the mentioned velocity field, that is

$$v_\varphi \equiv 0 .$$

As a result, v_ρ depending on t and ρ will be the only non-zero component of this velocity field.

Thus, the strain rate tensor will have only two non-zero physical components in the coordinate system used:

$$D_{\varphi\varphi} = \frac{v_\rho}{\rho}, \quad D_{\rho\rho} = \frac{\partial v_\rho}{\partial \rho} . \tag{1}$$

These components must be associated with the rates of change of the normal components of the stress tensor, based on the equations of state of a transversely isotropic elastic medium:

$$\frac{\partial \sigma_{\varphi\varphi}}{\partial t} = C_1 D_{\varphi\varphi} + C_2 D_{\rho\rho}, \quad \frac{\partial \sigma_{zz}}{\partial t} = C_2 D_{\varphi\varphi} + C_3 D_{\rho\rho}, \quad \frac{\partial \sigma_{\rho\rho}}{\partial t} = C_2 D_{\varphi\varphi} + C_4 D_{\rho\rho} , \tag{2}$$

where C_k are the elastic constants. The tangent components of the stress tensor in the coordinate system used will be zero at all points of the solid.

Naturally, the stress equilibrium equation and its corollary for the stress rates remain valid, so

$$\frac{\partial^2 \sigma_{\rho\rho}}{\partial \rho \partial t} + \frac{1}{\rho}\left(\frac{\partial \sigma_{\rho\rho}}{\partial t} - \frac{\partial \sigma_{\varphi\varphi}}{\partial t} \right) = 0, \quad R_0 < \rho < R(t) . \tag{3}$$

The rigid coupling of the first wound layer of material with the form surface is expressed by the boundary condition

$$v = 0, \quad \rho = R_0 . \tag{4}$$

In the here employed model of continuous growth, the outer surface of the being formed solid, onto which the layers of additional material are applied during the winding process, moves in space in the normal direction due to constant replenishment with new points, in the vicinity of which some initial stress state is predetermined:

$$\sigma_{\varphi\varphi} = \sigma_*(R(t)), \quad \sigma_{zz} = \sigma_{\rho\rho} = 0, \quad \rho = R(t). \tag{5}$$

This stress state is characterized by the function $\sigma_*(R)$ that sets the pre-tension in the circumferential direction of the applied layer of material at the moment when the outer radius of the solid being formed by winding is equal to R. Denoting this moment in time by $\tau_*(R)$, we will have the following identity fair for the entire cylindrical growing solid under consideration, that is, for all layers of radius ρ included in its composition:

$$R(\tau_*(\rho)) \equiv \rho.$$

On the basis of this identity, setting $t = \tau_*(\rho)$ in (5), we obtain

$$\sigma_{\varphi\varphi} = \sigma_*(\rho), \quad \sigma_{zz} = \sigma_{\rho\rho} = 0, \quad t = \tau_*(\rho). \tag{6}$$

It is obvious that initial conditions (6) unique to the deformable solid are equivalent to boundary conditions (5) on the moving growth surface.

3. Results and discussion

We can show (but omit the corresponding derivations here because of the limited paper) that specific initial conditions (6) considered together with the stress equilibrium equation lead to the condition on the moving growth surface that sets the initial rate of change of radial stresses at points attached to it:

$$\frac{\partial \sigma_{\rho\rho}}{\partial t} = -s(t), \quad \rho = R(t), \tag{7}$$

where the known function $s(t)$ in the right-hand side is determined by the particular programs of solid growth, $R(t)$, and of wound material pre-tension, $\sigma_*(R)$:

$$s(t) = \frac{\sigma_*(R(t))}{R(t)} \frac{\mathrm{d}R(t)}{\mathrm{d}t}.$$

The formulation of condition (7) closes the statement of the boundary value problem for evaluation of the rates of change of the stress-strain state characteristics for the solid under consideration during its continuous growth. The problem is given by differential equation (3) supplemented with relations (2) and (1), and by boundary conditions (4) and (7) on the stationary and moving boundaries of the growing solid in the considered coordinate system, respectively.

It turns out, it is possible to construct the exact analytical solution of the above non-classical mechanical problem (the constructing procedure is also to omit here). The following rates of change of all stress tensor non-zero components during the growth process will correspond to this solution:

$$\begin{aligned}
\frac{\partial \sigma_\varphi}{\partial t} &= \frac{[R(t)/R_0]^{\gamma+1} s(t)}{(C_2+\gamma C_4)[R(t)/R_0]^{2\gamma} - C_2 + \gamma C_4}\left[\frac{C_1 - \gamma C_2}{(\rho/R_0)^{\gamma+1}} - (C_1+\gamma C_2)(\rho/R_0)^{\gamma-1}\right] \equiv S_\varphi(\rho,t),\\
\frac{\partial \sigma_z}{\partial t} &= \frac{[R(t)/R_0]^{\gamma+1} s(t)}{(C_2+\gamma C_4)[R(t)/R_0]^{2\gamma} - C_2 + \gamma C_4}\left[\frac{C_2 - \gamma C_3}{(\rho/R_0)^{\gamma+1}} - (C_2+\gamma C_3)(\rho/R_0)^{\gamma-1}\right] \equiv S_z(\rho,t),\\
\frac{\partial \sigma_\rho}{\partial t} &= \frac{[R(t)/R_0]^{\gamma+1} s(t)}{(C_2+\gamma C_4)[R(t)/R_0]^{2\gamma} - C_2 + \gamma C_4}\left[\frac{C_2 - \gamma C_4}{(\rho/R_0)^{\gamma+1}} - (C_2+\gamma C_4)(\rho/R_0)^{\gamma-1}\right] \equiv S_\rho(\rho,t),
\end{aligned} \tag{8}$$

where $\gamma = \sqrt{C_1/C_4}$ is the dimensionless material constant which can be called, in its mechanical sense, the index of anisotropy of the composite material obtained as a result of winding (it would be $\gamma = 1$ if the material were fully isotropic).

After solving the problem given by (3), (2), (1), (4), and (7), it is possible to determine how the fields of technological stresses in the resulting composite will evolve during its winding. To do this, one should integrate expressions (8) in time taking into account initial conditions (6):

$$\sigma_\varphi = \sigma_*(\rho) + \int_{\tau_*(\rho)}^{t} S_\varphi(\rho,\tau)\mathrm{d}\tau, \quad \sigma_z = \int_{\tau_*(\rho)}^{t} S_z(\rho,\tau)\mathrm{d}\tau, \quad \sigma_\rho = \int_{\tau_*(\rho)}^{t} S_\rho(\rho,\tau)\mathrm{d}\tau. \tag{9}$$

The initial stresses in the applied elementary layers of additional material, caused by pre-tension of the reinforcing fibres, can be calculated in the proposed model of mechanics of continuous additive processes as follows. Let $F(R)$ be the pre-tension force acting on the fibre wound onto the surface of the being formed solid when this solid has an external radius R. Let $\kappa(R)$ be the packing density of reinforcing fibres (the number of fibres per unit area of the axial section) in the vicinity of this fibre. Then the function $\sigma_*(R)$ in boundary condition (5) can be set as

$$\sigma_*(R) = \kappa(R)F(R).$$

The analytical dependences obtained above make it possible to formulate various problems on technological control of the process stresses developing in filament-wound cylindrical composites. One of the goals of such controlling may be to minimize the risk of destruction of the composite both during its manufacture and in particular conditions of further work [18]. Constructed formulas (8), (9) also allow us to determine the evolution of the contact pressure from the wound layer of composite material onto the form used during its winding, which is extremely important for analyzing the stability and wear of the latter [19–23]. In addition, the developed model provides a simple apparatus for solving the problem on predicting the residual stresses in manufactured composites of the considered kind after their separation from the form having been used as technological tool. These stresses are unavoidable in every additive process of obtaining materials due to the very specifics of such processes and are explained by strain incompatibility of different material elements making up the solid being formed in a such process [24].

4. Conclusion

A non-classical mathematical model of solid mechanics has been developed to predict and control the evolution of technological stresses in cylindrical composites during their additive manufacturing by multi-layer winding of fibre-reinforced material. The model takes into account the geometric scheme of fibre packing and the possibility to arbitrarily variate the force of their pre-tension and the packing density. The corresponding boundary value problem in terms of the stress-strain state characteristics rates in the being manufactured composite solid is stated. This problem exact solution is constructed, on the basis of which the development of stress fields in the obtained composite is reconstituted throughout the entire manufacturing process for concrete programs of changing the control parameters of the simulated technology.

Acknowledgements

This study was carried out within the framework of the State Assignment Research (under Russian Ministry of Science and Higher Education Theme No. AAAA-A20-120011690132-4), and partially on the topic of the Russian Foundation for Basic Research Project (under grant No. 18-01-00920-a).

References

[1] Manzhirov A V 2017 Fundamentals of mechanical design and analysis for AM fabricated parts *Procedia Manufacturing* **7** pp 59–65

[2] Manzhirov A V 2017 Advances in the theory of surface growth with applications to additive manufacturing technologies *Procedia Engng.* **173** pp 11–6
[3] Manzhirov A V 2017 Mechanical design of AM fabricated prismatic rods under torsion *MATEC Web Conf.* **95** 12002
[4] Parshin D A 2017 Analytic solution of the problem of additive formation of an inhomogeneous elastic spherical body in an arbitrary nonstationary central force field *Mech. Solids* **52** (5) pp 530–40
[5] Manzhirov A V 2017 Mechanical analysis of an AM fabricated viscoelastic shaft under torsion by rigid disks *Proc. World Congr. on Engineering 2017 (London) Lecture Notes in Engng and Computer Sci.* **2230** pp 856–60
[6] Manzhirov A V and Gupta N K 2017 Fundamentals of continuous growth processes in technology and nature *IUTAM Symp. on Growing Solids 2015 (Moscow)* **23** 1–12
[7] Manzhirov A V 2018 Theory of continuous additive growth for solids *Materials of IUTAM Symp. on Mechanical Design and Analysis for AM Technologies 2018 (Moscow)* ed A V Manzhirov and E V Murashkin (Moscow: IPMech RAS) pp 27–9
[8] Manzhirov A V 2018 What is inside AM fabricated parts? The nature of residual stresses and shape distortion *Int. J. Advancements in Technology* **9** p 38
[9] Manzhirov A V and Parshin D A 2018 Analytical solution of the mechanical problem on additive thickening of aging viscoelastic tapers under nonstationary longitudinal end forces *Engng Letters* **26** (2) pp 267–75
[10] Manzhirov A V 2018 Some problems in mechanics of growing solids with applications to AM technologies *J. Phys.: Conf. Ser.* **991** 012056
[11] Manzhirov A V and Mikhin M N 2018 2D problems of surface growth theory with applications to additive manufacturing *J. Phys.: Conf. Ser.* **991** 012057
[12] Manzhirov A V and Parshin D A 2019 Mechanical modeling additive processes of rotational molding in case of viscoelastic behavior of the being manufactured articles *AIP Conf. Proc.* **2116** 380014
[13] Parshin D A 2019 Solving a problem of theory of viscoelasticity for continuously accreted solids with the satisfaction of static boundary conditions in the integral sense *J. Phys.: Conf. Ser.* **1203** 012006
[14] Parshin D A 2020 Mathematical modelling of the process of internal technological stress fields development in cylindrical solids being layer-by-layer made of ageing viscoelastic materials *J. Phys.: Conf. Ser.* **1479** 012145
[15] Ashkenazi E K and Ganov E V 1980 *Anisotropy of Structural Materials: Handbook* 2nd ed (Leningrad: Mashinostroenie) p 247
[16] Manzhirov A V and Parshin D A 2006 Modeling the accretion of cylindrical bodies on a rotating mandrel with centrifugal forces taken into account *Mech. Solids* **41** (6) pp 121–34
[17] Parshin D A 2018 Impact of the rate of the additive process of forming a heavy structure deforming in creep on the development of its technological stresses *AIP Conf. Proc.* **1959** 070029
[18] Hakobyan V N, Amirjanyan H A and Kazakov K Ye 2020 Axisymmetric stressed state of uniformly layered space with periodic systems of internal disc-shaped cracks and inclusions *Herald Bauman Moscow State Technical University: Nat. Sci. Ser.* **89** (2) pp 25–40
[19] Kazakov K E and Manzhirov A V 2017 Axisymmetric contact between an annular rough punch and a surface nonuniform foundation *Mech. Solids* **52** (4) pp 444–51
[20] Kurdina S P and Kazakov K E 2019 Axisymmetric contact problem for a punch and nonuniform foundation with rough surfaces *AIP Conf. Proc.* **2116** 380006
[21] Kazakov K E and Parshin D A 2019 On the indentation of a system of punches into a layered foundation *Engng Letters* **27** (1) pp 251–5
[22] Kazakov K E 2020 Wear of elastic tube with nonuniform coating by rigid bush *E3S Web Conf.* **162** 02002

[23] Kazakov K E and Kurdina S P 2020 Contact problems for bodies with complex coatings *Math. Methods Appl. Sci.* **43** (13) pp 7692–705

[24] Bychkov P S, Kozintsev V M, Manzhirov A V and Popov A L 2017 Determination of residual stresses in products in additive production by the layer-by-layer photopolymerization method *Mech. Solids* **52** (5) pp 524–29

A method of technological identification of metal-cutting machine

A V Sakharov

Mechanical Engineering Research Institute of the Russian Academy of Sciences, Moscow, 101990, Russia

klondikes@yandex.ru

Abstract. The paper demonstrates the disadvantages of formulations of the purposes of metal-cutting machine. The authors suggest to use the modules of the parts surfaces as the subject of machine tool production. The paper presents the method of determination the process capabilities of the machine tool and identifies the process capabilities of a milling machine.

1. Introduction

To form the manufacturing program of an enterprise, determine the possibilities of performance of individual orders for output of products, acquisition of machine tools as well as to develop the processes of parts manufacture, it is necessary to know the process capabilities of the enterprise machinery. The process capabilities of machinery consist of the process capabilities of each machine tool, which means the list of the subjects of production of particular range of sizes, accuracy and roughness with particular parts dimensions. The process capabilities of the machine tool should be reflected in the formulation of its purpose. The analysis of formulations of purposes of various types of machine tools has shown that they contain no information about process capabilities of the machine tools. As an example, the table 1 shows the formulations of purposes of three different types of machine tools which do not clearly set out, what kind of surfaces can be manufactured at these machine tools and what their characteristics will be.

Table 1. Formulations of purposes of machine tools.

Machine tool	Formulation of purpose
Screw-cutting lathe CDE6250A	Designed for performance of various turning works, including cutting of metric thread, British standard Whitworth thread, modular and pitch thread. It provides the use of the advanced high-speed processing modules with coolant supply, with the use of progressive tools and devices.
Vertical milling machine 6P12	Designed for milling, drilling and boring of parts of any shape of steel, pig-iron, non-ferrous metals, their alloys and other materials. The high-performance drives and improved rigidness of the structure of this reliable tool provides efficient utilization of the mills of high-speed steel as well as supplementary equipment provided with plates of rare and super-hard synthetic materials.
IS horizontal boring mill 2A636	Intended for complex mechanical processing of box-type parts weighing below 12 t.

This problem is especially important for multi-purpose machine tools, and, to a lesser degree, for specialized and operational tools.

The ambiguity in the description of process capabilities of the machine tools complicates the choice of machine tools during the designing of parts manufacturing processes, the acquisition of new machine tools, complicates the process audit of the company's machinery, assessment the possibility of adherence to manufacturing programs and orders.

In view of this fact, the purpose of determining the process capabilities is quite important. To reach the set purpose, the authors have formulated the task to elaborate a method of determining the process capabilities of the machine tool.

2. Methods

The problem of determining the process capabilities of the tool lies in the determination of the subject of production. In the specifications of the purpose of machine tools the subject of predication means a surface, an integrity of surfaces or a part. A part is characterized by almost unlimited diversity of possible structures and their characteristics which are impossible to cover. The integrity of surfaces is characterized by a greater diversity of the surfaces it consists of and their characteristics. The surfaces also feature with unlimited diversity both in geometrical shape and in size and dimension characteristics.

The above-mentioned disadvantages are possible to eliminate using the principle of modular technology, which means that any part can be presented by the integrity of functional surface modules (SM). SM means the combination of surfaces united by joint performance of the service function of the part [1]. The SMs' main advantage is their limited variety. As per to the classification, all SMs fall under three classes: basic, working and connecting ones. Each SM has its own number of standard designs, and each design has its own classification by size, accuracy and roughness.

Taking an SM as a subject of machine tool production, the formulation of process capabilities of the machine should contain the list of the produced SM designs with the ranges of size, accuracy and roughness of surfaces and overall dimensions of the parts manufactured.

3. Results

As a result of the work performed, the authors developed a method for determining the process capabilities of machine tools at the modular level. The initial data are the processing methods implemented using the machine tool, the machining tool used and the technical characteristics of the machine tool. The method consists of the following stages:

- determination of the surfaces manufactured using the machine;
- determination of SMs by the structure of the surfaces manufactured using the machine;
- determination of conformity of positions of surfaces in SM designs with their positions in working space of the machine in the process of production;
- determination of the ranges of SM sizes obtained using the machine;
- determination of achievable accuracy of SM production using the machine.

To test the presented method, the authors chose a vertical milling machine 6P12, widely used in machine-building enterprises.

First, a list of surfaces manufactured on the machine was established. They are determined by the machining methods, the diagrams of shaping movements (DSM) of the tool, and in some cases by the geometry of the machining tools used.

The working bodies of the vertical milling machine 6P12 perform the following shaping movements (figure 1): rotation of the spindle with the tool around Z-axis (R_Z^S); vertical movement of the spindle sleeve (M_Z^S); longitudinal (M_X^T), transverse (M_Y^T) and vertical (M_Z^T) movement of the table with the workpiece.

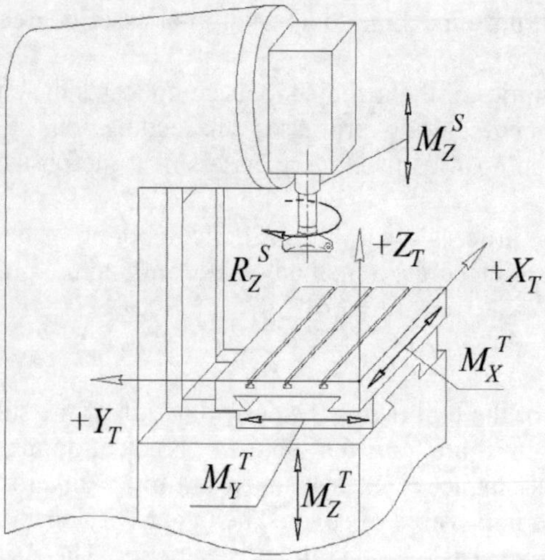

Figure 1. Shaping movements of vertical milling machine 6P12.

Table 2 presents the processing methods implemented at the vertical milling machine 6P12, machine DSM for every processing method, processing tool used and the manufactured surfaces.

From the list of surfaces produced at the machine tool (table 2), the authors determined SMs containing these surfaces: B11, B12, Б311, B321, B51, B52, R111, R112, R121, R21, P22, C111, C112, C121, C21 и C22.

Table 2. Surfaces manufactured at the machine 6P12.

Processing method	DSM	Machining tool	Surface manufactured
	R_Z^S, M_X^T	face mill, end mill	Flat outer surface
		mills: end mill, angle mill, T-slot cutters	Flat inner surface
1. Milling	R_Z^S, M_Y^T	face mill, end mill	Flat outer surface
		mills: end mill, angle mill, T-slot cutters	Flat inner surface
	R_Z^S, M_X^T, M_Y^T	end mill	Shaped closed-contour surface
2. Drilling	R_Z^S, M_Z^S	twist drill, countersink	Cylindrical inner surface
3. Boring	R_Z^S, M_Z^S	boring head	Cylindrical inner surface

At the next stage, the conformity of the surfaces in the SM structure to the surfaces they occupy in the machine tool workspace during the manufacture was determined.

To check the structure, the SM is conditionally placed on the diagram of the machine tool workspace so that one of the SM surfaces to occupy the position providing its manufacture. After that, the other SM surfaces are checked for compliance with the position necessary for its manufacture. If the position of at least one of the SM structure surfaces does not correspond to the necessary position, such SM structure cannot be manufactured at the machine tool.

The ranges of SM sizes obtained at the machine tool are determined from the diagram of the machine tool workspace diagram. The diagram specifies the boundaries of the machine tool workspace, the ranges of it working bodies displacement and, in some cases, the sizes of the working part of the machining tool. Figure 2 presents the diagram with the boundaries of the machine tool workspace (WS) and the ranges of displacement of the working bodies of the vertical milling machine 6P12.

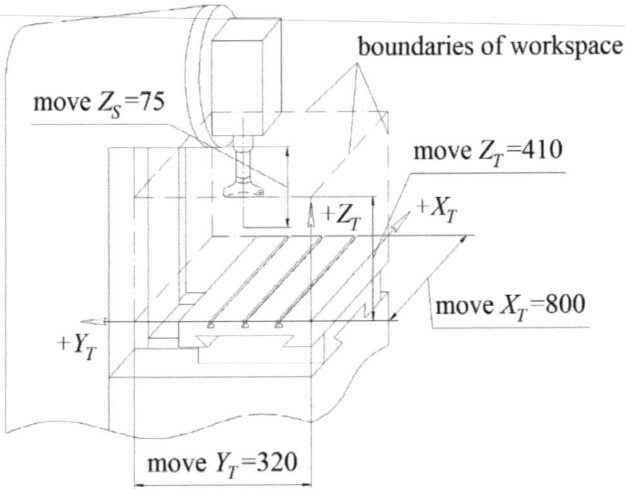

Figure 2. The diagram of the vertical milling machine 6P12 workspace.

For example, the ranges of dimensions of SP B12, that consists of flat outer surface, are determined by the values of the characteristics of the vertical milling machine 6P12, presented in the table 3.

Table 3. SP B12 dimensions.

Dimensions	Machine tool characteristics	Value (mm)
l	Maximal longitudinal displacement of table;	800
	Length of the table workspace.	1250
b	Maximal table transverse displacement;	320
	Width of the table workspace.	320

The next stage in the determination of the process capabilities of the machine tool is the establishment of achievable accuracy of SM production, which depends on geometrical precision of the machine tool, because at the finishing processing modes, when maximal accuracy is obtained, the effect of other factors in negligible. In turn, the geometrical accuracy of the machine tool is characterized by accuracy parameters governed by the corresponding standards for each machine tool type.

As per GOST 17734-88 [2], the accuracy of console milling machines is governed by eighteen accuracy parameters, including, but not limited to, the straightness of the table workspace, the straightness of guiding slot, axial runout of milling spindle, radial runout of conical hole of the milling spindle, etc.

SM accuracy is described by the accuracy of dimensions, the accuracy of surface shapes, the accuracy of relative position and the roughness of the surfaces. For example, B12 SM (figure 3) is described by:

- the accuracy of length (*l*) and width (*b*) of the flat outer surface;
- the shape accuracy: the deviation of flatness and the straightness of the flat outer surface;
- the roughness of flat outer surface (*Ra*).

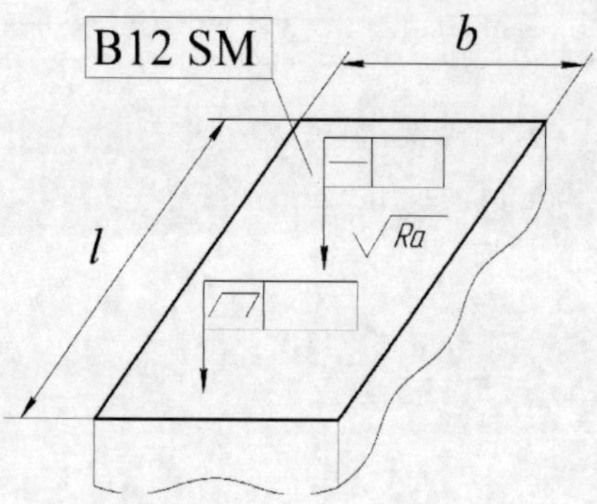

Figure 3. B12 SM draft with accuracy parameters.

It is necessary to establish the indicators of geometrical accuracy of the machine tool affecting the indicators of SM accuracy and to determine the value of their effect.

Table 4 shows the accuracy indicators of the machine 6P12, affecting each accuracy indicator of B12 SM.

Table 4. Machine tool and SM accuracy indicators.

B12 SM accuracy indicators	Machine tool accuracy indicators
1. Accuracy of dimensions	Accuracy of size adjustment of the tool.
2. Deviation from flatness of flat outer surface	Straightness of the table working surface in longitudinal and transversal sections.
3. Deviation from straightness of flat outer surface	Straightness and parallel alignment of trajectory of parallel displacement of the table against its working surface;
	Straightness and parallel alignment of trajectory of transversal displacement of the table against its working surface.
4. Roughness of flat outer surface, Ra	Axial runout of milling spindle;
	Radial runout of bearing end of the spindle.

The values of SM errors arising from the effect of geometric errors of the machine tool are determined by geometric constructions on the diagrams showing the effect of geometric errors of the machine tool on the accuracy indicators of the SM.

As a result, the wording of formulation of the purpose of the vertical milling machine 6P12 will sound as follows: “The machine tool is designed for milling, drilling and boring of the following SMs: B11, B12, B311, B321, B51, B52, R111, R112, R121, R21, R22, S111, S112, S121, S21, S22 on the parts of any shape, made of steel, cast iron, non-ferrous metals, their alloys and other materials”. Process capabilities of the machine should be executed as an appendix to the certificate of the machine tool, which provides a list of SMs manufactured at the machine tool with the ranges of characteristics obtained (dimensions, accuracy and roughness). For example, table 5 presents the characteristics of the B12 SM obtained at a vertical milling machine 6P12.

Table 5. B12 SM parameters.

SM parameters	Value
l (mm)	0-800
b (mm)	0-320
IT(l)	6-8
IT(b)	6-8
Ra (μm)	12.5-0.63
Deviation from flatness (μm)	25-60
Deviation from straightness (μm)	25-60

4. Conclusions

The considered example has shown that the methods suggest can be used for assessment of process capabilities of metal-cutting tools at modular level, reduce the labor intensity of solution of production tasks, related to process capabilities of the machine tools.

References

[1] Bazrov B 2001 *Modular technology in mechanical engineering* (Moscow: Mashinostroenie) p 368
[2] GOST 17734-88 Console milling machines. Standards of accuracy and rigidity (Moscow: Izdatel'stvo standartov)

Modelling and experimental validation of thermoset resin curing during pultrusion

A N Vedernikov, A A Safonov and I S Akhatov

Skolkovo Institute of Science and Technology, Center for Design, Manufacturing and Materials, Moscow, Russia

Aleksandr.Vedernikov@skoltech.ru

Abstract. Mathematical modelling of the pultrusion manufacturing process involves many parameters of the resin mixture. In this article thermo-chemical behavior of resin used for pultrusion of glass fiber/epoxy-vinyl L-shaped profiles was characterized. Profiles were manufactured at pulling speeds of 200, 400, and 600 mm·min^{-1}. Dependency of resin specific heat from temperature was determined by differential scanning calorimetry (DSC); the influence of temperature on resin's thermal conductivity was found; parameters of resin cure kinetics were described. Subsequently, the gained parameters were used for numerical simulation of the pultrusion thermo-chemical problem in ABAQUS software. The temperature and cure degree evolutions obtained from the experiments run at different pulling speeds, and those from the numerical model were shown to be correlated.

1. Introduction

Recently fiber reinforced polymers (FRPs) have been widely adopted as structural elements in the civil engineering and construction sector due to their better mechanical properties compared to those of traditional materials (steel, timber, concrete) [1-4]. As the most efficient manufacturing technique of composite materials, pultrusion allows the production of constant cross-section profiles with virtually unlimited length and least engineers' involvement [5,6]. Advantages of pultruded profiles are apparent: high strength-to-weight ratio [7-9] improved corrosion resistance [10-12] and durability [13-15], ease of transportation, installation, and maintenance [16].

The thermoset pultrusion process starts from pulling fiber reinforcement (unidirectional rovings, mats, woven fabric) through the bath, filled with the resin matrix [17,18]. Next, impregnated material is pulled to the preformer for the elimination of excessive resin [19]. Then preformed reinforcement is guided to the heated die for the initiation of resin polymerization process [20]. Cured composite is subsequently directed to the cutting saw by the system of puller units [21]. Nevertheless, the performance of the final pultruded product is very sensitive to processing parameters [22,23] and the choice of raw materials [24-26]. Aiming quality improvement of produced profiles by minimizing residual stresses [27], occurrence of cracks [28] and manufacturing induced shape distortions, mathematical modelling of the pultrusion process is a widely applied tool [29,30]. It allows to avoid time-consuming and expensive trial-error experiments [31,32]. Since the pultrusion manufacturing process incorporates thermo-chemical and mechanical problems, plenty of input parameters of composite raw elements are needed to achieve mathematical modelling reliable results [33,34].

The objective of the current article is to experimentally determine temperature-dependent parameters of resin mixture: specific heat and thermal conductivity. These parameters are subsequently coupled with resin kinetics and utilized to solve the pultrusion thermo-chemical problem in ABAQUS software. The computation procedure results are then compared to those detected during the pultrusion experiment performed at three different pulling speeds (200, 400, and 600 mm·min^{-1}).

2. Materials and methods

2.1. Pultrusion process setup and experimental determination of resin parameters

Resin mixture and pultruded profiles manufactured with Pultrex Px500-6T (Pultrex, UK) pultrusion machine were produced at the Laboratory of Composite Materials and Structures of the Center for Design, Manufacturing and Materials (Skolkovo Institute of Science and Technology, Moscow, Russia).

The matrix used for this study was epoxy vinyl-based Atlac 430 (DSM Composite Resins AG, Switzerland). To improve the quality of the resin mixture the following components were added: Trigonox C (Akzo Nobel Polymer Chemicals B.V., The Netherlands), Perkadox 16 (Akzo Nobel Polymer Chemicals B.V., The Netherlands), BYK-A555 (BYK Additives & Instruments, Germany), and Zinc Stearate (Baerlocher GmbH, Germany). Mentioned additives played the following roles: Trigonox C – initiator for (co)polymerization of ethylene, styrene, acrylonitrile, acrylates, and methacrylates; Perkadox 16 – initiator for suspension polymerization of acrylates and methacrylates; BYK-A555 – deaerator; Zinc Stearate – friction reduction. The amount of each component used in the resin mixture is indicated in table 1.

Table 1. Components of the resin mixture.

Component	Amount [kg]
Resin Atlac 430 (DSM Composite Resins AG, Switzerland)	25.07
BYK-A555 (BYK Additives & Instruments, Germany)	0.09
Trigonox C (Akzo Nobel Polymer Chemicals B.V., The Netherlands)	0.38
Perkadox 16 (Akzo Nobel Polymer Chemicals B.V., The Netherlands)	0.13
Zinc Stearate (Baerlocher GmbH, Germany)	1.00

The length of the die was 600 mm. Four heating platens were used. L-shaped profiles were pultruded at three different pulling speeds: 200, 400, and 600 mm·min^{-1}. Therefore, abbreviations of V-200, V-400, and V-600 are used in this paper. During the pultrusion process temperature evolution of the processing composite was recorded by wired thermocouples. They had a length of 7 m and were introduced within the profile before die entrance.

Independently from pultrusion process, the pure resin mixture was fully cured, and two samples were cut off to evaluate the temperature dependence of resin specific heat. Samples had a shape of cylinders with a radius of 4 mm and a thickness of 2 mm. For the purpose of this study Differential Scanning Calorimeter DSC 204 (NETZSCH-Gerätebau GmbH, Germany) was used. Experiments were carried out at the temperature range from 20°C to 100°C. Two samples were cut off to evaluate the temperature dependence of resin thermal conductivity. Samples had a shape of thin plates 10 × 10 mm and a thickness of 1 mm. For the purpose of this study Laser Flash Apparatus LFA 457 (NETZSCH-Gerätebau GmbH, Germany) was used. Experiments were carried out at the temperature range from 20°C to 100°C.

2.2. Heat-transfer model

The general 3D heat-transfer model describing the pultrusion process is presented in equation (1). Subscript "1" represents the pulling direction of pultrusion, while "2" and "3" are related to the transversal direction. Index "c" determines the belonging of the property to the entire composite, while "r" and "f" are related to resin and fiber respectively.

$$\rho_c C_{p\_c}\left(\frac{\partial T}{\partial t}+v\frac{\partial T}{\partial x_1}\right)=k_{x_1,c}\frac{\partial^2 T}{\partial x_1^2}+k_{x_2,c}\frac{\partial^2 T}{\partial x_2^2}+k_{x_3,c}\frac{\partial^2 T}{\partial x_3^2}+q \quad (1)$$

$$q=(1-V_f)\rho_r H_{tot}\frac{d\alpha}{dt} \quad (2)$$

Where ρ_c, ρ_r are densities of the composite and resin respectively; C_p represents specific heat; T is temperature; t is time; $k_{x_1,c}, k_{x_2,c}, k_{x_3,c}$ are the thermal conductivities in the corresponding directions; v is pulling speed; q represents internal heat generation due to exothermic reaction of the resin; V_f represents fiber volume fraction; H_{tot} represents total heat released during curing; $\frac{d\alpha}{dt}$ is rate of curing reaction.

Kinetic model of the nth order reaction with autocatalysis was used for the modelling of resin polymerization process [35]:

$$\frac{d\alpha}{dt}=Ae^{-\frac{E_a}{RT}}(1-\alpha)^n(1+K_{cat}\alpha) \quad (3)$$

Where α represents instantaneous degree of cure; t is time; A represents a preexponential factor; E_a is activation energy; R is universal gas constant; n is the order of the reaction; K_{cat} is activation constant. Parameters of the model are presented in the table 2 in accordance to those described in [22].

Table 2. Resin kinetic parameters.

$A[s^{-1}]$	$E_a\ [kJ\cdot mol^{-1}\cdot K^{-1}]$	$n[-]$	$K_{cat}[-]$
$10^{9.34}$	93.3	1.91	$10^{2.73}$

2.3. *Finite element simulations*

Mathematical modelling was performed at the FEM software ABAQUS. The following subroutines were used for the solution of the thermo-chemical problem: FILM (to define die temperatures and heat transfer coefficient as functions of position and time), USDFLD (to define cure degree of the composite as a field variable), HETVAL (to define a heat flux due to internal heat generation in the material during heat transfer analysis). The finite element model consisted of linear quadrilateral element (CPE4RT). The number of elements is over than 1000.

3. Results and discussions

The results of the temperature dependence of resin specific heat evaluation are presented in table 3. According to equation (4), the first-order polynomial was then used for the approximation of temperature dependent resin specific heat experimental results.

$$C_p=5.1T+1080 \quad (4)$$

Where T represents the instantaneous temperature of the resin; C_p is instantaneous specific heat of the resin.

The correlation between resin thermal conductivity and temperature is found to be weak. Therefore, further resin thermal conductivity is considered independent from the temperature and adopted as the average value of experimental results equal to 0.177 $W\cdot m^{-1}\cdot K^{-1}$.

Table 3. Temperature dependence of resin specific heat and thermal conductivity.

Temperature [°C]	Measured value of specific heat			Measured value of thermal conductivity		
	C_{p1} [J·kg^{-1}·K^{-1}]	C_{p2} [J·kg^{-1}·K^{-1}]	$C_{average}$ [J·kg^{-1}·K^{-1}]	λ_1 [W·m^{-1}·K^{-1}]	λ_2 [W·m^{-1}·K^{-1}]	$\lambda_{average}$ [W·m^{-1}·K^{-1}]
20	1181	1199	1190	0.1730	0.1761	0.1746
30	1227	1241	1234	0.1741	0.1718	0.1730
40	1277	1290	1284	0.1773	0.1704	0.1739
50	1333	1339	1336	0.1800	0.1769	0.1785
60	1386	1384	1385	0.1802	0.1787	0.1795
70	1430	1421	1425	0.1773	0.1784	0.1779
80	1476	1461	1468	0.1793	0.1765	0.1779
90	1538	1518	1528	0.1796	0.1785	0.1791
100	1628	1610	1619	0.1823	0.1817	0.1820

Figure 1 represents the evolution of composite temperature and cure degree as functions of coordinate measured in the pulling direction of pultrusion starting from the die entrance. Since the most significant part of the polymerization process happens within the die, only 600 mm length is considered. The solid line represents the results of mathematical modelling, while the dashed line is related to the experimentally obtained values. Good correlation between numerical and experimental results for all the pulling speeds is observed. Therefore, the thermo-chemical model of pultrusion presented above is considered to be experimentally verified.

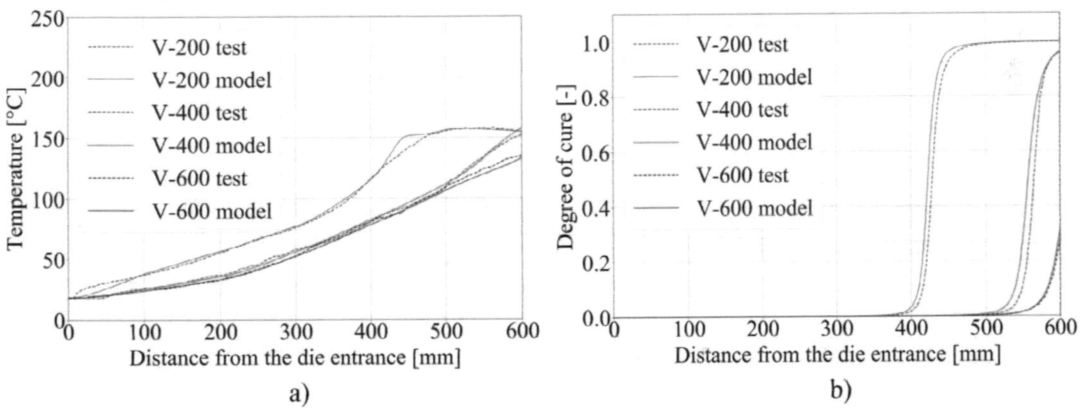

Figure 1. Comparison of modelling and experimental results. (a) Evolution of composite's temperature; (b) Evolution of composite's cure degree.

4. Conclusions

Both an experimental and a numerical study were performed to investigate thermo-chemical problem of pultrusion process in relation to the production of glass fiber/epoxy-vinyl L-shaped profiles. Tests on epoxy-vinyl based matrix were conducted to characterize the dependency of resin specific heat and thermal conductivity on temperature. Specific heat of the resin exhibited a linear temperature dependence, while thermal conductivity demonstrated weak correlation and was adopted as constant.

These results were then jointed to resin kinetic and used for the mathematical modelling of the thermo-chemical problem of pultrusion in ABAQUS software. Simulating the resin curing process, temperature and cure degree evolution profiles of composites were obtained for three pulling speeds: 200, 400, and 600 mm·min^{-1}. Modelling results demonstrated a good agreement with the results detected by thermocouples during the pultrusion experiment.

This paper is the first step of authors to experimentally characterize thermal and mechanical parameters of pultruded composites' constituent parts needed for the further development of the linear viscoelastic mathematical model.

References

[1] Vedernikov A, Safonov A, Tucci F, Carlone P and Akhatov I 2020 Pultruded materials and structures: A review *J. Compos. Mater.* **54** 4081–117

[2] Liu T, Feng P, Lu X, Yang J-Q and Wu Y 2020 Flexural behavior of novel hybrid multicell GFRP-concrete beam *Compos. Struct.* **250**

[3] Shimba Carneiro Vieira P, de Souza F, Taissum Cardoso D C, Domingos Vieira J and de Andrade Silva F 2020 Influence of moderate/high temperatures on the residual flexural behavior of pultruded GFRP *Compos. Part B Eng.* **200**

[4] Madenci E, Özkılıç Y O and Gemi L 2020 Experimental and theoretical investigation on flexure performance of pultruded GFRP composite beams with damage analyses *Compos. Struct.* **242**

[5] Vedernikov A N, Safonov A A, Gusev S A, Carlone P, Tucci F and Akhatov I S 2020 Spring-in experimental evaluation of L-shaped pultruded profiles *IOP Conf. Ser. Mater. Sci. Eng.* **747** 012013

[6] Fairuz A M, Sapuan S M, Zainudin E S and Jaafar C N A 2014 Polymer composite manufacturing using a pultrusion process: A review *Am. J. Appl. Sci.* **11** 1798–810

[7] Zhu R, Li F, Shao F and Zhang D 2020 Static and dynamic behaviour of a hybrid PFRP-aluminium space truss girder: Experimental and numerical study *Compos. Struct.* **243**

[8] Sutherland L S, Sá M F, Correia J R, Guedes Soares C, Gomes A and Silvestre N 2017 Impact response of pedestrian bridge multicellular pultruded GFRP deck panels *Compos. Struct.* **171**

[9] Liu T, Liu X and Feng P 2020 A comprehensive review on mechanical properties of pultruded FRP composites subjected to long-term environmental effects *Compos. Part B Eng.* **191**

[10] Xiong Z, Liu Y, Zuo Y and Xin H 2019 Experimental evaluation of shear behavior of pultruded GFRP perforated connectors embedded in concrete *Compos. Struct.* **222**

[11] Madenci E, Onuralp Özkılıç Y and Gemi L 2020 Buckling and free vibration analyses of pultruded GFRP laminated composites: Experimental, numerical and analytical investigations *Compos. Struct.* **254**

[12] Xin H, Mosallam A, Liu Y, Yang F and Zhang Y 2017 Hygrothermal aging effects on shear behavior of pultruded FRP composite web-flange junctions in bridge application *Compos. Part B Eng.* **110**

[13] Li C, Yin X, Wang Y, Zhang L, Zhang Z, Liu Y and Xian G 2020 Mechanical property evolution and service life prediction of pultruded carbon/glass hybrid rod exposed in harsh oil-well condition *Compos. Struct.* **246**

[14] Yang W-R, He X-J and Dai L 2017 Damage behaviour of concrete beams reinforced with GFRP bars *Compos. Struct.* **161**

[15] Gooranorimi O, Suaris W, Dauer E and Nanni A 2017 Microstructural investigation of glass fiber reinforced polymer bars *Compos. Part B Eng.*

[16] Liu T, Vieira J D and Harries K A 2020 Predicting flange local buckling capacity of pultruded GFRP I-sections subject to flexure *J. Compos. Constr.* **24**

[17] Bakis C E, Bank L C, Brown V L, Cosenza E, Davalos J F, Lesko J J, Machida A, Rizkalla S H and Triantafillou T C 2002 Fiber-reinforced polymer composites for construction - State-of-

the-art review *J. Compos. Constr.* **6**
[18] Vedernikov A, Tucci F, Safonov A, Carlone P, Gusev S and Akhatov I 2020 Investigation on the shape distortions of pultruded profiles at different pulling speed *Procedia Manuf.* **47** 1–5
[19] Starr T F 2000 *Pultrusion for engineers*
[20] Costa Dias R D C, Santos L D S, Ouzia H and Schledjewski R 2018 Improving degree of cure in pultrusion process by optimizing die-temperature *Mater. Today Commun.* **17** 362–70
[21] Baran I 2015 *Pultrusion: state-of-the-art process models* (Shropshire: Smithers Rapra)
[22] Vedernikov A, Tucci F, Carlone P, Gusev S, Konev S, Firsov D, Akhatov I and Safonov A 2021 Effects of pulling speed on structural performance of L-shaped pultruded profiles *Compos. Struct.* **255**
[23] Silva F J G, Ferreira F, Costa C, Ribeiro M C S and Meira Castro A C 2012 Comparative study about heating systems for pultrusion process *Compos. Part B Eng.* **43**
[24] Sorina T G, Safonov A A and Khairetdinov A K 2010 Pecularities of using carbon glass-reinforced plastic in pultrusion composite profiles for bridge engineering *J. Mach. Manuf. Reliab.* **39** 47–51
[25] Paciornik S, Martinho F, de Mauricio M H and d'Almeida J R 2003 Analysis of the mechanical behavior and characterization of pultruded glass fiber–resin matrix composites *Compos. Sci. Technol.* **63** 295–304
[26] Fairuz A M, Sapuan S M, Zainudin E S and Jaafar C N A 2016 Effect of filler loading on mechanical properties of pultruded kenaf fibre reinforced vinyl ester composites *J. Mech. Eng. Sci.* **10** 1931–42
[27] Yuksel O, Baran I, Ersoy N and Akkerman R 2019 Investigation of transverse residual stresses in a thick pultruded composite using digital image correlation with hole drilling *Compos. Struct.* **223**
[28] Safonov A, Gusev M, Saratov A, Konstantinov A, Sergeichev I, Konev S, Gusev S and Akhatov I 2020 Modeling of cracking during pultrusion of large-size profiles *Compos. Struct.* **235**
[29] Baran I, Tutum C C, Nielsen M W and Hattel J H 2013 Process induced residual stresses and distortions in pultrusion *Compos. Part B Eng.* **51** 148–61
[30] Safonov A A, Carlone P and Akhatov I 2018 Mathematical simulation of pultrusion processes: A review *Compos. Struct.* **184**
[31] Ding A, Wang J and Li S 2020 Understanding process-induced spring-in of L-shaped composite parts using analytical solution *Compos. Struct.* **250**
[32] Al-Dhaheri M, Khan K A, Umer R, van Liempt F and Cantwell W J 2020 Process induced deformations in composite sandwich panels using an in-homogeneous layup design *Compos. Part A Appl. Sci. Manuf.* **137**
[33] Baran I, Akkerman R and Hattel J H 2014 Material characterization of a polyester resin system for the pultrusion process *Compos. Part B Eng.* **64** 194–201
[34] de Cassia Costa Dias R, Costa M L, de Sousa Santos L and Schledjewski R 2020 Kinetic parameter estimation and simulation of pultrusion process of an epoxy-glass fiber system *Thermochim. Acta* **690**
[35] Worzakowska M 2006 Kinetics of the curing reaction of unsaturated polyester resins catalyzed with new initiators and a promoter *J. Appl. Polym. Sci.* **102** 1870–6

Influence of additional vibration loads on the processes of deformation and fracture of composite tubular samples

Valerii Wildemann and Oleg Staroverov

Perm National Research Polytechnic University, 29 Komsomolsky AVE, Perm 614000, Russia

cem_staroverov@mail.ru

Abstract. The work is devoted to an experimental study of the influence of additional vibration torsional effects on the deformation and fracture of composite tubular specimens under quasistatic tensile stress. A series of tests with various parameters of torsional vibration load for thin-walled CFRP tubular samples were performed according to the developed method. As part of the work, the methodological aspects of testing tubular specimens are considered. As a result, experimental data were obtained that reflect the effect of vibration effects on the behaviour of composite specimens during quasi-static tension. Additional vibrations contributed to the implementation of the deformation resources of the composite structural elements.

1. Introduction

During operation, structures made of composite materials, in addition to cyclic, low-speed impacts longitudinal and transverse loads, are subjected to vibration effects. The additional vibrational torsional influences have a stabilizing effect of the postcritical deformation process under tension conditions of solid cylindrical samples of structural steels [1, 2]. Based on this, it can be assumed that additional vibration effects, in addition to changing the residual strength, can also affect the processes of damage accumulation and destruction of composites.

The number of works devoted to the study of the processes of deformation and fracture of tubular specimens [3-6] is growing. The development of technologies for the production of composite materials and the improvement of test systems that make it possible to implement complex modes of mechanical loading, such as (in) proportional quasi-static and cyclic tension with torsion, all-round tension and hydrostatic compression, including internal pressure, etc. [7-9].

The main aim of this research was to study the regularities of the mechanical behavior of polymer composite tubular specimens under conditions of quasi-static tension with additional vibration loads.

2. Materials and equipment

The method of experimental study of the processes of deformation and fracture of polymer tubular composite specimens under additional torsional cyclic actions at different stages of quasi-static tension is schematically shown in figure 1, where:

P - maximum value of the tensile stress; u – maximum value of the displacement of the loading system of the specimen-traverse of the testing machine under tensile stress; P' – maximum value of the load under tensile with torsion; u' – maximum value of the displacement of the loading system (specimen-traverse of the testing machine) under tensile with additional torsional vibrations; φ_a – amplitude of the twist angle during torsional vibrations; γ – frequency of vibrational torsional loading.

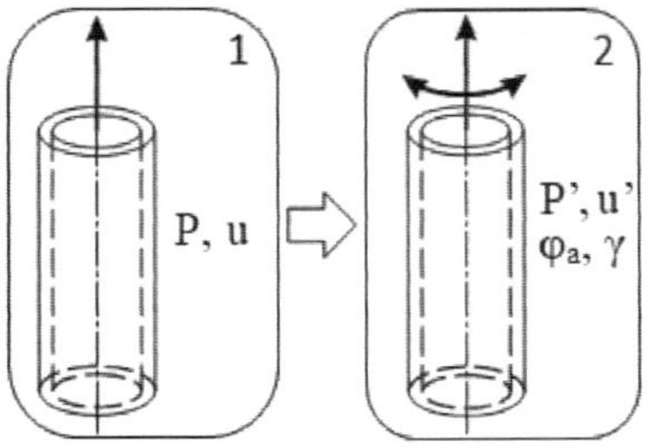

Figure 1. The research methodology of tensile tests with additional torsional vibration loads.

The technique includes the following tests: 1 – carrying out uniaxial tensile tests with the determination of the value of the maximum tensile forces at destruction P; 2 – tensile tests with additional torsional vibration loads with low amplitude and high frequency.

In order to study the influence of possible effects of additional vibration effects on the processes of deformation and destruction of carbon fiber reinforced polymer (CFRP) composite materials, a series of experiments were carried out on carbon-fiber (T700SC-12000-50C with Etal-Inject SL / M binder) tubular specimens with an outer diameter of 26 mm, wall thickness of 0.5 mm and concentrator 5 mm in diameter (figure 2).

Figure 2. CFRP tubular specimen.

Vibration load mean cyclic loading with a small amplitude of the twist angle φ_a and a high frequency ν.

To implement such a complex loading, it is necessary to take into account the rigidity of the loading test system. The Instron 8802 universal biaxial testing system is the most rigid and allows the implementation of complex loading conditions. The main advantage of biaxial test systems is associated with their design feature, which allows simultaneous application of axial load and torque to the specimen: ± 100 kN, ± 1000 N·m, 30 Hz (figure 3).

Figure 3. The Instron 8850 testing system.

Methodological aspects of testing tubular composite specimens under complex stress state conditions are reflected in [10]. In order to prevent crushing of the tubular specimen in the grips of the test system, cylindrical metal rods were pressed into the ends of the specimen by the length of the collet (figure 4).

Figure 4. Photo of a metallic rod.

3. Results

The study included experiments on quasi-static tension (figure 5) and tension with vibration torsional actions.

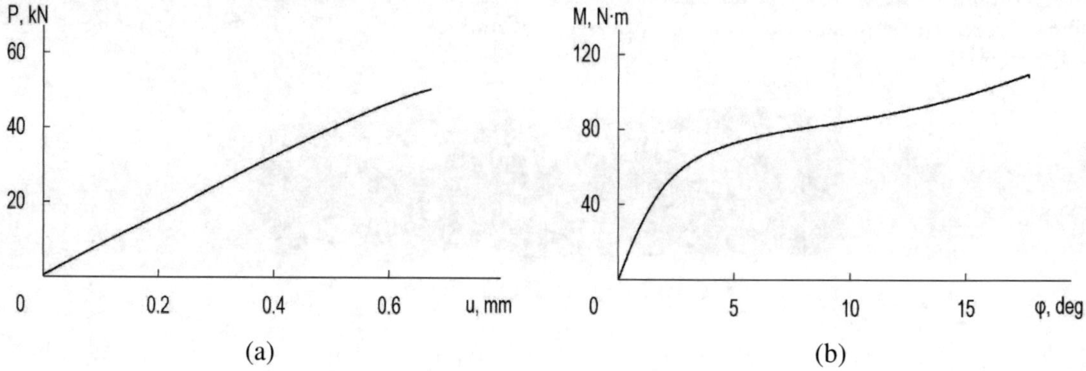

Figure 5. Loading diagrams of GFRP tubular specimens under quasi-static tension (a) and torsion (b).

Based on the results of quasi-static tests, the following parameters of the vibration torsional action were selected: $\varphi_a = 1$ °, $\nu = 20$ Hz; $\varphi_a = 0.5$ °, $\nu = 25$ Hz; $\varphi_a = 0.25$ °, $\nu = 30$ Hz.

As a result of the imposition of additional actions on the stress loading diagrams under tension, there appeared areas of stage failure and zones of equilibrium deformation (figure 6). For the convenience of analyzing the results, the diagrams are plotted in relative coordinates P'; u'.

$$P' = P \cdot P_{max}^{-1} \quad (1)$$

Where:
P – value of the compressive load, N;
P_{max} – value of the compressive load at destruction, N.

$$u' = u \cdot u_{max}^{-1} \quad (2)$$

Where:
u - sample displacement, mm;
u_{max} - displacement of the sample during destruction, mm.

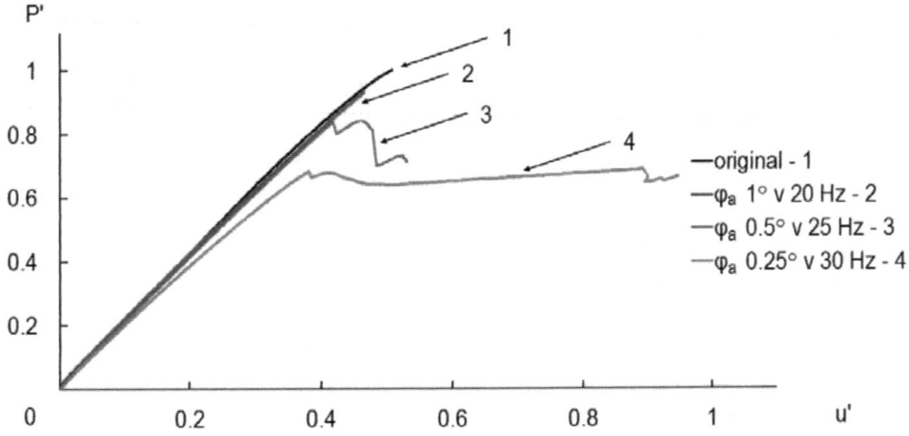

Figure 6. Loading diagrams of CFRP tubular specimens.

Additional vibration loads reduce the value of the maximum load P_{max} during specimen failure, but contribute to the realization of equilibrium loading sections. The obtained experimental dependencies can be used to create databases of materials, which are subject to increased requirements from the point of view of the safety of destruction and catastrophe. Photos destroyed specimens are shown in figure 7.

(a) (b) (c)

Figure 7. Photographs of fractured CFRP specimens after tests for tensile (a), torsion (b), tension with additional vibration effects $\varphi_a = 0.25$ °, ν = 30 Hz (c).

4. Conclusions

Within the framework of the study, a technique was developed for studying the processes of damage accumulation and destruction of composite tubular specimens under conditions of a plane stress state under joint tension with vibrational torsion using a two-axis servo-hydraulic testing machine. The study of the effect of additional vibration torsional vibrations on deformation and fracture of CFRP tubular specimens has been carried out. It was revealed that additional vibration loads contribute to the implementation of the deformation resources of the samples under study.

Acknowledgements

The work was carried out in the Perm National Research Polytechnic University with support of the Russian Science Foundation (Project 16-19-00069).

IOP Conf. Series: Materials Science and Engineering 1129 (2021) 012012 doi:10.1088/1757-899X/1129/1/012012

References

[1] Wildemann V E, Tretyakov M P and Lomakin E V 2016 Effect of vibration stabilization of the process of postcritical deformation *Doklady Phys.* **61 (3)** pp 147–51
[2] Kapica P L 1951 Mayatnik s vibriruyushchim podvesom UFN **44 (1)** pp 7–20
[3] Zhigun V I and Plume E Z 2017 Evaluation of carrying capacity of carbon-carbon composites in plane stress state *J. on Composite Mechanics and Design* **23 (1)** pp 25–40
[4] Abdewi E F, Sulaiman S, Hamouda A M S and Mahdi E 2008 Quasi-static axial and lateral crushing of radial corrugated composite tubes *Thin-Walled Structures* **46** (3) pp 320–2 doi: 10.1016/j.tws.2007.07.018
[5] Ogasawaraa T, Ontab K, Ogiharab S, Yokozekic T and Harad E 2009 Torsion fatigue behavior of unidirectional carbon/epoxy and glass/epoxy composites *Composite Structures* **90 (4)** pp 482–9
[6] Palanivelu S *et al* 2010 Comparative study of the quasi-static energy absorption of small-scale composite tubes with different geometrical shapes for use in sacrificial cladding structures *Polymer Testing* **29 (3)** pp 381–96 doi:10.1016/j.polymertesting.2010.01.003
[7] Tsvetkov S V, Kulish G G, Smerdov A A, Baryshev A N, Tashchilov S V, Magnitskiy I V and Ponomarev K A 2016 Experimental research of materials in triaxial tension *Herald of the Bauman Moscow State Technical University. Series mechanical engineering* **5** pp 76–88
[8] Shlyannikov V N, Tumanov A V and Zakharov A P 2014 The mixed mode crack growth rate in cruciform specimens subject to biaxial loading *Theoretical and Applied Fracture Mechanics* **73** pp 68–81
[9] Shlyannikov V N, Ishtyriakov I S and Yarullin R R 2014 Deformation characteristics for d16t alloy under combined loading by tension, compression, torsion, and internal pressure *Transactions of Academenergo* **3** pp 78–89
[10] Staroverov O A, Strungar E M, Tretyakov M P and Tretyakova T V 2017 Features of the experimental test of tubular specimens of composite materials under complex stress state conditions *PNRPU Aerospace Engineering Bulletin* **51** pp 104–14

The analysis of technologies and modeling for forming parts in isothermal conditions

Y I Yakovlev, D L Rakov and R Y Sukhorukov

Mechanical Engineering Research Institute of the Russian Academy of Sciences, Moscow, Russia

rdl@mail.ru

Abstract. One of the most promising ways to parts making is isothermal in a state of superplasticity. Mathematical modeling is an effective way to determine the necessary energy-force parameters of the process. Examples include the method of analysis and selection of technological processes of parts production in isothermal conditions. The technological processes and simulation of deformation-thermal treatment has been carried out. The simulation ensured the formation of the ultra-fine-grain structure in the blanks under the super-plastic roll-out of the shaft from the alloy EhK79. The modeling shows the effect of the microstructure of alloys on the physical-mechanical and operational properties of parts. Model used to predict microstructure evolution is the Johnson-Mehl-Avrami-Kolmogorov model.

1. Introduction

Among the main tasks in the development of the aircraft industry is the production of parts for Gas Turbine Engines (GTE) [1]. The parts work in aggressive environments at high temperatures and high mechanical loads. Appropriately, they are made from modern heat-resistant alloys and steels based on nickel, iron and titanium. The implementation of new high-performance technologies for obtaining complex parts in isothermal conditions is possible only if there is a structure in the blanks of Ultra-Fine-Grain (UFG) [2, 3]. Examples of large-sized UFG blanks from heat-resistant nickel alloys are presented in the works. Ensuring the conditions of formation determines the high quality of the resulting semi-finished products and parts [4, 5]. Further consideration the method of analysis and selection of processes in the form of parts in isothermal conditions. The technological processes of making discs and hollow shafts of isothermal rolling are considered, as well as modeling the effect of the microstructure of alloys on the physical-mechanical and operational properties of parts.

2. Selection of technological processes for forming heat-resistant alloys

The general principle of construction of the morphological analysis and its varieties is the creation of a Morphological Matrix (MM) based on essential attributes belonging to the entire class of the object under study. The morphological approach is a field for search of possible variants of the problem solution.

The stages of the morphological analysis procedure include [6, 7]:

- The exact wording of the problem to be solved.
- Division of essential morphological attributes of the MM.

IOP Conf. Series: Materials Science and Engineering 1129 (2021) 012013 doi:10.1088/1757-899X/1129/1/012013

- Search for possible options of execution of each feature individually and combining them into columns of the morphological matrix.
- Evaluation of all variants of solution of the problem obtained from the MM.
- Selection of the most rational and preferable variants of the problem solution.

To evaluate the variants of technological processes with the help of expert assessments, a morphological matrix was compiled [8, 9]. Any set of all elements is a possible variant of the technology of making hollow shafts of GTE. The main features are selected - the process of preparing the material of the blank, obtaining the required size of the structure of the material, the speed of deformation, etc. To reduce the size of the matrix, criteria were introduced on which the assessment was carried out. At the further stage each alternative was given a ballroom score for each criterion. Thus, 480 variants were generated, and then 120 variants were selected for further clustering. All variants were designed to measure similarity. Of the 120 generated and selected variants, 18 clusters containing similar technological solutions were formed (figure 1). The variants under study have high ratings, which makes it possible to conclude that there are high technical and economic indicators.

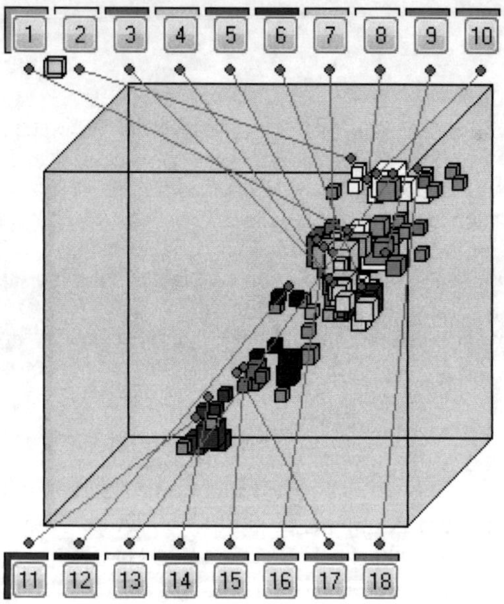

Figure 1. Location of technological processes clusters in the morphological solutions space.

3. Formation UFG structures

The mechanism of deformation of materials in structural superplasticity is grain boundary sliding. In order for grain boundary sliding to be the main mechanism of deformation, it is necessary that the grain size does not exceed 20 μm.

The main method of grinding grains to form an ultra-fine-grain structure in multiphase alloys is dynamic recrystallization. Due to the presence of secondary phases in the interphase space of the matrix, the thermal stability of the resulting structure significantly increases, allowing widespread use of dynamic recrystallization in both laboratory and industrial conditions.

In order to obtain UFG structures in the blanks of industrial nickel alloys, which usually have a large-grain structure, various types of deformation-thermal treatment are used, such as isothermal forging, punching, controlled rolling, equal channel angular pressing, high pressure torsion, as well as other methods of forming processes. To obtain a blank for isothermal roll-out of the shaft from the alloy EhK79 is proposed to use closed isothermal punching. The drawing of the blank under the rollout of the shaft is presented on figure 2.

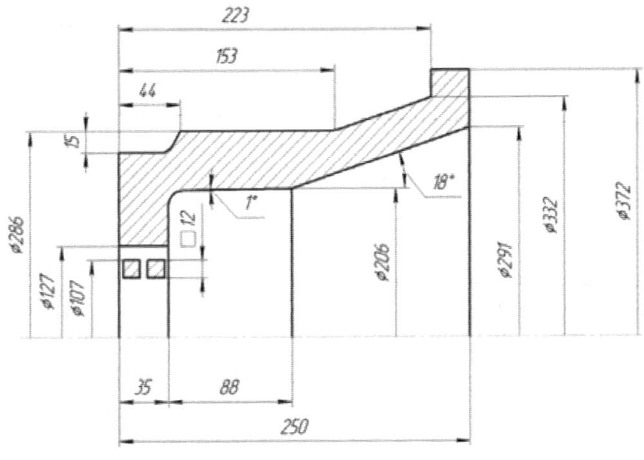

Figure 2. Preparation for the rollout of the shaft.

The kinetics of dynamic recrystallization and the size of the recrystallized grains obtained were described by the model Johnson-Mehl-Avrami-Kolmogorov (JMAK) [10]. The nature of the metal flow at punching practically does not depend on the temperature-speed parameters of deformation. The sequence of form-change is presented in figure 3. The deformation consists of two stages. In the first stage, the blank is deformed to the contact of the side surface of the blank with the matrix, and then there is a reverse squeezing until the final filling of the cavity of the stamp.

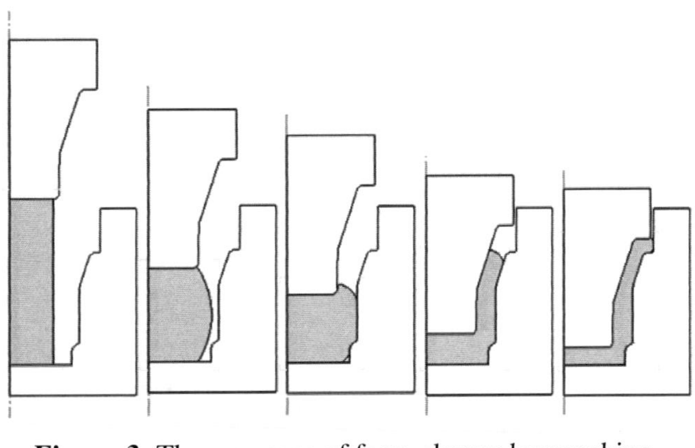

Figure 3. The sequence of form-change by punching.

The built model allows determining the proportion of recrystallized harvest volume, the average size of recrystallized grains and the average size of the grains of the harvest matrix depending on the degree, speed and temperature of the deformation of the alloy EhK79.

4. Johnson-Mehl-Avrami-Kolmogorov model to predict microstructure evolution

The same metals and alloys, with the same chemical composition, can have a different microstructure. This is due to different approaches and methods of forming parts and their heat treatment. The microstructure has a significant effect on the physical, mechanical and operational properties of parts.

The most common model used to predict microstructure evolution is JMAK model, which calculates the proportion of recrystallized volume and average grain size depending on the parameters of deformation and holding conditions.

The analysis underwent dynamic recrystallization. The simulation of grain growth is carried out right before the punching operation (for non-deformable metal). The simulation was carried out under isothermal conditions (the temperatures of work piece, environment, and tool were equal). The essence of the experiment is to perform the operation of hot forging "upset" (figure 4) of a cylindrical billet with dimensions: diameter 5 mm, height 8 mm.

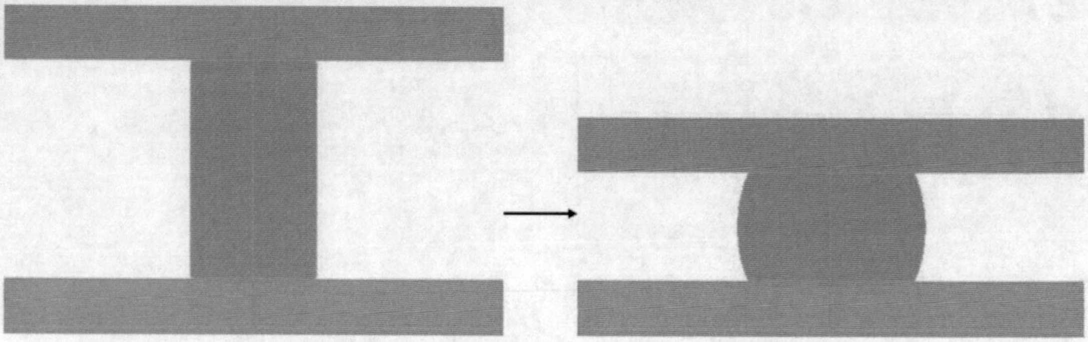

Figure 4. Upsetting of a cylindrical billet.

For a full comparative analysis, 12 simulations were calculated with various input data (table 1).

Table 1. Dynamic recrystallization simulation route.

Dynamic recrystallization					
№	Degree of deformation	Height, mm	Temperature, °C	Strain rate, $1 \cdot s^{-1}$	Deformation velocity, $mm \cdot s^{-1}$
1	0.7	3.97	970	0.01	0.058
2	0.7	3.97	970	0.10	0.575
3	0.7	3.97	970	1	5.753
4	0.7	3.97	1080	0.01	0.058
5	0.7	3.97	1080	0.10	0.575
6	0.7	3.97	1080	1	5.753
7	1	2.94	970	0.01	0.051
8	1	2.94	970	0.10	0.506
9	1	2.94	970	1	5.057
10	1	2.94	1080	0.01	0.051
11	1	2.94	1080	0.10	0.506
12	1	2.94	1080	1	5.057

As an example, the figures show the fields: the proportions of dynamically recrystallized volume (figure 5), average size of dynamically recrystallized grains (figure 6) and average size of grains (figure 7), obtained as a result of modeling the dynamic recrystallization of the Huang's model, according to input data presented in table 1. Three JMAK models proposed by Huang [11], Jong-Sang Na [12] and Chen Liging [13] were used to analyze grain growth.

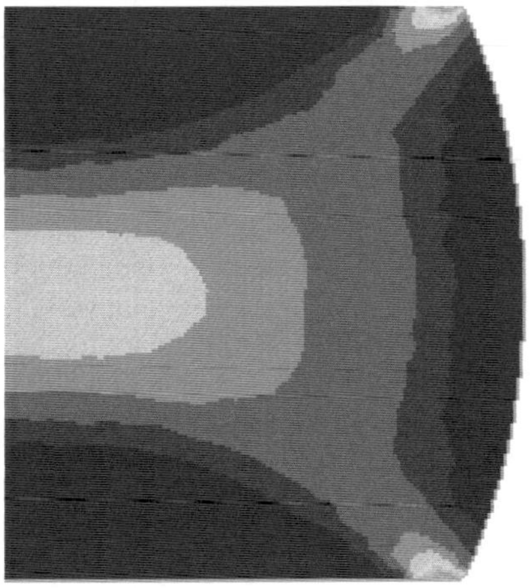

Figure 5. The proportions of dynamically recrystallized volume (JMAK Huang model).

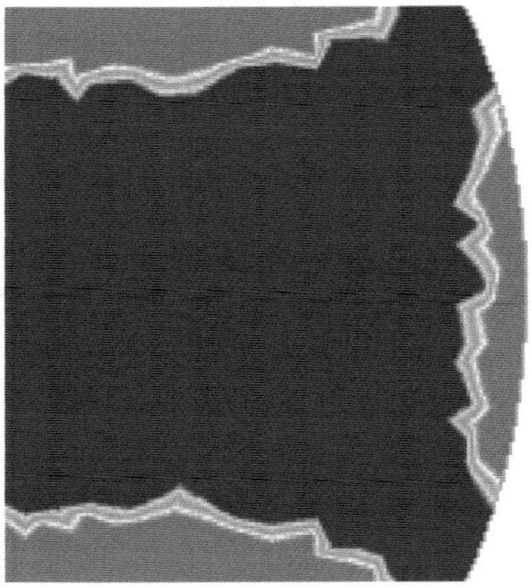

Figure 6. The average size of dynamically recrystallized grains (JMAK Huang model).

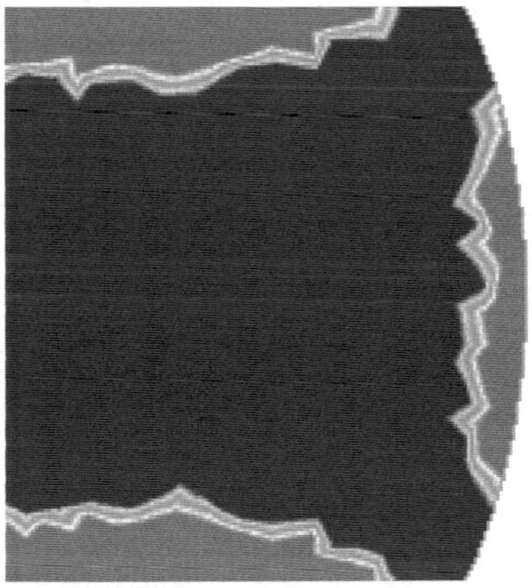

Figure 7. The average size of grains (JMAK Huang model).

For a full comparative analysis, 10 simulations were carried out with different input data (table 2).

Table 2. Grain Growth Simulation Route

	Grain growth	
№	T, °C	t, h
1	970	0.5
2	970	1
3	970	2
4	970	4
5	970	10
6	1080	0.5
7	1080	1
8	1080	2
9	1080	4
10	1080	10

Simulation of the calculation of the volume fraction of dynamically recrystallized grains showed that the value varies with a change in the temperature signification, the degree of deformation, and the strain rate. It appears to be a similar development trend: with an increase in the strain rate and temperature, the value of the average grain size decreases. This result was obtained after a deep analysis of calculation of the average size of dynamically recrystallized grains.

Modeling the calculation of the average grain size showed a similar development trend for all three models: everywhere, over time, the average grain size will increase. In the models of Chen Liging and D. Huang, the average grain size at 1080°C practically coincides, and the values of the third model proposed by Young-Sang Na, on average, are 2 times less than in the others.

5. Conclusion

In the article include the method of analysis and selection of technological processes in the form of parts in isothermal conditions. The technological processes and simulation of deformation-thermal treatment has been carried out. The simulation ensured the formation of the ultra-fine-grain structure in the blanks under the super-plastic roll-out of the shaft from the alloy EhK79. The modeling shows the effect of the microstructure of alloys on the physical-mechanical and operational properties of parts. As the model used to predict microstructure evolution is the Johnson-Mehl-Avrami-Kolmogorov model.

References

[1] Utyashev F Z, Sukhorukov R U, Nazarov A A and Potekaev A I 2015 The values of strain components and their role in formation of ultrafine-grained and nanosized structure in materials by means of severe plastic deformation *Russian Phys. J.* pp 64-71

[2] Valiev R Z, Enikeev N A and Murashkin M Yu 2010 Superstrength of ultrafine-grained aluminum alloys produced by severe plastic deformation *Dokl. Phys.* **55 (6)** pp 267-70

[3] Utyashev F Z and Raab G I 2006 The area of the surfaces of fragments, grains, and the sample upon large cold deformations of metals and the effect of these surfaces and the surface of the deformation zone on structure refinement *Fiz. Metal. Metalloved.* **101(3)** pp 311–22

[4] Valiev R Z , Enikeev N A and Murashkin M Y 2012 Using intensive plastic deformations for manufacturing bulk nanostructure metallic materials *Mech. Solids* pp 463–74

[5] Valiev R Z, Murashkin M Yu, Bobruk E V and Raab G I 2009 Grain refinement and mechanical behavior of the al alloy subjected to the new SPD technique *Mater. Trans.* **50(1)** pp 87–91

[6] Zwicky F 1948 The morphological method of analysis and construction *Courant Anniversary*

Volume (New York: Interscience Publishers) pp 461-70
[7] Zwicky F 1969 *Discovery, Invention, Research through the morphological approach* (Toronto: The Macmillan Company)
[8] Rakov D L, Sukhorukov R Y and Gavrilina L V 2015 Analysis and evaluation of technologies and equipment for the manufacture of long hollow shafts of gas turbine engines from heat-resistant nickel and titanium alloys on the basis of morphological approach *Probl. Mashinostr. Avtomatiz* pp 136-43
[9] Rakov D 2019 Okkam - advanced morphological approach as method for computer aided innovation (CAI) *MATEC Web of Conf. 298. ICMTMTE 2019, 00120*
[10] Wei X 2012 Hot compressive deformation behavior and microstructure evolution of HIPed FGH96 superalloy *Transactions of Nonferrous Metals Society of China* **22 (1)** pp 66–71
[11] Huang D, Wu W T, Lambert D and Semiatin S L 2013 Computer simulation of microstructure evolution during hot forging of waspaloy and nickel alloy 718 *Scientific Forming Technologies Corporation* (Columbus: OH 43220)
[12] Na Y, Yeom J, Park N *et al* 2003 Prediction of microstructure evolution during high temperature blade forging of a Ni–Fe based superalloy alloy 718 *Met. Mater. Int.* pp 15–19
[13] Yeom J T, Lee C S, Kima J H and Park N K 2007 Finite-element analysis of microstructure evolution in the cogging of an alloy 718 ingot *Materials Sci. and Engineering* pp 722–6

Fracture toughness experimental investigations of metal structures from low-alloyed steel

A V Chelnokov[1] and M N Erofeev[2]

[1] The Military Academy of Strategic Rocket Troops after Peter the Great, Moscow region, Balashikha, 143900, Russia
[2] Mechanical Engineering Research Institute of the Russian Academy of Sciences, Moscow, 101990, Russia

chelnokov-av@yandex.ru

Abstract. Presented are the results of experimental fracture toughness investigations of metal structures from low-alloyed steel (<0.12% C, 0.6-0.9% Cr, 0.8-1.1% Si, 0.5-0.8% Ni, 0.4-0.6% Cu and others), the influence of structural, technological and operational factors taken into account. The results of the investigations can be used to form a fracture toughness data bank of metal structures for mechanical engineering, to conduct of metal structure defectation, to assess of the brittle fracture risk due to the presence of crack defects, to improve the quality control system at the production stage and to solve other problems of ensuring the safety of machinery and equipment.

1. Introduction

Low-alloyed structural steel (<0.12% C, 0.6-0.9% Cr, 0.8-1.1% Si, 0.5-0.8% Ni, 0.4-0.6% Cu and others) is widely used in mechanical engineering for manufacturing welded metal structures and various parts, the requirements of high strength and operating temperature from -70 to 450 °C being taken into account. It is necessary to emphasize the favorable combination of mechanical characteristics (strength, ductility and impact strength) in this steel, good technological properties, which determines its wide application for manufacturing critical welded structures of technological equipment for rocket and space complexes, lifting equipment, bridges, ship hulls, etc. [1-3]. Under modern economic conditions a number of technical devices, machines and equipment are characterized by their intended use outside the standard / warranty periods of operation. In this regard, the issues of ensuring the reliability and safety of long-term operation of critical welded structures under conditions of material characteristics changing, the development of damage and defects, as well as the flow of other destruction processes become relevant. The greatest danger is the development of fatigue cracks, which can cause sudden destruction of structures, sometimes with disastrous consequences.

The aim of the research was to form a fracture toughness data bank of metal structures from low-alloyed steel (<0.12% C, 0.6-0.9% Cr, 0.8-1.1% Si, 0.5-0.8% Ni, 0.4-0.6% Cu and others) [4], to determine the characteristics of fracture toughness and fracture mechanism of welded steel, the influence of constructive, technological and operational factors being considered.

2. Materials and equipment

Many methods and criteria have been developed for assessing the structures fracture toughness [5]. Determination of fracture toughness characteristics under the static loading was carried out in accordance with the requirements of the State standard GOST 25.506-85 [6] based on the assessment of the stress intensity factor.

Fracture toughness was estimated incase of the maximum tensile stresses being perpendicular to the plane of the initial crack, i.e. for a crack of normal separation according to the I model of linear fracture mechanics.

Rectangular compact samples of type 3 with an edge crack were made from welded steel billets for off-center tension testing in accordance with the requirements of the State standard GOST [6]. The initial welded joints with symbols C8, C15, C17 are shown in table 1.

Table 1. Welded joints of steel billets for cutting samples.

Welded joint symbol	Cross-sectional shape of prepared edges	Cross-sectional shape of the weld
C8		
C15		
C17		

The samples were prepared to identify the dependence of fracture toughness on the thickness of the metal samples. Initiating cracks on the samples were applied to the base metal (BM), the weld (W), and the heat-affected zone (HAZ) of the welded joint.

Test samples were made from metal structures of lifting structures that fulfilled the standard service life to assess the destruction of low-alloyed steel elements under the influence of long-term operation [1-3].

The conditions were provided to ensure minimal hardening, residual stresses, as well as changes in the structure and phase composition in the area of application of the initiating incision when cutting and manufacturing the samples.

Experimental investigations planning and statistical processing of the results were carried out while taking into account the requirements of Standardization guidance document RD 50-705-91 [7].

A fatigue pulsator machine of the CDM PU-10t type were used to apply initial fatigue cracks on test samples, a TIRAtest 2300 test machine were used to load the samples to failure.

After the test, additional metallophysical investigations were carried out. There were fractographic analysis by the Tescan VEGA II LMU scanning electron microscope with INCA energy-dispersive and wave-dispersive microanalysis systems and metallographic analysis by the AxioVert.A1 microscope to determine the presence of non-metallic inclusions in accordance with the requirements of the State standard GOST 1778-70 [8].

3. Results and discussion

The tests have shown that the fracture toughness of the base metal and the zones of structures welded joints made of low-alloyed steel varies from 79 to 107 MPa·m$^{1/2}$ [3].

Evaluation of the influence of the structural factor (thickness) on the fracture toughness have shown that the fracture toughness of metal structures increases slightly in the thickness range from 4 to 20 mm (figure 1) [3].

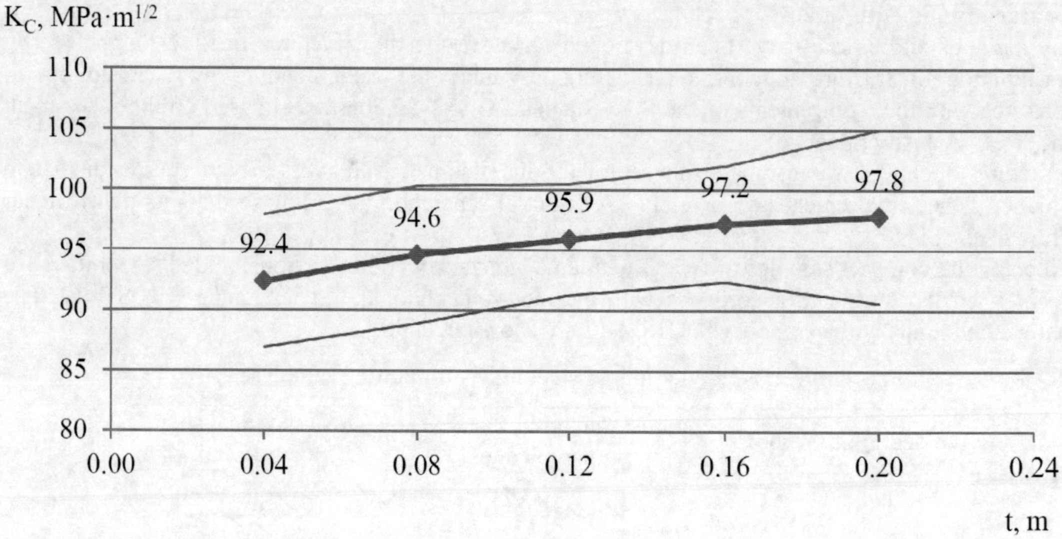

Figure 1. The average values and confidence interval of base metal fracture toughness depending on the thickness of the steel samples.

When analyzing the mechanics of welded structures destruction, features of the welded joints structure and defects should be taken into account [9].

The assessment of the influence of zones and types of welded joints on fracture toughness is shown in figure2 [3].

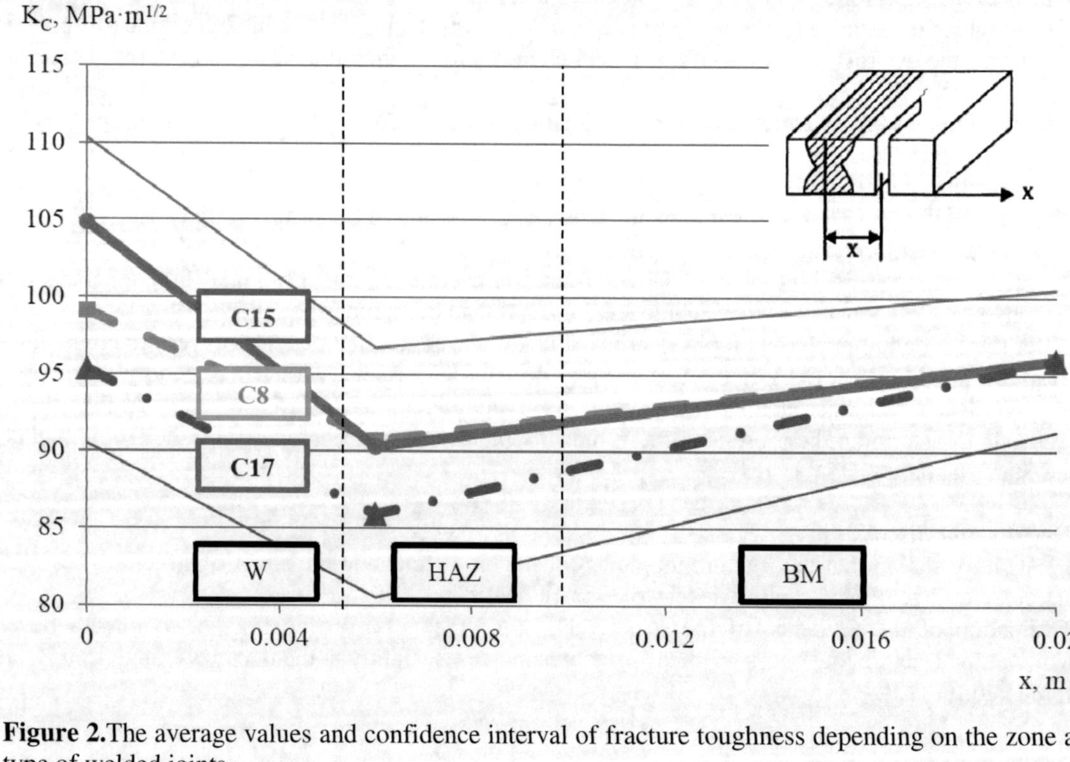

Figure 2. The average values and confidence interval of fracture toughness depending on the zone and type of welded joints.

The change in the structures fracture toughness after long-term operation is insignificant (about 3-6%) (figure 3) [1, 3].

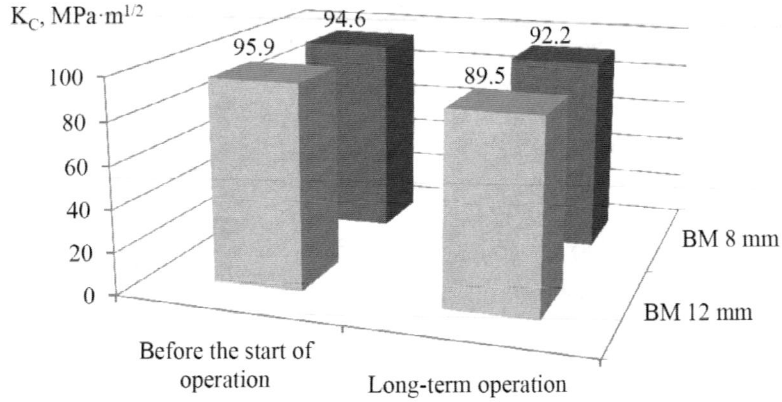

Figure 3. The average values of base metal fracture toughness taking into account the influence of long-term operation.

During the tests, the characteristic feature of the individual samples destruction was revealed. There was lamination, which contributes to a decrease in the base metal fracture toughness (figure 4). This feature was identified both for the structures before the start of operation (fracture toughness reduced by up to 10%) and for the long-term operation structures (fracture toughness reduced by up to 14%) [3].

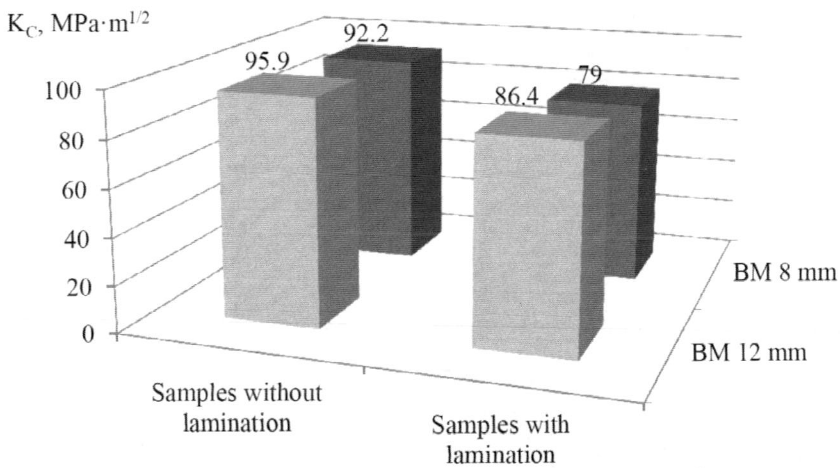

Figure 4. The average values of base metal fracture toughness taking into account the lamination.

The complex of metallophysical investigations was carried out in order to clarify the reasons for the lamination of samples. The results of fractographic analysis and chemical composition analysis allows us to understand the factors that contribute to the lamination of the samples during the tests:

- quantitative composition of chemical elements in the sample of low-alloyed steel (figure 5) meets the requirements of the State standard GOST [4], however, the sample is characterized by a high content of silicon and sulfur in the fracture zone near the lamination;

- the welded joint of a sample of low-alloyed steel (figure 6) has non-metallic inclusions and non-melting areas with a high content of manganese, silicon and sulfur in its structure.

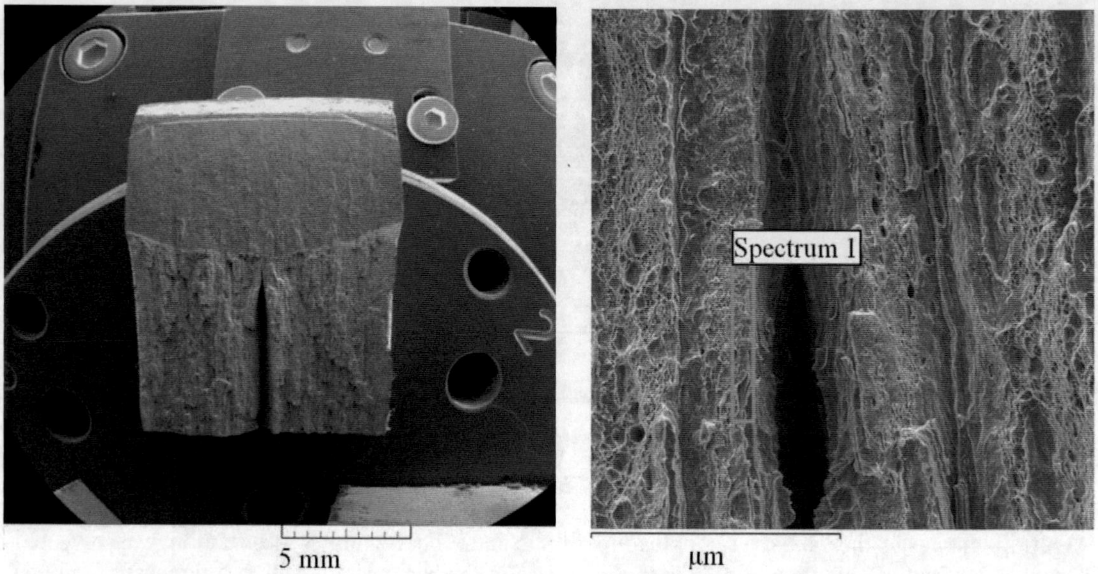

(a) (b)

Figure 5. Fracture surface of base metal sample with lamination: (a) – x12; (b) – x600.

(a) (b)

Figure 6. Fracture surface of base metal sample in the heat affected zone: (a) – x19; (b) – x600, non-metallic inclusion.

Metallographic analysis has shown that samples with lamination and reduced fracture toughness are characterized by an increased content of a number of non-metallic inclusions: non-deformable silicates (CH, figure 7), brittle silicates (CX, figure 7), and sulfides [3].

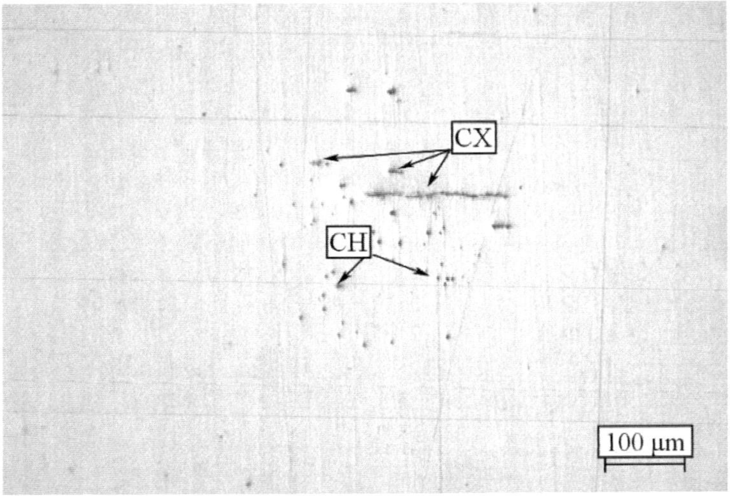

Figure 7. Contamination by non-metallic inclusions of a steel sample with lamination.

The results of metallophysical investigations of the destroyed samples show that, most likely, separate inconsistencies in the production technology (rolling) led to lamination. Inconsistencies in the technology are expressed in the increased content of non-metallic inclusions. A lamination of rolled steel is not allowed. Rolled products are checked for the absence of lamination by inspection of the edges and ends without the use of magnifying devices. In addition, the lamination of rolled products can be controlled using the ultrasonic method of non-destructive testing. However, ultrasonic testing showed that no fixed or unacceptable defects of the samples were detected (an ultrasonic flaw detector A1212 MASTER was used) [3].

It should be noted that the methods of rolled steel control do not include a mandatory quantitative and qualitative assessment of non-metallic inclusions, which probably contributed to the lamination of the metal. In this regard, it is advisable to conduct additional investigations in order to establish requirements for contamination of rolled products made of low-alloyed structural steel with non-metallic inclusions.

4. Conclusions

The complex of experimental investigations of metal structures from low-alloyed steel(<0.12% C, 0.6-0.9% Cr, 0.8-1.1% Si, 0.5-0.8% Ni, 0.4-0.6% Cu and others) was carried out. This involves determination of fracture toughness characteristics under the static loading and metallophysical investigations of destroyed samples.

The features of fracture toughness and the mechanism of welded metal structures destruction have been identified, the influence of structural, technological and operational factors taken into account.

The results of the investigations can be used to form a fracture toughness data bank of metal structures for mechanical engineering, to conduct of metal structure defectation, to assess of the brittle fracture risk due to the presence of crack defects, to improve the quality control system at the production stage and to solve other problems of ensuring the safety of machinery and equipment.

References

[1] Chelnokov A V and Pankin D A 2017 Assessment of the impact of long-term operation on thefracture toughness of rocket equipment metal structures *Deformation and destruction of materials and nanomaterials* (Moscow: A A Baikov IMET RAS) pp 748-9

[2] Agafonov K V, Gnevko A I, Polyanskiy V I and Chelnokov A V 2015 *Dual technology* **4 (73)** 60-6

[3] Chelnokov A V 2018 Fracture toughness investigation of lifting equipment metal structures *New materials and promising technologies* (Moscow: BukiVedi) pp 876-81
[4] State standard GOST 19281-2014 High strength rolled steel. General specification
[5] Zhu X – K and Joyce J A 2012 *Engineering Fracture Mechanics* **85** 1-46
[6] State standard GOST 25.506-85 Design, calculation and strength testing. Methods of mechanical testing of metals. Determination of fracture toughness characteristics under the static loading
[7] Standardization guidance document RD 50-705-91 Design, calculation and strength testing in mechanical engineering. Planning and statistical processing of static and fatigue test results
[8] State standard GOST 1778-70. Steel. Metallographic methods for the determination of nonmetallic inclusions
[9] Zerbst U, Ainsworth R A, Beier H Th, Pisarski H, Zhang Z L, Nikbin K, Nitschke-Pagel T, Münstermann S, Kucharczyk P and Klingbeil D 2014 *Engineering Fracture Mechanics* **132** 200-76

IOP Conf. Series: Materials Science and Engineering 1129 (2021) 012015 doi:10.1088/1757-899X/1129/1/012015

The hydroplane body development of design and topological optimization

G V Dolunts [1], L B Maslov[1,2] and M A Zhmaylo[2]

[1] Ivanovo State Power Engineering University, 34 Rabfakovskaya, Ivanovo, 153003, Russia
[2] Center of National Technological Initiative, Institute for Advanced Manufacturing Technologies, Peter the Great St. Petersburg Polytechnic University, 29 Politekhnicheskaya, St. Petersburg, 195251, Russia

doluntss@gmail.com

Abstract. This work is devoted to the study of the stress-strain state of the hydroplane body and its topological optimization. During laboratory tests, physical and mechanical properties of materials were identified, which are necessary for use in further calculations to study the stress-strain state of the hydroplane body, followed by reducing its weight, using topological optimization of the structure. The end result is a developed method for manufacturing the hydroplane body using additive technologies.

1. Introduction

Any aircraft is a compromise between lightness and strength, and the large range of materials available on the market allows you to print the body of an unmanned aerial vehicle (UAV) for a highly specialized task, without spending a lot of money for implementation. This paper presents a computer simulation of the hydroplane body based on studies of the physical and mechanical properties of materials used in the manufacture of the fuselage, and topological optimization of the body based on the study of the stress-strain state.

2. Materials and equipment

Tensile testing is the main and most common method for controlling the mechanical properties of materials [1]. The task is to select the optimal material for manufacturing the hydroplane fuselage using the additive method.

Reproducibility of the experiment results is achieved by repeated repetition. The minimum number of repetitions is 5 times. Experiments were performed for a plastic thread (the material before printing), a shell model (along and across the fibers), and a massive structure (the fibers are arranged at a 45-degree cross-to-cross angle). Figure 1 shows samples for tensile testing, according to ISO 527-2:2012 [2].

In the course of the experiments (table 1) it was found that PLA is the most optimal of the presented options of plastic threads. The tensile strength of the plastic thread is 45 MPa. The young's modulus for PLA plastic was also derived and differs from the passport data by 3%. The Poisson ratio is 0.39.

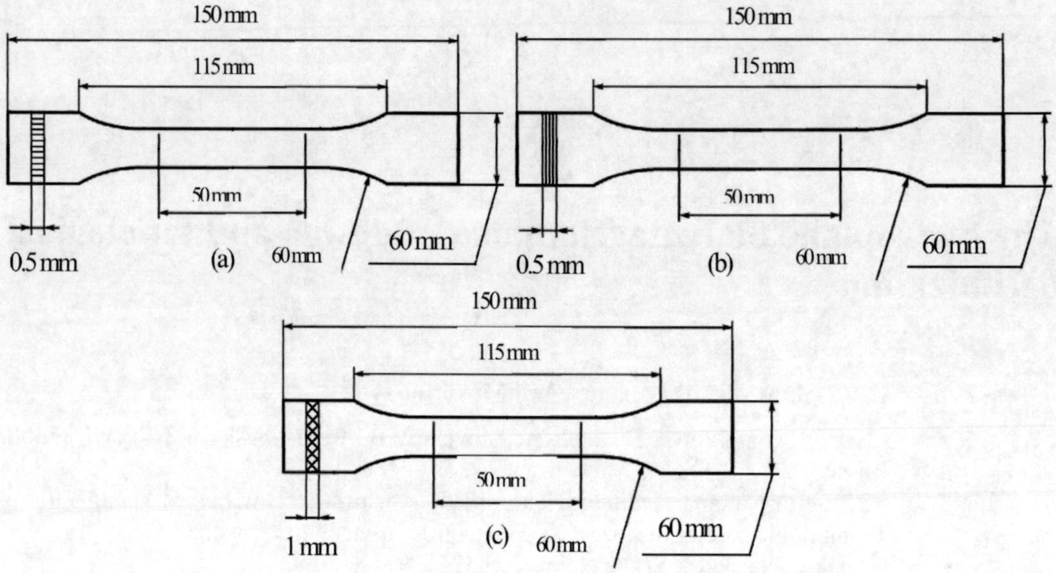

Figure 1. Standard for stretching ISO 527-2:2012. (a) – sample for the shell, fibers along the part, (b) – sample for the shell, fibers across the part, (c) – sample for a massive structure, fibers cross to cross at an angle of 45 degrees.

Table 1. Physical and mechanical properties of the materials.

Material	Maximal stress, MPa	Elongation, mm	Young's modulus, MPa	Poisson ratio
Aerotex (filament)	62.1	0.026	1557	-
PLA (filament)	45	0.05	1332	-
Total GF-30 (filament)	47.1	0.37	300	-
Longitudinal shell made of PLA	35	0.04	1447	-
Cross shell made of PLA	25	0.022	1032	-
Massive construction of PLA	45.2	0.067	1335	0.39

3. Computational model

The fuselage of a flying boat consists of a massive structure to which all the power elements are attached and where the payload is located. And the shell is necessary to improve the aerodynamic qualities of the aircraft, as well as for waterproofing and landing on the water surface (figure 2).

Figure 2. Scheme for developing the fuselage of a flying boat.

For further testing, it is necessary to select the design case where the maximum loads are applied to the UAV. There are landing, flight, and takeoff cases. The hydroplane must withstand loads, and the structure must not collapse during flight in all these cases. Consider the case during landing, when the maximum load occurs. The fuselage does not receive weight loads from the wing structure during flight, as it creates lift. During landing, the lift is noticeably reduced, thus the weight of the wing is transmitted to the fuselage through the connecting elements. Also, during a sharp loss of altitude, there is an overload that can reach up to 3g. The hydroplane has a wingspan of 1200 mm and a fuselage length of 758 mm. The maximum take-off weight with a payload is 2 kg. At point A in figure. 3, there is a pivotally fixed support; at points B, C, and D, there is a pivotally movable support. m1 = 0.2 kg (wing mass), m2 = 0.05 kg (stabilizer mass), m3= 1 kg (payload mass), L = 0.3 m, l= 0.04 m, M=1 $N{\cdot}m^{-1}$.

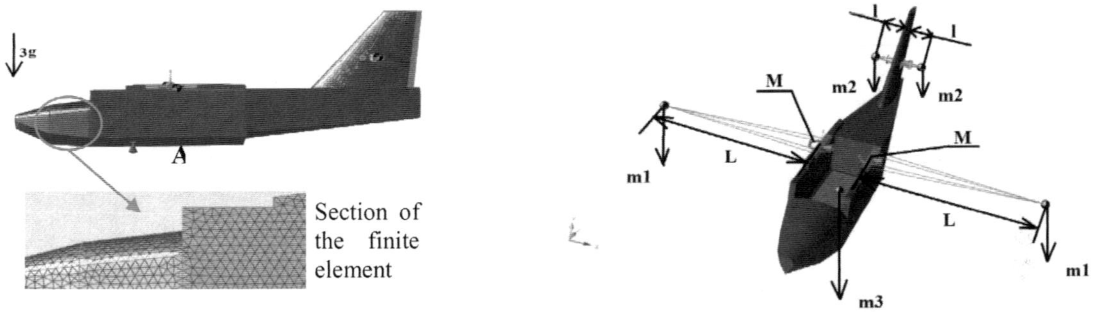

Figure 3. Loads applied to the hydroplane during take-off and landing.

The Von-Mises equivalent stresses of the massive structure and shell are shown below (figure 4).

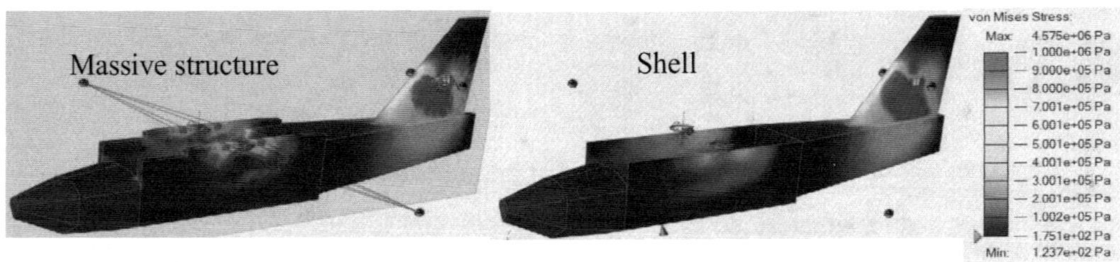

Figure 4. Mises equivalent stresses

4. Results and discussion

After performing the calculation to obtain the resulting stresses, we proceed to topological optimization [3], which is an iterative operation, at each step the design VAT is estimated. If the stress in some areas is significantly less than the critical ones, then this area of the material (continuous medium) is removed from the calculation (figure 5, 6).

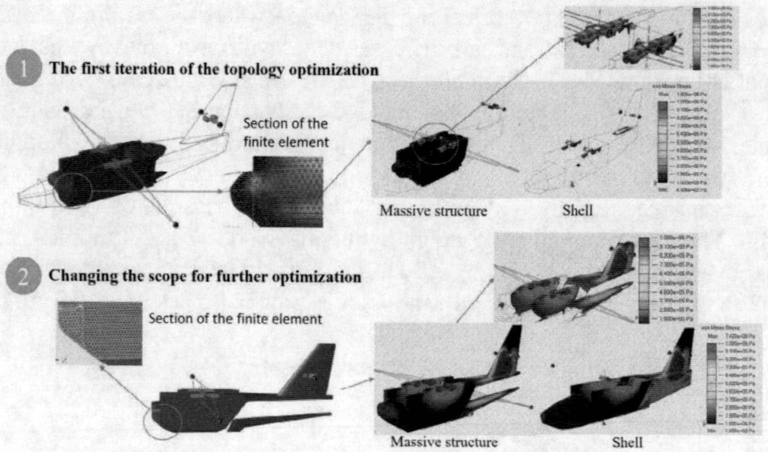

Figure 5. The first iteration of topological optimization of the hydroplane body.

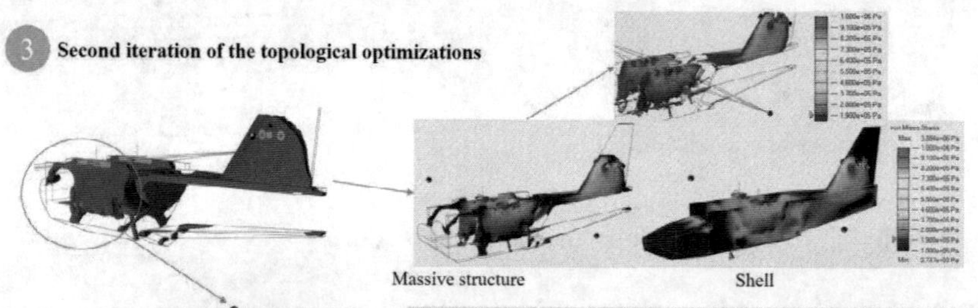

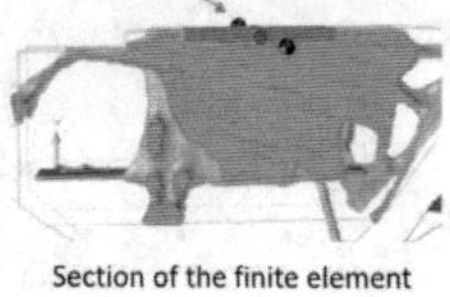

	Original case	1st iteration	Updated case	2nd iteration
The maximum von Mises stress (MPa)	4.6	2	7.4	3.9
The minimum von Mises stress(MPa)	$0.124\cdot10^{-3}$	$0.65\cdot10^{-3}$	$0.037\cdot10^{-3}$	$2.7\cdot10^{-3}$

Figure 6. The second iteration of topological optimization of the hydroplane body.

The design has a safety margin of more than 10 times. For aviation equipment, this is a huge margin and with the help of topological optimization, it was possible to reduce the weight by 70% without losing strength characteristics (figure 7).

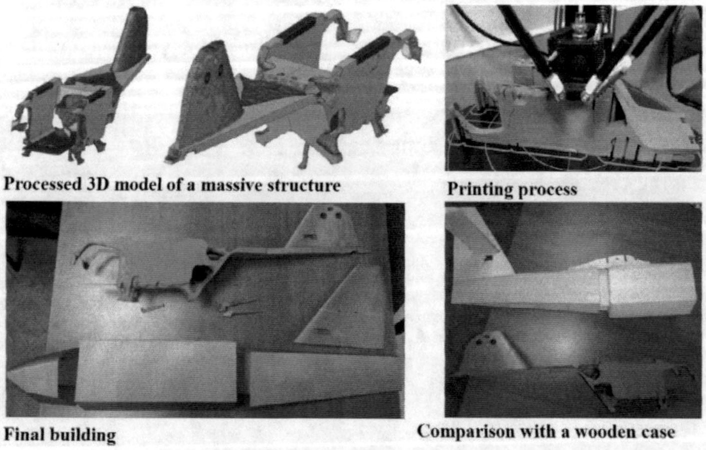

Figure 7. Post-processing and printing.

5. Conclusions

The physical and mechanical properties of materials for the massive structure and shell of the hydroplane body, as well as the plastic thread itself, are revealed. Thus, in the course of tensile tests, the strength limits, young's modulus, and Poisson's Ratio were obtained. For shells, the properties were studied from longitudinal and transverse fibers. Computer models of an unmanned aerial vehicle were developed in the SolidWorks software package.

The calculation of the stress-strain condition of the hull of a seaplane has been done. Thus, it was possible to create a method for manufacturing the UAV body using additive technologies and topological optimization.

References

[1] Birger I A and Mavlyutov R R 1986 *Resistance of materials* (Moscow: Nauka) p 560
[2] ISO 527-2:2012 Plastics – Determination of tensile properties
[3] Bashin K A, Torsunov R A and Semenov S V 2017 Methods of topological optimization of structures used in the aerospace industry *PNRPU Bulletin* (Perm National Research Polytechnic University) **51** pp 51-60

IOP Conf. Series: Materials Science and Engineering 1129 (2021) 012016 doi:10.1088/1757-899X/1129/1/012016

Application of UHMWPE in energy-absorbing road safety systems

A S Gruzdev, V S Nadezhdin, O V Titov and I S Prohanov

Moscow Automobile and Road Construction State Technical University (MADI), 64, Leningradsky Prospect, Moscow, 125319, Russia

shashki18@gmail.com

Abstract. The article considers the possibility of using a polymer material based on ultra-high molecular weight polyethylene (UHMWPE) in energy-absorbing structures of road safety systems. To determine the mechanical and energy absorbing characteristics of the material, bench tests carried out on the material and element of the designed structure in the form of a thin-walled cylinder. The possibility of using UHMWPE in the actual design of the road safety system checked by creating a validated digital twin of the impact attenuator. Based on the simulation results, polymer and metal structures compared.

1. Introduction

Currently, there are more than 30 million kilometers of roads worldwide, and the number of registered vehicles reaches more than a billion, and this number growing at a rapid rate every year. Due to the significant increase in traffic intensity, mass of cars, speeds, as well as the length of the road network, the number of fatal accidents (accidents), injuries, economic and other dangerous consequences increased significantly in recent years.

The problem of ensuring road safety and reducing road mortality is especially highlighted [1]. Analyzing the statistics of accidents, it can be seen that in 2019 there were about 150000 such incidents, as a result of which about 190000 people injured, and about 18000 died. One of the main reasons for this victims number is the lack of various security systems types on many sections of roads.

The structures of most types of barriers until recently represented either by steel materials (side road barriers, wire rope barriers, frontal road barriers) or concrete (parapet barriers) [2]. However, in connection with the development of the chemical industry, it seems promising to develop road barriers designs using new polymer materials with a number of advantages: high energy absorption capacity, corrosion resistance, chemical inertia, low weight. The use of modern polymer materials can significantly improve the quality of existing structures road construction systems. We pay special attention to the sections of roads on which the road lane divided, since in this case the direction of the vehicle impact during an accident does not occur tangentially (at an average angle of 200), but perpendicular to the road barrier, which leads to serious consequences (figure 1).

IOP Conf. Series: Materials Science and Engineering 1129 (2021) 012016 doi:10.1088/1757-899X/1129/1/012016

Figure 1. Vehicles hitting on barriers intended for splitting of lanes.

The way out of this situation is the installation of additional fences – frontal barriers, designed to absorb the energy of frontal impact with minimal risks of severe injuries for passengers. Currently, the following used as frontal traffic barriers:

1. Plastic enclosures filled with sand or water from the inside (figure 2). However, such a fence practically does not absorb the energy of the vehicle impact, and installed mainly to inform road users about the separation of the road lane.
2. Designs with energy-absorbing elements in the form of metal honeycombs have become widespread (figure 3).

The SMA (Safety Modular Absorber) energy-absorbing enclosures use high-strength steel, which absorbs impact energy by plastic deformation of cells (hexagonal cells) installed inside the system. The use of a honeycomb absorbent structure ensures uniform energy absorption during the entire enclosure deformation process. With a frontal impact on a barrier 2.8 m long, a passenger car weighing 1500 kg at a speed of 100 km·h^{-1}, energy-absorbing honeycombs compressed, due to this the impact energy extinguished, the side barrier panels folded. The impact energy is 532 kJ. However, this design also has a number of drawbacks: the price of a metal structure, the need for corrosion protection, high inertia during operation. A way out of this situation may be the use of energy absorbing elements from, for example, polymeric materials.

In works [3, 4] quite detailed analysis of foreign structures of frontal barriers carried out. From this analysis, you can select the Guard LMC (figure 4).

Figure 2. Dividing the Road Lane.

Figure 3. Safety Modular Absorber.

Figure 4. Frontal barrier Guard LMC.

Guard LMC has a length of 2.3 m. The guard consists of movable frames (diaphragms), which surrounded by a movable side panel and energy absorbing elements - elastomeric cylinders. During frontal impact, the cylinders and side panels compressed, so that the vehicle slows down smoothly to a complete stop. The impact energy is 532 kJ. In addition, the elastomeric cylinders almost completely recover their shape after they discharged, and can be reused.

2. Materials and equipment

UHMPE considered as the material of the energy absorbing element. The determination of mechanical and strength characteristics carried out using a test machine with a force limit of 100 kN and a maximum loading speed of 1000 mm·min^{-1}. Tensile stress loading carried out at different loading rates (20-500 mm·min^{-1}). As a result of the experiment, the elongation amount of the sample fixed, as well as the force amount of resistance to the sample elongation. Based on the test results mechanical and strength characteristics determined, which reflected in [5]. The remaining characteristics determined from studies [6-8].

Determined the main mechanical characteristics of the UHMWPE, it decided to develop a barrier design with an energy absorbing element in the form of a cylinder (figure 5).

Compression tests performed to determine the energy absorption of such an element. The test diagram is shown in figure 6.

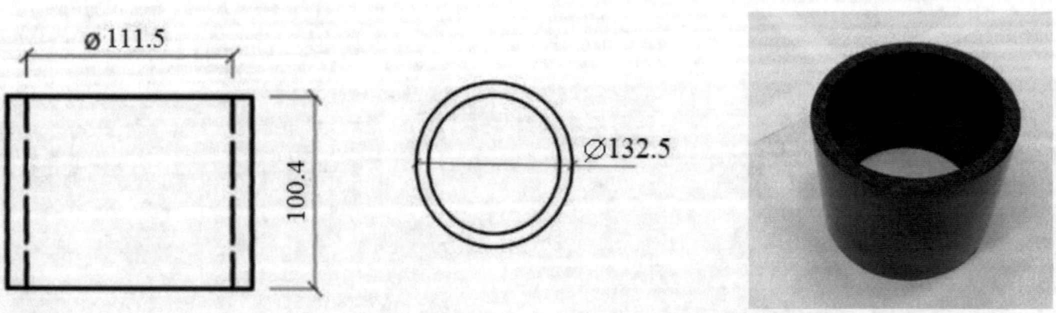

(a) (b) (c)

Figure 5. Sample of energy absorbing element from UHMWPE: (a) sketch, front view, (b) sketch, top view, (c) natural sample.

As a result of the full-scale experiment, a diagram of the force dependence on the movement of the active traverse of the test machine constructed (figure 7). Based on the results of experimental data on the diagram, the energy absorption amount was energy 0.326 kJ. The sample after the tests is shown in figure 8.

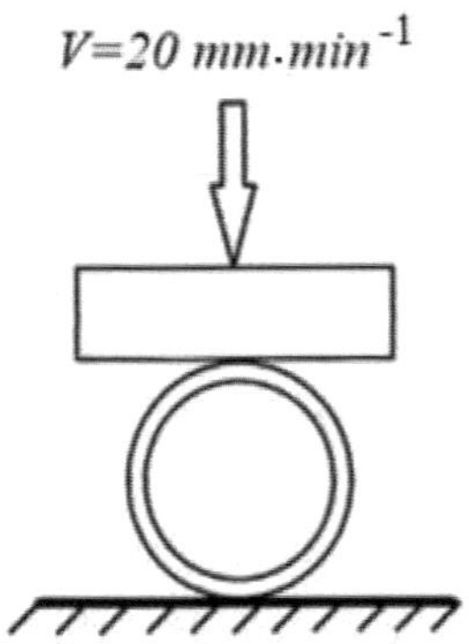

Figure 6. Diagram of tests for compression of sample of energy absorbing element from UHMWPE.

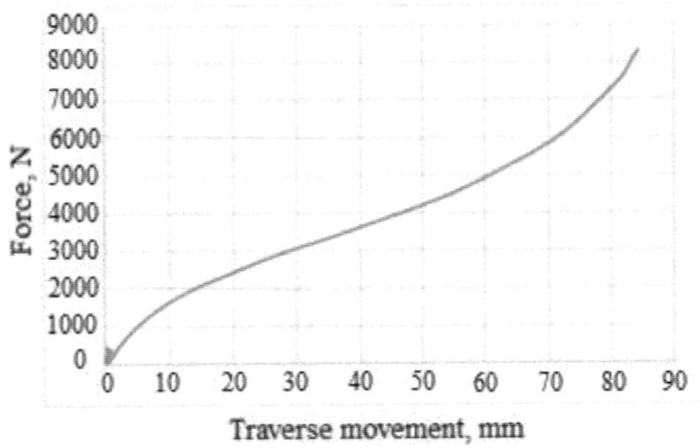

Figure 7. Graph of dependence of compression resistance force on cross-arm movement.

Analyzing the type of sample after the tests, noted that after three days the residual deformations were less than 10%, from which to conclude that with such a design, the energy absorbing element from the UHMWPE can be reused after significant deformations, unlike steel energy absorbing elements.

Figure 8. Sample energy absorbing element at the moment of full compression.

3. Model and methods

Virtual tests conducted on the basis of the finite element method (FEM) using the LS-Dyna multi-purpose software complex [9-12]. As a result of the virtual experiment, a graph of the force dependence on the displacement of the compressing plate plotted, which compared with the identical graph obtained during the full-scale experiment (figure 9). The process of compressing the energy absorbing element sample from the UHMWPE shown in figure 10.

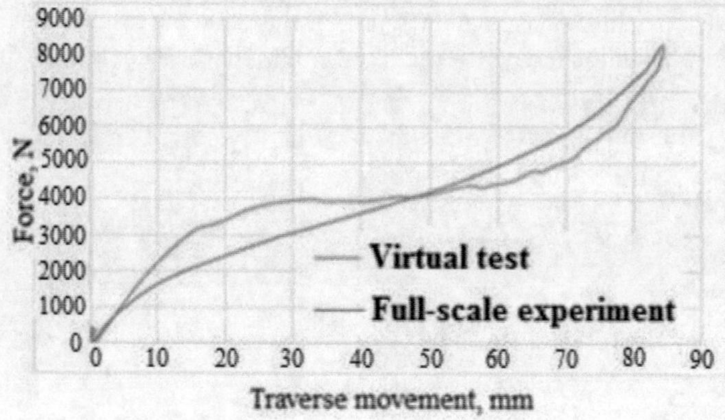

Figure 9. Graph of forces dependence on compressive element movements.

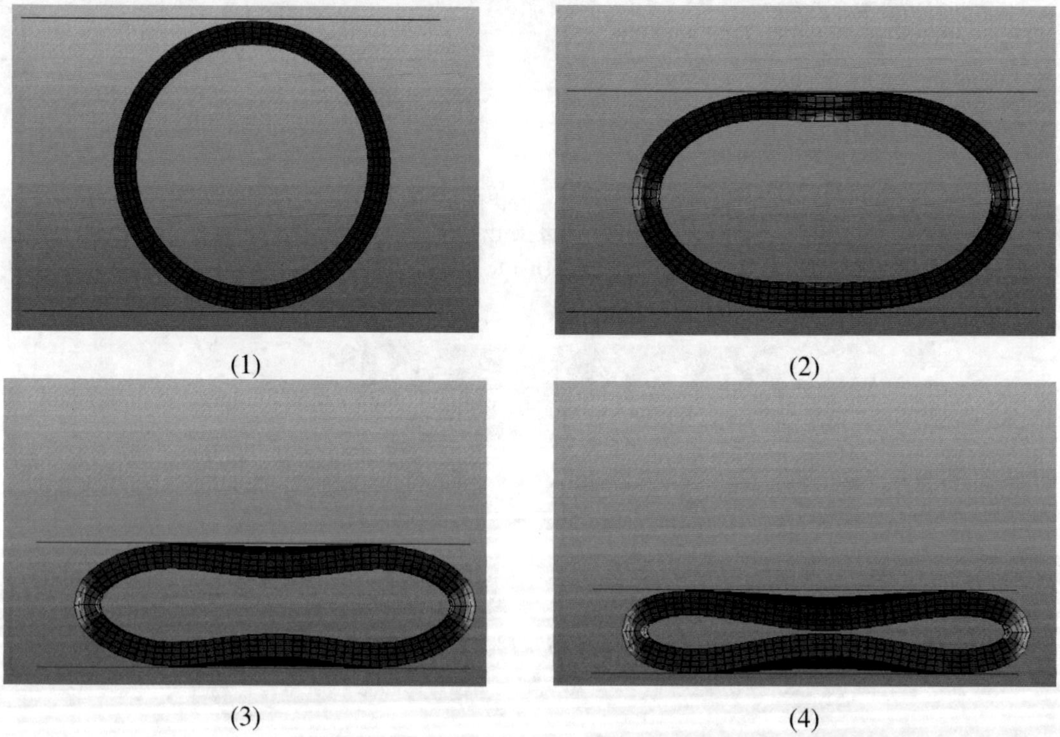

Figure 10. Compression process of FE model of energy absorbing element sample from UHMWPE.

Based on the results of the virtual experiment, the energy absorption of the sample from the UHMWPE was U=0.335 kJ.

Consider one of the possible design options for sections of energy absorbing elements in the form of cylinders from UHMWPE (figure 11). To compare energy absorption capacity, a model of energy absorption elements in the form of metal sections of honeycombs also constructed (figure 12).

Figure 11. Possible design of UHMWPE cylinders assembled.

Figure 12. Model of steel energy absorbing elements (honeycombs).

To determine the amount of energy absorption capacity, a virtual model of compression tests of energy absorption elements assembled built. The test diagram shown in figure 13. The loading process realized by means of an absolutely rigid plate moving at a speed of 20 mm·min^{-1}.

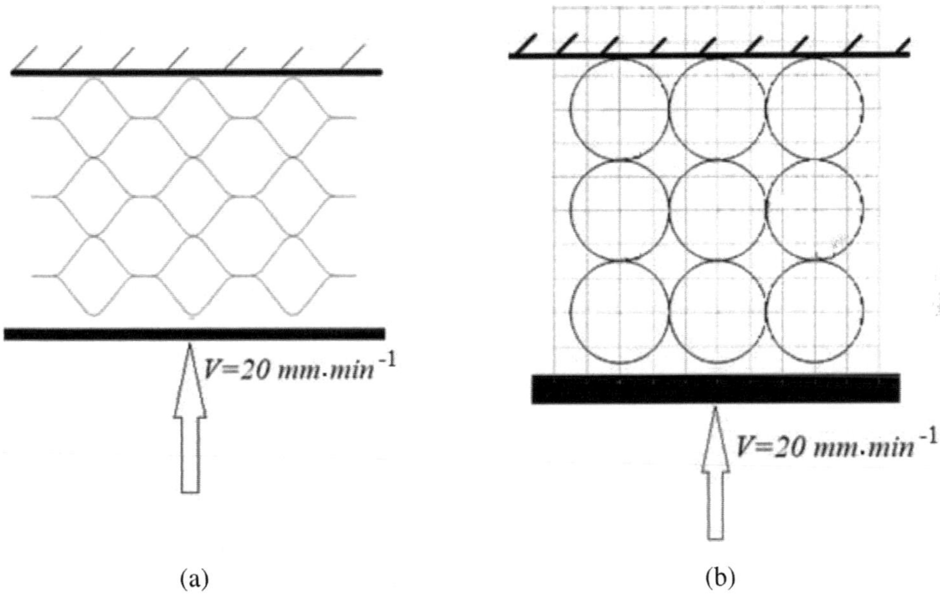

Figure 13. Virtual test schematic diagram: (a) steel energy absorbing elements, (b) UHMWPE cylinders.

4. Results

Based on the results of virtual tests, energy absorption capacity values obtained. The energy absorption diagrams shown in figure 14.

Table 1 shows the range of energy absorption values of sections from UHMWPE at an external diameter of 90 mm during loading, and the weight of the structure. For comparison, table 2 shows similar values for steel elements.

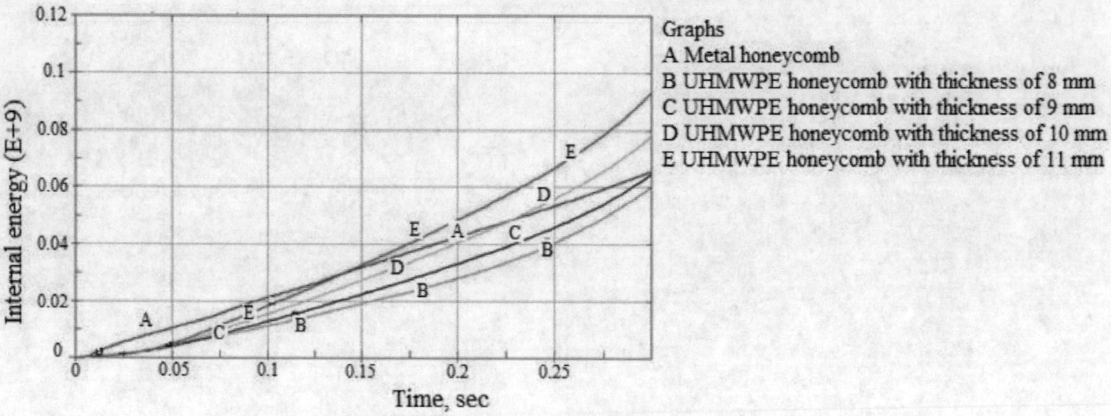

Figure 14. Energy absorption diagrams.

Table 1. Energy absorption and mass of structure from UHMWPE.

Parameter	Thickness 8 mm	Thickness 9 mm	Thickness 10 mm	Thickness 11 mm
Energy absorption	$6.0\cdot10^7$ J	$6.42\cdot10^7$ J	$7.73\cdot10^7$ J	$9.24\cdot10^7$ J
Weight	37.99 kg	42.49 kg	46.92 kg	51.32 kg

Table 2. Energy absorption and mass of structure from steel.

Parameter	Weight	Energy absorption
Steel honeycomb with 2 mm wall thickness	64.72 kg	$6.52\cdot10^7$ J

5. Conclusions

Based on the results obtained, it can previously concluded that the UHMWPE can used as a material for energy absorbing elements of the safety modular absorber. With an outer diameter of 90 mm and thicknesses of 8 mm to 11 mm, energy absorption closest to steel achieved.

Acknowledgment

Authors are grateful their scientific advisor, Ph.D., prof. I.V. Demiyanushko for guidance, comments and advice in carrying out this research work.

Financial support

The work carried out with funding under the Development - Digital Technologies Program with the support of the Fund for the Promotion of the Development of Small Forms of Enterprises in the Scientific and Technical Sphere (Innovation Assistance Fund) (№ 47GRCTC10-D5/56182).

References

[1] 2018 On National Goals and Strategic Objectives for the Development of the Russian Federation for the Period until 2024 *Decree of the President of the Russian Federation Assembly of Acts of the President and Government of the Russian Federation* **204**

[2] Barrier-type road fences Available at: http://nado-zabor.ru/article/dorozhnye-ograzhdeniya-barernogo-tipa

[3] American Association of State Highway and Transportation Officials. Roadside Design Guide, 4th Edition 2011 (Washington D.C.) p 356

[4] Bligh, Roger, Nauman Sheikh, Wanda Menges and Rebecca Haug 2005 Portable Traffic Barrier for Maintenance *Operations Report FHWA/TX-05/0-4692-1* (Texas Department of Transportation, Austin) p 70
[5] Titov O V, Demiyanushko I V, Nadezhdin V S and Gruzdev A S 2020 The effect of loading speed on the strength characteristics of ultra-high molecular polyethylene (UHMWPE) *XXXI Int. Innovation Conf. of Young Scientists and Students on Machine Science* pp 183-6
[6] Selyutin G E and Popova O E 2015 Application: 2013156379/05 Russian Federation MPK C08L 23/06 (2006.01) C08J 5/04 (2006.01) Composite material with improved damping properties based on ultra-high molecular weight polyethylene (UHMP) (Tuutin Applicant: Limited Liability Company NGO GELAR) **18**
[7] Mohseni Taromsari S, Salari M, Bagheri R and Faghihi Sani M A 2019 Optimizing tribological, tensile & in-vitro biofunctional properties of UHMWPE based nanocomposites with simultaneous incorporation of graphene nanoplatelets (GNP) & hydroxyapatite (HAp) via a facile approach for biomedical applications *Composites Part B: Engineering 175,107181*
[8] Xiao S, Lanceron C, Wang P, Soulat D and Gao H 2019 Mechanical and thermal behaviors of ultra-high molecular weight polyethylene triaxial braids: the influence of structural parameters *Textile Research Journal* **89(16)** pp 3362-73
[9] LS-DYNA® 2016 Keyword User's Manual. Livermore Software. Technology Corporation (Livermore, CA)
[10] Demiyanushko I V, Karpov I A, Zharikova S V and Mukhametova A A 2017 Study of the model of MAT26 material for modeling cellular elements for the LS-DYNA program *Collection of works of the MIKMUS conference* pp 379-82
[11] Büyük M, Atahan A O and Kurucuoğlu K 2018 Impact performance evaluation of a crash cushion design using finite element simulation and full-scale crash testing *Safety* **4** 48
[12] Wright L and Davidson S 2020 How to tell the difference between a model and a digital twin *Adv. Model. and Simul. in Eng. Sci. 7:13*

IOP Conf. Series: Materials Science and Engineering 1129 (2021) 012017 doi:10.1088/1757-899X/1129/1/012017

Research of the strength of structural elements (wing console) for aviation purposes obtained using 3d printing

A D Latypov, G V Moskvitin, O V Naumov and M S Pugachev

Mechanical Engineering Research Institute of RAS (Russian Academy of Sciences), Bardina street 4, Moscow, 101990, Russia

naumov57.on@gmail.com

Abstract. 3D printing technology allows you to obtain parts with a number of characteristics that have great importance for a number of areas, in particular small aircraft. The article presents a statistical and modal analysis of the naca0012 configuration wing console made of ABS and PLA plastics. These calculations allow us to draw conclusions about the possibility of using printing technology in the development and design of aircraft.

1. Introduction

Presently much attention is given to using 3D printing technology in the course of aircraft manufacturing. The peculiarity of this technology consists in layer-by-layer application. Different printing technologies have appeared at the moment, which make it possible to produce intricate-configuration parts that are as good as already existing ones. The application of such technology as 3D printing in some fields will help gain significant advantages. Specifically, aviation industry is worthwhile noting, where special attention is paid to the issues of weight-strength combinations. The usage of 3D printing technology helps solve problems in the course of building and designing aircraft, viz., get a required configuration of elements, units, reduced structure weight (as compared with the steel parts) and prime cost.

The purpose of this work is to analyze a possibility of using parts made by 3D printing when designing aircraft. In order to solve the formulated problem, the static and modal analyses of *naca* configuration wing have been carried out with the use of AutoCAD Inventor 2020 Professional software system. The grades of plastics are ABS (acrylonitrile-butadienestyrene) and ABS with polycarbonate. The data for calculation have been taken from the acquired results of static tests of samples produced by printing using 3D printer [1, 2].

2. Problem definition

The acquired mechanical characteristics of ABS plastic have been entered into materials base of Autodesk Inventor 2020 Professional design program, where modal and static analyses of outer wing panel of symmetrical airfoil *naca0012 (n0012-il)* for unmanned aerial vehicle have been carried out (figure 1). A mesh has been plotted for the panel, its parameters are shown in table 1. A choice of straight wing is substantiated by low speeds, however, the results of calculations of outer wing panel made of different plastics and comparison of the results thereof are of immediate interest. Dimensions: wing chord – 0.3 m, wing span – 2 m, sweep – 0.

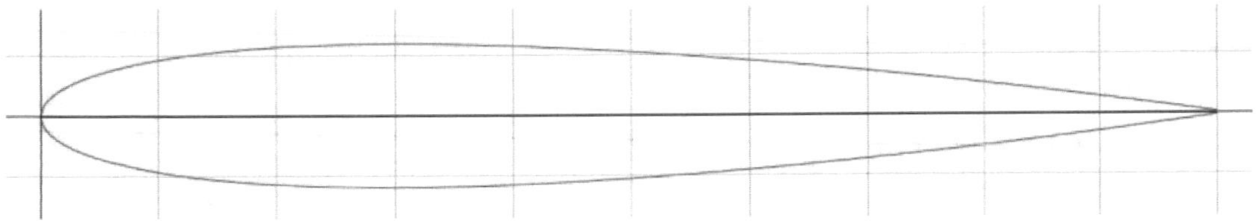

Figure 1. Outer wing panel of airfoil-shaped profile naca0012 (n0012-il).

Table 1. Network configuration.

Average dimension of elements, units	Average dimension of elements, units	Dissimilarity coefficient	Maximum turn angle
0.030	0.200	1.500	60.00

Pressure of 5 kPa has been set over entire panel area and one of its ends has been rigidly fixed for static analysis of outer wing panel made of plastic. The load has been selected so for the proof von Mises stress that to match closely the material yield limit. Theoretically the plastic material begins to get damaged, where von Mises stresses become equal to the limiting ones.

The same load has been set mainly for comparing displacements and safety factors for ABS with polycarbonate featuring somewhat surpassing mechanical characteristics as compared with pure ABS.

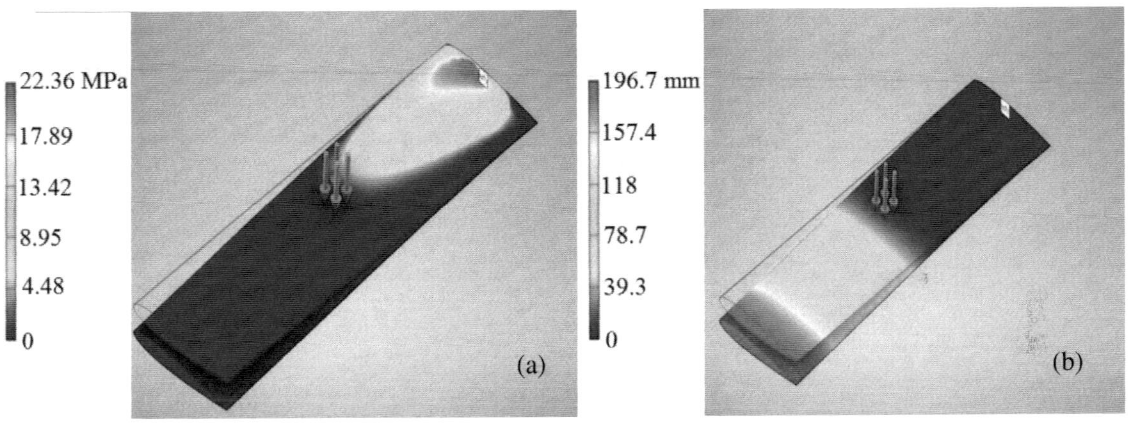

Figure 2. ABS: (a) – von Mises stresses; (b) – displacement simulation result.

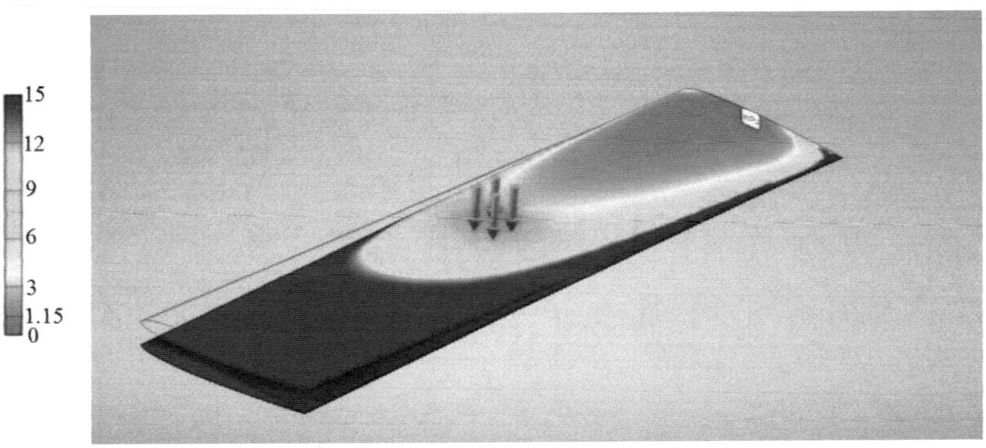

Figure 3. ABS safety factor.

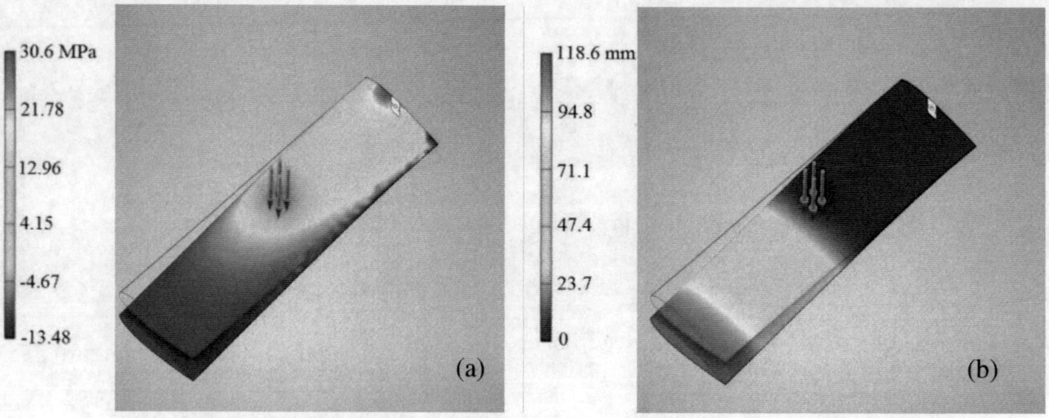

Figure 4. ABS with polycarbonate: (a) - von Mises stresses; (b) – displacement simulation result.

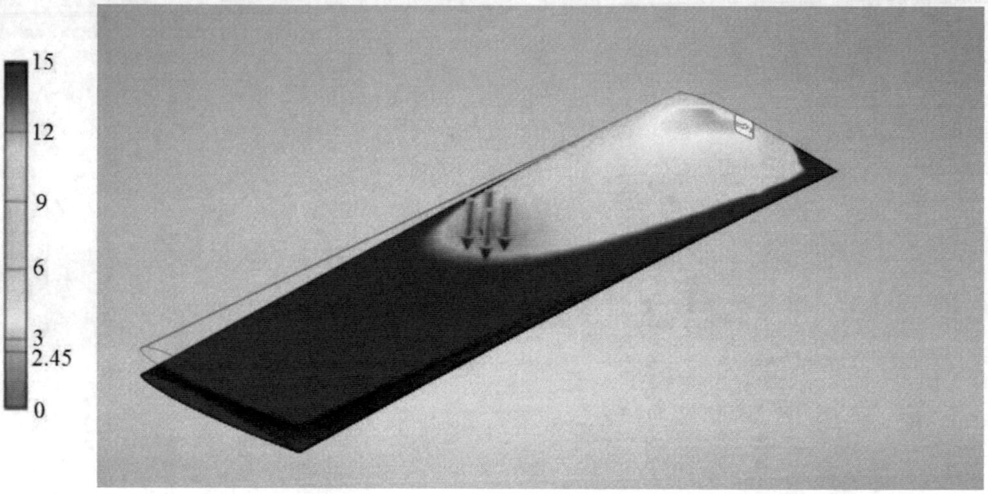

Figure 5. ABS with polycarbonate safety factor.

The modal analysis makes it possible to assess natural vibrations frequency, which is one of the important parameters when estimating a possibility of using material for aircraft [3]. Figures 6, 7 show the results of investigations by modes giving the values of frequencies with limiting values for every type of plastic.

The results of modal analysis show the presence of frequencies, when the displacement maximum values appear being localized in different parts of the wing. Table 2 is given for comparative analysis of two plastics.

Table 2. Calculation results.

Material	Density, g·cm^{-3}	Panel weight, kg	Yield limit, MPa	Young modulus, GPa	Displacement (max), mm	Safety factor (min)
ABS	1.06	7.84	25.8	1.686	196.7	1.15
PC/ABS	0.35	2.64	54.4	2.78	118.7	2.43

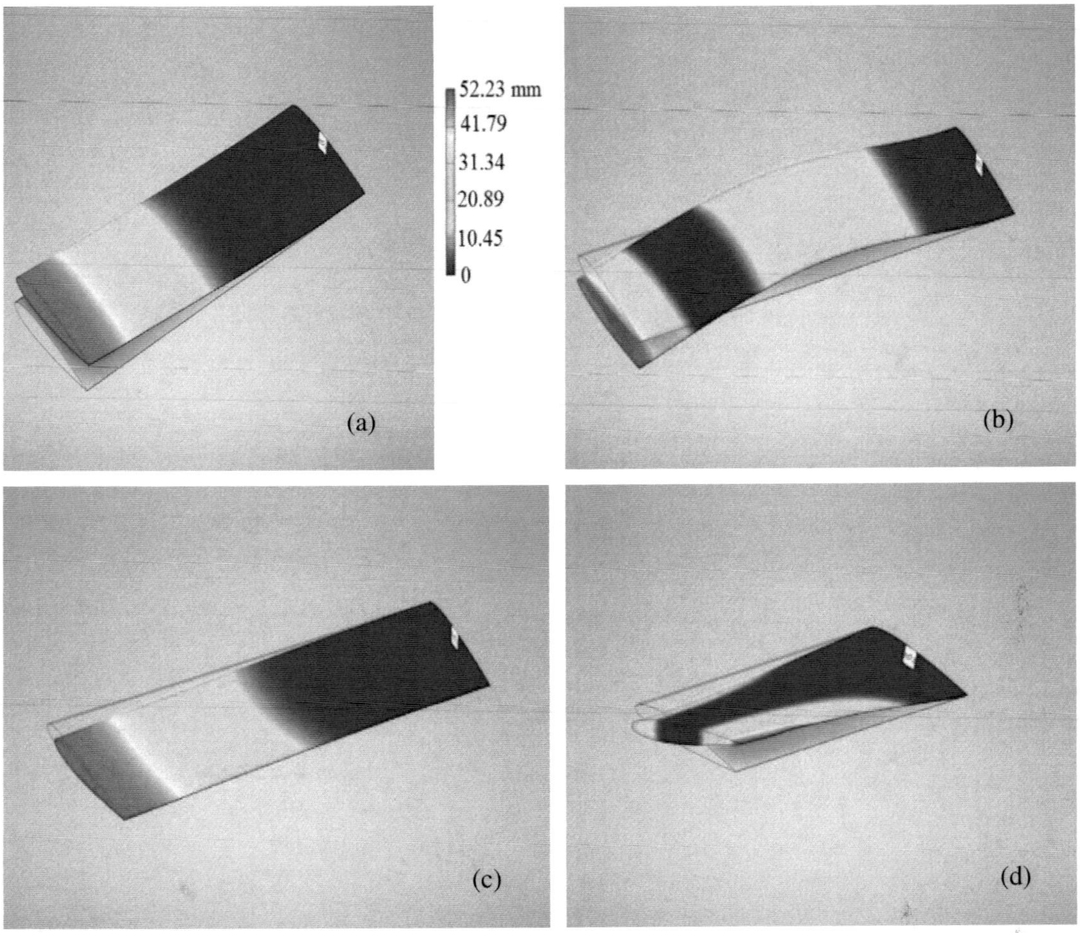

Figure 6. ABS: (a) – the first mode - 6.28 Hz, (b) – the second mode - 39.09 Hz, (c) – the third mode - 48.14 Hz, (d) – the fourth mode - 48.86 Hz.

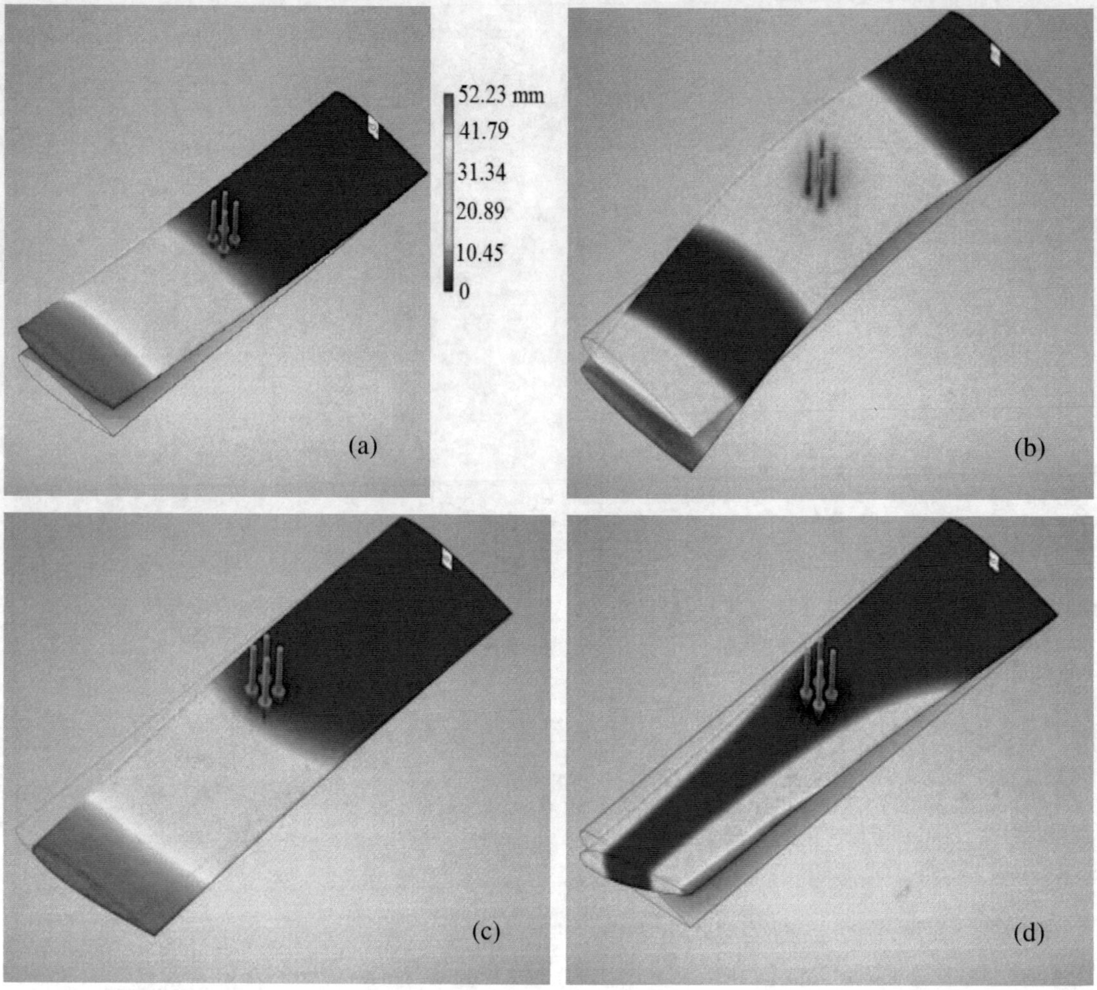

Figure 7. ABS with polylactide: (a) - the first mode - 13.95 Hz, (b) – the second mode - 86.81 Hz, (c) – the third mode - 106.56 Hz, (d) – the fourth mode - 107.31 Hz.

The acquired results make it possible to come to a conclusion: plastic of ABS type with polycarbonate notably surpasses ABS with the assumption that its density is 3 times lower.

3. Conclusions

Proceeding from the performed investigations we arrive at the following conclusions:

- It is apparent proceeding from the test results that ABS plastic with polycarbonate features higher strength, but lower plasticity, while ABS plastic features strength 25% lower, at that, its plasticity is 14% higher.
- The calculation results have demonstrated that the panel weight made of ABS features weight 60% higher as compared with the same made of ABS with polycarbonate. A difference of displacements also amounts to around 45% in favour of smaller displacements of a panel made of ABS with polycarbonate. The safety factor with ABS with polycarbonate is 47% higher than that with ABS without admixtures.
- The parts manufactured with the use of technology of 3D printing of plastics can be used in aviation industry, where it will help reduce the weight of aircraft. It is also worthwhile noting the application of this technology in manufacturing unmanned aerial vehicles.

IOP Conf. Series: Materials Science and Engineering 1129 (2021) 012017 doi:10.1088/1757-899X/1129/1/012017

Acknowledgments
The authors would like to thank colleagues for important advices and assistance in preparing this article.

Reference

[1] Naumov O V, Moskvitin G V, Pugachev M S and Polyakov A N 2017 *Investigation of stress-related and strength properties of ABS plastic when printing parts using 3D printer* (Institute of Machines Science of Russian Academy of Sciences) pp 57– 60

[2] Naumov O V, Moskvitin G V, Pugachev M S and Polyakov A N 2016 *Experimental estimate (investigation) of strength of parts manufactured at 3D printer* (Institute of Machines Science of Russian Academy of Sciences) pp 461 – 4

[3] Leontiev N V 2006 Application of ANSYS system for solving tasks of modal and static analysis *Educational learning material for refresher training program Information systems in mathematics and mechanics* (Nizhny Novgorod) p 101

Estimation of the additive and multiplicative error of the standard algorithm of acoustic emission sources linear location

N A Makhutov[1], I E Vasil'ev[1], D V Chernov[1], I V Mishchenko[2] and D S Moskovskaia[2]

[1] Mechanical Engineering Research Institute of the Russian Academy of Sciences (IMASH RAN), Moscow, Russian Federation
[2] Moscow Power Engineering Institute (MPEI), Moscow, Russian Federation

chernovdv@inbox.ru

Abstract. The paper deals with the results of the standard acoustic emission (AE) source linear location algorithm, which evaluates the location of developing damage in a steel sample under static load test to destruction. The linear location algorithm error is due to many factors: acquisition threshold of the AE signals and the dispersion properties of recorded signals, in particular, the dependence of the AE signal propagation velocity on its amplitude. According to the test results, the additive error in determining the time difference of arrival (TDA) of the AE signals to the transducers array is 52.8 μs and the average multiplicative error level is 0.67.

1. Introduction

Currently, a great number of dangerous industrial facilities with extended life are being in operation. The continuous monitoring systems based on physical methods of non-destructive testing are used to provide uninterrupted operation of such facilities. The most commonly used monitoring systems are the acoustic emission systems that provide the assessment of the level of the acoustic signal sources hazard that occurs in the process of generation and growth of the defects [1-4]. The standard linear location algorithm is used to identify the location of the developing damages. The use of the standard linear location algorithm consists in the calculation of the developing damages location based on the acoustic emission (AE) signal velocity (V) in the material of the test object and the time difference of arrival (Δt) of the AE signals to the antenna array transducers [5-8]. It should be noted that the standard algorithm does not consider the anisotropy of acoustic properties of the test objects and the dispersion properties of the registered AE signals. The acoustic properties anisotropy results in the dependence between V parameter and the direction of AE signal propagation. During the acoustic wave propagation, the nonlinear changes of signal frequency characteristics occur in the dispersion medium, which is accompanied by high-frequency modes attenuation in the AE source near zone. This physical phenomenon results in tailing of the leading edge of the recorded AE signals and the increase of the time parameter Δt. Therefore, the joint effect of acoustic properties anisotropy and dispersion produces significant impact on the values of Δt and V parameters, used for calculation of AE sources coordinates. Figure 1 presents the diagram of determination of AE source coordinate at linear location.

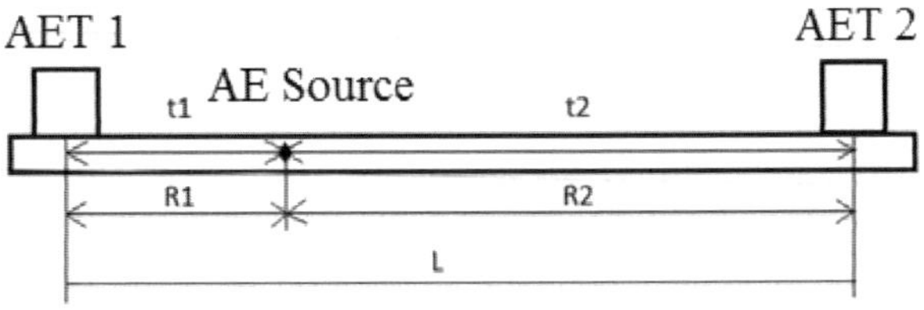

Figure 1. Diagram of determination of AE source coordinate using standard linear location algorithm.

The standard linear location algorithm is based on calculating the coordinates (R_1 and R_2) of the AE source from the time difference of arrival of the AE signals to the antenna array transducers ($\Delta t = t_1 - t_2$) and its propagation velocity V. The AE source coordinates are calculated as follows:

$$\begin{cases} t_1 = \dfrac{R_1}{V} \\ t_2 = \dfrac{R_2}{V} \\ t_1 - t_2 = \Delta t \end{cases} \quad (1)$$

$$R_1 = \frac{L - V \cdot \Delta t}{2} \quad (2)$$

2. Materials and equipment

To determine the accuracy of the AE sources coordinates using the standard linear location algorithm, laboratory tests on the static tension of a steel sample to destruction have been performed. The figure 2 presents the general view of testing the sample at electromechanical testing system Instron-5982.

Figure 2. The sample (1), installed in the grips (2 and 3) of electromechanical testing system Instron-5982 with the transducers VS 150-RIC (4) fixed with the clamps (5).

During the sample testing, the load was automatically increased at the rate of the testing system active traverse displacement of 1 mm·min^{-1}. The overall dimensions of the tested samples were 500x40x4 mm. Two holes 5 mm in diameter were drilled as stress concentrators at a distance of 110 mm from the end surfaces of the sample. The resonant acoustic emission transducers (AET) of *VS 150-RIC* were installed at the distance of 40 mm from the concentrators so that the linear location size was L = 360 mm. Before loading the steel sample, the measurement devices were adjusted. The acoustic signal recording was performed using four-channel AE measurement system *Vallen AMSY-6*. The AE signal discrimination threshold was selected equal to u_{th} = 44 dB, digital filter band pass corresponded to frequency range Δf= 100-850 kHz. To determine the AE signal propagation velocity, the standard calculation method described in the regulatory document PB 03-593-03 was used. The group velocity value determined using Su-Nielsen source (0.3 mm clutch pencil lead fracture) constituted V_g= 4500 m·s^{-1} at the level of AE signals maximal amplitude u_m=94-96 dB.

3. Results and discussion

Synchronously with Ae monitoring, the video camera *EOS 60D* recorded the process of material deformation in the concentrators area. In the process of loading the samples, some asymmetry in the degree of damages accumulation was observed in the concentrators area and, correspondingly, the asymmetry of the loading processes in the areas of location of upper and lower hole which correspondingly impacted on the AE events accumulation diagrams recorded by the acoustic emission systems. The earlier development of the main cracks accompanied by the loss of the bearing capacity and the sample destruction also took place in one of the concentrators. Figure 3 presents the pattern of material deformation and the sample destruction.

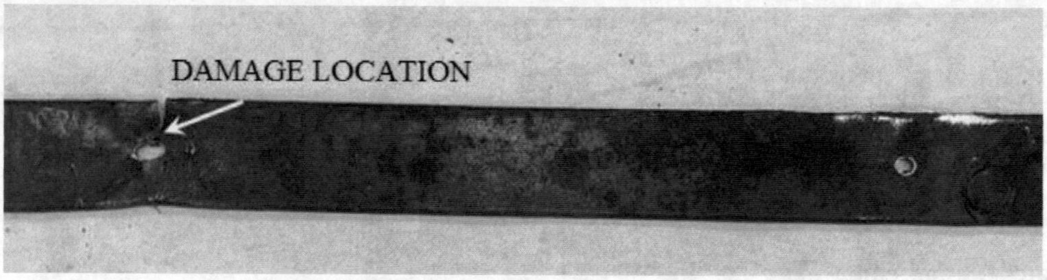

Figure 3. The pattern of deformation and destruction of the steel sample of grade St3.

Figure 3 shows the areas of concentrators where the most intensive local material structure destructions took place. It is known, that the maximal density of AE events is to be registered at the area of the sample material destruction. However, as we can see from the experiments conducted, in the event of presence of stress concentrators in the near zone of the receiving transducers (at the distance of 40 mm from the holes), the maximal AE events density recorded by *Vallen AMSY-6* system, shifted correspondingly by 70 mm from the center of the holes. The diagrams of figure 4 present the change dynamics of locational signals maximal amplitude (*a*) and the coordinate location of AE events (*b*), recorded in the course of AE diagnostics of the steel sample during tension.

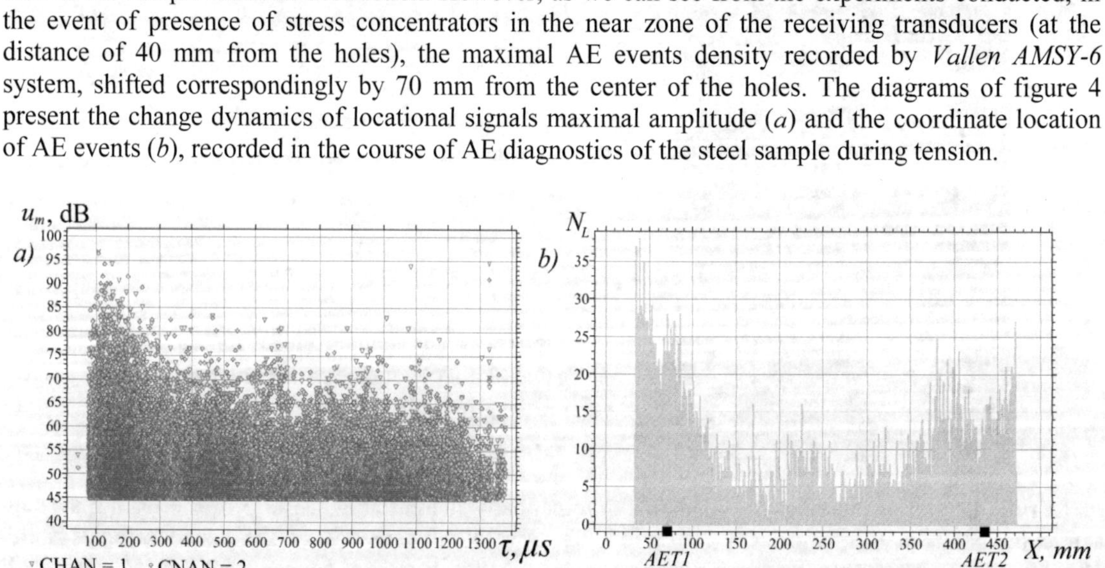

Figure 4. The change dynamics of locational signals maximal amplitude (a) and the coordinate location of AE events (b), recorded in the process steel sample testing.

As follows from the figure 4 (a), the arrays of locational signals with the maximal amplitude of 44 to 97 dB were registered in the process of AE testing. At the period from 85th to 400th second of AE monitoring, the maximal level of the registered signal amplitude reached 97 dB. The maximum of AE events distribution was registered at the area of #1 AET installation, and the sample destruction occurred in the hole area at the distance of 40 mm from it. According to the linear location results, presented in figure 4 (b), the maximum density of AE events, reaching 39 units·mm^{-1}, was observed at the coordinate x_{max} = 40 mm, which corresponds to the absolute error $\Delta = |x_{max} - x_t| = |40 - 110| = 70$ mm, where x_t is the true coordinate of the destruction place. A specific feature of the standard location algorithm application is the use of the group velocity value $V_g = 4500$ m·s^{-1} determined for Su-Nielsen source with the maximal amplitude level u_m=94-96 dB, while the amplitude of the main array of the recorded location signals was in the range of u_m=45–65 dB. Therefore, the use of the standard method for locating AE sources oriented to the signal amplitude level from a Su-Nielsen source results in significant errors in determination the coordinates of AE sources located in AET near zone, as follows from figure 4 (b).

To assess the impact of the acoustic signals dispersion properties on the value of the group velocity of AE signal propagation V_g, the additional experimental studies have been performed. In this case, the AE signals were generated using the electronic simulator "*Interunis*" with the connected broadband transducer UT-1000, situated outside the location area at the distance of 40 mm from # 1 AET. During the testing of the group velocity V_g, the electronic simulator voltage was selected so that to change the maximal amplitude of the recorded signals (u_m) in the range from 55 to 100 dB in 5 dB steps. Based on the results of the studies performed, figure 5 shows the dependence of the change in the group velocity of AE signals propagation from the level of the maximal source amplitude (u_m).

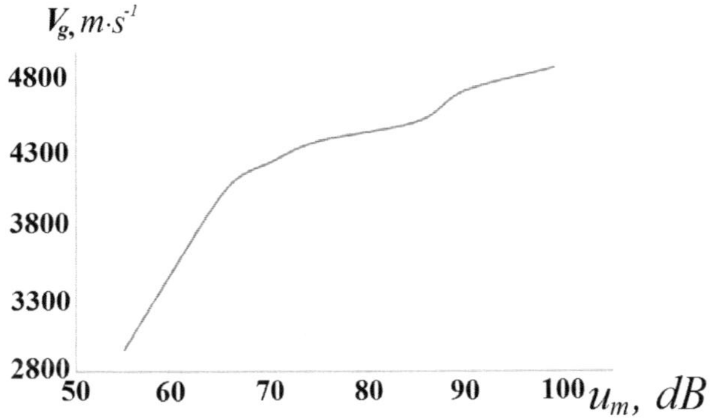

Figure 5. The dependence of the velocity of AE signals propagation on the acoustic signals source amplitude.

As we can see from figure 5, the value of the AE signals propagation velocity largely depends on the maximum amplitude of the recorded signals. The use of the Su-Nielsen source makes it possible to generate 90-95 dB AE signals. The average propagation velocity of such signals corresponds to V_g=4500 m·s^{-1}. When using an electronic simulator, it is possible to change the values of the AE signals amplitude within a wide range. For example, for 55 dB signals, the TDA value was equal to $\Delta t = 122$ μs, which corresponds to $V_g = L/\Delta t = 360 \cdot 10^{-3}/(122 \cdot 10^{-6}) = 2950$ m·s^{-1}. In this case, for 65 dB signals, the velocity value reached as much as 3800 m·s^{-1}. Therefore, even the minor changes in the AE signals amplitude result in significant scatter of V_g values, and subsequently, to linear location building error. The greatest effect of the nonlinearity of $V_g(u_m)$ diagram and the dispersion properties of acoustic signals is observed when the AE source is located in the near zone of one of the antenna array transducers at a distance of $R_i \leq 100$ mm. Depending on the recorded signals amplitude, the absolute error level can exceed the true AE source position by more than 1.5 times.

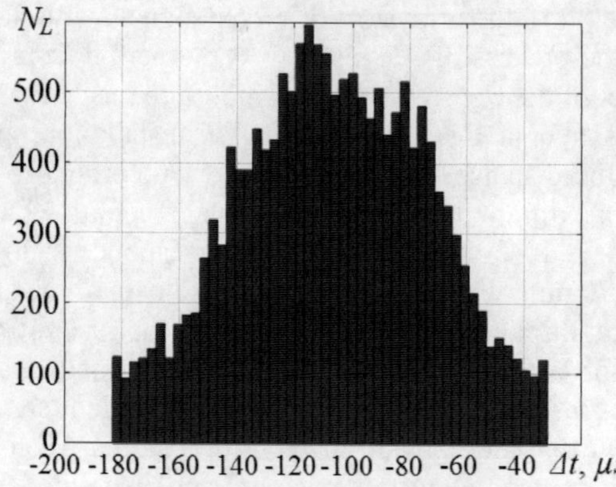

Figure 6. Distribution of the amplitude of AE signals recorded near the destruction area.

As is seen from figure 6, in the process of the steel sample loading 5553 AE signals of the amplitude below 60 dB was registered in the destruction area, which corresponds to 75% of the total number of the located signals. According to the results presented in figure 5, the maximal velocity of propagation of AE signals of the amplitude below 60 dB does not exceed $V_g = 3300$ m·s^{-1}. The average velocity value for high-amplitude signals corresponds to $V_g = 4500$ m·s^{-1}. Such V_g values changes are due to the effect of the dispersion properties and the threshold AE signals discrimination algorithm in the time parameter value Δt. As follows from the publications [4, 5], the threshold registration of AE signals characterized with the oscillating front edge form, results in temporary delay and the TDA increase by the permanent value Δt_a, correlating with the additive error component. The temporary delay is due to the fact that at the initial period of the signal distribution, the main energy of the wave packet is born by the high-frequency components possessing the highest value of the propagation velocity. As the AE source moves off from the transducers, the high-frequency modes quickly attenuate which reduces the TDA measurement error. The change of the frequency characteristics in the process of acoustic signals attenuation significantly impacts the AE signals propagation velocity. The reduction of the AE signal maximal amplitude results in the reduction of the energy of high-frequency components of the spectrum, and, therefore, the increase of Δt measurement error. Therefore, the effect of AE signals dispersion properties results in multiplicative error of TDA measurement. The above-listed factors significantly affect the accuracy of AE events location in the transducer near zone. For numerical estimation of additive and multiplicative error components, figure 7 shows the distribution of Δt values of AE events, registered in the #1 AET near zone at the distance of ±40 mm from the center of the holes where the sample destruction took place.

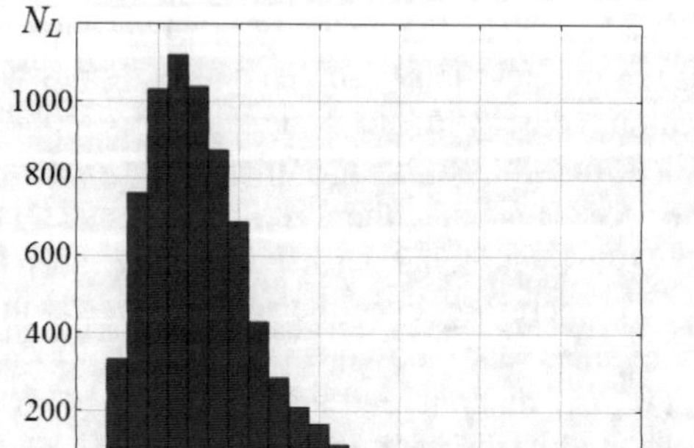

Figure 7. The distribution of TDA values of AE events, registered in the #1 AET near zone at the distance of ±40 mm from the center of destruction (hole).

Figure 7 shows the distribution of the values of the temporal parameter Δt of AE events recorded by the antenna array transducers in the steel sample destruction area. Due to the fact that the destruction area is located at a distance of 40 mm from #1 AET and 320 mm from #2 AET, $t_1 < t_2$ and $\Delta t = t_1 - t_2 < 0$. The maximum of TDA distribution corresponds to the value of Δt_m = -115 μs. To determine the additive error component, the TDA parameter was calculated according to the standard algorithm $\Delta t_s = \frac{2\cdot(x_t - x_1) - L}{V_g} = \frac{2\cdot(110-70)-360}{4.5} = -62.2$μs. Therefore, the additive component of TDA measurement error was $\Delta_A = |\Delta t_m - \Delta t_s| = |-115 + 62.2| = 52.8$ μs. For numerical estimation the multiplicative error, the average scatter of Δt values was calculated relative to the TDA maximum distribution coordinate, depending on the number of recorded AE events at the distance of ± 40 mm from the center of destruction (hole): $\gamma(N_L) = \Delta_\gamma \cdot (N_L/N_\Sigma)$, where N_Σ is the total number of registered AE events. For Δt = -115 μs, the deviation from Δt_m corresponds to Δ_γ = 0 μs, which means that the multiplicative error component is equal to zero. For example, for Δt = -135 μs, the TDA values deviation corresponds to $\Delta\gamma = \Delta_\gamma = |-135+115| = 20$ μs, and the multiplicative error is $\gamma(N_l) = 20\cdot(439/16407) = 0.535$. Based on the statistical processing of the data presented in figure 7, it was determined the average value of the multiplicative error, equal to $\gamma(N_L) = 0.67$, indicating the average deviation of the TDA values by 67% relative to the median of Δt distribution.

4. Conclusions

The effect of the additive and multiplicative error components on the accuracy of the linear location of AE sources located in the near zone at a distance of 40 mm from AET has been studied. To assess the coordinate location accuracy, calculated by the standard algorithm, a tensile test of a steel sample with stress concentrators in the form of holes was performed. According to the studies, when the standard algorithm is used, the absolute error in the coordinate location of AE events, that occurs when the source is located in the AET near zone, can differ from the true position by 1.5 times. Based on the studies performed, the dependence between the group velocity of AE signal propagation and its amplitude was recorded. A numerical estimation of the effect of the additive and multiplicative error components on determination of TDA parameter on the coordinate location accuracy was performed. To improve the accuracy of determining the coordinate location of AE events, it is necessary to take into account the dependence of the wave packet propagation velocity on the signal amplitude, as well as the effect of the additive and multiplicative error components in determining the TDA.

Acknowledgments

The author acknowledges the support of the Russian Science Foundation (project No. 20-19-00769).

References

[1] PB 03-593-03 2003 *Regulations on organization and execution for acoustic emission inspection of vessels, boilers, apparatus and process pipping* (Moscow: NTTs Industrial Safety) p 53
[2] GOST R ISO 22096-2015. *Condition monitoring and diagnostics of machines. Acoustic emission method* (Moscow: Standartinform) p 8
[3] Ivanov V I and Barat V A 2017 *Acoustic emission diagnostics* (Moscow: Spectrum) p 368
[4] Matvienko Yu G, Vasilev I E, Bubnov M A and Chernov D V 2020 The effect of the size and shape of process cutouts on the accuracy of the acoustic emission sources location *Defektoskopiya* **2** pp 3-11
[5] Makhutov N A, Vasilev I E, Chernov D V, Ivanov V I and Elizarov S V 2019 The effect of frequency filter band pass on acoustic emission pulse parameters *Defektoskopiya* **3** pp 3-14
[6] Kosel T, Grabec I and Kosel F 2005 Intelligent location of simultaneously active acoustic emission sources *Aerospace science and technology* **9** pp 45-53
[7] Matvienko Yu G, Ivanov V I, Vasilev I E, Chernov D V and Mishchenko I V 2020 Determination of velocity of wave packet distribution in composite materials *Pribory i tekhnika eksperimenta*. **1** pp 115-20

[8] Makhutov N A, Vasilev I E, Chernov D V and Mishchenko I V 2019 The improvement of accuracy of acoustic emission sources linear location *XXXI Int. Innovative Conf. of Young Scientists and Students on Machine Engineering Problems MIKMUS 4-6 December, Moscow* pp 669-72

Optimization of open-hole variable stiffness composite plates under tensile loading using curved continuous fibers

A V Malakhov

Mechanical Engineering Research Institute of the Russian Academy of Sciences, 4 Maly Kharitonyevsky Pereulok, Moscow, 101990, Russia

anmaviko@gmail.com

Abstract. This paper describes optimization of variable stiffness composite plates (VSCPs) with open-holes under tensile loading. Curved continuous fibers are used to optimize distributions of fiber orientation and the fiber volume fraction during the optimization for the VSCPs. The modeling of VSCPs is performed by means of fiber trajectories, which are aligned in the direction of maximum principal stress. The progressive failure of the VSCPs is modeled by a material property degradation method, taking into account the variable distributions of material properties to predict the ultimate load. It is demonstrated that the ultimate load for the composite plates is increased using the change in a reinforcement structure and a transition from unidirectional to curvilinear reinforcement of fibers. Thus, application of VSCPs with curved fibers compared with unidirectional composites results in more effective use of composite materials in various fields of industry.

1. Introduction

Fiber reinforced polymers (FRPs) are used in various fields of industry especially where it is necessary to minimize weight due to the high specific stiffness and the high specific strength of composite materials. The most common composite structures are laminates consisting of unidirectional (UD) layers. The laminates are optimized by variations of fiber orientations of the UD layers to maximize the efficiency of FRPs. To optimize the laminates, fiber orientations are aligned in the direction of loads. However, the optimization is effective only if there are not stress concentrations.

To improve performance of composite structures with stress concentrations, shape optimization [1] and local reinforcements [2] are used. However, since stiffness in UD FRPs within a layer is constant, the variation of design variables during the optimizations is limited. To change stiffness locally within a layer of laminates and enlarge design variables, fiber orientation and the fiber volume fraction can be varied along the layer coordinates. Similar composite structures are observed in nature, namely in wood where curvilinear fibers are curved in the vicinity of a knot. There are different design methods to model composite structures with curved fibers. For example, fiber orientation can be aligned in the direction of fluid streamlines [3], load paths [4, 5] and maximum principal stress [6-11]. Besides, fiber orientations can be varied depending on the objective function during optimization [12-16].

There are various advanced manufacturing technologies such as automated fiber placement, tailored fiber placement [2, 10] and 3D printing [11, 17-27], which are used to make composite structures with curved fibers. It is shown in tests [4, 5, 9-11, 24-27] that the change in a composite structure from UD to curvilinear reinforcement of fibers leads to increases in the efficiency of FRPs.

Thus, it is possible to design and make FPRs with greater efficiency using curved fibers compared to standard UD composite structures.

2. Definition of the problem

Variable stiffness composite plates (VSCPs) with holes were modeled to estimate the effect of curved fibers on the ultimate load of the VSCPs. Dimensions of the plates are shown in figure 1a. Two types of reinforcements were considered for the plates, namely, curvilinear reinforcement and unidirectional reinforcement. Curvilinear reinforcement was modeled in the VSCPs where curved fiber trajectories are aligned in the direction of the maximum principal stress. Fiber orientation and the fiber volume fraction are changed along coordinates of the VSCPs depending on the distribution of fiber trajectories. The maximum fiber volume fraction of the VSCPs is 57.7%. Fiber orientation in the direction of the *x*-axis (figure 1) and the fiber volume fraction (V_f) of 57.7% were used to model the UD plates. The fourth part of the plates, namely the left and top part of the plates ($0 \leq x \leq L/2$ and $0 \leq y \leq W/2$), were only modeled since the plates are symmetrical in the geometry, material and boundary conditions. The boundary conditions for the plates correspond to the tensile test where loading is directed along the *x*-axis. All types of the considered plates with different dimensions and reinforcements are given in table 1.

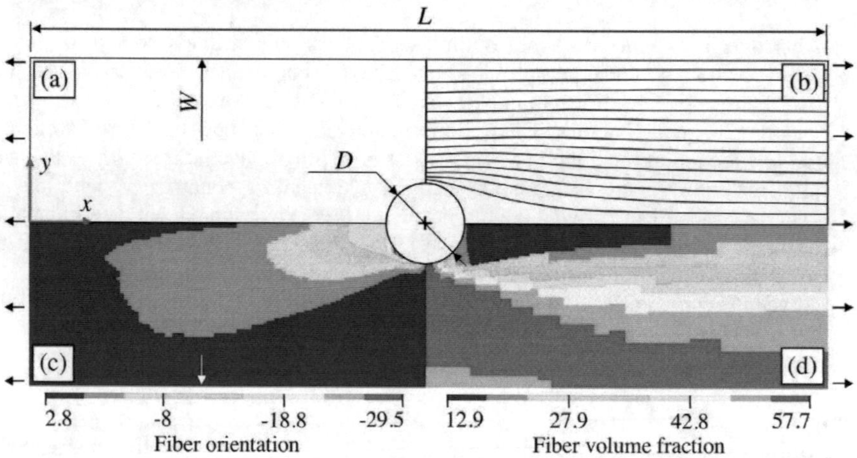

Figure 1. Dimensions (a) and distributions of fiber trajectories (b), fiber orientations [°] (c) and the fiber volume fraction [%] (d) for the plate. The distributions correspond to C-3.

Table 1. Labels, dimensions and reinforcement structures of composite plates.

Label	*D* (mm)	*L* (mm)	*W* (mm)	Reinforcement of composite structures
C-1	6	60	9	Curvilinear reinforcement of fibers
C-2	6	60	15	Curvilinear reinforcement of fibers
C-3	6	60	24	Curvilinear reinforcement of fibers
C-4	6	60	36	Curvilinear reinforcement of fibers
UD-1	6	60	9	Unidirectional reinforcement of fibers
UD-2	6	60	15	Unidirectional reinforcement of fibers
UD-3	6	60	24	Unidirectional reinforcement of fibers
UD-4	6	60	36	Unidirectional reinforcement of fibers

To model the composite structures, carbon FRP of IM7/8552 is used. Properties of carbon fiber of IM7 and matrix of 8552, as well as ultimate strengths of the IM7/8552 with V_f = 57.7%, are given in tables 2 and 3, where E_1 and E_2 are Young's moduli in the longitudinal and transverse fiber direction, G_{12} is shear modulus in the 1–2 plane, μ_{12} is the Poisson's ratio, and ε_T is ultimate tensile strain. X and Y are longitudinal strength and transverse strength, respectively. In this case, the subscripts T and C for X and Y denote tension and compression. S_{12} and S_{23} are the shear strengths in the 1–2 and 2–3 planes.

Table 2. Material properties of IM7 fiber and 8552 matrix [8].

Material properties	E_1 (GPa)	E_2 (GPa)	G_{12} (GPa)	μ_{12}	X_T (MPa)	ε_T (%)
Fiber	276	19.5	70	0.28	5654	1.9
Matrix	4.76	4.76	1.74	0.37	121	1.7

Table 3. Ultimate strengths for the IM7/8552 with V_f = 57.7% [8].

Ultimate strengths	X_T (MPa)	X_C (MPa)	Y_T (MPa)	Y_C (MPa)	S_{12} (MPa)	S_{23} (MPa)
Values	2524	1690	63.4	285.7	101.1	107.6

3. Methods

3.1. Modeling method of VSCPs

Initially, an isotropic material (E = 200 GPa, μ = 0.3) is used to model the VSCPs and compute stress. On the basis of the stress fields, fiber trajectories are modeled, which are aligned in the direction of the maximum principal stress [28]. Then variable material properties along coordinates of the plates are simulated by the finite element method (FEM). Each element of the VSCPs is assigned to its own material properties depending on the distribution of fiber trajectories. Fiber orientation of each element is aligned in the direction of the maximum principal stress. The change in distance between fiber trajectories is modeled by the variable V_f, where the maximum V_f is located in the zones with the minimum distance between fiber trajectories. The V_f decreases with increasing distance between fiber trajectories. Material properties depend on fiber orientation and the fiber volume fraction. The modeling of variable distributions of fiber orientations and the fiber volume fraction, as well as the change in material properties, are described in more detail in [8].

After material properties are assigned to the VSCPs, the distribution of fiber trajectories is changed due to redistribution of stresses. Thus, the modeling process is iterative, and it is completed when the effective stress concentration factor (K_t') will not exceed 1% between iterations. The calculation of the K_t' is performed by equation (1) where σ_1 and V_f are stress in the fiber direction and the fiber volume fraction, and σ_0 and V_{f0} are the averaged values of σ_1 and V_f along the left side of the plates. Distributions of fiber trajectories, fiber orientations and the fiber volume fraction are shown in figure 1b, 1c and 1d for the VSCP with label of C-3 after finishing the simulation.

$$K_t' = (\sigma_1 \cdot V_{f0})/(\sigma_0 \cdot V_f) \quad (1)$$

3.2. Progressive failure modeling of VSCPs

To model the progressive failure of the composite structures, a material property degradation method (MPDM) is used. According to the MPDM, if failure occurs, material properties of FRPs are decreased with increasing load in the zone with the failure. To fix the beginning of failure, the Hashin criterion is used. The decrease of material properties is performed by the degradation factors [29, 30]. If the zone of fiber failure crosses from the hole to the free edge of the plate, total failure of the composite structures occurs.

Since material properties are changed in the VSCPs, ultimate strengths are assigned to each element in accordance with the fiber volume fraction. To calculate the ultimate tensile strength in the

IOP Conf. Series: Materials Science and Engineering 1129 (2021) 012019 doi:10.1088/1757-899X/1129/1/012019

fiber direction, component properties of the IM7/8552 (table 2) and the modified rule of mixtures [31] are used. Other ultimate strengths are constant and they are not changed depending on the fiber volume fraction. The progressive failure modeling is described in more detail in [8].

4. Results and discussion

The stress concentration factor and the effective stress concentration factor are calculated for all types of plates and shown in figure 2. It is worth noting that since $V_f = V_{f0}$ for the UD plates, $K_t = K_t'$ for these plates. The effective stress concentration factor significantly decreases with the change in composite structure and a transition from UD reinforcement to curvilinear reinforcement.

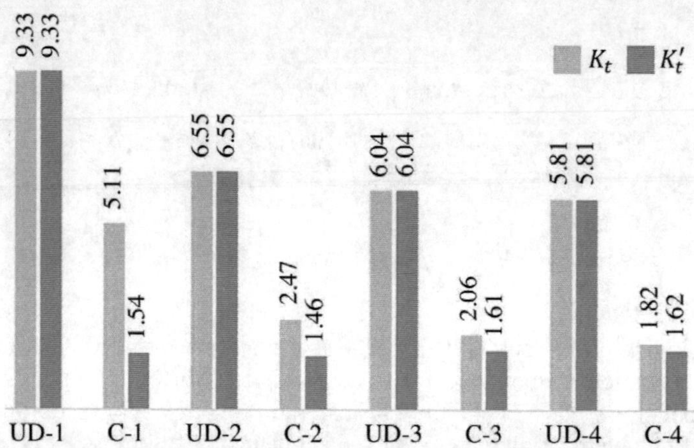

Figure 2. The stress concentration factor (K_t) and the effective stress concentration factor (K_t') for all types of plates.

Transverse stresses and shear stress are shown in figure 3. The stresses are given for loading when tensile stress is 1 MPa. As seen from figure 3, transverse stresses are close to each other for the same geometry of plates, but shear stress for the VSCPs is always lower than it is for the UD plates. Since fiber trajectories are aligned in the direction of the maximum principal stress, in fact, shear stress is minimized. The difference in shear stress is especially noticeable between UD-4 and C-4 where shear stress decreases by 8.8 times using the transition from the UD-4 plate to the C-4 plate.

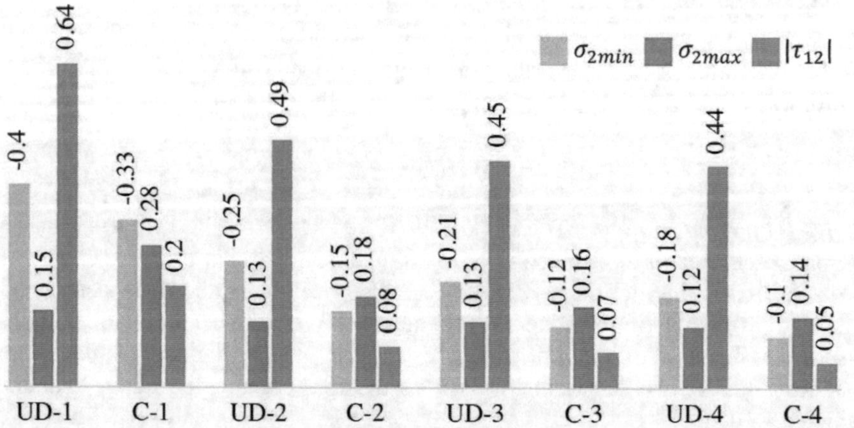

Figure 3. Minimum transverse stress (σ_{2min}) (MPa), maximum transverse stress (σ_{2max}) (MPa) and shear stress (τ_{12}) (MPa).

Tensile stress values (MPa) are shown in figure 4, at which the failures of the matrix (MF) and fiber (FF) begin as well as the proportional limit (PL) being reached for the composite plates with different reinforcements. The results show that load with the initial fiber failure increases by about 4 times if the reinforcement structure is changed from UD fibers to curvilinear fibers for the plates. Since the failures are achieved at higher load for the VSCPs than for the UD plates, the VSCPs can be exploited with greater efficiency and safety. To understand how much the efficiency can be increased, the proportional limit is used. It has been established in tests [32, 33] that the number of the acoustic emission counts grows only after exceeding the proportional limit on the stress–displacement curves. When the load is higher than the proportionality limit, irreversible damage of the composite structures already begins. Thus, it is necessary to exploit composite structures up to the proportional limit to ensure their safe use. The proportional limits and the stress–displacement curves (for C-3 and UD-3) are shown in figures 4 and 5, respectively.

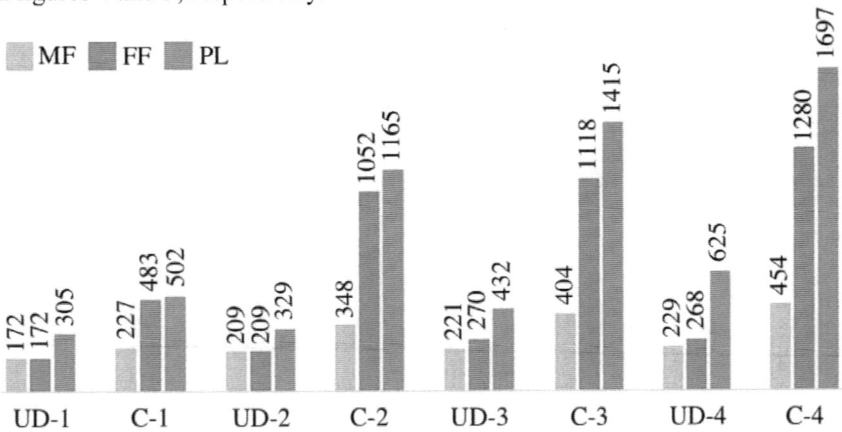

Figure 4. Failure of the matrix (MF) and fiber (FF) as well as the proportional limit (PL) for different reinforcement structures.

As seen from figure 4, the proportional limit for the VSCPs is higher by about 3 times compared to the UD plates. The maximum values of the stress–displacement curves (ultimate loads) are given in table 4. Thus, the results show that the change in reinforcement structure significantly increases the efficiency of composites.

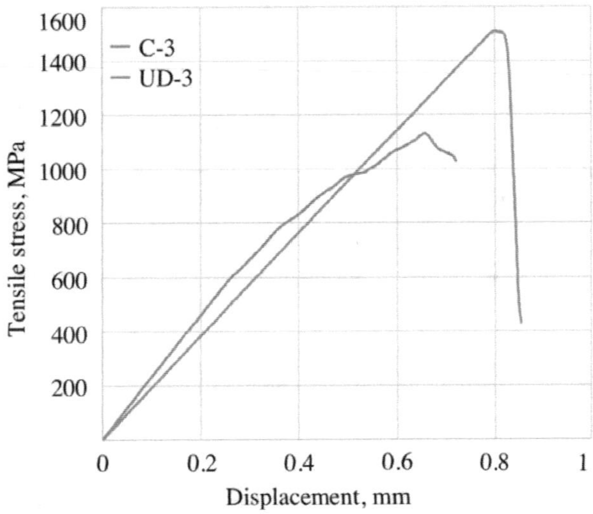

Figure 5. The stress–displacement curves for C-3 and UD-3.

Table 4. Ultimate loads for all types of plates.

Label	UD-1	C-1	UD-2	C-2	UD-3	C-3	UD-4	C-4
Ultimate loads (MPa)	569	559	827	1238	1130	1509	1535	1748

5. Conclusions

Different composite structures with both UD fibers and curved fibers were modeled. A material property degradation method was used to model the progressive failure and calculate the proportional limits and the ultimate loads for all types of considered plates. As the results showed, the proportional limit of composites significantly increases with a transition from UD composite structures to curvilinear composite structures. Composites with curved fibers can be made by advanced manufacturing technologies where fiber orientation is controlled. Thus, composites can be used with even greater efficiency, which will lead to safer and more reliable application of composites in various industries.

Acknowledgments

This work was carried out with financial support from the Program of Basic Scientific Research of State Academies of Sciences for 2013-2020.

References

[1] Pedersen P 2000 *Structural and Multidisciplinary Optimization* **19** 169–82
[2] Katagiri K et al 2019 *Composite Structures* **207** 658–64
[3] Khan S, Fayazbakhsh K, Fawaz Z and Nik M A 2018 *Additive Manufacturing* **24** 378–85
[4] Li R, Kelly D, Crosky A, Schoen H and Smollich L 2006 *J. of Composite Materials* **40** 1645–58
[5] Crosky A, Kelly D, Li R, Legrand X, Huong N and Ujjin R 2006 *Composite Structures* **76** 260–71
[6] Malakhov A V and Polilov A N 2016 *Composites Part A: Applied Science and Manufacturing* **87** 23–8
[7] Hyer M W and Charette R F 1991 *AIAA Journal* **29** 1011–5
[8] Malakhov A V, Polilov A N, Zhang J, Hou Z and Tian X 2020 *Applied Composite Materials* **27** 185–208
[9] Tosh M W and Kelly D W 2000 *Composites Part A: Applied Science and Manufacturing* **31** 1047–60
[10] Almeida Jr J H S, Bittrich L and Spickenheuer A 2020 *Composites Science and Technology* **185** 107889
[11] Sugiyama K, Matsuzaki R, Malakhov A V, Polilov A N, Ueda M, Todoroki A and Hirano Y 2020 *Composites Science and Technology* **186** 107905
[12] Cho H K and Rowlands R E 2009 *J. of Composite Materials* **43** 1177–98
[13] Huang J and Haftka R T 2005 *Structural and Multidisciplinary Optimization* **30** 335–41
[14] Honda S, Igarashi T and Narita Y 2013 *Composites Part B: Engineering* **45** 1071–8
[15] Ferreira R T L and Ashcroft I A 2020 *Structural and Multidisciplinary Optimization* **61** 2155–76
[16] Yu H, Hong H, Cao S and Ahmad R 2020 *Applied Sciences* **10**(3) 943
[17] Hou Z, Tian X, Zhang J, Zhe L, Zheng Z, Li D, Malakhov A V and Polilov A N 2020 *Composite Structures* **237** 111945
[18] He Q, Wang H, Fu K and Ye L 2020 *Composites Science and Technology* **191** 108077
[19] Dutra T A, Ferreira R T L, Resende H B and Guimaraes A 2019 *J. of the Brazilian Society of Mechanical Sciences and Engineering* **41** 133
[20] Hou Z, Tian X, Zheng Z, Zhang J, Zhe L, Li D, Malakhov A V and Polilov A N 2020 *Composites Part B: Engineering* **189** 107893
[21] Shi B et al 2020 *Matter* **2**(6) 1594–604

[22] Azarov A V, Antonov F K, Golubev M V, Khaziev A R and Ushanov S A 2019 *Composites Part B: Engineering* **169** 157–63
[23] Ueda M, Kishimoto S, Yamawaki M, Matsuzaki R, Todoroki A, Hirano Y and Duigou A L 2020 *Composites Part A: Applied Science and Manufacturing* **137** 105985
[24] Dickson A N, Ross K-A and Dowling D P 2018 *Composite Structures* **206** 637–43
[25] Li N, Link G, Wang T, Ramopoulos V, Neumaier D, Hofele J, Walter M and Jelonnek J 2020 *Composites Part B: Engineering* **182** 107612
[26] Hou Z, Tian X, Zhang J, Zheng Z, Zhe L, Li D, Malakhov A V and Polilov A N 2021 *Composites Science and Technology* **201** 108502
[27] Zhang H, Dickson A N, Sheng Y, McGrail T, Dowling D P, Wang C, Neville A and Yang D 2020 *Composites Part B: Engineering* **186** 107835
[28] Malakhov A V and Polilov A N 2013 *J of Machinery Manufacture and Reliability* **42** 306–11
[29] Tan S C 1991 *J. of Composite Materials* **25** 556–77
[30] Tan S C and Perez J 1993 *J. of Reinforced Plastics and Composites* **12** 1043–57
[31] Lee C and Hwang W 1998 *J. of Materials Science Letters* **17** 1601–3
[32] Xiao Y and Ishikawa T 2005 *Composites Science and Technology* **65** 1022–31
[33] Zhang H, Li C, Dai W, Liu Y, Tian S, Huang W, Jia D, He D and Zhang Y 2020 *Composite Structures* **234** 111667

IOP Conf. Series: Materials Science and Engineering 1129 (2021) 012020 doi:10.1088/1757-899X/1129/1/012020

Design issues of energy-absorbing truck mounted attenuators

P S Mikheev, I A Karpov and A A Mukhametova

Moscow Automobile and Road Construction State Technical University (MADI), 64, Leningradsky Prospect, Moscow, 125319, Russia

psmikheev95@gmail.com

Abstract. The work is devoted to the issues of finite element modeling of energy-absorbing truck mounted attenuators (TMA) used on the shadow vehicles in the zones of stationary and mobile road and construction works to protect workers from collisions with vehicles, as well as to reduce the severity of an accident for the driver who collided. Discusses various designs of TMAs, the main problems in design and defines the criteria for the development of structures based on mathematical modeling by the finite element method.

1. Introduction

The One of the most dangerous types of road accidents is a collision with a shadow vehicle. This type of road accidents is the fourth most common (6.88%) and is as traumatic as head-on collisions of vehicles. Truck mounted attenuators (TMAs) fixed on shadow vehicles are used to reduce the severity of accidents [1, 2] (figure 1). The use of such security systems can reduce the number of victims from 44% to 35%, and the death rate from 19% to 4% [3]. Currently, persistent efforts are underway to create local TMA designs, but engineers are faced with a large number of difficulties due to the lack of knowledge of these devices mechanics. Digitalization is one of the main trends in the Russian economy, therefore, the work was based on digital models, which make it possible to predict the behavior of a structure with high accuracy throughout the entire life cycle. This article, based on a digital model, examines the TMAs mechanics of one of the types, highlights the typical problems during operation and formulates requirements for design and testing safety. The work was carried out as part of the GOST R development "Public roads. Truck mounted attenuators. General requirements. Test methods" with the Federal Road Agency (FRA) support.

(a) (b)

Figure 1. Collision with a shadow vehicle, covering road works: (a) without an impact attenuation system, (b) with an impact attenuation system.

2. The operation principle of the impact attenuation systems

The main task of impact attenuation systems like TMA is to absorb the kinetic energy of an oncoming vehicle, then stop or redirect it along a safe path.

Figure 2 shows some TMA designs by foreign manufacturers [4, 5]. The oncoming vehicle impact attenuation in these structures is carried out by the work of deformation, friction forces or destruction of specially designed TMAs structural elements [6, 7, 8] (figure 3).

(a) 

Figure 2. Truck mounted TMAs designs: (a) TTMA-100, (b) SST trailer.

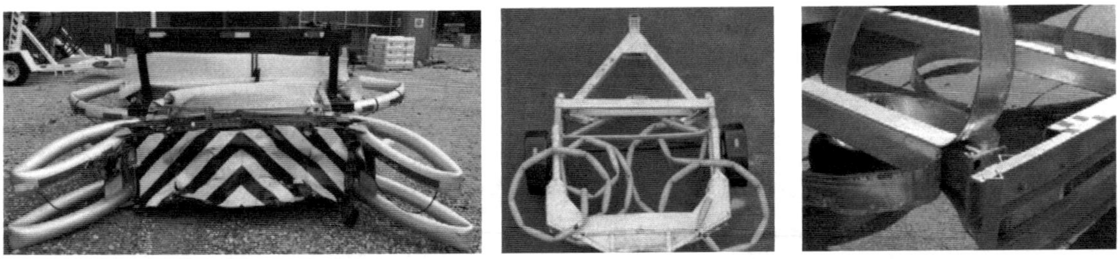

(a) (b) (c)

Figure 3. An example of energy attenuation at the impact to a TMA due to: (a) deformation, (b) friction, (c) destruction.

The behavior of the vehicle and the risk of injury upon impact can be described by the following parameters: roll and pitch angles (rotation around the longitudinal and transverse axes of the vehicle), speed and the driver and passengers head overload, the vehicle center of gravity (CG) overload. These characteristics can be obtained by a number of tests [9, 10]: direct impact over the TMA axis, direct impact with displacement, and impact at an 10° angle to the TMA axis with displacement.

However, in the real conditions it is impossible to predict the initial impact parameters, therefore, the general operation principle of the impact attenuation system should be maintained regardless of the collision conditions.

The vehicle and a structure collision process is fast (not more than 0.5 s), with complex boundary conditions in the form of contacts and restrictions on displacements, accompanied by large displacements and deformations in the system; therefore, to solve the problem, the finite element method was used in an explicit formulation. All calculations have been performed using the LS-DYNA software package. Next, we will consider the impact attenuation system structure, and, based on it, will formulate the basic design principles of these systems.

3. Study of the impact attenuation system structure by the finite element method (FEM)

Figure 4 shows a TMA based on the deformation principle of energy absorption due to the formation of plastic hinges. The main energy-absorbing element is the steel frame to which the draw-bar, fairing frame and rear bumper reinforcement are attached. The structure is mounted on a single-axle suspension and is attached to the shadow vehicle hitch with a steel loop.

The structure consists of rectangular sections and sheet stamped and bent parts with a thickness of 2.8 - 5 mm, therefore, the developed finite element (FE) model was built with shell elements. The shell elements size is 5 mm. In the FE model, an elastoplastic model of the material

MAT_024_PIECEWISE_LINEAR_ PLASTICITY was used with the specification of the full deformation curve and the constants of the effect of strain rates according to the Cooper-Symonds model [11].

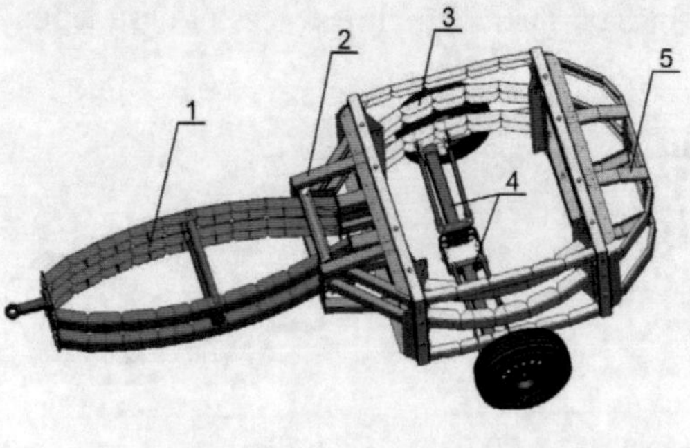

Figure 4. The main components of the TMA structure are: 1 – draw-bar, 2 – fairing frame, 3 – frame, 4 – wheels with axles, 5 — rear bumper reinforcement.

The structure has connections that should be disconnected during the operation of the attenuators. They connect the draw-bar arch with the crossbar (figure 5 (a)), and parts of the coupling (figure 5 (b)). Based on this, these connections were modeled using FE models of bolts. In other bolted connections, were used CNRB elements, rigidly connecting the parts to each other.

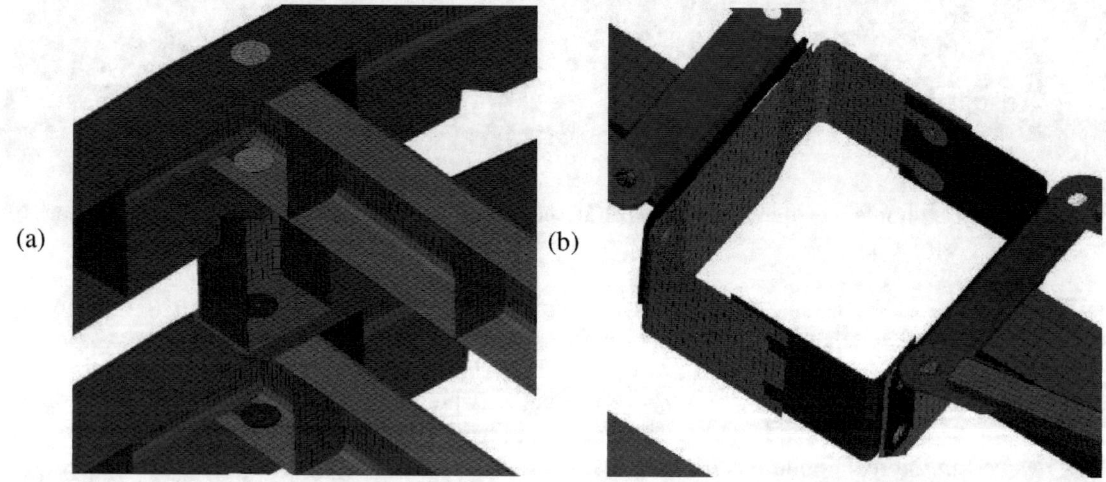

Figure 5. Collapsing joints during the TMA operation are: (a) the arc and the draw-bar connection, (b) the coupling connecting parts.

Figure 6 shows the complete FE model of a virtual test, consisting of a shadow vehicle with an TMA attached to it, and a driving vehicle weighing 1500 kg at a speed of 70 km·h^{-1}.

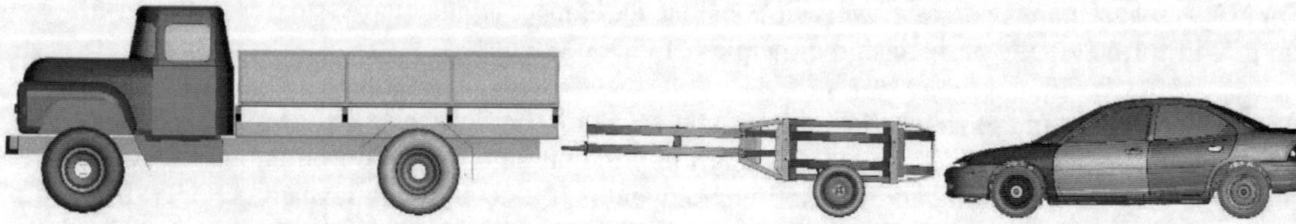

Figure 6. FE model of a virtual test with a collision along the axis of the TMA device.

4. Structural analysis based on the FEM calculation

Figure 7 shows the TMA after the impact. The longitudinal axis of the main impact attenuation part of the structure (frame) deflected from the initial position and turned out in a vertical position. This was due to the resulting bending moment and loss of the attenuators stability, since the point of impact was significantly lower than the central axis of the impact attenuation system structure, which the draw-bar could not withstand due to lack of rigidity. Figure 8, (a) and (b) shows the process of the draw-bar deformation, and figure 8 (c) — energy-absorbing frame after impact.

Figure 7. The TMA after the impact.

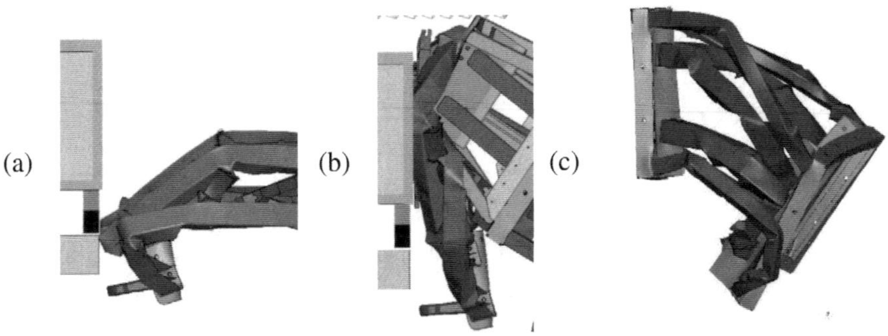

Figure 8. TMA structural elements deformation: (a), (b) draw-bar, (c) frame.

Consider in more detail the work of the structure on the internal energy of the main parts (figure 9). The impact kinetic energy was 283.56 kJ. The graph of the absorbed energy shows that the structure elements do not work sequentially and the energy absorption is interrupted (the graph is parallel to the time axis), in contrast to the most optimal case, in which energy is absorbed by the elements from the impact point to the draw-bar sequentially along the links.

The frame and the truck parts absorbed the same amount of energy during the impact: 58.9 kJ and 57.8 kJ, respectively. The draw-bar absorbed the greatest energy (67.9 kJ), which is not acceptable, since the metal consumption of the draw-bar is less than that of the frame, which is the main energy-absorbing element of the TMA structure. The TMA absorbed 180 kJ of energy in total.

Also, the work of the structure can be assessed by the speed graph of the vehicle's CG: the smoother the graph, the more efficiently energy is absorbed. However, figure 10 shows the period of time in which the vehicle does not decelerate, and the path of the vehicle in which deceleration could occur is not used.

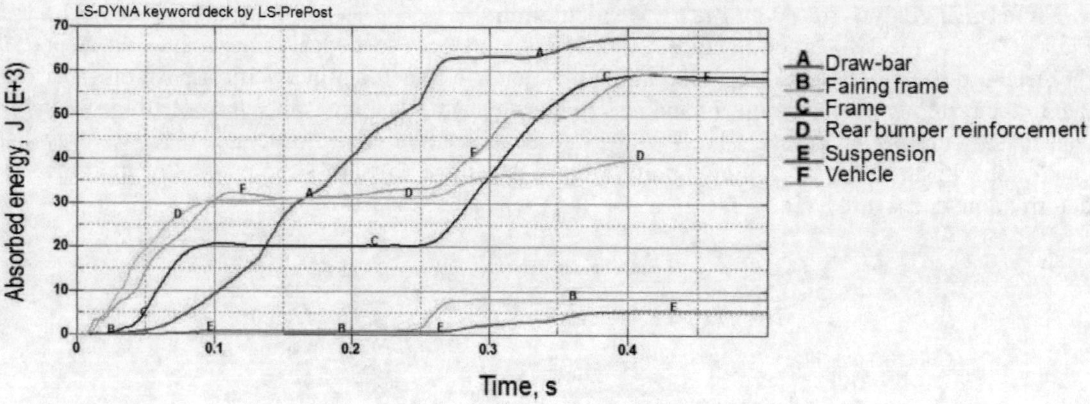

Figure 9. The time dependence graph of the absorbed energy by the TMA parts and the truck.

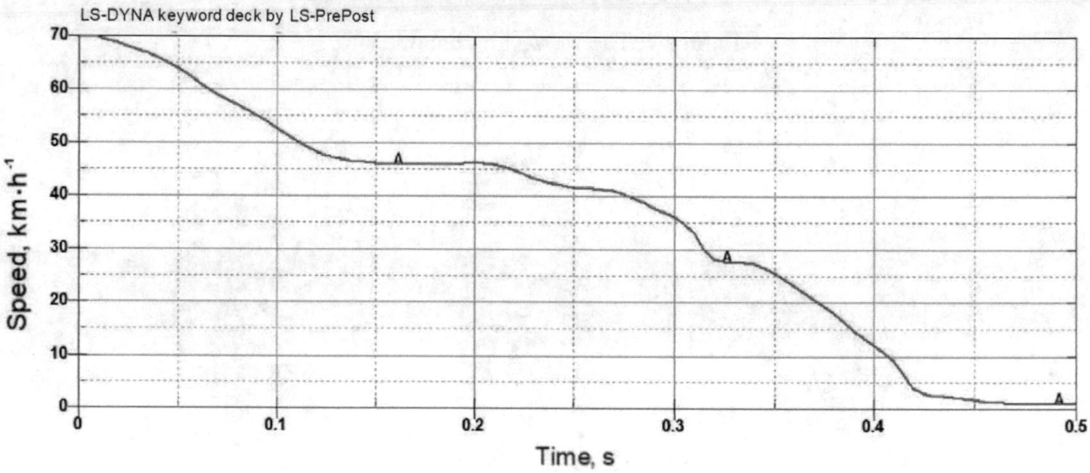

Figure 10. The graph of the vehicle speed change over time.

The collision case considered above is similar to the case shown in figure 11. The beam is hinged on one side (tow hitch), in the middle the point cannot move downward (suspension). The force is applied longitudinally, below the geometric CG. The result is an eccentric compression of the attenuators.

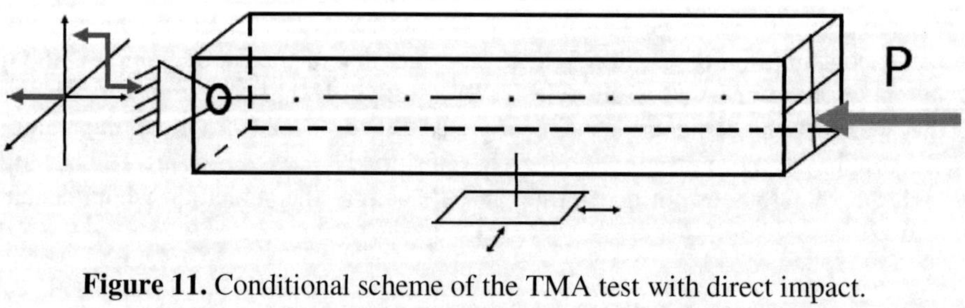

Figure 11. Conditional scheme of the TMA test with direct impact.

In the case of an angular impact with a displacement (figure 12), the impact point is displaced relative to the TMA center of gravity not only downward, but also sideways, and also applied at an angle. As a result, there is a compression and a bending moment in two planes, which tends to deflect the test structure up and to the side (figure 13). The TMA did not pass the virtual test: the vehicle passed under the attenuators and was not stopped. The speed of the approaching vehicle after contact with the TMA was 40 km·h^{-1}, after which there was a collision with the rear of the shadow vehicle. This result, as in the first case of impact, was due to the draw-bar low bending rigidity.

(a)

(b)

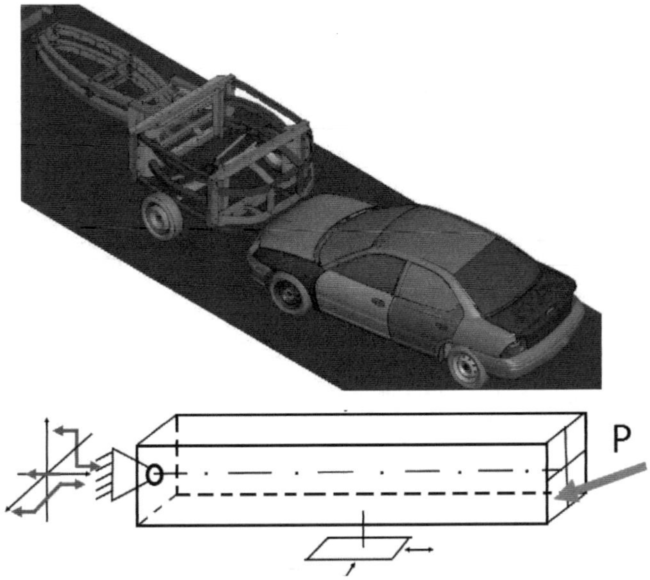

Figure 12. Testing the TMA at an 10° angle of to the axis with displacement: (a) FE model, (b) schematic diagram.

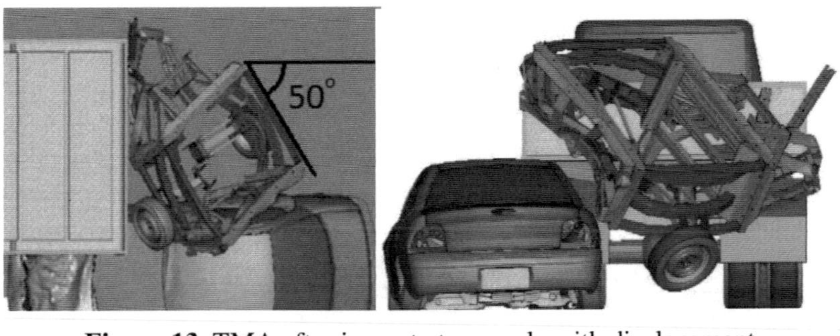

Figure 13. TMA after impact at an angle with displacement.

5. Design principles

When designing TMAs, it is necessary to strive to structurally ensure the impact redistribution of load evenly over the entire cross section, so that there is only longitudinal compression.

In the event of a direct impact, a safe uniformly slow motion of the vehicle is achieved by selecting the rigidity of the main energy-absorbing elements and the geometry of the TMA. The height is so selected that the oncoming vehicle does not run under or onto the TMA during the impact, which can lead to serious consequences, and the TMA CG should be at the impact level.

When impact with displacement or at an angle, the TMA structure is subjected to eccentric compression, therefore, the bending stiffness must also be correctly selected, but this will only partially solve the problem, since in the hinge of the hitch remains the rotational degree of freedom. Therefore, draw-bars with a telescopic structure (figure 14 (a), (b)) or hydraulic dampers [12] (figure 14 (c)) are used in the TMA designs, which make it possible to prevent the TMA from turning during the impact.

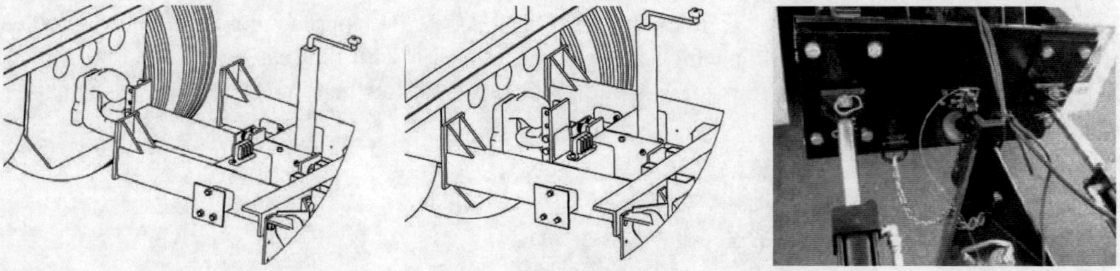

(a) (b) (c)

Figure 14. The TMA anti-rotation design of the draw-bar: (a), (b) telescopic before and after impact, (c) with a hydraulic damper.

6. Conclusions

The study, based on virtual digital analysis, made it possible to study the mechanics of operation and the processes that occur as a result of a vehicle hitting a truck mounted TMA. Using the energy method as an example, it is shown how to evaluate the efficiency of the structure for the successful fulfillment of the assigned task — the safe stop of an oncoming vehicle. The constructed conditional schemes of the considered impact problem — eccentric compression with loss of stability and oblique impact, made it possible to formulate the basic principles of these safety structures designing. The performed research is the initial stage of work on the design of a domestic truck mounted TMA.

Acknowledgment

The authors are grateful to their scientific advisor, professor, doctor of technical sciences I.V. Demiyanushko for advice and attention to work.

References

[1] Demiyanushko I V, Karpov I A, Tavshavadze B T and Mikheev P S 2020 Improving road safety during repairs *World of roads* **131** pp 102-7
[2] Theiss L and Bligh R 2013 *Compare trailer-mounted attenuators vs truck-mounted attenuators protection for workers* 70 Available at: https://static.tti.tamu.edu/tti.tamu.edu/documents/0-6707-1.pdf
[3] Silyanov V V and Sidorova A S 2012 Organization of traffic in places where road works are performed *Science and technology in the road industry* **2** pp 2-5
[4] Trinity Highway Products, LLC 2020 SST® Trailer *Product Description Assembly Manual* Available at: https://trinityhighway.com/
[5] Gregory Industries 2010 TTMA-100 User's Manual Available at: https://www.gregorycorp.com
[6] Rohde J R, Reid J D, Sicking D L and Mak K K 2008 Trailer truckmounted attenuator *Int. J. of Crashworthiness* **13 1** pp 77-87 DOI 10.1080/13588260701731708
[7] TrafFix Devices, Inc 2018 Scorpion II® Trailer Attenuator Assembly Manual and Mounting Instruction Guide Available at: www.traffixdevices.com
[8] Trinity Highway Products, LLC 2020 Vorteq® Trailer *Product Description Assembly Manual* Available at: https://trinityhighway.com/
[9] GOST R development Public roads. Truck mounted attenuators. General requirements. Test methods Available at: http://niimech.ru/
[10] MASH 2016 2nd Edition Available at: bookstore.transportation.org/Item_details.aspx?id=2707
[11] Hallquist J 2006 LS-DYNA theory manual Available at: https://www.dynasupport.com/
[12] Trinity Highway Products, LLC 2020 SMT™ *Product Description Assembly Manual* Available at: https://trinityhighway.com/

Exploration of strength characteristics quadrocopter frame structure obtained using 3d printing technology

R A Navrotsky, G V Moskvitin, O V Naumov and M S Pugachev

Mechanical Engineering Research Institute of the Russian Academy of Sciences, 4 Maly Kharitonyevskiy Pereulok, Moscow, 101990, Russia

naumov57.on@gmail.com

Abstract. This article discusses the use of 3D printing technology for manufacturing the frame structure of a quadrocopter. The strength of the plastic frame structure is calculated. The results obtained allow us to draw conclusions about the possibility of using this technology for the production and operation of aircraft of the quadrocopter type.

1. Introduction

The application of 3D printing technology affords ample opportunities for different fields of application; one of such fields implies small aircraft building. The prototyping, printing of units and parts makes it possible to obtain new configurations of devices bringing about the improvement of existing devices.

The quadcopters – the small remote-controlled aircraft have become widespread over the last few years. The quadcopter design includes the following elements - motors, speed controller, propellers, board, sensors (GPS, gyroscopes, etc.), frame [1]. Depending on the applied tasks and required fixturing there are quadcopters with different frame designs (figure 1).

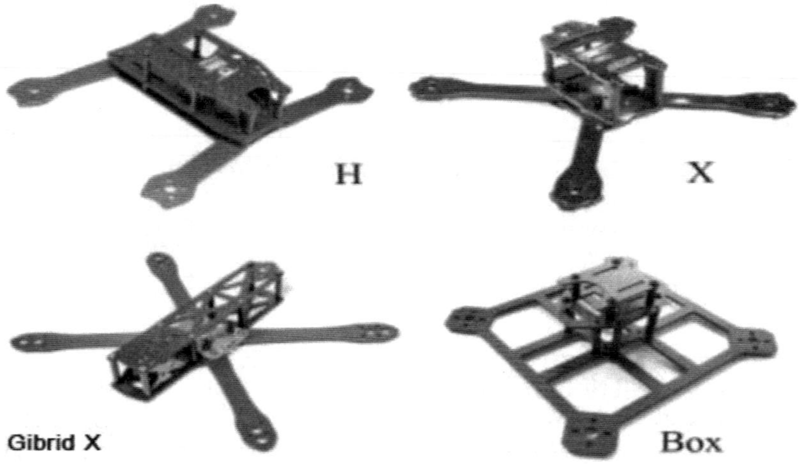

Figure 1. Configurations of quadcopters H, X (cross), hybrid X, quadrate (box).

The principle of quadcopter operation is based on the counter rotation of propellers, owing to this the flight becomes smooth, stabilized, and a vehicle is well-controlled (figure 2).

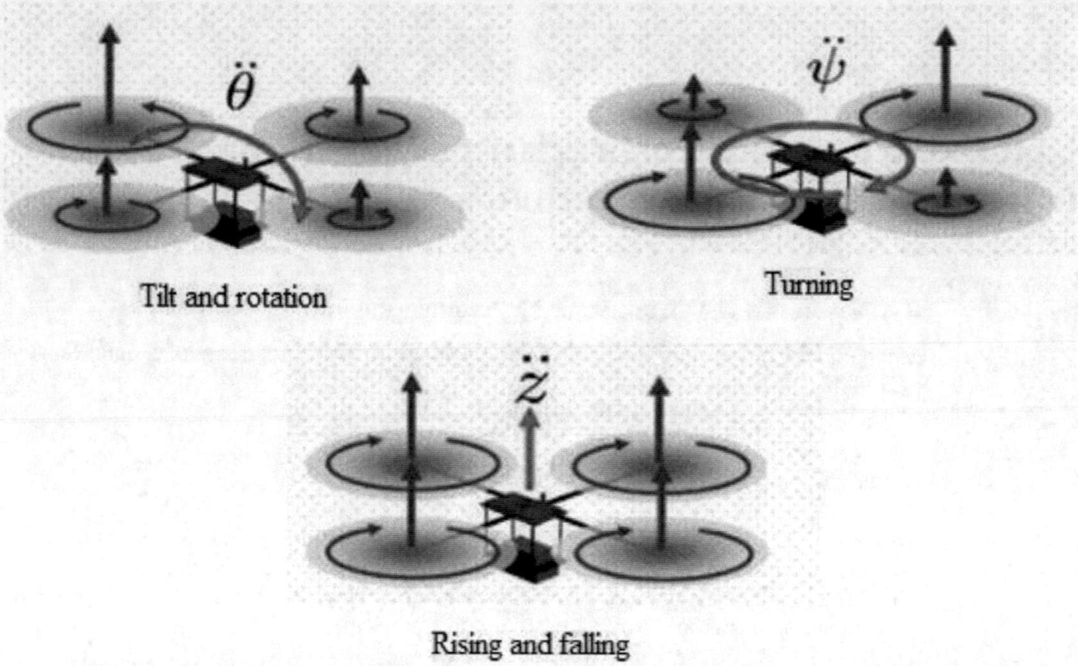

Figure 2. Capabilities of changing space attitude of a quadcopter.

Taking into account peculiarities of spatial movement and load created by equipment, the following requirements are produced to the frame structure:

- ruggedness – a well-designed frame should be rugged, hold up well in case of accidents. If the bars become bent or warped, the copter will be prone to vibrations and fluctuations. As a result it will be difficult to adjust and calibrate it. The flight controller will also suffer, it will be more difficult for it to stabilize the flight, finally, the motors may become overheated;
- weight – reduction of frame weight without decreasing mechanical characteristics of the structure will help carry out further structure improvements and increase the time of aircraft operation;
- longevity;
- robustness.

Printing of parts using 3D printer helps solve technological tasks and meet the produced requirements to the structure. In particular, for production of parts with preset geometrical parameters, the non-recurrent blanks are cheaper than flow-line production. But in spite of an important advantage of printing an issue of structure strength remains less than fully investigated, in particular, the strength characteristic of aircraft of certain configurations. The strength analysis of quadcopter frame structure for static strength has been conducted in this work.

2. Investigation methods

The investigation of strength characteristic of quadcopter frame structure has been carried out in Free Cad software system with the use of finite elements method. A common frame type has been selected as an analytical model – a flat cross with holes for motors and a platform for flight controller and storage battery (figure 3). Frame overall dimensions: distance between opposite motors - 180 mm, dimension – distance between neighboring motors: 127.28 mm. The necessary values pertaining to PLA plastic have been taken from works [2, 3] which are obtained using [4, 5].

PLA plastic (polylactide) has been chosen as the material. The properties are given in table 1.

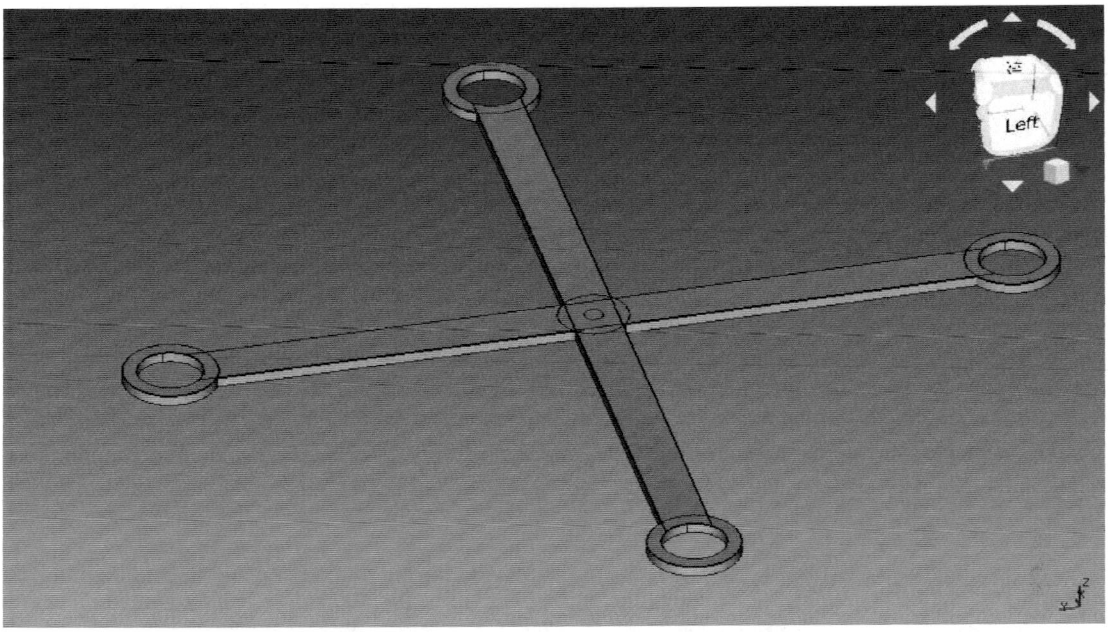

Figure 3. Model of quadcopter frame of x-shaped configuration.

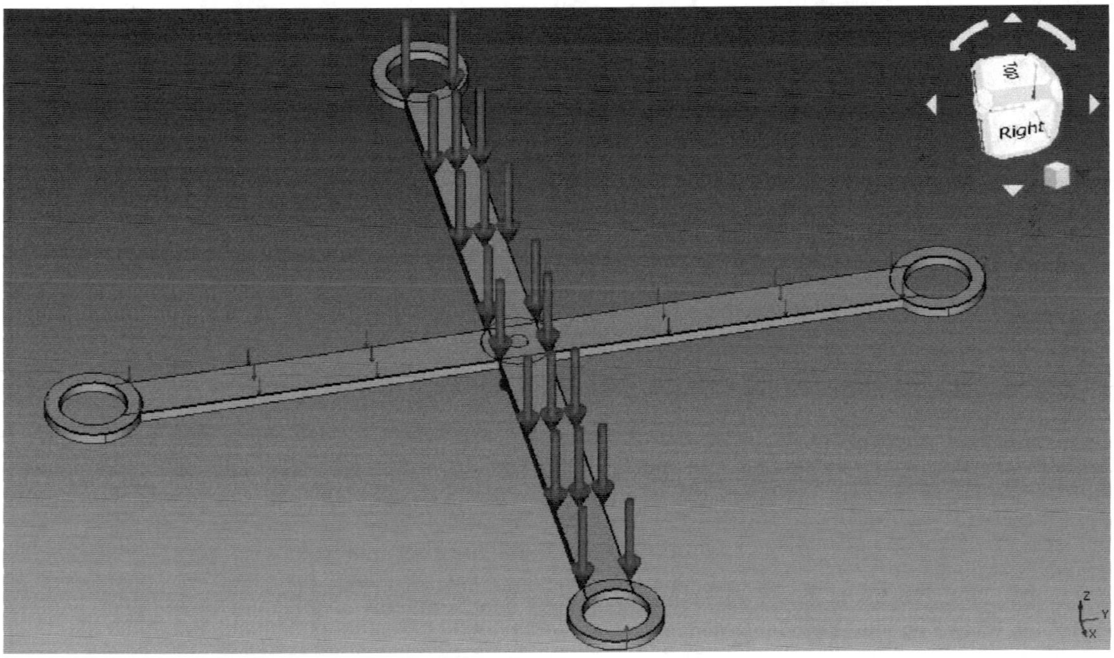

Figure 4. Distributing forces applied vertically to the frame upper planes, simulated gravity force and propelling force of motors.

Table 1. PLA plastic properties.

Density	1.240 kg·m^{-3}
Modulus of elasticity	3.640 MPa
Thermal conductivity factor	0.13 W·(m·K)$^{-1}$
Expansion coefficient	41 C^{-1}
Specific heat capacity	1.800 J·(kg·K)$^{-1}$

The forces applied vertically to the frame upper planes have been distributed, gravity force and propelling force of motors have been simulated. The propelling force of one motor should equal 50% of quadcopter weight. This implies that the total propelling force of all four motors should exceed the weight of the copter proper twice. The conventional values of applied forces have been selected taking into account this condition. It stands to mention that the forces simulating weight include the frame weight and the weight of such elements as storage battery, flight controller and motors.

3. Results

Results of static calculation are shown in figure 5 and table 2.

Figure 5. Visual result of static calculation.

Table 2. Calculation results.

	Minimum value	Average value	Maximum value
Absolute displacement, mm	10.00	16.24	26.84
Main stresses (von Mises), MPa	0.008	6.5	39.39
Shearing stresses, MPa	0.005	3.32	20.06

Comparing the calculation data with the earlier-acquired results of testing samples [2, 3], it is possible to conclude that the printed frame structures can be used in the course of building and developing quadcopters without any loss of the main characteristics of small aircraft.

4. Conclusions

Proceeding from the performed investigation it is possible to make the following findings:

- frame structures made with the use of technology of 3D printing from PLA plastics can be used in building small aircraft (quadcopters);
- values of absolute displacement at the preset design load are admissible (limiting displacement value for frame of x-shaped configuration equals 45 – 50 mm);
- values of design stresses are below the values received experimentally (41.8 MPa) [2].

Acknowledgments

The authors would like to thank colleagues for important advices and assistance in preparing this article.

References

[1] Radio-controlled models Available at: https://rc-like.ru/kvadrokopter-chto-eto-takoe
[2] Naumov O V, Moskvitin G V, Pugachev M S and Polyakov A N 2017 *Investigation of stress-related and strength properties of ABS plastic when printing parts using 3D printer* (Institute of Machines Science of Russian Academy of Sciences) pp 57– 60
[3] Naumov O V, Moskvitin G V, Pugachev M S and Polyakov A N 2016 *Experimental estimate (investigation) of strength of parts manufactured at 3D printer* (Institute of Machines Science of Russian Academy of Sciences) pp 461–4
[4] STATE STANDARD 15139-69 Plastics Methods of density (bulk weight) identification
[5] STATE STANDARD 14359-69 Plastics Methods of mechanical tests General requirements

IOP Conf. Series: Materials Science and Engineering 1129 (2021) 012022 doi:10.1088/1757-899X/1129/1/012022

Biomimetic design of fibrous composite structures

A N Polilov and N A Tatus'

Mechanical Engineering Research Institute of the Russian Academy of Sciences, 4 Maly Kharitonyevsky Pereulok, Moscow, 101000, Russia

polilovan@mail.ru

Abstract. The paper is devoted to the analysis of the use of Nature's methods in the design of composite structures, in particular, with the use of optimal curvilinear paths for laying fibers. In the first part models of different structural levels in macro-, mini-, micro-mechanics of composites are noted. The second part includes analysis of the optimal elastic-strength properties of wood and composites for crack stopping by weak interfaces. The third part is devoted to methods for modeling curvilinear fibers trajectories "flowing around the holes". In the fourth part the technologies for manufacturing joints are analyzed in which holes are formed using curved fiber trajectories. The fifth part describes the "bio-inspired" principles for the optimal design of composite pipe structures which are similar to multilink bamboo stems. In the sixth part examples of the effective use of fiber composites in elastic elements are considered. The seventh part is devoted to the intensively developing additive technologies of three-dimensional printing of parts made from fiber composites with laying the fibers along the calculated trajectories. The role of composite technologies in creating structures in open space is analyzed.

Introduction

There are numerous examples of the use of Natural Solutions in engineering, and particularly in the design of composite structures. The terms biomimetics and «bio-inspired method» are widely used. For example, a question arises: how is the branch-barrel connection arranged? You can break a branch, but no effort can tear the branch off the trunk. The connection is stronger than the part itself. So far, it is only possible to dream about such methods of fastening composite parts.

1. Modeling at different structural levels

Scientific approaches to studying fibrous composites are very diverse. With the same object (composite construction), they can differ in the subject of study which means they can be based on models of different levels. The tasks analyzed in this report relate to composite models that are placed in an intermediate position between "microstructural" and "macromechanical" models.

Micromechanics of wood and composites

Wood, bamboo stem have complex, hierarchical structure [1]. Without going into the details of micromechanical models, we note only a series of scientific works by G.A. Vanin (Fan Fo Fa) [2] and other authors, which set out approaches to the construction of accurate solutions to the problems of the theory of elasticity for periodic fibrous structures. However, the use of gradient models of the moment theory of elasticity of the Kosser-Mindlin-Lurie type in the design of composite structures is extremely limited, since existing technologies do not allow the production of fibers and binders with a specially created bio-similar hierarchical structure.

2. Mini-midi mechanics. Optimal properties of components for crack stopping by interface

The main advantages of fibrous composites are not only high strength along the fibers, but also toughness due to the presence of fragile fiber-matrix interfaces.

Figure 1 shows the crack stopping mechanisms due to the complex stress state near its apex. The refined analysis [3] allows corrections (figure 1 (b)) to the known Gordon scheme (figure 1 (a)):

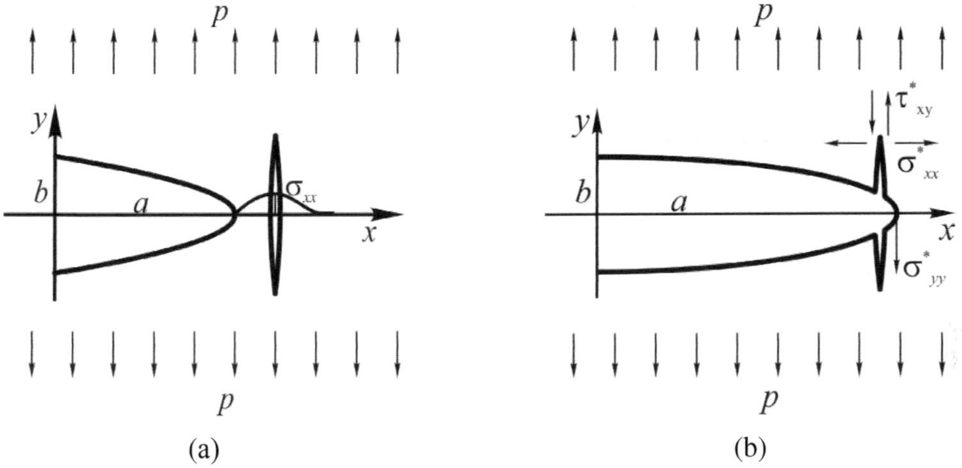

Figure 1. Schemes for stopping the crack by the weak interface: a - according to Gordon - splitting before the crack; b - updated diagram - splitting on the loop.

- The ratio of normal stresses $\frac{\sigma_{xx}^{\max - x}}{\sigma_{yy}^{\max}}$ equal in the isotropic case $\frac{1}{3\sqrt{3}} = 0{,}192 \approx \frac{1}{5}$, is not a constant but depends on the anisotropy of the material.
- The largest values of all stress components are achieved on the hole contour, and therefore splitting always begins on the hole or crack contour, not in front of it.
- The main role in splitting is played by tangent stresses, since they are significantly more transverse, especially in highly anisotropic materials such as wood or unidirectionally reinforced plastics, which allows us to formulate the condition for the occurrence of splitting before fracturing of fibers in a simple form:

$$\frac{\tau_{xy}^{max}}{\sigma_{yy}^{\max}} \geq \frac{\tau^*}{\sigma_y^*}. \tag{1}$$

where τ^*, σ_y^* are composite shear and tensile strengths along the fibers.

The ability to stop cracks due to the splitting of weak interfaces is one of the main advantages of natural and polymer composites, which allows to simultaneously increase strength and crack resistance.

It is interesting to note that equation in (1) is reasonably fulfilled for strong wood varieties: for oak $\frac{\tau_{xy}^{\max}}{\sigma_{yy}^{\max}} = 0.095;\ \frac{\tau^*}{\sigma_y^*} = 0.091$, for pine oak the corresponding ratios are 0.14 and 0.093. For unidirectional fiberglass according to equation (1), the shear strength should be about 150 MPa, and the real strength of the polymer matrix is three times lower, so unidirectional composites are split near stress concentrators long before they reach limit stress.

3. Methods for constructing biosimilar curvilinear fiber paths

Nature never uses linear reinforcement. Observations of Nature have led to the idea of the need to use an optimal, curvilinear reinforcement structure in composite design, when the fiber trajectories are consistent with stress fields. A large number of works [4, 5 et al.] is devoted to the construction of optimal trajectories of fibers "flowing around" a circular hole.

In [5], a new approach to the construction of fiber paths coinciding with the trajectories of the main tensile stresses is developed. Direct, iterative construction of continuous fiber trajectories makes it possible to directly switch to additive technologies that provide a rational structure of curvilinear reinforcement in the zone of holes or connecting units. The trajectories of the fibers flowing around the hole resemble the structure of wood in the knot area (figure 2).

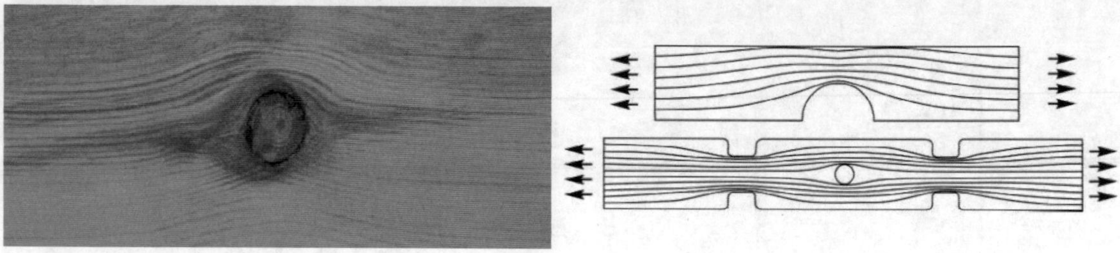

(a) (b)

Figure 2. (a) structure of pine wood in the knot area, (b) equally stressed structure of fibre laying along trajectories of greatest main stresses near holes and recesses in stretched plate.

The main conclusion of the analysis is that the maximum stress "per fiber" in the optimally reinforcement structure becomes about 3-4 times less than with a unidirectional fiber placement which means that the effective stress concentration coefficient decreases from 5 to 1.3, and the fiber overload at the bottom of the hole is only about 30%. At the same time, it is extremely important that shear stresses disappear, causing splits near the holes.

It becomes clear that damaging effect of stress concentration in composite joints can be reduced precisely by means of a special reinforcement structure consistent with the stress field.

4. Biomechanical principles of composite parts attachment units design

One of the main fundamental tasks of composite design is the development of connection methods that implement high fiber strength.

The fundamental disadvantages of all known connection methods make it necessary to turn to the Nature for experience, and the "design" of the knot can suggest optimal fiber paths in the connection area through a bolt or rivet (figure 3).

Rational reinforcement leads to a significant reduction in local stresses per fiber, elimination of splits and an increase (at least 50%) in the load capacity of the connection.

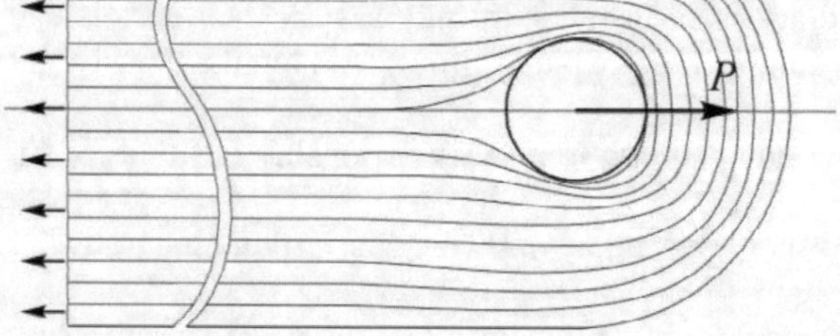

Figure 3. Rational fibers trajectories enveloping the loaded hole.

5. Bamboo and composite pipe failure mechanisms

One of the most amazing and perfect works of Nature - the fast-growing stem of bamboo - attracts not only as an indispensable material for scaffolding in Hong Kong, but also as an optimal prototype of composite pipes [6]. The energy criterion of splitting [3, 7] - in the form of a "Chinese flashlight" during compression (figure 4) and with deplanation of the section (figure 5) during torsion of thin-walled unidirectional composite pipes - allows you to indicate rational distances between the rings of a multi-link composite pipe, qualitatively consistent with the dimensions of the bamboo stem.

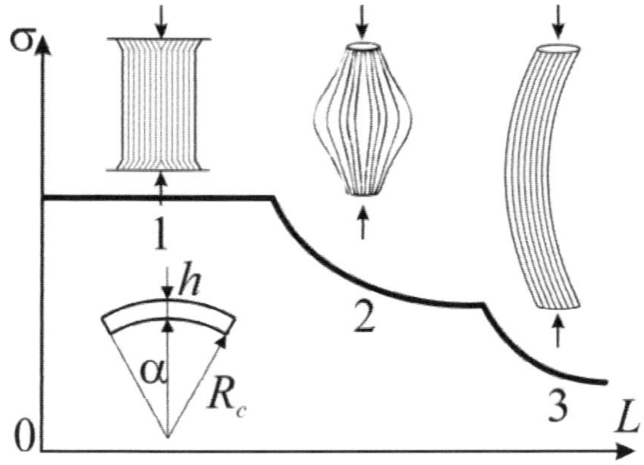

Figure 4. Dependencies of critical stresses on length L during compression of composite pipes for three types of destruction: 1 - local crushing; 2 - "Chinese lantern"; 3 – macro buckling.

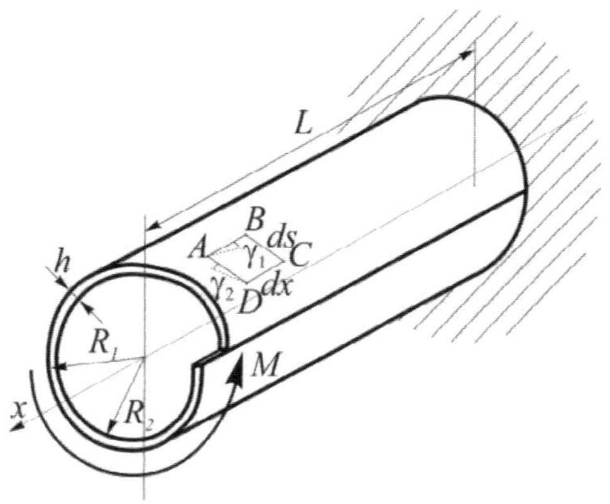

Figure 5. Torsion split diagram of a thin wall pipe with section deplanation and illustration of the calculation of total shear strain $\gamma = \gamma_1 + \gamma_2 = R_c\theta + \dfrac{\partial u}{\partial s}$.

Increasing torsional stiffness after constrained splitting leads from the energy criterion to an increase in critical torque and to the appearance of significant axial stresses:

$$M_0^+ = \left(\frac{4\pi\gamma^* G_{xy} R_c^3}{h^{-2} - [3R_c^2(1-2pB_2/\lambda)]^{-1}} \right)^{0.5}; \quad \sigma_{\max} = 0.76 \frac{MB_2}{R_c h^2} \left(\frac{E_x}{G_{xy}} \right)^{0.5}. \tag{2}$$

In (2) p - sealing factor (p = 0 - non-shear torsion, p = 1 - absolutely rigid sealing);

$$B_2 = \frac{\mathrm{ch}\,\lambda - 1}{\mathrm{sh}\,\lambda},$$

E_x, G_{xy} are Young's and shear modulus,

$$\lambda = \frac{L}{k}$$

$$k = \frac{R^2}{h} \left(\left(\pi^2 - 6\right) \frac{E_x}{G_{xy}} \right)^{0.5}$$

It can be understood from (2) that there is such a small length of the pipe link L* at which the denominator turns to zero, that is, splitting cannot occur at any torque M+. This is approximately the length of the bamboo link that Nature "chooses" to avoid premature splitting.

6. Design of bio-similar branching and profiled composite elastic elements

The secrets of the tree crown structure for many centuries have attracted the attention of researchers. Leonardo da Vinci in his notes expressed the following statement: "The sum of the squares of the diameters of the branches is the same before and after branching." The Leonardo Rule suggests how to create branched (figure 6) and profiled (figure 7) composite elastic elements [8-11].

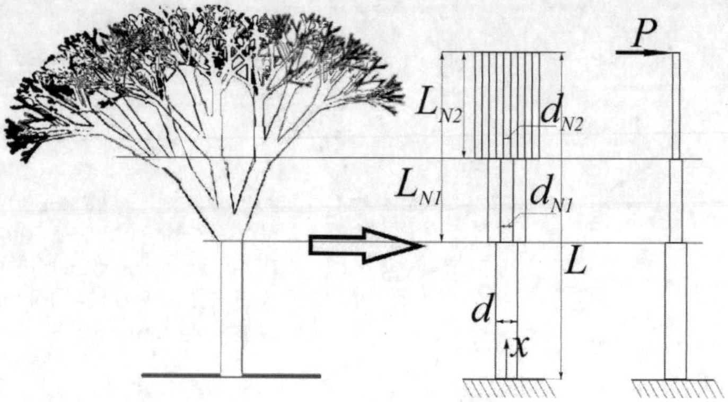

Figure 6. Diagram of equal-strength branch of rod with circular section.

When acting on a perfectly branching or profiled beam with a variable bending moment:

$$\mathrm{M}(\bar{x}) = M(1)\bar{x}^{\gamma};\ \bar{x} = x/l. \tag{3}$$

in order to maximize the stored elastic energy, sequential branching on $1+2\gamma$ of the branches is advantageous, and surprisingly, the maximum possible mass reduction coefficient is also equal to $1+2\gamma$.

Let us note that in (3) the load of the end force P corresponds to $\gamma = 1$, $\gamma = 2$ for the uniformly distributed load, $\gamma = 3$ for the linearly varying load.

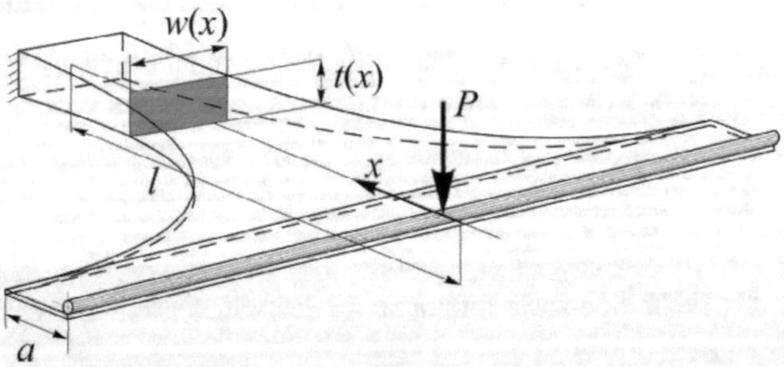

Figure 7. Constant-area beam with constant cross-sectional area.

In [11], the reduction in mass of the band spring from the shaped half-rings was analyzed, and it turned out that in the ideal case it is possible to double it, which corresponds to $\gamma = \frac{1}{2} < 1$, since the change in the bending moment in the diametrically compressible half-ring is proportional to the sine of the polar angle, i.e. it is "slower" than the linear one.

Branching equally stressed composite elastic elements are as effective as profiled beams of the constant area type, but the advantage of branching compared to profiling is the preservation of

IOP Conf. Series: Materials Science and Engineering 1129 (2021) 012022 doi:10.1088/1757-899X/1129/1/012022

continuous linear reinforcement and the possibility of reducing dimensions when collecting "branches" into a bundle.

7. Additive Biomimetic Technologies

The development of additive 3D printing technologies extends the possibilities of creating optimal structures with curved fiber paths [12, 13]. In this way, it is possible not only to design and create products of different shapes, but also to manufacture bio-similar joints [14], which are much more effective than traditional "metal-like" connections.

Since such technologies do not require high temperatures and considerable effort, in the future it is possible to create technological areas for the manufacture and repair of composite elements directly in orbit [15].

Conclusion

In our opinion, the three most promising areas in the mechanics of composites are biomechanics of strength, computer modeling of optimal structures and technological mechanics of composites. It is the modeling of biotechnologies and biomaterials structures that can provide a breakthrough in the creation of composite structures.

References

[1] Wegst U G K, Bai H, Saiz E, Tomsia A P and Ritchie R O 2015 Bioinspired structural materials *Nature Materials* 14 pp 23-36
[2] Vanin G A Moment mechanics of composites *Mechanics of composite materials* 2001 **36** pp 621-54
[3] Polilov A N and Tatus' N A 2018 *Strength biomechanics of fibrous composites* (Moscow: FIZMATLIT) p 328
[4] Papapetrou V S, Patel C and Tamijani A Y 2020 Stiffness-based optimization framework for the topology and fiber paths of continuous fiber composites *Compos. Part B-Eng.* 183 107681
[5] Malakhov A V and Polilov A N 2016 Design of composite structures reinforced curvilinear fibres using FEM *Compos. Part A* **87** pp 323–8
[6] Zou M, Xu S, Wei C, Wang H and Liu Z 2016 A bionic methods for the crashworthiness design of thin walled structures inspired by bamboo *Thin Walled Structures* **101** pp 222-30
[7] Plitov I S and Polilov A N 2015 Rational dimensions of segments of bamboo stems and composite tubes subjected to compression, flexure, and torsion *J. of Machinery Manufacture and Reliability* **44** (3) pp 239-48
[8] Polilov A N, Tatus N A, Arutjunova A S and Tian X 2019 Equistrong branchy composite beams with a constant total area of variable elliptic cross sections *Mechanics of Composite Materials* **55** (3) pp 325-36
[9] Polilov A N, Tatus N A and Tian X 2019 Analyzing the correctness of equal strength composite profiled beam bending problems *J. of Applied Mechanics and Technical Physics* **60** (1) pp 144-55
[10] Polilov A N, Tatus' N A and Tian X 2019 Analysis of efficiency of uniform-strength composite leaf springs under various loading conditions *J. of Machinery Manufacture and Reliability* **48** (5) pp 431-9
[11] Polilov A N, Tatus' N A, Zhavyrkin V V and Tian X 2020 Analysis of the efficiency of annular and tape springs made of fiberglass *Problems of machine building and machine reliability* **3** pp 79-93
[12] Khan S, Fayazbakhsh K, Fawaz Z and Nik M A 2018 Curvilinear variable stiffness 3D printing technology for improved open-hole tensile strength *Addit. Manuf.* **24** pp 378-85
[13] Velasco Hogan A, Xu J and Meyers M A 2018 Additive manufacturing as a method to design and optimize bioinspired structures *Adv. Funct. Mater.* **30** 1800940

[14] Malakhov A, Polilov A, Zhang J, Hou Z and Tian X 2020 A modeling method of continuous fiber paths for additive manufacturing (3D printing) of variable stiffness composite structures *Applied Composite Materials* **27** pp 185-208
[15] Misochenko A A, Prozhega M V, Salamandra K B, Tatus' N A and Shokhin A E 2015 Orbital technological complex (ORTeK) for the composite structures production in outer space *In the book: Vostochny Cosmodrome and Prospects for the Development of Russian Cosmonautics* pp 64-5

Acknowledgments
This work was supported by the Russian Foundation for Basic Research (grant 18-08-00372-A).

On the strength of chemical anchors

O V Titov[1], V S Nadezhdin[1], A S Gruzdev[1] and A V Kuprin[2]

[1] Moscow Automobile and Road Construction State Technical University (MADI), 64, Leningradsky Prospect, Moscow, 125319, Russia
[2] BIT United Ltd

shashki18@gmail.com

Abstract. The work presents the strut attachment unit study of the bridge barrier road rack to the reinforced concrete slab (strength class of concrete B35) using adhesive (chemical) compositions and anchor-studs under the dynamic load. Bench static tests results and virtual experiment results based on study of digital models presented. The required depth of anchor-stud deepening determined.

1. Introduction

Reducing the number of road accidents, and therefore improving road safety, is a priority for any country. One of the most difficult road sections for safety is road bridging. This is primarily due to the fact that in an accident, the bridge barrier obliged to keep the vehicle from leaving the pedestrian part and/or beyond the bridge. However, the selection of the road barrier limited by the width of the road bridge and the difficulty of installing the road barrier to the load-bearing system of the bridge. One of the modern methods of securing the safety system to the bridge bearing system under limited deepening conditions is the use of chemical anchor consisting of a metal element and an adhesive composition. The purpose of this study is to determine the reasons for breaking of anchor studs [1] during full-scale bench impact tests [2].

As a result of bench impact tests, which carried out at the stand of FSUE "NAMI" the post using BIT-NORD adhesive composition and anchor pins for attaching the strut to the foundation [2], the front anchor pins torn out during loading (figure 1).

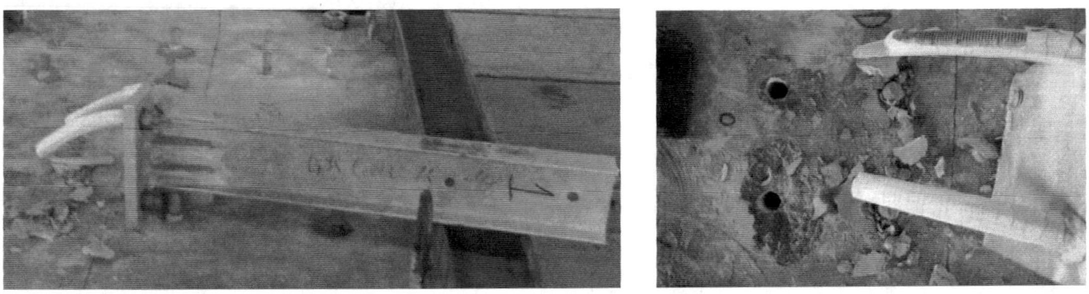

Figure 1. Removal front anchor pins as a result of bench impact tests.

IOP Conf. Series: Materials Science and Engineering 1129 (2021) 012023 doi:10.1088/1757-899X/1129/1/012023

Bench impact tests carried out in accordance with the terms of reference and Standart R 52721. Bench impact test diagram shown in figure 2.

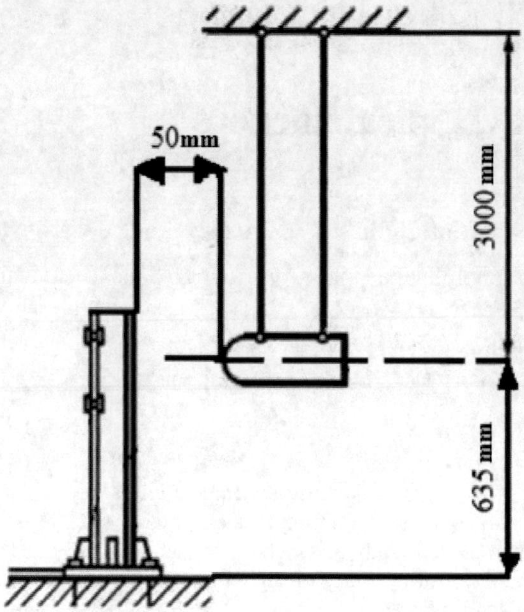

Figure 2. Bench impact test.

2. Materials and equipment

To identify the causes of chemical anchors breaking out of the support surface, the study divided into two stages. At the first stage, the ultimate force determined by conducting bench static tests for breaking out of anchor pins (figure 3), which attached to a reinforced concrete slab, with geometric dimensions of 1200x1400x350 and strength class of concrete B35, using adhesive compositions. Bench tests carried out for two installation depth of studs: 170 mm (stud-170) and 210 mm (stud-210) on two types of adhesive compositions BIT-NORD and BIT-200 [1].

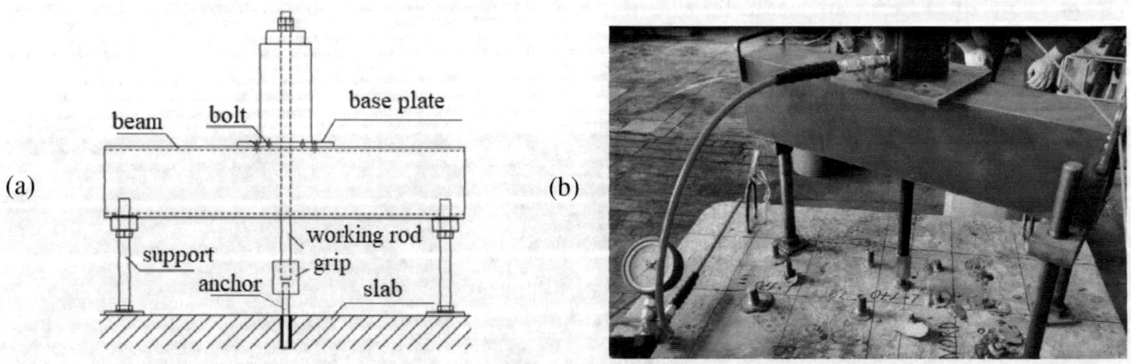

Figure 3. Bench static tests for removal of anchor pins: (a) circuit of the accessory, (b) device for the experiment.

As a result of bench tests of stud-170 and stud-210 on adhesive compositions BIT-NORD and BIT-200, the values of the maximum load in the connection between the stud and the reinforced concrete slab obtained, depending on the depth of sealing, and two main types of destruction also recorded:

1) destruction of the body of the anchor pin (figure 4 (a));

2) destruction of the adhesive joint (figure 4 (b)).

The results of bench tests to determine the ultimate force of the BIT-NORD and BIT-200 adhesive compositions shown in table 1.

(a) (b)

Figure 4. Types of destruction: (a) destruction of the body of the anchor pin, (b) destruction of adhesive joint.

Table 1. Bench test results.

adhesive compositions	installation depth	stud	maximum force	destruction
BIT-NORD	170 mm	Stud-170	15.5 t (152 kN)	adhesive compositions
BIT-NORD	170 mm	Stud-170	15 t (147 kN)	adhesive compositions
BIT-NORD	210 mm	Stud-170	18.5 t (181 kN)	stud
BIT-200	170 mm	Stud-210	18.5 t (181 kN)	stud
BIT-200	170 mm	Stud-210	17.5 t (172 kN)	stud
BIT-200	210 mm	Stud-210	17.8 т (175 kN)	stud

Analyzing the results (table 1), it concluded that with an installation depth of 170 mm and the use of BIT-NORD adhesive composition, the anchor stud-210 breaks out of the reinforced concrete slab (destruction of the adhesive connection between the stud and the reinforced concrete slab) without breaking the anchor stud. When the installation depth increased to 210 mm, the adhesive joint between the stud and the reinforced concrete slab using the BIT-NORD composition was not destroyed, while the body of the anchor-stud was destroyed stud-210 with a force corresponding to the fracturing force determined by previously performed laboratory bench static tests [3].

When using adhesive composition BIT-200 for attachment of anchor stud-170 and stud-210 with installation depth of 170 mm and 210 mm, respectively, in both cases the body of anchor stud destroyed with forces corresponding to those determined earlier during laboratory bench static tests [3]. At the same time, the loss of the load-bearing capacity of the adhesive connection of the anchor-stud with the reinforced concrete slab did not occur.

3. Modeling

At the second stage, the behavior of chemical anchors during dynamic loading was simulated. Virtual tests conducted on the basis of the finite element method (FEM) using the LS-Dyna multi-purpose software complex [4]. According to the design diagram (see figure 2), the digital test model shown in figure 5 constructed.

Figure 5. Virtual dynamic test model.

Virtual bench dynamic (impact) tests carried out for anchor stud-170 on the adhesive (chemical) composition of BIT-NORD on the reinforced concrete base of the bridge post. Figures 6 and 7 show the developed digital models of the core elements used in virtual impact tests. The post is a thin-walled profile, so its modeling carried out by shell finite elements. Stud-170 modeled with three-dimensional finite elements [5].

(a) (b)

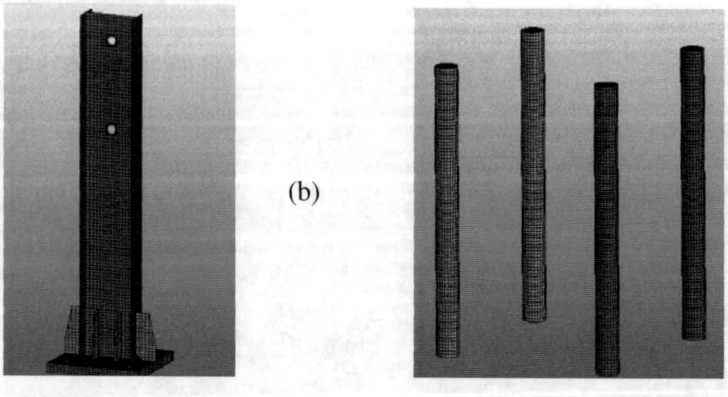

Figure 6. Digital models of the main elements: (a) post, (b) anchor stud-170.

(a) (b)

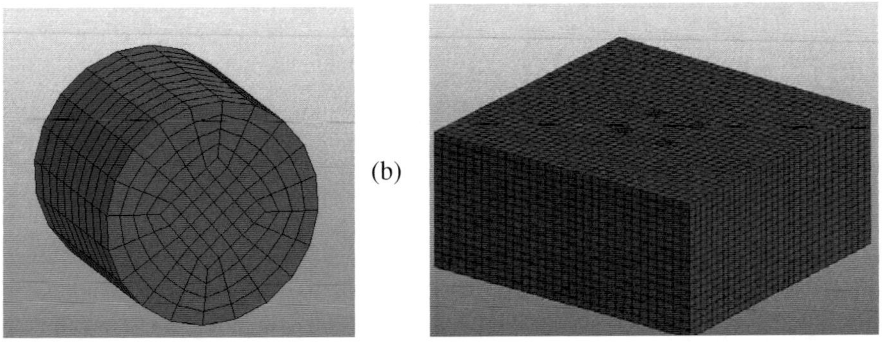

Figure 7. Digital models of the main parts: (a) loading element (striker), (b) reinforced concrete slab.

The main mechanical characteristics of the post material according to the quality certificate given in table 2. The true stress-strain curve of the post material used in the virtual bench impact test model is shown in figure 8. In the digital model, an elastoplastic model of the material MAT_024_PIECEWISE_LINEAR_ PLASTICITY used with the specification of the full stress-strain curve and the constants of the effect of strain rates according to the Cooper-Symonds model [6]. For loading element and concrete slab used rigid material MAT_020_RIGID.

Table 2. Main mechanical characteristics of the post material

Yield strength, MPa	351
Tensile strength, MPa	473
Relative deformation at failure, %	31
Elastic modulus, MPa	200000
Poisson ratio	0.31

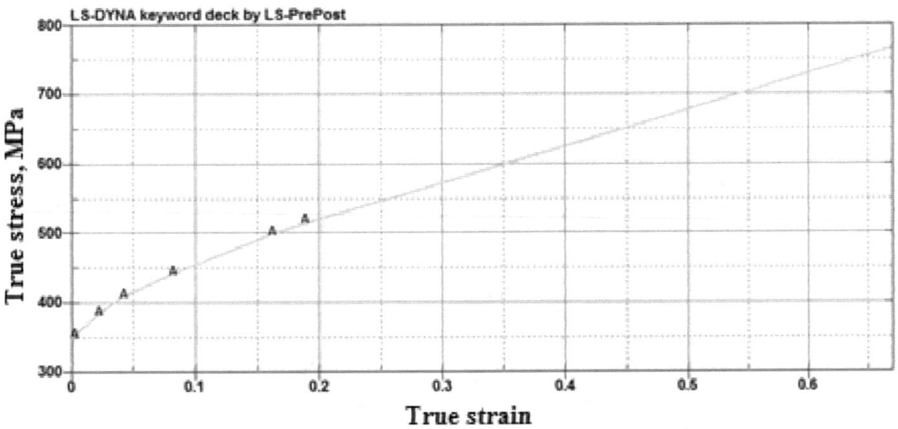

Figure 8. Stress-strain curve of the post material used in digital model.

The main mechanical characteristics of the material of anchor stud-170, obtained in previous studies [5], shown in table 3. The true stress-strain curve of the anchor-stud material used in the virtual model of bench shock tests shown in figure 9.

Table 3. Main mechanical characteristics of the anchor stud material

Yield strength, MPa	479
Tensile strength, MPa	713
Relative deformation at failure, %	7

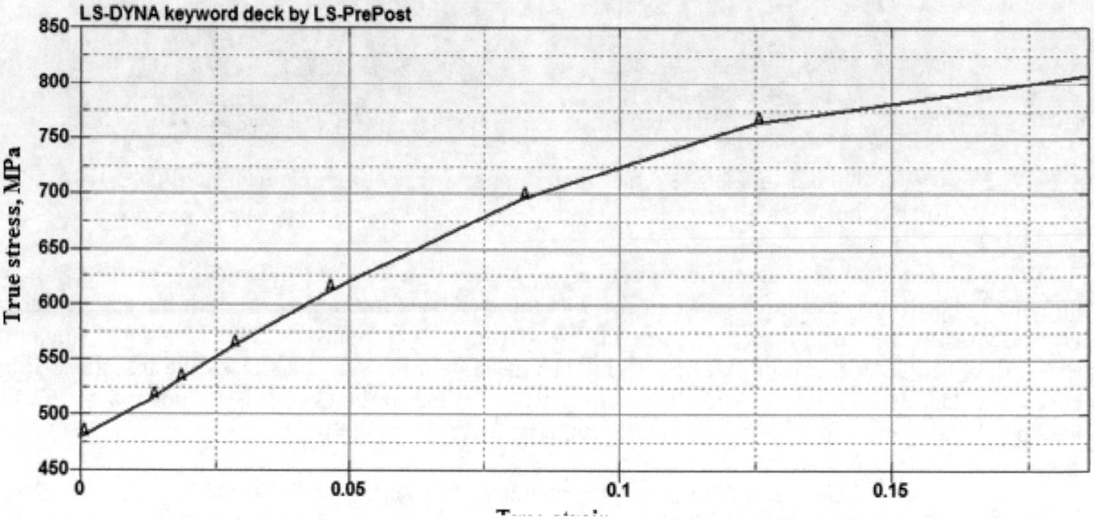

Figure 9. Stress-strain curve of the post material used in digital model.

4. Results

The loading element and reinforced concrete slab in the virtual model of bench shock tests presented as absolutely rigid [7-9]. The connection between the reinforced concrete slab and the anchor studs carried out by means of an indestructible adhesive contact [10, 11]. During dynamic loading (figure 10), the impact energy was 29.43 kJ. The values of the resulting forces in contact between the anchor studs and the reinforced concrete slab shown in figure 11.

(a)

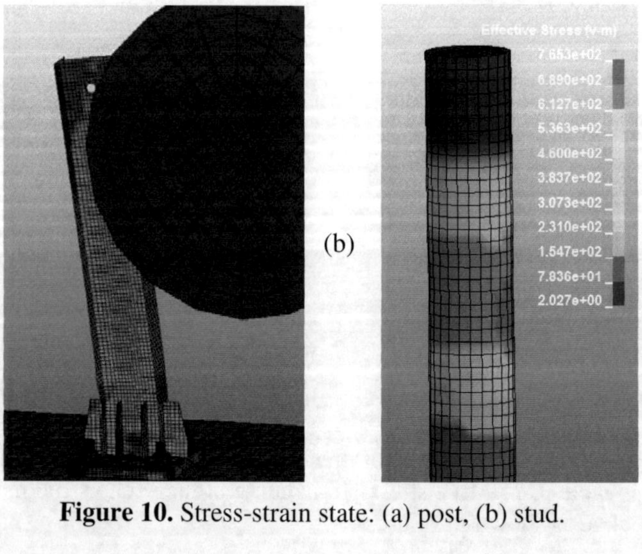

Figure 10. Stress-strain state: (a) post, (b) stud.

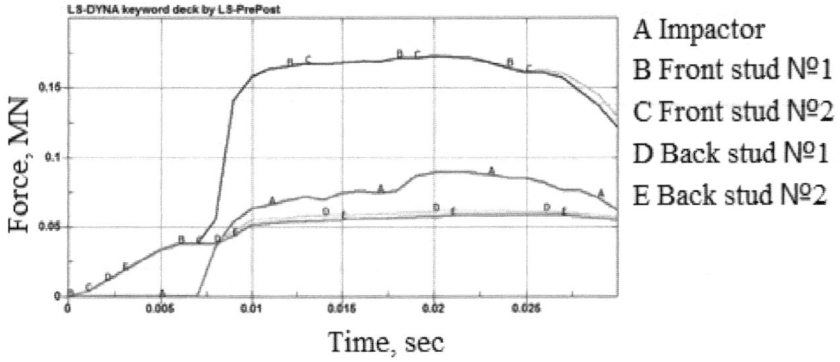

Figure 11. Resultant forces in contact between the anchor studs and the reinforced concrete slab.

5. Conclusions

Based on the results of virtual impact tests, it concluded (figure 11) that the maximum force that occurs in contact between the anchor stud and the reinforced concrete slab is 174 kN, which exceeds the amount of force required to pull out the anchor stud-170 with an installation depth of 170 mm using the BIT-NORD adhesive composition (table 1). This fact explains the reason for breaking out the anchor stud during bench impact tests and the destruction of the adhesive joint. It is worth noting that the obtained maximum force that occurs in contact between the anchor pin and the reinforced concrete slab, according to the results of virtual impact tests, does not exceed the amount of force required to pull out the anchor stud-210 with a depth of 210 mm and using BIT-NORD adhesive composition (table 1) conducting virtual experiments based on the use of digital twins [12] is a modern tool for solving research, development and engineering problems with a significant reduction in both time and material costs.

Acknowledgment

Authors are grateful their scientific advisor, Ph.D., prof. I.V. Demiyanushko for guidance, comments and advice in carrying out this research work.

Financial support

The work carried out with funding under the Development - Digital Technologies Program with the support of the Fund for the Promotion of the Development of Small Forms of Enterprises in the Scientific and Technical Sphere (Innovation Assistance Fund) (№ 47GRCTC10-D5/56182).

References

[1] Standart STO 96043391-001-2018 Chemical anchorages for road barriers of the barrier type of the bridge group. Technical conditions LLC Bit-united

[2] Protocol FSUE NAMI No. 1928/C0/W/W/24-20 bench impact test to determine the strength of anchor stud-170 on the BIT-NORD adhesive composition for fixing the rack of the road fence of the bridge group

[3] Scientific and technical report of MIP research Institute of Mechanics and quality problems on the topic: Research of chemical anchor systems BIT for road barrier fences (under the contract no. M2019-17 dated 26.06.19 between LLC Bit-united and LLC MIPK)

[4] LS-DYNA® 2016 Keyword User's Manual. Livermore Software. Technology Corporation (Livermore, CA)

[5] Demiyanushko I V, Nadezhdin V S, Karpov I A, Tavshavadze B T and Titov O V 2019 *2nd Int. Conf. on Computational & Experimental Methods in Mechanical* (Greater Noida: India IOP Conference Series) vol 691

[6] John O Hallquist 2006 *LS-DYNA theory manual* Available at: https://www.dynasupport.com
[7] Burzyński S, Chróscielewski J and Pachocki L 2018 Finite element method simulations of various cases of crash tests with N2/W4/A steel road barrier *MATEC Web of Conferences* (Gdansk: Poland) vol 231
[8] Demiyanushko I, Karpov I and Tavshavadze B 2017 *Mathematical and Numerical Aspects of Dynamical System Analysis Int. Conf. «Dynamical systems theory and application»* (Lodz) pp 165-74
[9] Demiyanushko I V and Karpov I A 2017 *J. world roads* **102** 49-5
[10] Boumakis I, Marcon M, Ninčević K, Czernuschka L M and Wan-Wendner R 2018 *Engineering StructuresVolume* **175** 790-805
[11] Sasmal S, Thiyagarajan R, Lieberum K H and Koenders E A B 2018 *Computers and Concrete* **22** 395-405
[12] Wright L and Davidson S 2020 *Adv. Model. and Simul. in Eng. Sci.* **7** 13

Features of accounting machining process influence on fiber reinforced composites characteristics

D D Vlasov[1], **N A Tatus'**[1], **A N Polilov**[1] **and E Kh Akhmedshin**[2]

[1] Mechanical Engineering Research Institute of the Russian Academy of Sciences, 4 Maly Kharitonyevsky Pereulok, Moscow, 101990, Russia
[2] Moscow Polytech, Bolshaya Semenovskaya str, 38, Moscow, 107023, Russia

davlasov98@gmail.com

Abstract. The results of the experimental study of machining effect on the fiberglass specimens' tensile strength realization are presented. According to the data obtained, the size of the damaged zone, which does not carry applied loads and reduces the actual net cross-section of the specimen, has been estimated.

1. Damaged zone thickness estimation method for machined FRP composite specimens

Fibrous composite materials have a heterogeneous structure where strong and brittle fibers, which are easily damaged during machining, play a major bearing role. The damage may not be visible, but if for metals machining process leads to local hardening and microplastic deformations that have little effect on strength, then for glass and carbon plastics machining damage can have a significant negative effect [1]. Due to this fact machining on the thickness of composite plates is not allowed, and the technology and cutting modes of linear specimens have been carefully worked out.

The need to cut composite materials along curved trajectories required the acquisition and adjustment of a modern, imported CNC 6040Z milling machine, which allows you to direct the cutting tool along any trajectory, and not only straight.

1.1 Setting the task

When working out cutting modes, several questions arise: How does cutting technology affect the damage of material near the cutting line? How can we estimate the thickness of damaged layer of material that actually goes out of bearing part of specimen?

To answer these questions, at the first stage, it was decided to test strip specimens (figure 1.1) with different widths cut out using a single technology for stretching. This made it possible to estimate the width of the damaged zone - the "edge," which is formally the part of the cross section of the specimen but does not bear any load. It is considered that in a strip specimen cut from the plate with a width of b and a length of L the width of damaged zone δ (figure 1.1) remains constant.

It should be noted that the width of the "edge" must be constant for a given size and type of cutting tool, as well as for fixed cutting modes (rotation and feed speed). Therefore, with an increase of the size of the structure (specimen width), the relative width of the damaged edge and its effect on strength should be reduced.

At the same time, changing the diameter of the cutting tool leads to change of size of the damaged zone. A larger diameter cutter will damage the specimen more heavily, while the effect of smaller diameter cutters should not greatly affect the resulting strength values. A great result would be to get a clear dependence of size of the damaged zone on the size of the tool used which is relevant for all types of cutters.

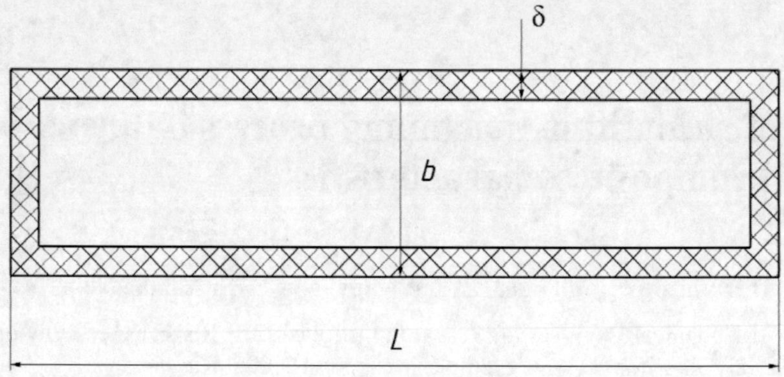

Figure 1.1. The damaged zone of specimen after machining process.

To estimate the width of the damaged "edge" the research has been carried out at the Mechanical Engineering Research Institute of the Russian Academy of Sciences. Several series of specimens of various widths have been produced and tested with different size of tool and constant cutting modes.

1.2 Specimen fabrication and testing

Since the tests have been comparative and finding exact values of material characteristics has not been the main goal, a separate test procedure has been developed that is different from used standards. The fiberglass plate with the thickness of h = 2 mm has been made with the vacuum infusion technology. The plate has been molded from 6 layers of glass fabric with a weave type of "web" and [0/90] lay-up.

Specimens with the length of 250 mm with different widths have been cut from a plate on the CNC milling machine using a corn cutter tools (figure 1.2) with a diameter of 1 mm and 1.5 mm. Horizontal feed rate: 250 mm·min^{-1}, inverter current frequency: 160 Hz. The cutting mode for all specimens has not changed.

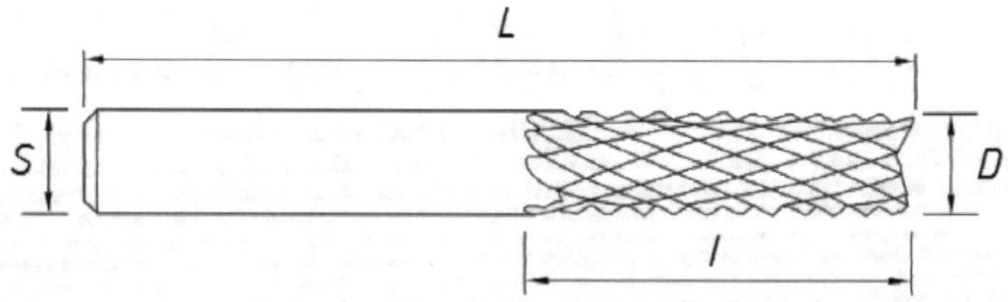

Figure 1.2. Corn type cutter tool.

For the test implementation with 1.5 mm cutter tool 7 series of three specimens with different widths have been made: 2.5; 3; 4; 5; 7.5; 10 and 30 mm. Series number i = 1, 2,..., 7. And with 1 mm cutter tool 5 series of three specimens have been made: 2.5; 5; 7.5; 10 and 40 mm. Series number j = 1, 2,..., 5.

The specimens have been tensile tested on the universal electromechanical test machine INSTRON 6025 in self-tightening grips. All specimens have collapsed in the working area, which indicates the correctness of the test results.

1.3 Test results

According to the obtained experimental data, the dependence of strength of the specimen on its width has been built. Figure 1.3 shows that for 1.5 mm tool when the width of the specimen increases to 10 mm, the conditional critical stresses increase sharply, and then they go to the "plateau" with a value of about 400 MPa. This behavior of the diagram shows a strong effect of the damaged zone on the conditional strength of small-width specimens and a much smaller effect for wider ones. A similar situation can be traced for a mill with the diameter of 1 mm. Since the diameter of the cutter has decreased, the damaged zone practically does not affect the strength of the specimens at a width of 7.5 mm, while for wider specimens the strength also reaches about 400 MPa, which indicates a very little influence of the tool.

To estimate the width of the damaged zone, we use simple formulas for conditional and real ("true") strength.

Conditional strength (critical load to full cross section):

$$\sigma_{i,j}^{eff} = \frac{F_{i,j}}{b_{i,j} \cdot h} \tag{1}$$

Real strength from test (load to unknown net cross section):

$$\sigma_{i,j}^{real} = \frac{F_{i,j}}{b'_{i,j} \cdot h} \tag{2}$$

where $b'_{i,j} = b_{i,j} - 2\delta -$ is the effective (actually resisting applied force) width.

For approximate calculations, the maximum value of the conditional strength of this material can be used as the real ("true") strength. On the basis of the diagram (figure 1.3), the "true" strength has been taken as the value of the conditional strength obtained on specimens with the width of 30 mm since at this width the damaged zone no longer has a significant impact on the implementation of the strength.

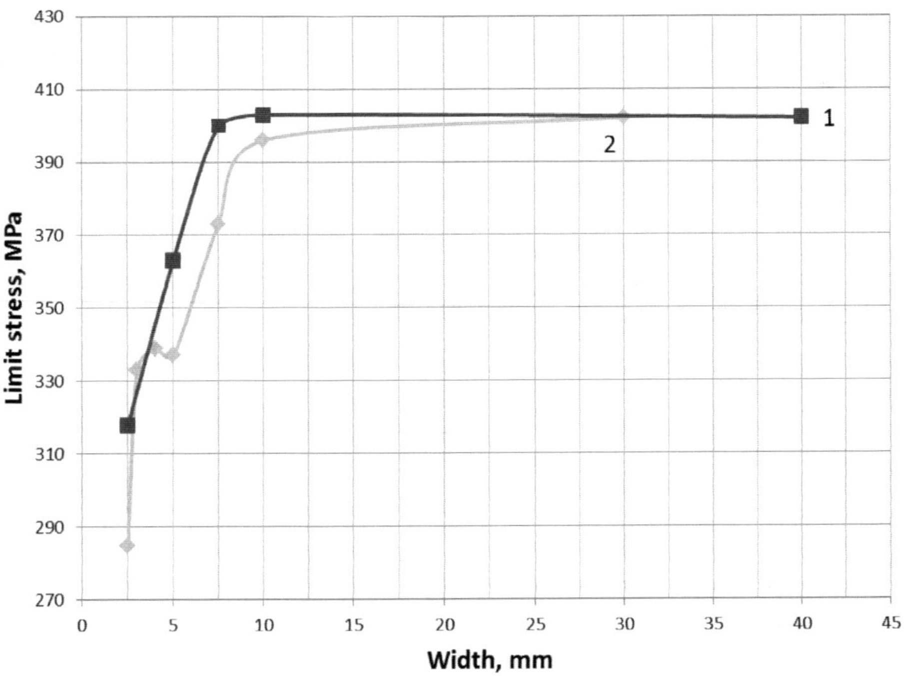

Figure 1.3. Conditional strength on specimen width dependence. 1 – 1 mm cutter tool; 2 – 1.5 mm cutter tool.

Changing $\sigma_{i,j}^{real}$ to $\sigma^* = \sigma_7^{eff} = \sigma_5^{eff} = 402MPa$ in (2) (indexes «7» and «5» mean the width of specimens b_7 = 30 mm and b_5 = 40 mm) we estimate from the experimental values of the critical load $F_{i,j}$ the width of the damage zone δ, which turns out to be slightly different for specimens of different widths:

$$\sigma^* = \frac{F_{i,j}}{b'_{i,j} \cdot h} \tag{3}$$

$$\delta = \frac{1}{2}\left(b_{i,j} - \frac{F_{i,j}}{\sigma^* \cdot h} \right) \tag{4}$$

The results of the evaluation according to the formula (4) of the size of the "edge" for specimens of different widths (i = 1…5 and j=1…3), taking into account the spread, are given in figure 1.4. The initial assumption that the width of the damaged zone should not depend on the width of the specimen was not confirmed, instead, the results showed a linear dependence. The main reason for this can be the effect on the strength of the specimen of a much larger number of factors in addition to the type and size of the tool, as well as machining modes. In this regard, it is difficult to talk about an accurate determination of the size of the damaged zone. However, the average value has been used for practical evaluation of the effect of machining on the strength of the material in both series. As shown in figure 1.4, for a 1.5 mm diameter cutter, the delta is 0.24 mm and for a 1 mm cutter 0.17 mm, which in both cases is about a sixth of the tool diameter.

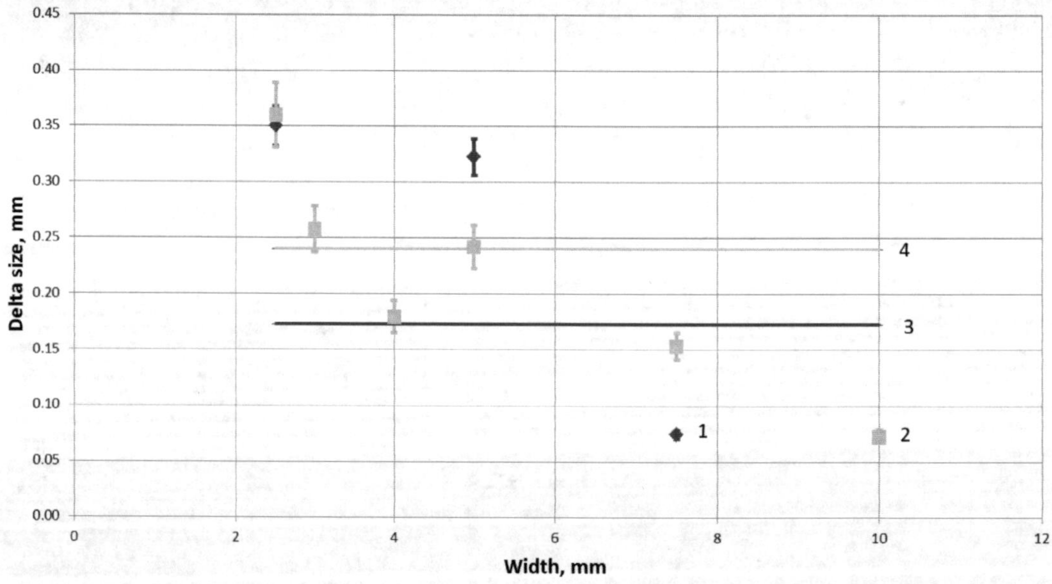

Figure 1.4. Damaged zone size estimation. 1 – 1 mm cutter tool; 2 – 1.5 mm cutter tool; 3 – average for series 1; 4 – average for series 2.

1.4 Results of chapter 1

1. The showed method can become the basis for estimating the effective width of the damaged zone when cutting composite parts. For the given data, $\delta_{1.5} \approx 0.24$ mm and $\delta_1 \approx 0.17$ mm, which is about a sixth part of the diameter of the cutting tool.
2. Strictly speaking, obtained values correspond only to the tool sizes and type used in the experiments at the selected horizontal feed rate and rotation speed. When changing the tool type and cutting modes, the effective width of the damage zone may be different.
3. The nominal size of the damaged zone is small and can be ignored for big parts, but making small parts from fibrous composites, the effect of damaged zone should be taken into account.

2. The scale effect of strength in composite specimens with hole

Strength properties of structural elements made of polymer fibrous composites are closely associated with the scale effect. It is well known that multiple scaling of a structural element or a specimen leads to reduction in limit stresses. Because of this feature of composite materials scale modeling and the creation of small-sized models of composite structures for testing are very difficult.

The article is devoted to the analysis of dependence of the size of specimen on the scale effect rate. The presence of stress concentrators in composite parts, such as holes, increases the scale effect - the larger the specimen with the hole, the more bearing fibers are turned off in cross section. It is therefore obvious that in anisotropic fibrous composites the constant ratio d/b (figure 2.1) will not guarantee the same strength in specimens of different size precisely due to the scale effect. In order to verify this assumption, the experimental part of the work has been built. The research has been carried out at the Mechanical Engineering Research Institute of the Russian Academy of Sciences.

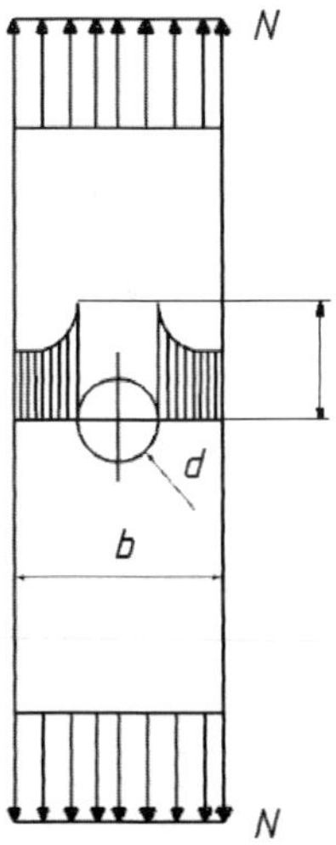

Figure 2.1. Composite specimen with hole during the tensile test.

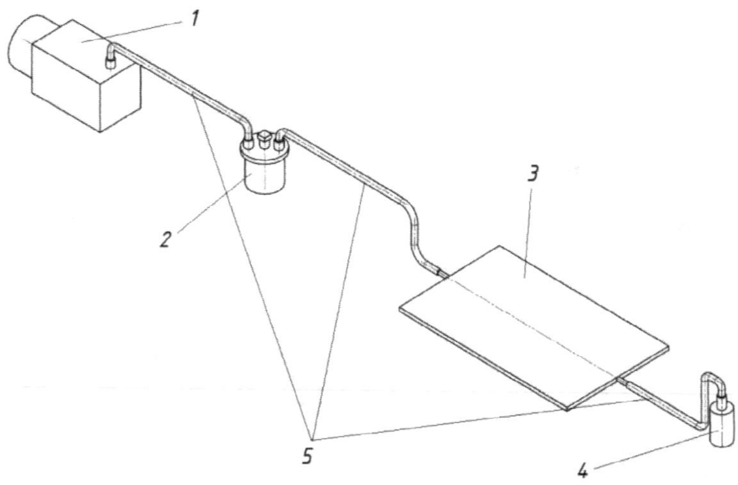

Figure 2.2. Vacuum infusion principle diagram. 1 - vacuum pump, 2 - resin trap, 3 - composite bag, 4 - degassed binder, 5 - vacuum pipeline.

2.1 Specimen fabrication and testing

The specimens with length of 250 mm and different widths have been produced for tests (table 2.1). The ratio d/b =0.1 has been constant for all dimensions. Holes have been drilled.

Table 2.1. Specimen size and hole diameter.

Specimen width (mm)	Hole diameter (mm)
5	0.5
10	1
15	1.5
20	2
25	2.5
30	3

Strip specimens have been cut from a 2.5 mm thick fiberglass plate produced by vacuum infusion method (figure 2.2). The plate consisted of 8 layers of fabric with [0/90] lay-up.

Strip specimens with and without holes have been tensile tested on the universal electromechanical test machine INSTRON 6025 machine in self-tightening grips. All specimens failed in the working area so the test results are correct. The results obtained for the hole specimens have been compared with the results obtained on the reference specimens without holes.

2.2 Processing of test results

As it was expected, specimens with holes showed lower net stress limits than solid specimens (figure 2.3). Moreover, with the increase of the size (width) of specimen, the hole effect is more noticeable, which means the smallest limit stresses has been shown by the widest specimens with big holes.

The strength reduction factor based on the obtained data has been calculated for each width value:

$$K_\sigma = \frac{\sigma_0^*}{\sigma_n^*} \tag{5}$$

where σ_0^*, σ_n^* – stress limits of solid specimen and similar specimen with hole. After that, the dependence of the coefficient on the width of the specimen has been built (figure 2.4).

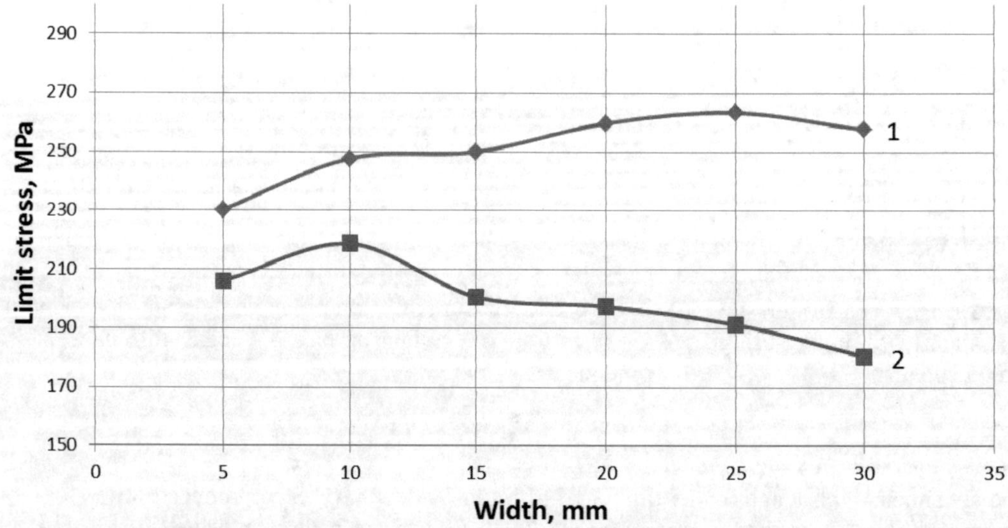

Figure 2.3. Specimen stress limit on width dependence. 1 – specimens without hole; 2 – specimens with hole

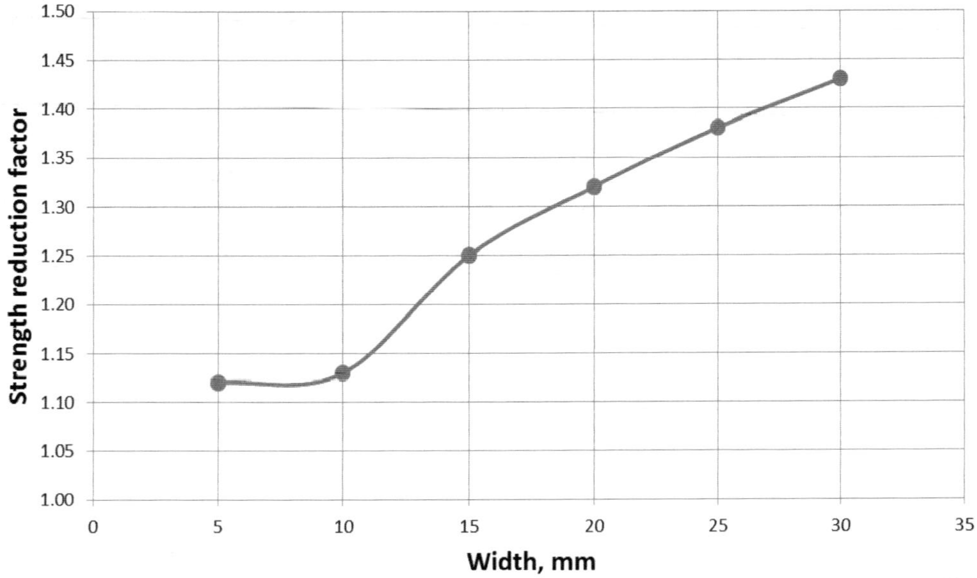

Figure 2.4. Strength reduction factor on width dependence.

It is also interesting to note that limit stresses increase with increasing width of solid specimens. Most likely, this is due to the effect of damage from the tool during the machining process of specimens, which was described in chapter 1. At the same time, the results and their graphical representation are slightly different from those shown in figures 1.3 and 1.4. As noted in chapter 1, the size of the damaged zone and therefore its effect on the strength of the specimen are individual for each type of tool and processing modes. In this experiment the specimens have been cut on a vertical milling machine using a disk cutter made of fast-cutting steel, but the trend of strength increasing with the increase of specimen width is still relevant.

2.3 Results of chapter 2

1. Definitely the stress concentration factor for composite specimens with holes with the same ratio of hole diameter to width is not a constant so traditional equations for isotropic materials are not suitable.

2. As the diameter of the hole and the width of the specimen increase, the strength reduction factor (effective stress concentration factor) increases. The larger specimens are most affected by the hole.

3. Analyzing the slope of the strength reduction curve (figure 2.4) for wide specimens the authors should say that there should be a plateau which the strength of specimens with holes is approaching. The test results suggest that the maximum reduction in strength for wide specimens of orthogonal fiber-reinforced composites at the ratio of d/b=0.1 would be about 50%.

3. Conclusion

This paper presents data from the experimental study of the effect of various types of machining on the strength of parts (specimens) made of fibrous polymer composite materials: the results clearly describe the phenomenon of strength reduction in composite specimens with drilled holes and in milled specimens. Thus, the strength of the composite parts, in addition to the accuracy of the process of creating the material, is also affected by the machining process, which is undesirable, however, often mandatory when moulding products.

References

[1] Savitskiy R S and Veshkin E A 2017 Effect of mechanical processing of specimens at cutting on composites testing *Izvestia of Samara Scientific Center of the Russian Academy of Sciences* pp 214-19
[2] Feodosyev V I 2018 Theory of strength of materials *Publishing House Bauman Moscow State Technical University* p 543
[3] Polilov A N 2014 Mechanisms of stress concentration reduction in fiber composites *J. of Applied Mechanics and Technical Physics* pp 154-63
[4] Polilov A N and Tatus N A 2018 *Biomechanics of the strength of fiber composites* (Moscow: FIZMATLIT) p 328

Influence of running-in in dry and wet conditions on the interfacial adhesion of tire tread rubber

A V Morozov[1], Yu Yu Makhovskaya[1] and K S Kravchuk[2]

[1] Ishlinsky Institute for Problems in Mechanics of the Russian Academy of Sciences, Moscow 119526, Russia

[2] Technological Institute for Superhard and Novel Carbon Materials, Troisk 142190, Russia

morozovalexei@mail.ru

Abstract. To study the running-in effects on the adhesion properties of tire tread compounds, an approach is proposed based on theoretical analysis of experimental load-distance curves. The approach uses the Maugis model of adhesion in an axisymmetric elastic contact. Classical AFM tests with a ball of 450 nm glued to the probe are performed for two different surfaces of the tire tread rubber after running-in in dry and wet friction regimes and for the original rubber surface. It is shown that running-in in wet conditions decreases the surface energy of rubber. Also, running-in in both wet and dry conditions leads to softening of a surface layer of rubber.

1. Introduction

It is known that the friction of elastomers has two basic mechanisms – adhesion and hysteresis – and thus depends on the adhesion and mechanical properties of elastomers [1, 2]. But the process of friction, in turn, influences the properties of elastomer surfaces, depending on many factors, such as type of materials, regime of friction, surface relief. In this study, a highly elastic tribotechnical material – tire tread rubber – is tested in sliding contact with a regular rigid surface having macro- and microtexture in the conditions of dry and wet friction. Under the influence of friction, wear, and liquid medium (soap solution in distilled water) during running-in, the surface of rubber changes its adhesion properties. To evaluate these changes, we propose an approach based on theoretical analysis of experimental load-distance dependencies obtained by AFM before and after running-in.

Usually, to extract the adhesion properties of materials from load-distance curves, the classical models of Johnson-Kendall-Roberts (JKR) and Derjaguin-Muller-Toporov (DMT) are used. These models give the specific work of adhesion w by the following simple formulas, which use only one value obtained from the load-distance curve – the pull-off force F_1 [3]:

$$w_{JKR} = 2F_1 \cdot (3\pi R)^{-1}, \quad w_{DMT} = F_1 \cdot (2\pi R)^{-1} \tag{1}$$

where R is the radius of curvature of the probe tip. But these simplified models cannot be used in the entire range of parameters. In the JKR model, the radius of adhesion force action is assumed to be infinitely small. The DMT model does not take into account the influence of adhesion on deformation of the interacting surfaces. Later, more exact models of adhesion were developed, both numerical and analytical, which can be used in wider ranges of parameters [4-6]. In particular, the Maugis model suggested for the description of the sphere-plane elastic adhesive contact [4] and generalized for axisymmetric bodies of power-law shape [5] takes into account both finite radius of adhesion force action and mutual influence of adhesion and contact deformation. This model is used as a basis for the approach suggested in the present study for the calculation of the adhesion properties of elastomers.

As a measure of adhesion of rubber, we use the specific work of adhesion at the silicon-rubber interface. This quantity is defined by the surface energy of rubber which plays the key role in friction of elastomers against smooth rigid surfaces.

2. Theoretical model

Consider the interaction between the AFM probe tip, which is assumed rigid and whose shape is described by the parabolic function $f(r)=r^2\cdot(2R)^{-1}$, where R is the radius of tip curvature, and a rubber specimen. The probe is acted on by the external load F, the distance between the surfaces d being registered (figure 1 (a)).

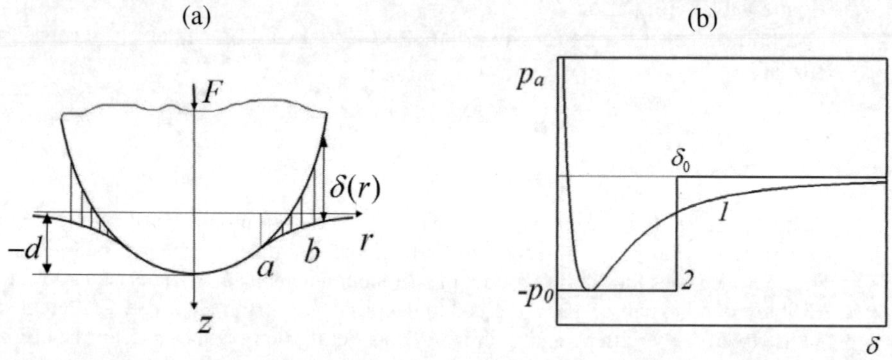

Figure 1. Illustrations for the theoretical model – scheme of contact between the probe tip and a specimen (a) and the adhesion pressure as a function of gap between the surfaces (b).

Let (r,φ,z) be the cylindrical system of coordinates whose origin lies on the surface of the specimen and the z - axis coincides with the axis of symmetry of the probe tip. The surfaces are in contact over the region $r\le a$. In the region $a<r\le b$, the adhesion attraction occurs between the surfaces defined by the adhesion pressure p_a as a function of the gap δ. This function follows from the potential of intermolecular interaction. For Lennard-Jones potential, it is given by curve 1 in figure 1 (b). In the present study, we use the Maugis model in which this function is approximated by a piecewise-constant function (curve 2 in figure 1 (b)) and is expressed by the relation [4]:

$$p_a(\delta)=\begin{cases}-p_0, & 0<\delta\le\delta_0\\ 0, & \delta>\delta_0\end{cases}, \tag{2}$$

where the characteristic radius of adhesion force action δ_0 and adhesion pressure p_0 are considered the adhesion properties of the interface. The specific work of adhesion w is the work done by the adhesion force when two surfaces are moved apart to infinity. In the case of model (2), the specific work of adhesion is given by the relation

$$w=\int_0^{+\infty} p_a(\delta)\,d\delta=p_0\delta_0 \tag{3}$$

Mechanical properties of the rubber specimen are described by the model of an elastic half-space under axisymmetric loading, for which the normal displacement of the boundary $u(r)$ is related to the normal pressure $p(r)$ by the expression [7]

$$u(r) = \frac{4}{\pi E^*}\int_0^b p(r')\mathbf{K}\left(\frac{2\sqrt{rr'}}{r+r'}\right)\frac{r'dr'}{r+r'}, \qquad 0 \le r \le b \tag{4}$$

where $\mathbf{K}(x)$ is the complete elliptic integral of the first kind, $E^* = E\cdot(1-\nu^2)^{-1}$ is the reduced modulus of elasticity of the rubber specimen. The conditions at the boundary $z=0$ of the elastic specimen follow from the contact conditions and relation (2):

$$\begin{aligned} &u(r) = -r^2\cdot(2R)^{-1} - d, && 0 < r < a \\ &p(r) = -p_0, && a \le r \le b \\ &p(r) = 0, && r > b \end{aligned} \tag{5}$$

Also, the condition following from the adopted model of adhesion (2) is satisfied:

$$\delta(b) = \delta_0 \tag{6}$$

where the gap between the surfaces δ is defined by the expression (see figure 1 (a)):

$$\delta(r) = r^2\cdot(2R)^{-1} + u(r) + d \tag{7}$$

Besides, the condition of equilibrium for the forces acting on the probe tip is satisfied:

$$F = 2\pi\int_0^b rp(r)dr \tag{8}$$

The solution of the adhesion contact problem (4)-(8) problem was constructed in [5] for the case of the power-law function of shape, $f(r) = Ar^{2n}$. By setting $n=1, A=(2R)^{-1}$, we get the following relations for the load applied to the probe

$$F = \frac{4E^*a^3}{3R} - \frac{2p_0a^2}{\gamma^2}\left(\arccos\gamma + \gamma\sqrt{1-\gamma^2}\right) \tag{9}$$

distance between the probe tip and the specimen

$$d = -\frac{a^2}{R} + \frac{2p_0a}{E^*\gamma}\sqrt{1-\gamma^2} \tag{10}$$

and the equation for the determination of the contact radius a:

$$\left[\frac{a}{R}\left(\frac{1}{\gamma^2}-2\right) + \frac{4p_0}{E^*\gamma}\sqrt{1-\gamma^2}\right]\arccos\gamma + \frac{a}{R\gamma}\sqrt{1-\gamma^2} - \frac{4p_0}{E^*}\left(\frac{1}{\gamma}-1\right) - \frac{\pi\delta_0}{a} = 0 \tag{11}$$

Relations (9)-(11) specify the dependence of the load F on the distance d via the parameter $\gamma = a\cdot b^{-1}$. These relations are applicable in the entire range of variation of the Tabor parameter [8]

$$\mu_T = \left(\frac{Rw^2}{E^{*2}\delta_0^3}\right)^{1/3} \tag{12}$$

Unlike the classical simplified models which are its limit cases for $\mu_T >> 1$ (JKR limit) and $\mu_T << 1$ (DMT limit) [4, 8]. The disadvantage of relations (9)-(11) is their complex parametric form which does not allow one to obtain simple computational relations similar to (1).

In this work, we propose a method based on using two values obtained from the load-distance curves – the pull-off force F_1 and difference $d_2 - d_1$ between the distance at which the force is zero and the pull-off distance (figure 2). These values are determined from the retraction AFM curves.

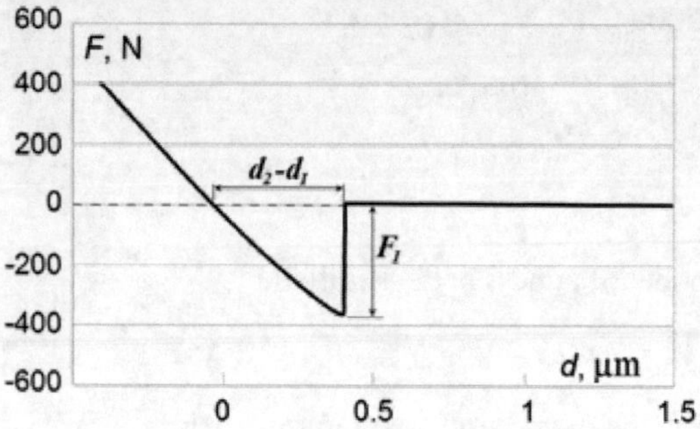

Figure 2. A retraction segment of the load-distance curve with denoted values used for the calculation.

We solve equation (11) for a, choose the positive root of two, and substitute it into equations (9) and (10). As a result, the load and distance are expressed as functions of three unknown parameters - $F = F(\gamma, p_0, \delta_0)$ and $d = d(\gamma, p_0, \delta_0)$. These functions are applied for two points of the load-distance curve – the pull-off point and the point of zero load. At the pull-off point, the load attains its extremum value F_1, i.e., the following relations are satisfied:

$$\partial F(\gamma, p_0, \delta_0) / \partial \gamma \big|_{\gamma=\gamma_1} = 0 \tag{13}$$

$$F(\gamma_1, p_0, \delta_0) = F_1 \tag{14}$$

where γ_1 is the unknown ratio of the radii of the contact and adhesion regions at the pull-off point. At the point of zero load, we have

$$F(\gamma_2, p_0, \delta_0) = 0 \tag{15}$$

where γ_2 is the unknown ratio of the radii of the contact and adhesion regions at the point of zero load. Also, the following equation for the distance difference $d_2 - d_1$ is satisfied:

$$d(\gamma_2, p_0, \delta_0) - d(\gamma_1, p_0, \delta_0) = d_2 - d_1 \tag{16}$$

The system of equations (13)-(16) is to be solved numerically by the iteration method for unknowns $\gamma_1, \gamma_2, p_0, \delta_0$, provided that the values R and E^* are known. After the adhesion properties p_0 and δ_0 are determined, the specific work of adhesion w is calculated as their product in accordance with (3).

Unlike formulas (1) of the classical models, the method suggested takes into account the influence of the modulus of elasticity on the pull-off force, and it does not assume any limitations on the value of the Tabor parameter μ_T (12).

3. Materials and equipment

As a material for specimens, we choose a tire tread rubber of car winter tire, which is produced from a compound based on mix of natural rubber (NR) and cis-butadiene rubber (CBR). Three groups of rubber specimens are tested. The first group (N1) of rubber specimens are tested in the conditions of wet friction in 20% solution of liquid soap in distilled water. The second group (N2) of specimens are subjected to dry friction tests in the conditions similar to those of the first group. The third group (N3) of specimens is the original rubber after vulcanization. Rubber specimens are produced as rings of 7 mm height with internal radius 41 mm and external one 55 mm. The rings are glued to a steel substrate. As a counterbody for friction tests, a laminated plywood is used. The photograph of the counterbody, its surface microroughness, and geometric characteristics of texture are presented in figure 3.

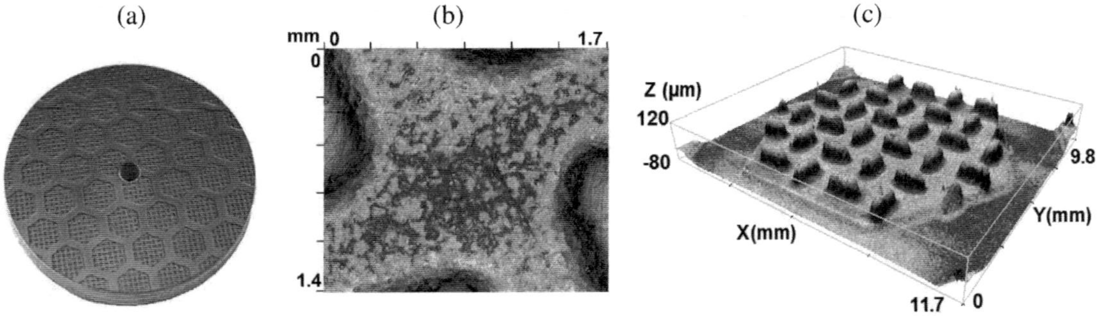

Figure 3. Photograph of the counterbody (a), its surface roughness with the parameters S_a = 1.4 μm, S_z = 6.2 μm (b), and 3D texture of a typical area of the plywood surface (c).

3.1. AFM technique

Experimental curves of approach and retraction were obtained by the atomic force microscope Integra Prima (NT-MDT, Russia). The photographs of the microscope and a rubber specimen under investigation are presented in figure 4.

We use the commercially available silicon cantilevers FMG01_Bio with the stiffness 1.5 N·m^{-1}. At the end of the probe (cantilever), a spherical ball of silicon oxide with 900 nm diameter is attached. The force sensor calibration was performed by a series of shots into the polished amorphous sapphire whose stiffness is orders of magnitude higher that the stiffness of the rubber surface under investigation. It is assumed that when the sapphire is indented at large depth, the stiffness of the cantilever coincides with the stiffness of the sapphire. The depth of indentation of the ball into a rubber specimen is 2000 nm, and it varies slightly because of surface roughness of the rubber, particularly that after tribological tests.

Figure 4. Photographs of the AFM (a) and its probe over a specimen being tested (b).

Figure 5. Photograph of a rubber sample 1, which is fixed in the tribometer holder, where 2 is the counterbody.

3.2. Tribological tests technique

Tests were carried out on the UMT-2 Tribometer (CETR Inc., USA) with ring-disk scheme of contact. A friction pair is shown in figure 5. A ring-shaped rubber specimen 1 is fixed in the self-adjustment holder of the tribometer. The disk-shaped counterbody 2 is fixed axisymmetrically on the stage rotated by the stepper motor. For tests in liquid medium, a lubricant container is set upon the stage, inside which the counterbody is fixed by a three-pin system. The working principle of the tribometer and the experimental technique are given in detail in [9]. Tests were performed under specific loads $P = 0.1\ldots0.5$ MPa and sliding velocities $V = 0.01\ldots100$ mm·s^{-1} at room temperature $T = 23 \pm 3$ °C.

4. Results and discussion

Results of the tests are presented in figure 6 as plots of the coefficient of friction μ vs. sliding velocity V at various nominal pressures ranging from 0.1 to 0.5 MPa for two friction regimes. Analysis of the obtained experimental data shows that the curves have a peak, more noticeable for the dry friction regime (figure 6 (a)), which is accounted for by the adhesion mechanism of friction [10].

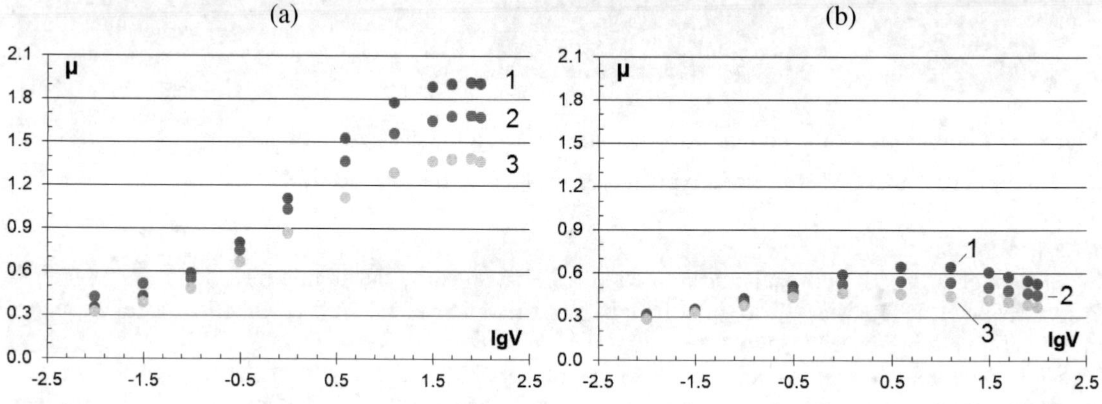

Figure 6. Coefficient of friction μ vs. sliding velocity V for dry (a) and wet (b) regimes of friction, where 1 is for $P = 0.1$ MPa, 2 is for $P = 0.2$ MPa, and 3 is for $P = 0.5$ MPa.

In the considered range of loads, the coefficient of friction decreases as the normal load increases. This is accounted for by the adhesion properties of rubber surface. It was shown in that in the presence of adhesion in dry contact, the coefficient of friction decreases as the load increases [11, 12]. In wet contact, the contribution of adhesion into friction is reduced, therefore the effect of the load on the coefficient of friction is less significant. Still, peaks associated with adhesion are seen in figure 6 (b), which is due to the artificial texture applied on the surface of the plywood. The texture makes the lubricant flow out from the interface, and thus the adhesion mechanism works even in wet conditions.

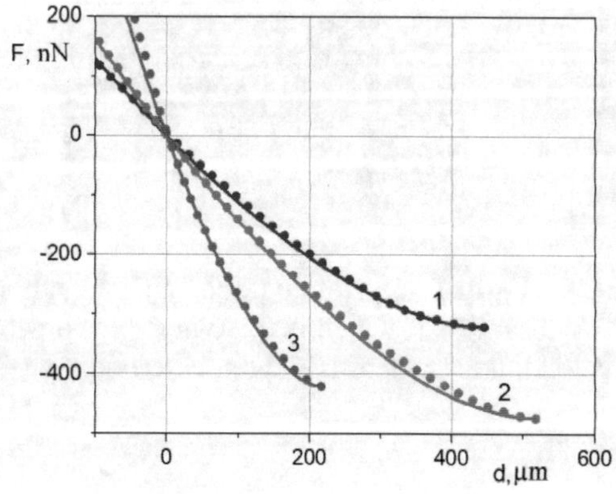

Figure 7. Experimental curves of retraction of the probe from the specimens surface (dots) and their approximation by calculation (solid lines), where curve 1 corresponds to the surface after wet friction, curve 2 to that after dry friction, and curve 3 to the original surface after vulcanization.

AFM tests of the rubber surfaces were carried out before and after tribological tests in wet and dry regimes. Each specimen was subjected to not less than 10 tests on different spots located at 50 μm from each other. In figure 7, typical load-distance retraction curves are shown for three tested surfaces of rubber specimens in the range of the probe displacements from $d=-100$ to the value of d at which the maximum tensile force F is registered. From these curves, the values of F_1 and d_2-d_1 were determined to be used as input data for solving the system of equations (13)-(16) to calculate the adhesion properties p_0 and δ_0.As one more input parameter, we first used the Young modulus of rubber $E=6.9$ MPa obtained by the classical test on uniaxial compression. But the results lead to high discrepancy between the calculated load-distance curves and experimental data for negative loads, which could not be reduced by any selection of the adhesion properties p_0 and δ_0 but it was eliminated well by choosing the modulus of elasticity *E*. This is supposedly explained by nonlinear elastic properties of rubber. However, our calculations show that in the range of small positive (≤ 100 nm) and negative depths of indentation of a ball with radius $R=450$ nm, the rubber properties can be approximated well by linear elasticity with its own modulus, which we consider particularly justified for comparative tests. Thus, we calculate the modulus of elasticity by approximating experimental data with the parametric function (9)-(11) by the least square method at each iteration of the solution of system (13)-(16). In doing so, the elastic (E) and adhesion (p_0, δ_0) properties are calculated simultaneously. The calculated curves are presented in figure 7. Note that simultaneous extracting of the elastic and adhesive characteristics from load-distance curves was first suggested in [13].

Three experimental curves were selected for each rubber specimen, the corresponding values of F_1 and d_2-d_1 presented in table 1 (specimen N1 is after wet friction, N2 is after dry friction, N3 is the original surface). Also, the calculated adhesion pressure p_0, radius of adhesion force action δ_0, and modulus of elasticity E are presented. The specific work of adhesion w averaged for three curves is calculated. For comparison, the specific work of adhesion determined by the JKR and DMT formulas (1) is given, as well as the Tabor parameter calculated by formula (12)

Table 1. Results of calculation of the p_0, δ_0, E and w for the three rubber specimens.

NN	Measured values		Calculated adhesion and elastic properties			Averaged work of adhesion	Work of adhesion by JKR and DMT models		Tabor parameter
	F_1, nN	d_2-d_1, nm	p_0, MPa	δ_0, nm	E, MPa	w, J·m^{-2}	w_{JKR}, J·m^{-2}	w_{DMT}, J·m^{-2}	μ_T
	320.4	445.1	7.768	19.39	1.530				
1	280.6	465.5	6.101	21.64	1.253	0.1399	0.1413	0.1054	8.35
	293.0	439.3	7.562	18.13	1.428				
	463.1	702.3	8.205	25.00	1.117				
2	473.5	517.2	10.020	22.25	1.807	0.1946	0.2081	0.1562	8.54
	415.1	670.3	8.156	23.97	1.368				
	463.1	702.3	19.761	9.98	5.935				
3	473.5	517.2	19.040	9.49	5.575	0.1919	0.1927	0.1445	8.17
	415.1	670.3	20.151	9.82	5.636				

Since the surface energy is directly related to the specific work of adhesion w at the interface, it can be concluded that after dry friction, the surface energy of rubber increases insignificantly, while after wet friction it considerably decreases in comparison with the surface energy of the original surface. These results are in good correlation with the results of the tribological tests which show the decrease in the friction losses when the rubber is tested in 20% solution of soap in water (figure 6). Thus, the coefficient of friction in wet conditions decreases not only due to a decrease in the real contact area but also due to a decrease in the surface energy of the rubber. This effect reduces the contribution of adhesion mechanism, which in most cases dominates the friction of elastomers.

The calculated values of the modulus of elasticity E correspond to the range of small compressive and tensile deformations occurring in a thin surface layer of a specimen, and they differ from its bulk Young modulus $E = 6.9$ MPa. Thus, running-in in both dry and wet regimes leads to a surface layer of rubber becoming effectively softer. The results of comparative tests suggest that it is softening of a surface layer rather than change in roughness or nonlinearity of the material. Also, the effective radius of adhesion force action δ_0 becomes larger than that of the original surface.

The calculated value of the specific work of adhesion w lies between of the values obtained by JKR and DMT models but it is closer to the JKR limit. This result is explained by high values of the Tabor parameter μ_T which correspond to soft materials with relatively high surface energy [8].

5. Conclusion

Series of tribological tests were carried out in the conditions of dry and wet friction. The results show that the dependence of the coefficient of friction on the sliding velocity has a distinguished peak associated with the adhesion mechanism of friction. Introducing the lubricant (liquid soap solution in water) into contact reduces the contribution of adhesion into friction. Increasing of the normal load leads to decreasing of the coefficient of friction, irrespective of the sliding velocity and conditions of lubrication. This behavior is typical for molecular mechanism of friction of elastomers.

To study the adhesion properties of the rubber surface after running-in in different conditions, an approach was proposed to evaluate the adhesion characteristics of materials based on the load-distance curves obtained by atomic force microscopy. The method uses the Maugis model of adhesion in contact of elastic axisymmetric bodies and it is applicable in the entire range of the Tabor parameter. The results obtained by AFM and their theoretical analysis shows that friction in water solution of liquid soap considerably decreases the surface energy of rubber, while friction against the dry surface slightly increases the surface energy of rubber in comparison with that not subjected to friction tests.

Since the suggested approach, unlike the classical JKR and DMT formulas, takes into account the modulus of elasticity as a model parameter, the change of stiffness of a surface layer of rubber due to running-in was also possible to evaluate. It was obtained that as a result of running-in, the rubber surface softens, the presence of lubricant having no significant influence on its level.

The obtained values of the specific work of adhesion for rubber are compared with the results obtained by JKR and DMT models. The approach suggested allows one to obtain not only more exact values of the specific work of adhesion, but also to evaluate the characteristic values of the adhesion pressure and radius of the adhesion force action, which can be used as input parameters in modelling the adhesion component of friction between rubber and a rigid surface at the microlevel [12, 14].

Acknowledgement

The work was carried out under the financial support of the Russian Foundation for Basic Research (grants 18-08-00558 and 19-01-00231).

References

[1] Grosch K A 1963 *Proc. R. Soc. Lond. A.* **274** 21–39
[2] Moor D F 1980 *Wear* **61** 273–82
[3] Bhushan B 1999 *Handbook of Micro/Nanotribology* (Boca Raton etc.: CRC press) p 859
[4] Maugis D 1991 *J. Colloid Interface Sci.* **150** 243-69
[5] Goryacheva I G and Makhovskaya Y Y 2001 *J. Appl Math Mech.* **2** 273-82
[6] Soldatenkov I A 2012 *J Appl Math Mech.* **5** 597-603
[7] Johnson K 1985 *Contact Mechanics* (Cambridge University Press)
[8] Tabor D 1977 *J. Colloids Interface Sci.* **2** 13
[9] Morozov A V and Petrova N N 2016 *J. of Friction and Wear* **2** 124–8

[10] Gent A N and Walter J D 2005 *The Pneumatic Tire* (Washington DC 20590) p 707
[11] Goryacheva I G and Makhovskaya Y Y 2017 *Friction* **3** 339–50
[12] Goryacheva I G and Makhovskaya Y Y 2015 *Mechanics of Solids* **4** 439–50
[13] Borodich F M and Galanov B A 2008 *Proc. R. Soc. Lond. A.* **464** 2759-76
[14] Morozov A V and Makhovskaya Y Y 2019 *Lecture Notes in Mech. Engineering* 1029–37

Analysis of the algorithm for determining the meshing forces in a spline coupling with misalignment

I V Nikolaev and M K Leontiev

Moscow Aviation Institute, Volokolamskoe highway 4, Moscow, 125993, Russia

nikolka831@mail.ru

Abstract. In modern engine buildings spline couplings have an important role. These connections can have a serious impact on the dynamic of the system, especially in misalignment conditions. Therefore, assessment its influence on the system is an important task. In this paper the algorithm for determining meshing forces in spline couplings and the influence of various parameters of connections on the meshing forces are analysed. Also the possibilities of further development this algorithm are considered.

1. Introduction

A misalignment in spline coupling it's a common problem in the rotor of aircraft engine. This process can lead to an increase the loads on the support, noise level and the vibration level [1]. Therefore, the analysis of a spline coupling with a misalignment is an important task for simulating dynamic system. One of the software packages that allows you to take into the consideration the misalignment in a dynamic system is the DYNAMICS R4. It provides the ability to model a system with angular or radial misalignment. The misalignment effects are modeled as external non-linear loads; the element is a link between two section with 12 degree of freedom. However, in this software package one of the important condition for the simulating of dynamic model is the additional modeling of an equivalent stiffness model of the connection. For the simulating a such equivalent stiffness model the theory and algorithm presented in [2] was used. The main task of this work is to assess the applicability of this algorithm for its further modification and use.

2. Model

An algorithm for determining the meshing forces for a spline coupling is considered. Conditions of the model are presented below:

- The spline connection has rectangular tooth;
- The spline is a cantilever beam with an elastic base;
- The distributed force from the torque is represented as a concentrated force acting at the end of the spline;
- The place of contact is considered as a point.

The geometric parameters of the spline are shown in figure 1.

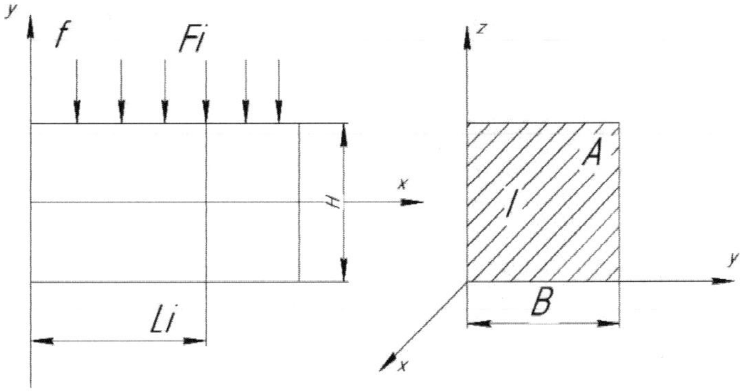

Figure 1. Parameters of a spline.

The total elastic deformation of the spline is found as:

$$\delta_i = \delta_{Bi} + \delta_{Si} + \delta_{Mi} \tag{1}$$

δ_{Bi} is a bending deformation, δ_{Si} is a shear deformation and δ_{Mi} is deformation of elastic foundation, which are defined as follows:

$$\delta_{Bi} = \frac{F_i L_i^3}{3E_e I} \tag{2}$$

$$\delta_{Si} = \frac{12 F_i L_i (1+\nu)}{5E_e A} \tag{3}$$

$$\delta_{Mi} = \frac{F_i}{E_e}\left(\frac{5.306}{BH^2} L_i^2 + \frac{2(1-\nu-2\nu^2)}{(1-\nu^2)BH} L_i + \frac{1.534}{B}\right) \tag{4}$$

where $I = \frac{BH^3}{12}$, $Q = \frac{2(1-\nu-2\nu^2)}{(1-\nu^2)BH}$, $X = \frac{12(1+\nu)}{5A}$, $A = BH$.

The stiffness of a single spline is defined a follow expression:

$$K_i = \frac{E_e}{(\frac{1}{3}I)L_i^3 + (\frac{5.306}{BH^2})L_i^2 + (Q+X)L_i + \frac{1.534}{B}} \tag{5}$$

When the system doesn't have misalignment, the centers of the spline coupling (male and female) are superposed and the distance between the splines in mesh is the same for each pair. In condition of misalignment in the X, Y directions, the eccentricity will be determined as follows:

$$e = ((x + e_0 \cos\varphi_0)^2 + (y + e_0 \sin\varphi_0)^2)^{1/2} \tag{6}$$

$$\cos\varphi = \frac{x + e_0 \cos\varphi_0}{e} \tag{7}$$

$$\sin\varphi = \frac{y + e_0 \sin\varphi_0}{e} \tag{8}$$

For the simplify assume, that misalignment appear only in X direction. Taking the positive direction of the splines clockwise, the angle between each spline and the positive direction of the X axis is defined as follow expression:

$$\varphi_i = \frac{2\pi(i-1)}{z} \tag{9}$$

In a misalignment state the meshing in the two parts of the spline coupling is also different. The distance between the splines for each pair in misalignment conditions is determined as:

$$L_i = L_0 - e\cos\varphi_i \tag{10}$$

In a real system all splines are rigidly meshed due to the transmission of a large torque:

$$T = \sum_{i=1}^{z}(F_{Ti}(R+L_i)) \tag{11}$$

where R is a radius of the root of the teeth, ϕ is the torsion angle (the same for the all splines) and F_{Ti} is the force transmitted by each spline.

$$F_{Ti} = \phi L_i K_i \tag{12}$$

Based on the above, the angle ϕ can be defined as:

$$\phi = \frac{T}{\sum_{i=1}^{z}(L_i K_i (R+L_i))} \tag{13}$$

The meshing force for each spline in misalignment conditions will be determined as follows:

$$F_i = (\phi L_i + e' \sin\varphi_i)K_i \tag{14}$$

$$e' = \sqrt{x^2 + y^2} \tag{15}$$

From the expression above we can conclude that the meshing force in the splines has the biggest values for the splines from 1 to $z/2$ and also cannot be negative. In addition, for the case when it's absence the torque, the meshing force is 0 (even if the misalignment is). Therefore, the expression for the determining the meshing force for each spline will take the form:

$$F_i = \begin{cases} (\phi L_i + e'\sin\varphi_i)K_i, & \phi L_i + e'\sin\varphi_i > 0 \\ 0, & \phi L_i + e'\sin\varphi_i \leq 0 \end{cases} . \tag{16}$$

The components of the above force:

$$F_{xi} = F_i \cos\theta_i \tag{17}$$

$$F_{yi} = F_i \sin\theta_i \tag{18}$$

$$\theta_i = \varphi_i + \frac{\pi}{2} \tag{19}$$

where θ_i is the angle between the positive direction of the X-axis and the meshing force.

The above analysis refers to the case of misalignment in only one direction. For real system the meshing forces are defined as follows:

$$f_x = F_x \cos\varphi - F_y \sin\varphi \tag{20}$$

$$f_y = F_x \sin\varphi + F_y \cos\varphi \tag{21}$$

3. Results

To assess the applicability of the algorithm, the spline coupling with the parameters presented in [2] were calculation. Table 1 shows the parameters of the studying spline connection.

Table 1. Parameters of the spline coupling.

Parameter	Value
Number of splines, Z	14
Spline width B, mm	100
Spline thickness H, mm	16
Spline length L_0, mm	5.07
Poisson's Ratio, ν	0.3
Elastic modulus E, Pa	$2.12 \cdot 10^{11}$
Radius of spline root R, mm	69.5
Transmitting torque T, Nm	87000

Figure 2 shows the dependence of the meshing force from the value of the initial displacement. For comparison, the dashed lines in the graphs represent the results obtained in [1].

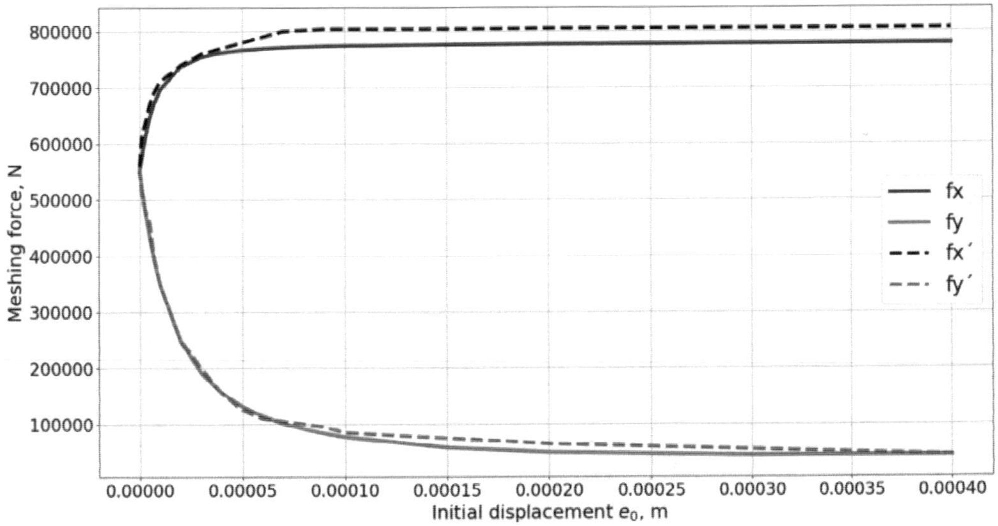

Figure 2. Dependence of the meshing force from the initial displacement.

Figure 3 shows the dependence of the meshing force from the misalignment.

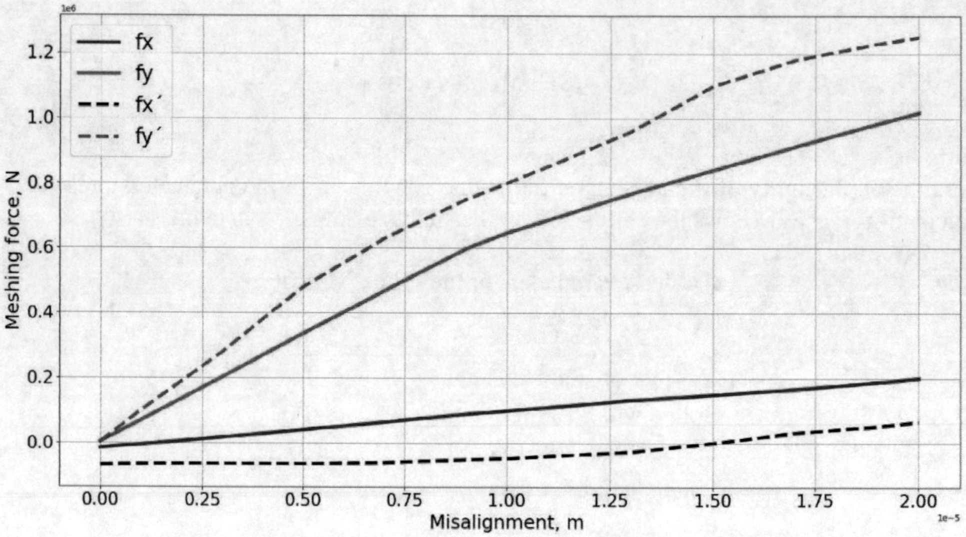

Figure 3. Meshing force versus misalignment.

The figures below show the dependence of the meshing force from the spline parameters: in figure 4 - from the spline thickness, in figure 5 - from the spline length.

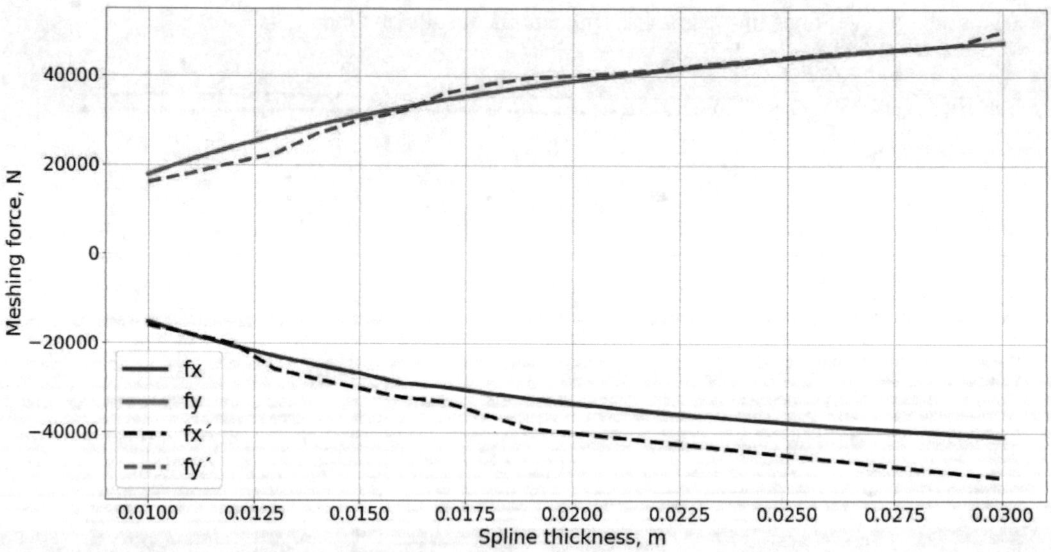

Figure 4. Meshing force versus spline thickness.

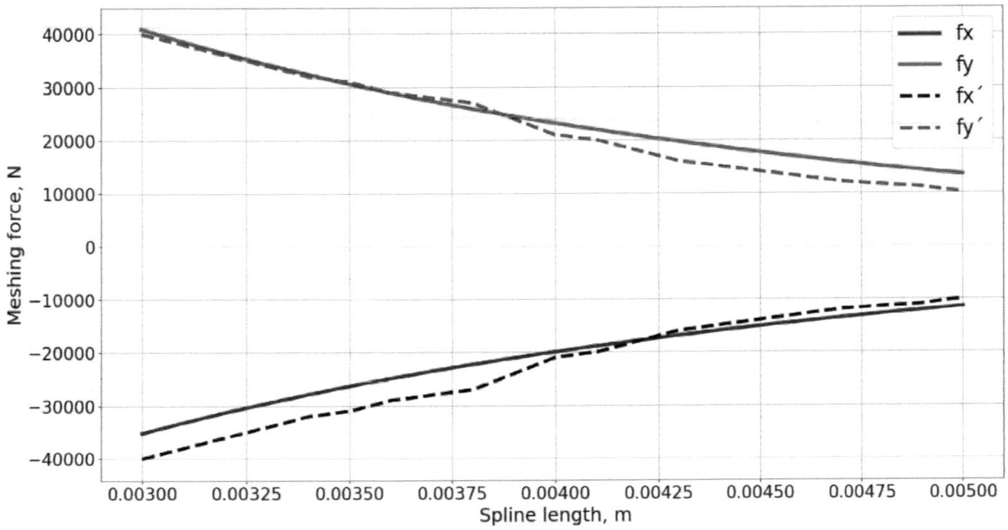

Figure 5. Meshing force versus spline length.

4. Conclusions

1. The dependences of the meshing force from the various parameters of spline coupling has been obtained: an increase of the thickness tooth leads to the increase of meshing force and an increase the tooth length leads to the decrease of meshing force; with an increase of the misalignment the meshing force increase too, the dependence of these parameters close to linear.

2. The research results presented in [2] are reproduced only qualitatively. It is necessary to confirm the calculation results by solving the problem using alternative methods (for example, the FE method).

3. The development of this algorithm seems possible in the transition from the model of rectangular tooth to the models of involutes and triangular spline couplings, taking into account such parameters as the contact patch, clearances (like in [3], [4]), and also subsequent use in the DYNAMICS R4 software package.

References

[1] Guang Z, Zhansheng L and Feng C 2008 Force of misaligned spline coupling and the influence on rotor system *Int. J. of Rotating Machinery* ID 321308

[2] Guang Z, Zhansheng L, Yongliang W and Jinfeng Z 2009 Study on stabilities of rotor-bearing systems with different types of couplings *Proc. ASME Turbo ExpoGT2009* 59751

[3] Crease A B 1977 Forces generated by gear couplings Michael Neale and Associates

[4] Al-Hussain K M 2003 Dynamic stability of two rigid rotors connected by a flexible coupling with angular misalignment *J. of Sound and Vibration* **266** p 217

IOP Conf. Series: Materials Science and Engineering 1129 (2021) 012027 doi:10.1088/1757-899X/1129/1/012027

Investigation of the contact fatigue strength of high quality carburised steel

S A Pakhomova, R S Fakhurtdinov, E Zhavoronkova and K Zinkovich

Bauman Moscow State Technical University, 2-d Bauman Street, 5, Russia, Moscow, 105005

mgtu2013@yandex.ru

Abstract. The parameters of contact fatigue strength during vacuum carburization and ion nitriding of high quality steel (*HQS*) *HQS-7* and *HQS-10* are considered. Information about the modes of chemical and heat treatment is given and the technological process of hardening for *HQS-7* and *HQS-10* steels is considered. The characteristics of the contact fatigue strength of steels in the statistical aspect at different levels of contact stresses are obtained. A graphical analysis of the obtained values after testing was carried out, the optimal stresses for research were determined and the optimal version of chemical heat treatment was selected, which determines the best indicators for mechanical characteristics and operational requirements.

1. Introduction

The gears of modern engines are subject to high power and temperature loads. They are made mainly from carburised or nitrided complex-alloyed heat-resistant steels [1, 2]. One of the main criteria for the performance of gears is the contact fatigue strength of the material. However, data on the contact fatigue strength of heat-resistant steels are fragmentary and sometimes contradictory [3–5]. Therefore, the study of the contact fatigue strength of high-loaded parts is an important task in solving the current problem of ensuring the reliability of aircraft engine gears.

The vacuum carburization process is increasingly used, especially in mass production. Vacuum carburization has a number of advantages over conventional carburization. Vacuum carburization was carried out in acetylene at a temperature of 940°C for 5 hours according to cyclic modes selected in accordance with the recommendations [6, 7]. To ensure a high limit of contact fatigue strength, the acetylene supply mode was used, which includes two groups of cycles: the first group of cycles provides active saturation of the surface with carbon, the second group – diffusion redistribution of the carbon concentration and active saturation of the near-surface zone of the layer.

The process of nitriding in a glow discharge is carried out on the final processed parts [2, 8–10]. Ion nitriding includes two stages: surface cleaning by cathode sputtering and nitriding itself, which is carried out at a temperature of 500…600°C in an atmosphere of 95% N_2 + 5% H_2. The first stage provides activation of the surface and destruction of oxide films, and the second stage – heating the surface to the diffusion temperature and saturation with nitrogen.

To increase the contact fatigue strength of gears, other methods of surface hardening are also used. For example, various methods of deformation processing are widely used [11–14]. One of the most effective methods is also electro–mechanical treatment [15–17], the combination of which with chemical heat treatment (*CHT*) is insufficiently studied and also the methods of laser carburization of low-alloy tool steels [18].

The purpose of this work is to analyze the contact endurance during vacuum carburization and ion nitriding of *HQS-7* and *HQS-10* steels used in the manufacture of gears. For this purpose, a number of fracture tests were performed and the optimal stresses for research were determined.

2. Materials and equipment of the experiment

The objects of research made of *HQS*-7 and *HQS*-10 steels that had undergone the entire complex of chemical heat treatment, including vacuum carburization (or nitriding), quenching and tempering. The table 1 below shows the chemical composition of *HQS*-7 and *HQS*-10 steels.

Table 1. Chemical composition of the studied steels.

Steel grade	Content of elements, % (wt.)								
	C	Cr	Ni	Mn	W	Si	V	Nb	Al
HQS-7	0.14–0.18	1.8–2.2	2.7–3.0	0.4–0.6	0.2–0.3	0.17–0.37	0.1–0.2	0.1–0.2	0.02–0.07
HQS-10	0.10–0.15	3.0–3.4	2.7–3.0	1.9–2.3	0.3–0.6	0.2–0.5	0.17–0.37	0.05–0.15	0.02–0.04

Note: no more than 0.015% of sulfur and no more than 0.025% of phosphorus.

Chemical and heat treatment was performed on *SECO/WARWICK* equipment [19]. For comparison, two methods of *CHT* were studied: vacuum carburization (*VC*) and ion nitriding (*IN*). For *HQS-7* and *HQS-10* steels, pre–heat treatment was performed before nitriding, which includes two operations: quenching and high tempering (table 2). This treatment provides high viscosity and strength of the part core [20, 21].

Table 2. Chemical and heat treatment of the studied steels.

Steel grade	The chemical heat treatment mode	The surface layer structure
HQS-7	vacuum carburization t = 880°C, τ = 5 h; high tempering t = 660°C, τ = 3 h; quenching t = 1010°C; cold treatment t = –70°C, τ = 0.5 h; tempering t = 240°C, τ = 2 h	fine-needlepoint martensite with inclusions of fine carbides, HR15N 89…92, the layer thickness of 3.2 mm
HQS-7	quenching t = 950°C; high tempering t = 640°C, τ = 3 h; ion nitriding t = 650°C, τ = 4 h	nitrogenous sorbitol with inclusions of fine carbonitrides, HR15N 87…89, the layer thickness of 0.5 mm
HQS-10	vacuum carburization t = 880°C, τ = 5 h; high tempering t = 660°C, τ = 3 h; quenching t = 1010°C; tempering t = 510°C, τ = 3 h; cold treatment t = –70°C, τ = 0.5 h; tempering t = 510°C, τ = 2 h	fine-needlepoint martensite with a small amount of fine carbide-inclusions, HR15N 89…92, the layer thickness of 3.0 mm
HQS-10	quenching t = 950°C; high tempering t = 670°C, τ = 3 h; ion nitriding t = 650°C, τ = 4 h	nitrogenous sorbitol with inclusions of fine carbonitrides, HR15N 87…89, the layer thickness of 0.5 mm

Samples that were not saturated with carbon or nitrogen were subjected to typical hardening heat treatment. It included quenching t = 950°C for both steels and tempering (τ = 3 h) with t = 650°C and t = 670°C, for *HQS-7* and *HQS-10*, respectively.

It is known [1–3] that after carburization it is impossible to obtain the maximum hardness. Therefore, the final hardening heat treatment was carried out, combining quenching and different tempering modes. The purpose of high tempering is to give the core of the parts maximum viscosity while maintaining sufficient resilience and strength limit of the metal. In this case, the residual austenite of the surface layer decays. The purpose of quenching: a) grind the core grain to increase the viscosity required by the core; b) break the carburization grid on the surface to eliminate its brittleness; c) complete dissolution of excess phases. There is an increase in hardness. Cold treatment is condact to ensure the decomposition of residual austenite, increase the properties (mainly hardness) and stabilize the size of parts. Last tempering to relieve residual stress. As a result of such complex processing, parts receive the necessary values of physical and mechanical properties: high hardness, wear resistance, contact fatigue strength, etc.

Contact fatigue strength tests were realized on a modernized *MKV-K* installation at contact stresses σ_{Zmax} = 780–1020 MPa. The contact fatigue specimen is a roller of 30.2 mm in diameter and 18.5 mm in length. This specimen is rolled reciprocally with a fatigue tester with a slip ratio of 20 %. Tests consist of running a sample between two test disks, one of which is a drive disk and the other is a pressure disk. The speed of the drive disk is 100 s^{-1}. The load on the sample is applied by means of a pressure test disc from the spring loading mechanism through a system of levers. The samples were lubricated by the drip method with synthetic oil with a complex of additives *IPM–10 (TU 38.10112 99-2006).*

The number of samples in one batch was 10–14 pieces to ensure high accuracy of test results. For the production of samples, rods with a diameter of 15 mm were taken from the studied steels.

Metallographic analysis of micro-glyphs was performed on a *Neophot-32* light optical microscope. For the study, steel samples were made after etching in a 5 % solution of HNO_3 in ethyl alcohol. Microhardness was measured on a *PMT-3* device at a load of 0.1 N.

3. Experimental results and discussion

Metallographic studies have shown that the diffusion layer after vacuum carburization has a structure characteristic of the carburised layer. The microstructure and distribution of carbon across the layer are shown in figure 1. It can be seen that the near-surface layer is sufficiently saturated with carbon and contains a well-developed carbide phase.

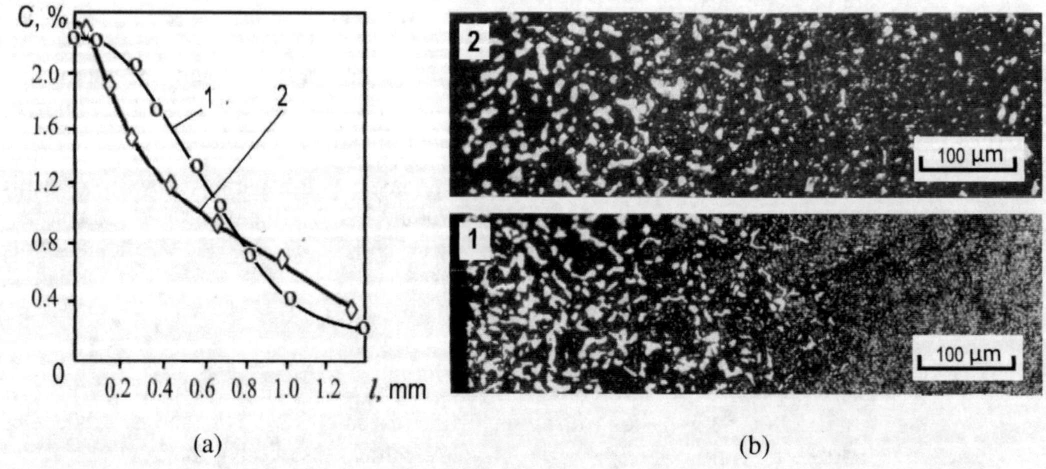

Figure 1. Carbon saturation (a) and microstructure of the carburised layer (b) of *HQS-10* (1) and *HQS-7* (2) steels after vacuum carburization.

The main research was related to contact endurance tests. Samples which had sharply distinguished (abnormal) number of cycles were screened out. Thus, gross errors are excluded from calculations. When processing the data, we used the obtained mathematical expectation (the arithmetic mean of the observed values of a random variable, and the sample average). It is proved that if certain conditions are met, the sample average tends to the true value of the mathematical expectation of a random variable when the sample size (the number of observations, tests, measurements) tends to infinity [22].

After plotting the dependence of the number of cycles on the various stresses applied to the samples, the optimal values required for testing were found (figure 2).

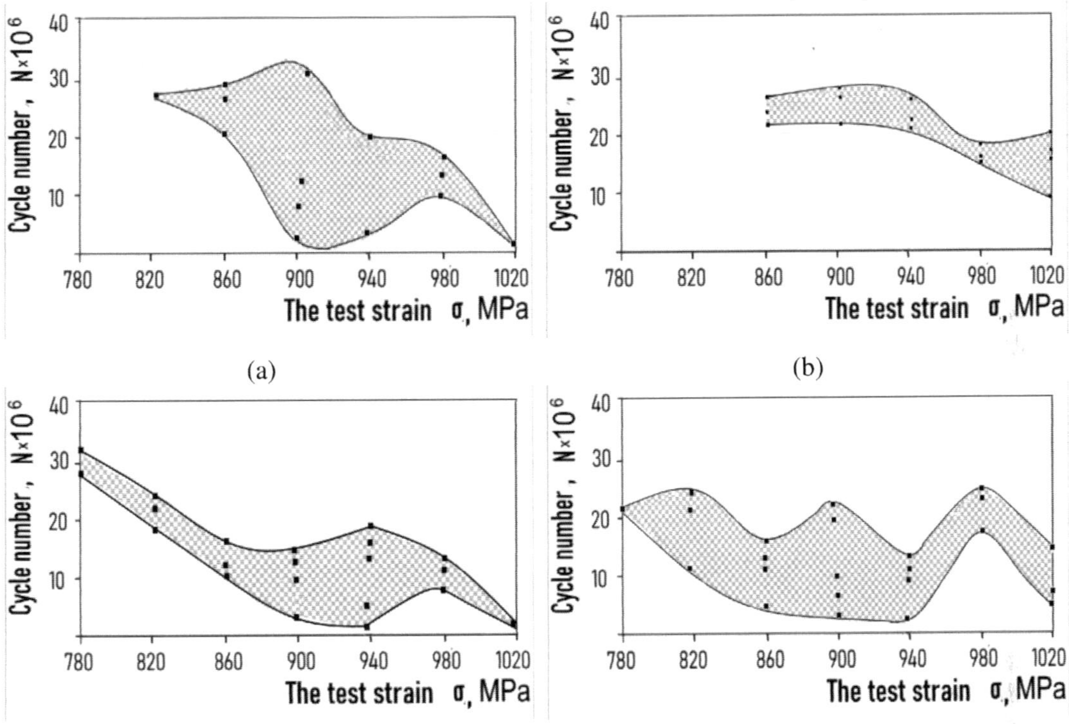

Figure 2. Contact fatigue strength of *HQS-7* (a), (b) and *HQS-10* (c), (d) steels after vacuum carburization (a), (c) and ion nitriding (b), (d) depending on the stresses σ during the test.

The graphs presented are inaccurate, but they give an idea of the maximum stresses at which further testing of samples should be performed. The inaccuracy of the graphs is due to the following reasons:

- heterogeneity of the material structure (large/small grain in the area of the neck of the gap/the place of fixing the sample or Vice versa), the concentration of impurities, etc.;
- anisotropy of the sample properties that increases with the number of loading cycles;
- hardening and accumulation of dislocations (especially in the area of the neck of the rupture);
- anisotropy of the sample properties, increasing with increasing number of loading cycles;
- systematic and static measurement error;
- non-systematic errors due to insufficient number of measurements.

Analysis of the results of measurements of the contact endurance of diffusion layers allows us to conclude that the optimal operating stresses are 980 MPa for testing the contact endurance of *HQS-7* and *HQS-10* steels, since they are the most indicative. After testing with stresses less than $\sigma = 940$ MPa, a lot of samples are taken without damage, and at σ = 1020 MPa — a very large variation in the final values.

Electronographic examination of surfaces with fractures (pitting) confirmed the typical fatigue nature of the fracture. The vast majority of the resulting microcracks are located at an angle of 30…45° to the surface, and on a number of transverse microshifts made from samples, primary cracks were observed under the surface, preferably in the zone of maximum shear stresses. A comparative cartogram of the studied steels after testing for contact endurance with operating stresses σ = 980 MPa is shown in figure 3.

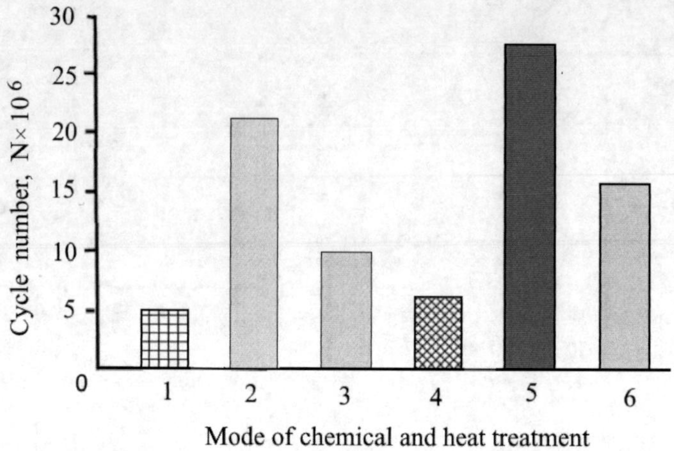

Figure 3. Contact endurance of *HQS-7* (1–3) and *HQS-10* (4–6) steels after *CHT* including vacuum carburization (2 and 5) or ion nitriding (3 and 6); without both of carburization and nitriding (1, 4) (tests were performed at σ = 980 MPa).

Tests have shown that after vacuum carburization, the durability of *HQS-7* and *HQS-10* steels is higher than after ion nitriding by about 23%, which is due to the greater length of the diffusion layer. It can be seen that both processes, such as carburization and nitriding, contribute to a significant increase in the durability of the studied steels. So, thanks to vacuum carburization, the durability becomes approximately 3.5 times more, and thanks to ion nitriding – 2 times.

In optical and electron microscopic studies, the greatest changes in the structure were observed on the run-in surface and in layers very close to the surface. As you move away from the surface, the degree of martensite distortion decreases and, starting from a depth of 30–40 μm, the structure does not differ from the original one. Figure 4 shows the distribution of microhardness and microstructure of the diffusion layer of *HQS-10* steel after vacuum carburization and hardening heat treatment, which provides maximum contact fatigue strength.

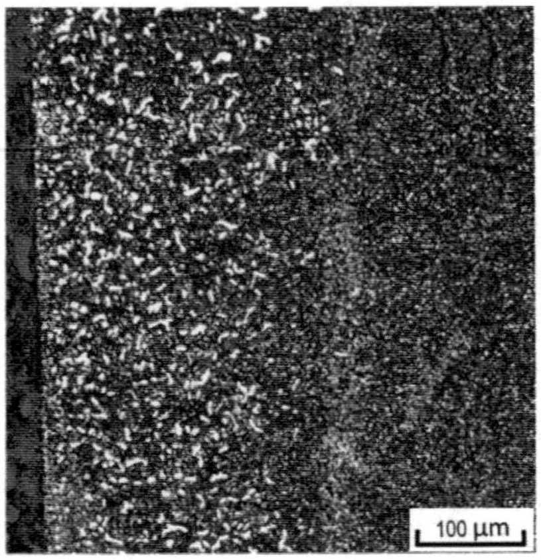

(a)

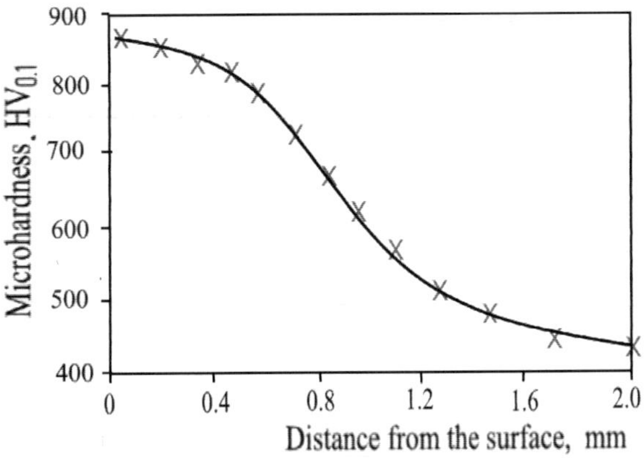

(b)

Figure 4. Microstructure (a) and distribution of microhardness $HV_{0.1}$ (b) of diffusion layer *HQS-10* steel after vacuum carburization.

Analysis of the microstructure of the carburised layer shows the following: the surface part is composed of martensite and carbides, the thickness of the carbide zone is about 0.5 mm, and the carbides are located discretely and have a favorable globular shape.

Measurement of microhardness in the depth of cross sections showed that the extended carbide zone and the dense arrangement of particles in it provides increased surface hardness and smooth distribution over the layer thickness (see figure 1 (b)).

4. Conclusions

1. It is shown that the maximum stresses during tests for contact fatigue strength of samples made of *HQS-7* and *HQS-10* steels should be 980 MPa.

2. It was found that samples after chemical heat treatment, including vacuum carburization, have on 23 % greater durability compared to ion nitriding.

3. Carburization and nitriding processes contribute to a significant increase in the durability of the *HQS-7* and *HQS-10* steels. So, due to vacuum carburization, the durability becomes approximately 3.5 times greater, and due to ion nitriding – 2 times.

References

[1] Kablov E N, Bakradze M M, Gromov V I, Voznesenskaya N M and Yakusheva N A 2020 New high-strength structural and corrosion-resistant steels for aerospace engineering developed by «VIAM» (review) *Aviation materials and technologies* **58** pp 3-11

[2] Suslov A G 2008 *Engineering of surface parts* (Moscow: Mashinostroyeniye) p 320

[3] Zinchenko V M 2001 *The Engineering of the Surface of Gear Wheels by the Methods of Chemical Heat Treatment* (Moscow: MGTU N E Baumana Press) p 303

[4] Fakhurtdinov R S, Ryzhova M Y and Pakhomova S A 2017 Advantages and commercial application problems of vacuum carburization *Polymer Science, Series D* **10** pp 79-83 DOI:10.1134/S1995421217010063

[5] Silkin A A, Linnik A A, Pankratov A S, Kurganova Y A, Kobernik N V and Mikheev R S 2016 Formation of the structure of the weld metal upon the introduction of nanoparticles into the weld pool *Russian metallurgy (Metally)* **13** 1253-6 DOI: 10.1134/S0036029516130206

[6] Fomina L P 2014 Influence of chemical heat treatment on the endurance of teeth when bending *Metal Technology* **1** pp 11-4

[7] Pakhomova S and Karpukhin S 2020 Technological bases for increasing the durability of aviation parts by chemical heat treatment *IOP Conf. Series: Materials Sci. and Engineering* **963** 012006 DOI:10.1088/1757-899X/963/1/012006
[8] Semenov M Yu 2013 Management of the structure of carburization layers of heat-resistant steels *Metal Sci. and Heat Treatment* **5** pp 31-8
[9] Kuksenova L I, Gerasimov S A, Alekseeva M S and Gromov V I 2018 Influence of vacuum chemical heat treatment on wear resistance of VKS7 and VKS10 steels *Aviation materials and technologies* **50** pp 3-8
[10] Kula P, Olejnik J and Kowalewski J 2001 New vacuum carburizing technology *Heat treatment progress* **1** pp 57–65
[11] Polyakov S A, Kuksenova L I, Lapteva V G and Alekseyeva M S 2016 Analysis of microplastic deformation processes of surface layers of nitrided structural steels *Inorganic Materials: Applied Research* **7** (4) pp 618-23 DOI: 10.1134/S2075113316040298
[12] Fomina L P 2018 Influence of high-temperature hardening processes on the deformation of aviation engine gears *Metal technology* **6** pp 28-31
[13] Lashnev M M 2011 Process control of vacuum carburizing (carbonitriding) heat-resistant steels *Metallurgy of Machinery Building* **4** pp 45-6
[14] Pakhomova S A and Manayev O I 2018 Effect of heat shotblast treatment exerted on the contact fatigue of carburised heat-resistant steel C0.12Cr2NiWV *Inorganic Materials: Applied Research* **9** pp 732–5 DOI: 10.1134/S2075113318040251
[15] Morozov A, Fedotov G, Fedorova L, Musharapov D and Khabieva L 2019 The providing durability of the movable square-sided spline joints by electromechanical treatment of the working surfaces *MATEC WEB of Conf. ICMTMTE 2019* **298** 00117 DOI:10.1051/matecconf/201929800117
[16] Fedorova L, Fedorov S, Sadovnikov A, Ivanova Y and Voronina M 2018 Abrasive wear of hilong botn hardfacings *IOP Conf. Series: Materials Sci. and Engineering* **307** 012038 DOI: 10.1088/1757-899X/307/1/012038
[17] Fedorova L V, Fedorov S K, Ivanova Y S and Voronina M V 2017 Increase of wear resistance of the drill pipe thread connection by electromechanical surface hardening *International J. of Applied Engineering Research* **18** 7485-9
[18] Marinin E A, Chirkov A M, Gavrilov G N, Fetisov G P, Chernyshov D A and Kurganova Yu A 2018 Experimental evaluation of the methods of laser cementation of low-alloy tool steels *Russian metallurgy (Metally)* **13** pp 73-7 DOI: 10.1134/S0036029518130153
[19] Fakhurtdinov R S, Pakhomova S A and Ryzhova M Y 2017 On the problems of modernizing equipment for vacuum carburization *J. of Machinery Manufacture and Reliability* **46** (2) pp 187–92 DOI: 10.3103/S1052618816060066
[20] Kuksenova L I, Polyakov S A, Lapteva V G and Alekseyeva M S 2015 Mechanical properties of surface layers of structural steels after nitriding and possibilities of adaptation of their nanostructure to contact deformation *Metal Sci. and Heat Treatment* **57** (7-8) pp 436-42 DOI: 10.1007/s11041-015-9901-7
[21] Kalashnikov A S, Morgunov Yu A and Kalashnikov P A 2018 Advantages of gas vacuum carburization and gas quenching under high pressure in processing a hypoid gear of the shaft type *Handbook Engineering journal* **8** pp 3-7 DOI: 10.14489/hb.2018.08.pp.003-007
[22] Livshits M Yu, Derevyanov M Yu and Yakubovich E A 2017 Optimal control of an object with distributed parameters on the example of vacuum carburization *Mathematical methods in engineering and technology* **12** pp 24-30

IOP Conf. Series: Materials Science and Engineering 1129 (2021) 012028 doi:10.1088/1757-899X/1129/1/012028

Influence of the composition of polymeric materials on the tribological characteristics under abrasive wear

A N Prince and V P Biryukov

Mechanical Engineering Research Institute of the Russian Academy of Sciences, Moscow, 101990, Russia

Laser-52@yandex.ru

Abstract. The paper presents the results of tribological tests of thermoplastic polyurethanes (TPU), UHMWPE, polypropylene and rubber based on nitrile butadiene rubber (BNR) on fixed silicon carbide and free abrasive. The influence of pressure and sliding speed on the coefficients of friction is shown. For PP2015 polypropylene, the coefficient of friction and mass loss were 0.3-0.35 and 0.003 g, which is 4 - 10 times lower than for BNR.

1. Introduction

Currently, rubber materials are used in the friction units of railway rolling stock. These include oil seals, cuffs, various seals, corrugations, shock absorbers, etc. Until now, the issue of wear of free and fixed abrasive parts made of rubber, and the possibility of replacing them with other polymer materials, has not been sufficiently studied.

In [1], polymer samples of polyamide and polypropylene PA6/PP composites filled with GR PA66/PP graphite and a polymer composite with the introduction of solid particles of nano-clay NC - PA66/PP with shore hardness of 64, 66 and 71, respectively, were made for testing for abrasive wear according to the US ASTM G-65 standard. Quartz sand and quartz with a particle size of 200 – 250 μm were used as an abrasive. The tests were carried out at a constant load of 36 N, the friction path was 150 - 600 m. The greatest wear is observed in samples with graphite filler PA66/PP – $0.0105 \cdot 10^{-9}$ $m^3 \cdot N^{-1} \cdot m^{-1}$, and the lowest wear is $0.0055 \cdot 10^{-9}$ $m^3 \cdot N^{-1} \cdot m^{-1}$ for the PA66 / PP mixture when tested with quartz abrasive. When using quartz sand as an abrasive, the relative wear rate is lower. Wear for graphite-filled PA66/PP and unfilled PA66/PP is $0.0085 \cdot 10^{-9}$ $m^3 \cdot N^{-1} \cdot m^{-1}$. and $0.004 \cdot 10^{-9}$ $m^3 \cdot N^{-1} \cdot m^{-1}$, respectively. For samples of NC-PA66/PP and GR PA66/PP composites, the specific wear rate increases with increasing friction path and decreases for the PA66/PP mixture. There is an increase in the wear of GR - PA66/PP samples by 196% and 167% compared to NC - PA66/PP and PA66/PP composites, respectively.

In [2], the effect of white soot (colloidal silicon dioxide and precipitated silicon dioxide) on the tribological properties of silicone rubber was studied. Friction and wear tests were carried out according to the scheme disk (silicone rubber with a diameter of 29 mm, height of 12.5 mm) - ball (GCr15 steel with a diameter of 6 mm, hardness 59-61 HRC) on a universal tribotester with a CFT-1 microcontroller, China. During the test, the upper steel ball GCr15 remained stationary, and the silicone rubber disk rotated at a frequency of 500 min^{-1}. The rubbing surface of each silicone sample was sanded with sandpaper to ensure even contact. All experiments were performed with a load of 10 N, three times to ensure repeatability of the results. The coefficient of friction of rubber with colloidal silicon dioxide constantly increased and reached a maximum value of 0.633, with further testing it stabilized at a value of 0.54. Silicone rubber with deposited silicon dioxide had a maximum coefficient of friction of 0.89, then it decreased and stabilized at a value of 0.65.

The wear value was 0.00190 g of samples with colloidal silicon dioxide, which is 1.5 times less than for samples with deposited silicon dioxide.

In [3], samples from thermoplastic polyurethane (TPU) and a mixture of TPU and ultra-high-molecular polyethylene (UHMWPE) were used 95/5, 90/10, 85/15, 80/20, 75/25 and 70/30%. The friction and wear test was performed on the MRH-3 installation, China using the disk (steel with a diameter of 49.22 mm) - a finger (a sample of TPU or TPU/UHMWPE with a thickness of 4 mm) glued to a steel rod with a thickness of 8 mm. Dry friction tests were performed at a normal load of 150 N, with a rotation speed of 100 min^{-1} at room temperature with a relative humidity of 20-50%. In addition, water in the amount of 55-65 drops fed to the friction zone was used as a lubricant under similar test conditions. The coefficient of friction and the rate of wear of samples under dry friction conditions decreased sharply, with the addition of 15% UHMWPE. The coefficient of friction decreased from 1.2 for TPU to 0.45 for TPU/UHMWPE 85/15%, and the wear rate decreased by 15%. For mixtures of TPU/UHMWPE with a higher UHMWPE content, the wear rate during water lubrication is comparable or even slightly higher than for dry friction.

The aim of this work is to determine the tribological characteristics of polymers for replacing rubber products in the friction units of rolling stock.

2. Research materials and methods

For friction and wear tests, thermoplastic polyurethanes (TPU), UHMWPE, and polypropylene (PP) were selected. The technical characteristics are shown in table 1 for comparison with the polyurethanes of the selected rubber based on butadiene-nitrile rubber (BNR). Each of the 5 materials studied was a plate with dimensions of 20×70×2 mm, which was glued to plywood with dimensions of 20×70×12 mm.

Table 1. Technical characteristics of TPU.

Material brand	Hardness, Shor	Density, $g{\cdot}cm^{-3}$	Ultimate strength, MPa	Elongation at break, %
TPU 50D2S10	52 D	1.18	45.0	450
TPU 60D2S10	60 D	1.18	55.0	320
UHMWPE	64 D	0.93	25	50
PP 02015	70 D	0.9	29	400
BNR	75 A	1.3	25	425

A steel ring mandrel with sandpaper made of silicon carbide with a grain size of 120 μm pasted on its end was used as a counter-tile. Tests for friction and wear with a fixed abrasive were performed according to the "plane (test sample) – ring" scheme. The sliding speed and pressure on the sample varied discretely in the range of 0.1–0.3 $m{\cdot}s^{-1}$ and 0.1–0.5 MPa, respectively. Wear tests with a free abrasive were performed when a flat sample was rubbed against the forming surface of a rubber disk. Quartz sand with a particle size of 0.2-0.6 mm was fed into the friction zone. The tests were performed at normal atmospheric pressure and temperature. The test load was 15 N, and the time was 5 minutes. The weight loss of the samples was determined on a vibra HT-220CE analytical balance. The wear value was determined as the arithmetic mean for three samples.

3. Results of experimental studies

Figure 1 ((a), (b)) shows the results in the form of dependences of the coefficient of friction on the normal pressure and sliding speed when tested on a fixed abrasive. As the pressure increases, the coefficients of friction decrease for all types of polymers and BNR. The obtained values of the friction coefficients can be arranged as they increase in the following order: 0.3 – 0.36 PP2015, 0.33 – 0.44 UHMWPE, 0.38 – 0.55 50D2S10, 0.4 – 0.58 60D2S10 and 0.85 – 0.98 BNR. An increase in the sliding speed from 0.12 to 0.23 $m \cdot s^{-1}$ led to a slight increase in the friction coefficients for the studied polymers. So for PP2015 and UHMWPE, it increased from 0.32 to 0.36. With a further increase in speed to 0.3 $m \cdot s^{-1}$, the coefficient of friction was 0.5 and 0.49 for PP2015 and UHMWPE, respectively. The coefficient of friction of BNR - 0.8 - 0.9 increased proportionally over the entire range of sliding speeds.

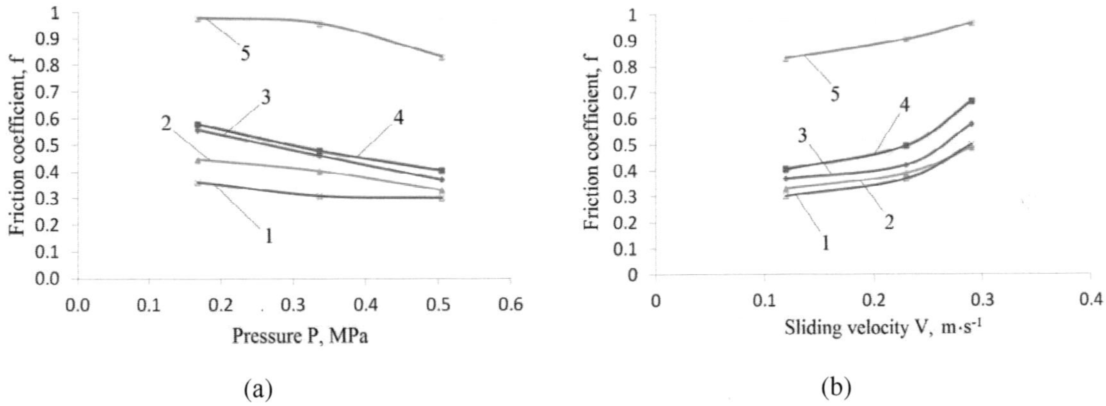

Figure1. Dependence of polymer friction coefficients on pressure (a) and sliding speed (b): 1 – PP2015, 2 – UHMWPE, 3 – 50D2S10, 4 – 60D2S10, 5 – BNR.

Figure 2 shows the mass loss of samples from pressure.

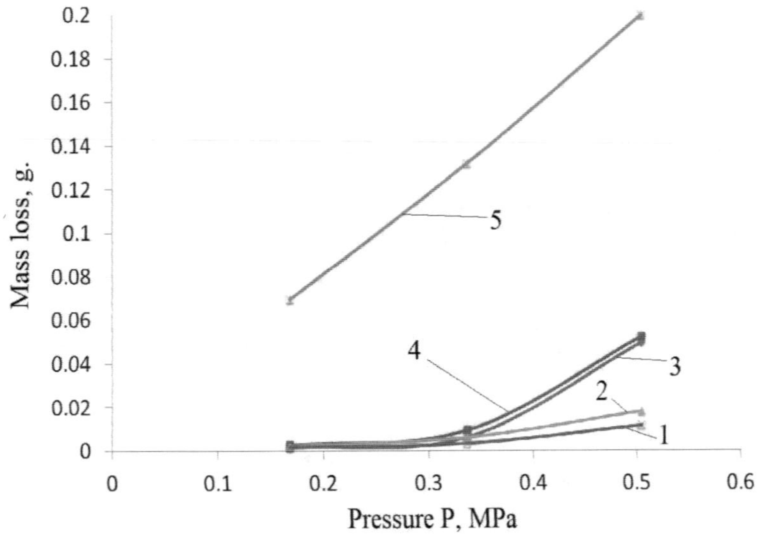

Figure 2. Dependence of polymer mass loss on pressure: 1 – PP2015, 2 – UHMWPE, 3 – 50D2S10, 4 – 60D2S10, 5 – BNR.

For BNR, the mass loss increased in proportion to the pressure. The wear of TPU samples at low pressures up to 0.34 MPa is insignificant, with a further increase in pressure up to 0.54 MPa for 50D2S10, 60D2S10 increases from 0.01 to 0.04 g. The minimum mass loss was obtained for PP2015 polypropylene. Figure 3 shows the morphology of the friction surfaces of the samples.

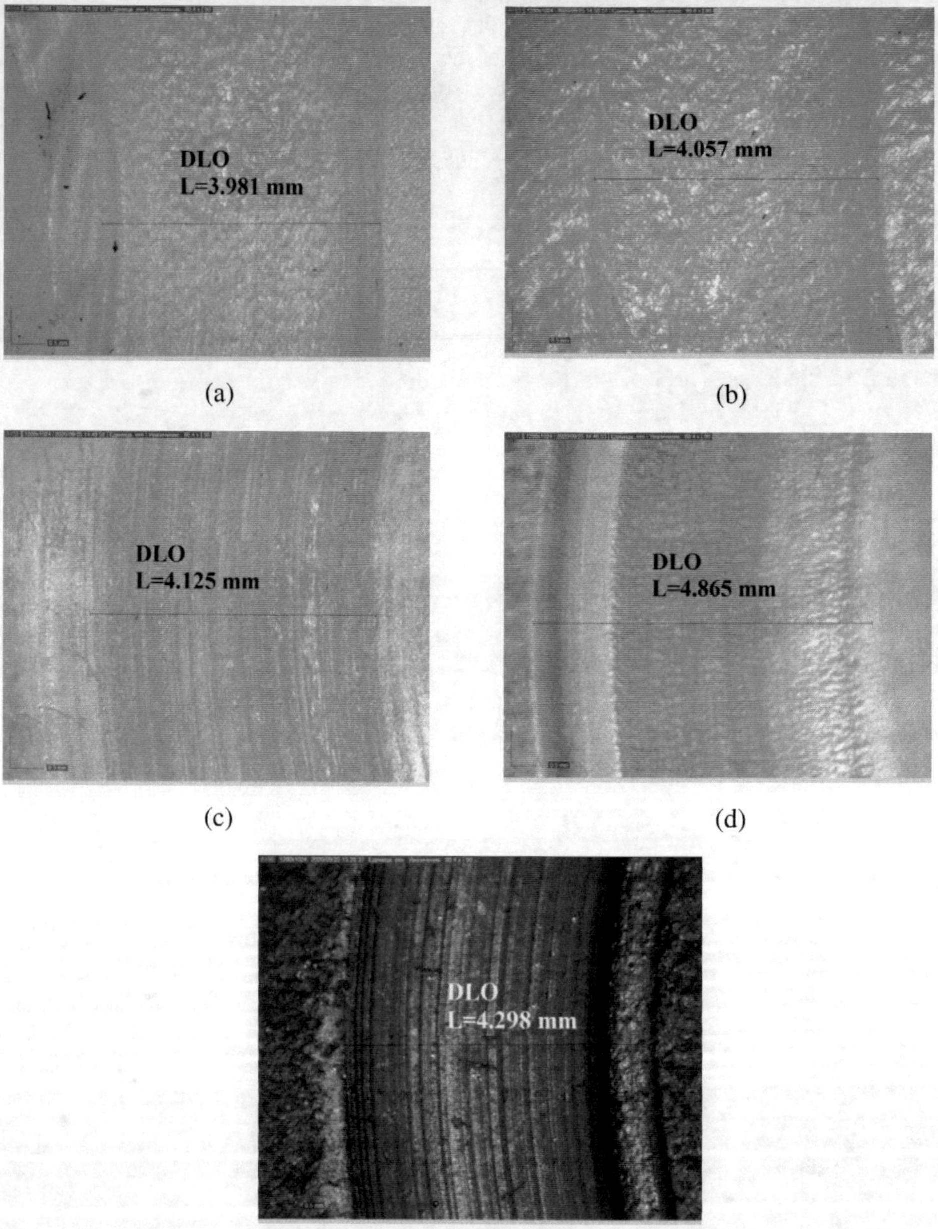

(a) (b) (c) (d) (e)

Figure 3. Morphology of the friction surfaces of polymers and BNC rubber: (a) - PP2015, (b) – UHMWPE, (c) - 50DS210, (d) - 60DS2010, (e) – BNR.

The friction surfaces of PP02015 and UHMWPE polypropylene with a minimum wear value of 0.003 and 0.004 g, respectively, did not have deep ridges or grooves in the sliding direction of the fixed abrasive grain. The undulation of the surface indicates a gradual accumulation of damage and separation of wear particles by the fatigue mechanism. The friction surfaces of the 50DS210 samples and the BNR throughout had risks in the sliding direction. It can be assumed that the prevailing mode of wear was micro-cutting. For the 60DS2010 samples, the wave height was minimal, and the wear pattern can be attributed to fatigue when friction occurs on a fixed abrasive grain.

Figure 4 shows the profiles of wear holes of polymer materials and BNR when tested with loose abrasive.

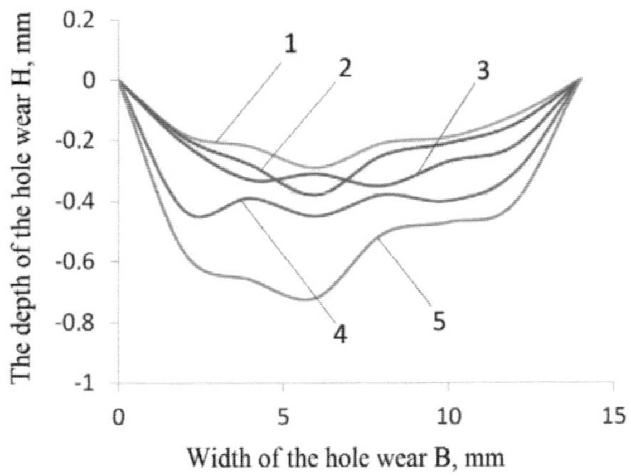

Figure 4. Profiles of material wear holes: 1 – 50DS210, 2 – 60DS2010, 3 – UHMWPE, 4 – PP2015, 5 – BNR.

In contrast to the tests with fixed abrasive, the samples TPU50DS210 and 60DS2010 showed minimal weight loss. Samples of UHMWPE has moved into third place. PP2015 and BNR samples showed the greatest weight loss, respectively. Figure 5 shows fragments of wear holes of materials during friction on a free abrasive.

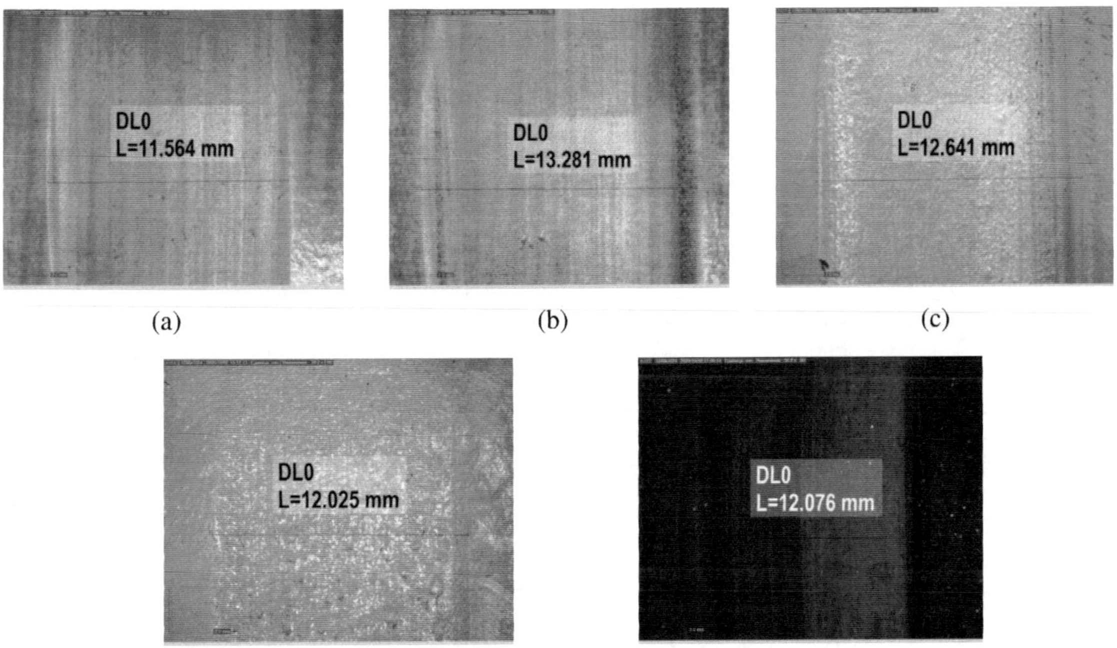

(a) (b) (c) (d) (e)

Figure 5. Morphology of the friction surfaces of polymers and BNC rubber when tested for free abrasive: (a) – 50DS210, (b) – 60DS2010, (c) – UHMWPE, (d) – PP2015, (e) – BNR.

The friction surfaces of 50DS210, 60DS2010 and BNR had longitudinal risks in the direction of sliding of the abrasive grain and the wear mechanism can be attributed to micro-cutting. The undulation of the surface of UHMWPE and PP2015 SAMPLES indicates a gradual accumulation of damage and separation of wear particles by the fatigue mechanism. Various results of friction tests with free and fixed abrasive grain indicate that the sandpaper was coated with polymer particles due to adhesion during the testing of TPU PP2015, and the wear rate and coefficient of friction were reduced. For polymers 50DS210, 60DS2010, the minimum mass loss due to friction with a free abrasive indicated that their hardness is higher than that of a rubber disk and some of the abrasive particles were simply pressed into the forming surface of the disk. Despite the different results, both test methods complemented each other and gave a complete description of polymer materials used in natural conditions during the operation of machines and mechanisms, when it is possible to affect the friction surface with fixed and free abrasive grain.

4. Conclusions

Regularities of changes in the coefficients of friction and mass loss of samples from pressure are obtained. The coefficient of friction of PP 2015 polypropylene had values of 0.3-036, while for BNR was 2.5 times higher and was 0.85-1.0 when friction on the fixed abrasive. A stable result in relative weight loss was shown by PTFE samples when tested with a free and fixed abrasive. The samples PP2015 and 50DS210 and 60DS2010 had the least wear during friction on the fixed abrasive and free abrasive, respectively. Materials such as TPU 50DS210, 60DS2010 and UHMWPE can be used in the future in friction units as a replacement for rubber products for the manufacture of cuffs, seals, gaskets and corrugations. PP 2015 polypropylene can be used instead of rubber in control panels, rotary mechanisms and devices inside the rolling stock.

References

[1] Ravi Kumar B N, Suresha B and Venkataramareddy M 2012 *Instrumentation Science and Technology* **17** pp 113-26
[2] He Q, Li A, Zhang Y, Liu S, Guo Y and Kong L 2018 *Tribology Materials Surfaces and Interfaces* **12** pp 9-16
[3] Wang X, Mu B and Wang H 2015 *Polymer Composites* **36** pp 897-906

Solid lubricating coatings tribological test rig development

M V Prozhega and E O Reschikov

Mechanical Engineering Research Institute of the Russian Academy of Sciences (IMASH RAN) 4 Maly Kharitonyevsky Pereulok, Moscow, 101990, Russia

egorreschikov@yandex.ru

Abstract. Solid lubricant coatings are commonly used under difficult operating conditions. Compositions containing graphite and molybdenum disulphide are the most used worldwide. Friction and wear tests are carried out to evaluate solid lubricant quality. Tribological properties are characterised by friction coefficient. The following paper presents the design of a ball-on-disc scheme testing rig. The normal load is applied by weights and remains constant during the entire testing time. The sliding speed is controlled by computer and changes smoothly. A strain gauge is installed to friction force determination. The sensor provides direct measurement of the bending strain and also the torque. The latter to be reduced for higher accuracy. To evaluate the design stiffness and contact point displacement the finite element analysis was done. The test rig is developed in accordance with ASTM G99.

1. Introduction

Lubrication in dry conditions is achieved by the means of solid lubricants. The same as liquid lubricants or greases solid lubricants decrease friction and wear. These materials are best in high or extremely low temperatures, vacuum, high contact pressure or other specific environmental conditions [1].

Solid lubricants are highly efficient at high loads, close to the yield strength of contacting materials, at low sliding speeds when oils cannot form a hydrodynamic wedge [2].

Solid lubricants are added in oils, greases and self-lubricating composites. Various materials are used as solid lubricants, but the most common are molybdenum disulfide, graphite and polytetrafluoroethylene (PTFE) [1-3].

There are various methods of applying solid lubricants on the friction surface: from the simplest staining with a suspension containing a solid lubricant to the formation of thin films by chemical and physical vapor deposition [2].

Staining with a suspension is the most common way to obtain a coating with a solid lubricant on a product surface. Solid lubricating coating is a composition of antifriction filler, solvent and film-forming substances. Antifriction filler provides a low coefficient of friction, a solvent is necessary to maintain a suspension in a liquid form suitable for application, and film-forming substances, which are polymer binders, are responsible for the coating high adhesion to the surface and bind solid lubricant particles into a continuous film [1].

The combination of solid lubricants makes the special properties of the coating. Compositions of MoS_2 with graphite or silver are corrosion resistant. MoS_2 combined with PTFE increases the service life of the antifriction coating [1-2].

Various compositions of solid lubricating coatings produced by different manufacturers are widely represented on the market. At the same time, despite the attention paid by researchers to this issue [1-2, 4-6], the composition of the coating is selected mainly experimentally [2].

The solid lubricating coating operates in different conditions. Hence its quality is assessed according to various criteria: the ability to withstand high contact pressure, high or low temperatures, work in a vacuum, etc. However, it is possible to evaluate the performance of the coating in general by the friction coefficient. It should represent the certain low values throughout the entire period of operation under specified conditions.

Tribological tests are necessary both for evaluation the coating quality and for composition developing, and therefore are in demand by manufacturers and consumers of solid lubricating coatings. However, the equipment for friction tests available on the market are too expensive. Thus, researchers are forced to develop their own tribological test rigs [1, 7-8].

It is important to use standard test designs so that results obtained by researchers can be compared. Not all types of test installations are equally suitable for modeling the operation of friction units, since the different nature of the relative motion can cause differences in experimental results [9].

The self-lubricating composites low friction coefficient depends on test conditions: load and relative velocity. Therefore, when studying the tribological properties of solid lubricating coatings, it is necessary to ensure the same experimental conditions [2].

Pin-on-Disk or Ball-on-Disk test scheme corresponding to the ASTM G99 is widespread all over the world. The Ball-on-Disk kinematic scheme is preferable for studying friction at high contact loads. The ball-shaped indenter is capable of self-centering, in contrast to the cylindrical or square shape pin. This ensures good reproducibility of results [8, 10].

The purpose of this work is to develop a tribological test rig for materials and coatings in accordance with international standards.

The equipment being developed must meet the following requirements. Test scheme - Ball-on-Disk in accordance with ASTM G99. Friction coefficient measurement data are displayed on a computer screen and recorded in digital form for further processing.

2. Test rig design

According to solid lubricating coatings properties and friction test equipment analysis technical characteristics of installation developed are determined (table 1).

Table 1. Test rig technical characteristics.

Parameter	Value
Kinematic scheme	Ball-on-Disk
Measurement data	Friction coefficient
Friction coefficient range	0.05…1 (±1.5%)
Sliding speed ($m \cdot s^{-1}$)	0.1…1
Maximum normal load (N)	25
Ball diameter (mm)	6
Disk diameter (mm)	40
Maximum torque (Nm)	1
Engine power (kW)	0.3
Size (mm)	360x220x600
Operating environment	Air
Operating temperature (°C)	20…25

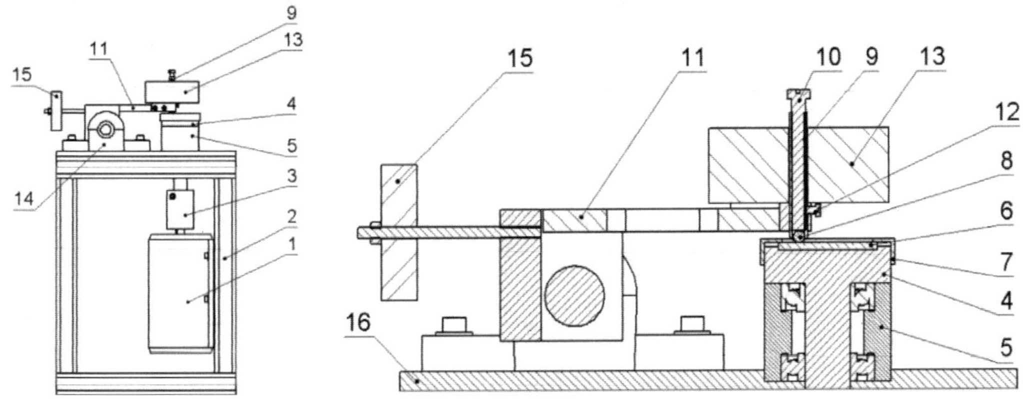

Figure 1. Test rig functional parts: 1 – electric motor, 2 – frame, 3 – coupling, 4 – chuck, 5 – support, 6 – sample, 7 – pressure washer, 8 – ball, 9 – indenter, 10 – screw, 11 – strain gauge, 12 – fixing screw, 13 – weight, 14 – support, 15 – balancing weight, 16 – plate.

The test rig consists of the following functional parts (figure 1).

The electric motor 1 is installed on the frame 2. The rotation of the motor shaft is transmitted by means of the coupling 3 to the cartridge 4, which is installed in the support 5. The support, which is a housing with an angular contact bearing and radial bearing, perceives the axial load and ensures the sample rotation axis is vertical. A sample 6 is fixed with a clamping washer 7 in the cartridge. A metal ball 8 is set at the end of the indenter 9. The ball contacts with the sample and its rolling is limited by a screw 10. The indenter is installed in the strain gauge mounting hole 11. The indenter ability to move vertically allows to test different height samples. It is fixed with a screw 12 in the working position. Normal force is applied to the sample by means of weight 13. Supports 14 allow the load cell to move freely in the vertical plane, bending stresses caused by the weight do not act on the sensor, and thus all normal load is transferred to the sample. To ensure that the weight of the sensor and the indenter does not affect the sample during the test, a balancing weight 15 is provided. All units are mounted on plate 16 to provide the exact mutual position of the indenter and the sample.

The test rig developed 3D model is shown in the figure 2.

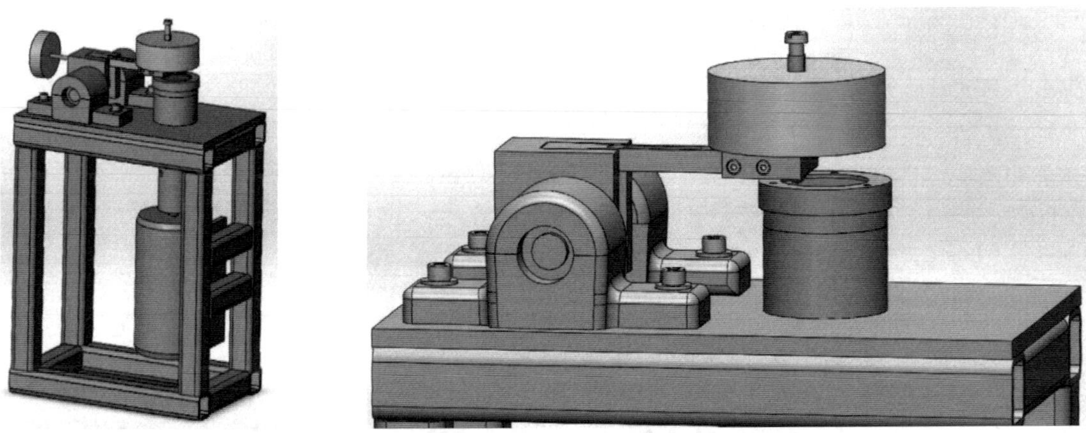

(a) (b)

Figure 2. Test rig model.

The installation is equipped with a strain gauge for continuous friction coefficient measurement during testing. The signals from the strain gauge are processed by the data acquisition system, displayed on the computer screen and written to a file. The test rig launch, the sliding speed and initial test conditions are set using the control system. Sample speed smooth control is achieved by using a frequency converter. The data acquisition and test rig control system is implemented through the National Instruments controller and LabView software (figure 3).

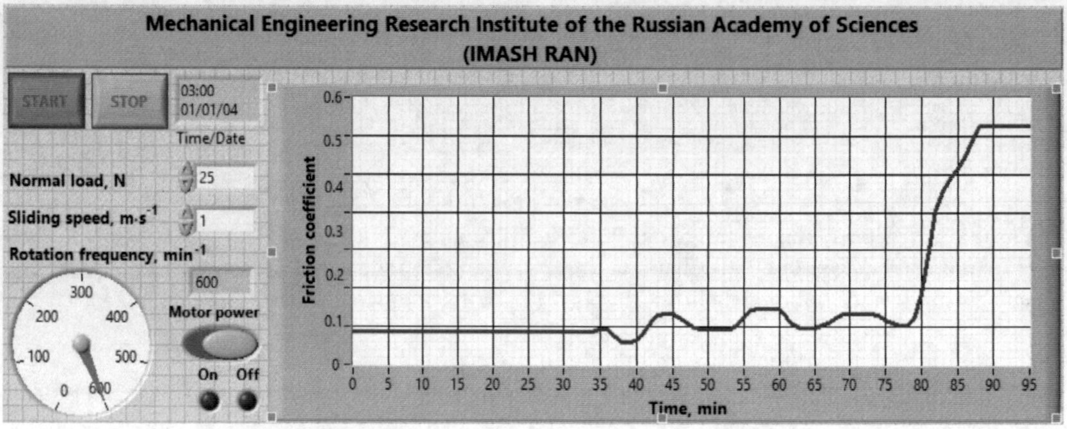

Figure 3. Tribological test LabView program window.

The friction force is measured by the deformation of a single point beam-type strain gauge. For the correct operation, it is necessary that the force measured is directed perpendicular to the sensor. Sensor deformation is caused by the combined action of bending moment and torque. To reduce the measurement error, the sensor is installed as close as possible to the sample surface in order to reduce the shoulder of the friction force.

3. Stress and strain simulation

To evaluate the design performance under operating loads the finite element method analysis of the stresses and strains in the test rig was done (figures 4 and 5). The SolidWorks Simulation software was used. The supports lower surface was fixed in the simulation. The force was applied to the ball and disk contact point. Disk surface was supposed to be flat and perpendicular to the indenter axis. The force 25 N is equivalent to the friction force arising at the maximum normal load and the coefficient of friction (table 1).

Figure 4. Finite element test rig model.

Figure 5. Test rig stress distribution.

The deformations and stresses caused are concentrated mainly on the sensor, and the support system has sufficient stiffness. The maximum stresses turn up in places where the geometry of the sensor changes. The middle sensor area, where the strain gauge is situated, has even stress distribution. At a frictional force of 25 N the maximum stress generated is 30 MPa which is lower than the yield stress of sensor material.

The contact point displacement due to system deformation was also measured. Ball offset leads to wear track cross section area increasing. The warn face shape obtained by profile measurement tends to be more elliptical than spherical. The maximum displacement calculated value seems to be less than 0.13 mm. We expect friction coefficient range obtained in vast majority of studies 0.05…0.2. Then friction force is 1.25…5 N and the real contact point is set 0.008…0.032 mm from theoretical place. This is necessary to consider while wear rate is calculated by the mean of profiling.

4. Conclusion

Lubrication in extreme operating conditions such as high contact loads is usually solid lubricant field. Solid lubricating coatings development and quality evaluation requires friction and wear tests. The Ball-on-Disk configuration is used for high contact loads friction studying.

The Ball-on-Disk tribological test rig model has been developed. The installation developed is suitable for tribological tests according to international standard ASTM G99. Friction coefficient measurement and writing in digital form is available. The design finite element analysis shows the measurement accuracy.

References

[1] Furlan K P, Biasoli de Mello J D and Klein A N 2018 *Tribology Int.* **120** pp 280-98

[2] Khopin P N 2018 Comprehensive assessment of tribotechnical indicators of interfaces with solid lubricating coatings *Doctoral dissertation* (Moscow) p 313

[3] Prozhega M V, Albagachiev A Y, Smirnov N I and Smirnov N N 2018 *J. Frict. Wear* **39** pp 335–40

[4] Khopin P N 2018 *BMSTU J. of Mechanical Engineering* **694** pp 84-93

[5] Ke Q, Qing F, Haoming L, Shuai Z, Mingyuan D and Fuchuan H 2020 *IOP Conf. Series: Earth and Environmental Sci.* **508** 012219

[6] Morgunov A P, Masyagin V B, Derkach V V and Matveev N A 2017 *IOP Conf. Series: J. of Physics: Conf. Series* **858** 012019

[7] Zagidullina D R, Ibatullin I D and Yemel'yanov S G 2016 *Izvestia RAS SamSC* **4(2)** pp 285-90
[8] Hoic M Hrgetic M and Deur J 2016 *Mechatronics* **40** pp 220-32
[9] Red'kin V E Tkachenko Yu S, Sukhodaev P O and Lyamkin A I 2020 *Industrial laboratory. Diagnostics of materials* **86** pp 66-71
[10] Chen Z *et al* 2017 *Tribology Int.* **116** pp 208-16

IOP Conf. Series: Materials Science and Engineering 1129 (2021) 012030 doi:10.1088/1757-899X/1129/1/012030

Solving one dimensional time-space fractional vibration string equation

Temirkhan S. Aleroev [1], **Asmaa M. Elsayed** [1,2,*] **and Elsayed I. Mahmoud** [2,3]

[1]NRU MGSU, Yaroslavskoe Boulevard, 26, Moscow, Russia.

[2]Faculty of Science, Zagazig University, Zagazig 44519, Egypt.

[3] Nikolskii Mathematical Institute, Peoples Friendship University of Russia, Moscow 117198, Russia.

E-mail: asmaa_mahdy1991@yahoo.com

Abstract. The article presents a solution for the one-dimensional space- time fractional vibration equation (FVE) by the separation of variables method (Fourier method). We describe the fractional derivatives in the sense of Caputo and Riemann-Liouville operators. Our method performs in the extreme well in terms of simplicity and efficiency. A sample of the problem of structural mechanics has been considered. This sample allows the demonstration of some advantages of the application of the suggested approach to solve the fractional vibration equation.

1. Introduction

Recently, fractional partial differential equations play an increasingly important role in many fields of science and engineering [1, 2] such as physics,fluid dynamics, signal processing, viscoelasticity, mechanical engineering, electrochemistry focus, mathematical biology and so on. The advantages of the analytical methods in explaining the basics of physical connotation and mechanical engineering problem, by which it is easier to study a different physical and mechanical engineering problem, as well as, it is less time consuming compared with the numerical method. However as is known that it is difficult to obtain the exact solutions of space-time fractional partial differential equations FPDEs by using the existing analytical methods [3, 4]. Thus, different numerical methods have been used to solve space-time FPDEs, see [4, 5, 6]. Also, many numerical methods, such as homotopy perturbation method(HPM), homotopy analysis method(HAM), finite difference method, finite element method, Adomian decomposition method(ADM), etc. are constructed and analyzed to obtain the numerical solutions for fractional partial differential equations.

In this paper, we will consider in the region $D = \{0 < x < L \ , \ 0 < t < T\}$ the following one-dimensional time-space fractional vibration string equation

$$a\frac{\partial^2 V(x,t)}{\partial t^2} = \frac{\partial^2 V(x,t)}{\partial x^2} + c_0 D^{\beta}_{0t} V(x,t) + c_1 D^{\alpha}_{0x} V(x,t), \tag{1}$$

constrained by boundary conditions

$$V(0,t) = V(L,t) = 0, \tag{2}$$

and initial conditions

$$\begin{aligned} V(x,0) &= \varphi(x), \\ V_t(x,0) &= \psi(x), \end{aligned} \tag{3}$$

here, $1 < \alpha, \beta < 2$, are the viscoelasticity parameters of the medium a, c_1 and c_0-are the mass of the granule and the viscosity modulus of resin respectively. D^{β}_{0t} and D^{α}_{0x} denote the temporal Caputo derivative with order β and spatial Riemann-Liouville derivative of order α respectively, V(x,t) is the displacement.

Where the Caputo fractional derivative of $h \in C^m_{-1}$ with order $\beta > 0$, and the Riemann-Liouville fractional derivative of $f \in C^m_{-1}$ with order $\alpha > 0$, for $m \in N \cup \{0\}$ are defined respectively as (see [7]),

$$D^{\beta}_t h(t) = \frac{1}{\Gamma(m-\beta)} \int_0^t (t-\tau)^{m-\beta-1} h^{(m)}(\tau) d\tau, \quad m-1 < \beta < m, m \in N, \tag{4}$$

$$D^{\alpha}_x f(t) = \frac{1}{\Gamma(m-\beta)} \frac{d^m}{dx^m} \int_0^x (x-\xi)^{m-\alpha-1} f(\xi) d\xi, \quad m-1 < \alpha < m, m \in N, \tag{5}$$

which, In particular, equation (1) is used to characterize the vibration of a string taking into account the friction in a fractal geometry medium. This equation is used in this work to model changes in polymer concrete's deformation-strength characteristics under loading. Intrinsically, equation (1) is a particular form of multi-term space -time fractional wave equations. Many works on analytical and numerical methods for space-time fractional wave equations have been existed, see [4, 5, 8]. The rest of the paper is organized as follows. In section 2, we apply the separation of variables scheme based on Fourier method for solving equation (1). Section 3, is devoted to illustrate some numerical example on mentioned method. In section 4, we give the conclusion.

2. Vibration string equation and Fourier method

In this section, by using the method of separating variables, we solve the vibration string equation.(1). Assume $V(x,t) = \chi(x)T(t)$ and replacing $V(x,t)$ in equation (1), we obtain on the following differential equation for $\chi(x)$:

$$\chi''(x) + c_1 D^{\alpha}_{0x}\chi(x) + \lambda\chi(x) = 0; \chi(0) = 0, \chi(L) = 0, \tag{6}$$

the parameter λ is an eigenvalue of the equation (6), if and only if λ is the zero of the function

$$\omega(\lambda) = \chi + \sum_{n=1}^{\infty} (-1)^n \sum_{k=0}^{n} \frac{\binom{n}{k} c_1^k \lambda^{n-k}}{\Gamma(2n+2-k\alpha)} \tag{7}$$

We find the first five eigenfunctions of the boundary value problem (6) numerically using the high-level technical computing language Wolfram Mathematica, and plot their graphs, taking $\alpha = 1.5$ and $c_1 = 0.5$, where the first five eigenvalues [8] are written in table 1.

Now to seek a solution of equation.(6), substitute from equation (5) in (6) and integrate equation(6) from 0 to x, we have

$$\int_0^x \chi''(t)dt + \int_0^x c_1 D^{\alpha}_{0x}\chi(t)dt = \int_0^x \lambda\chi(t)dt = 0; \tag{8}$$

$$\chi'(x) + \frac{c_1}{\Gamma(1-\alpha)} \int_0^x \chi(t)(x-t)^{-\alpha}dt = \lambda \int_0^x \chi(t)dt + \chi'(0); \tag{9}$$

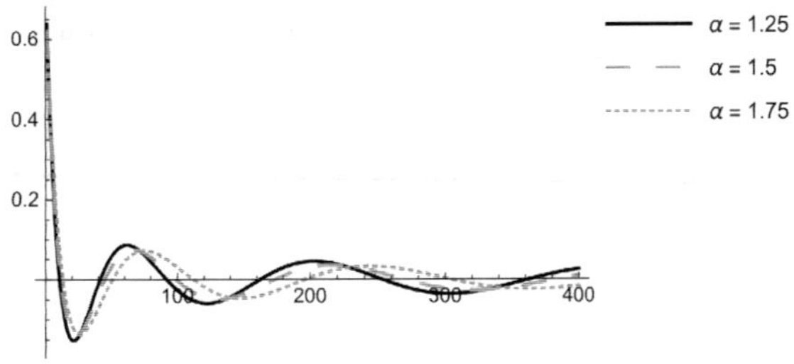

Figure 1. The graph of the function $\omega(\lambda)$ for $c_1 = 0.5$, at different $\alpha = 1.25, 1.5, 1.75$.

Table 1. The first five eigen values of equation (6) at $c_1 = 0.5$ and $\alpha = 1.5$

λ_1	λ_2	λ_3	λ_4	λ_5
11.63	44.93	99.01	173.73	268.93

integrate both sides again from 0 to x,

$$\chi(x) + \int_0^x \frac{c_1}{\Gamma(2-\alpha)}(x-t)^{1-\alpha} - \lambda(x-t)\chi(x)dt = x.A + B, \tag{10}$$

where, $\chi'(0) = A$, and $\chi(0) = B$. Notice that the the solution of equation(10) is

$$\chi(x) = x_0(x) + ... + x_n(x), \tag{11}$$

such that,

$$x_0(x) = x.A + B, x_n(x) = \int_0^x \left[\frac{-c_1}{\Gamma(2-\alpha)}(x-t)^{1-\alpha} + \lambda(x-t)\right] x_{n-1}(t)dt, \tag{12}$$

integrating equation(12) and substituting in equation (11), then we get a solution of equation (6)

$$\chi_j(x) = A\left[x + \sum_{n=1}^{\infty}(-1)^n \sum_{k=0}^{n} \frac{\binom{n}{k} c_1^k \lambda_j^{n-k}}{\Gamma(2n+2-k\alpha)} x^{2n+1-k\alpha}\right]; j = 1, 2, ..., 5. \tag{13}$$

The graphs of the first five eigenfunctions $\chi(x)$ of equation(6) are presented in figure 2 at $c_1 = 0.5$, $\alpha = 1.5$.

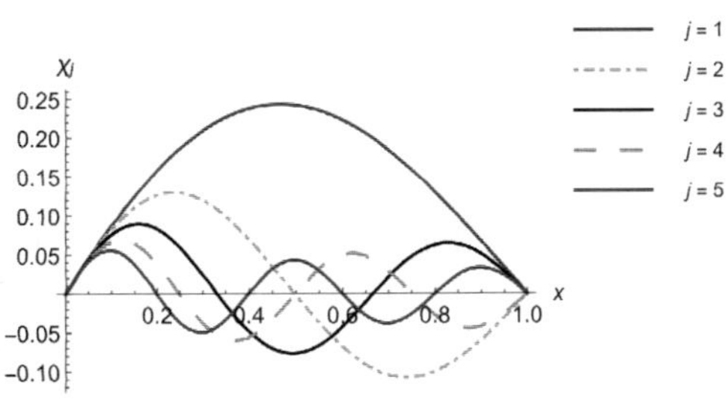

Figure 2. The graphs of the eigenfunctions $\chi_j(x); j = 1, 2, .., 5$ at $c_1 = 0.5$, $\alpha = 1.5$.

IOP Conf. Series: Materials Science and Engineering 1129 (2021) 012030 doi:10.1088/1757-899X/1129/1/012030

The system of eigenfunctions (13) is complete [9] anyhow, not orthogonal. Thus, we arise a system that will be biorthogonal to the system (13); therefore, we consider the operator $A(f) = f'' + c_1 D^{\alpha}_{0x} f$, so assume that the operator A^* is conjugated with the operator A, such that $\langle A(f), h\rangle = \langle f, A^*(h)\rangle$, where we have the scalar product $\langle f, h\rangle = \int_0^L f(x).h(x)dx$.

Let us assume that the functions f and g satisfy the boundary conditions (6), consider

$$\langle f, h\rangle = \langle f'' + c_1 D^{\alpha}_{0x} f, h\rangle, \tag{14}$$

we use the self-adjoint operator and linear differential operator

$$\langle f'' + c_1 D^{\alpha}_{0x} f, h\rangle = \langle h'', f\rangle + c_1.\langle D^{\alpha}_{0x} f, h\rangle, \tag{15}$$

further,

$$\langle D^{\alpha}_{0x} f, h\rangle = \int_0^L \frac{1}{\Gamma(2-\alpha)}\{\frac{d^2}{dx^2}\int_0^x (x-\xi)^{1-\alpha} f(\xi)d\xi\}h(x)dx, \tag{16}$$

let

$$V(x) = \int_0^x (x-\xi)^{1-\alpha} f(\xi)d\xi,$$

then

$$\begin{aligned}\int_0^L V''(x)h(x)dx &= V'(x)h(x)|_0^x - \int_0^L V'(x)h'(x)dx \\ &= -V(x)h'(x)|_0^x + \int_0^L V(x)h''(x)dx \\ &= -V(L)h'(L) + V(0)h'(0) + \int_0^L V(x)h''(x)dx, \end{aligned}\tag{17}$$

hence, equation (16) can be written as,

$$\begin{aligned}\langle D^{\alpha}_{0x} f, h\rangle &= \tfrac{1}{\Gamma(2-\alpha)}\left(\int_0^L V(x)h''(x)dx - V(L)h'(L)\right) \\ &= \tfrac{1}{\Gamma(2-\alpha)}\int_0^L \{\int_0^x (x-\xi)^{1-\alpha} f(\xi)d\xi\}\, h''(x)dx \\ &\quad - \tfrac{h'(L)}{\Gamma(2-\alpha)}\int_0^L (L-\xi)^{1-\alpha} h(\xi)d\xi, \end{aligned}\tag{18}$$

by change the limits of integration in the double integral, so equation(18), can be written as,

$$\langle D^{\alpha}_{0x} f, h\rangle = \frac{1}{\Gamma(2-\alpha)}\int_0^L \left\{\int_\xi^L (x-\xi)^{1-\alpha} h''(x)dx - h'(L)(L-\xi)^{1-\alpha}\right\} f(\xi)d\xi, \tag{19}$$

by using the relationship between Caputo fractional derivative, and Riemann-Liouville fractional derivative, we obtain:

$$\frac{(-1)^n}{\Gamma(n-\alpha)}\frac{d^n}{dt^n}\int_\xi^L \frac{h(s)}{(s-t)^{\alpha-1}}ds = \frac{1}{\Gamma(n-\alpha)}\int_\xi^L \frac{h^{(n)}(s)}{(s-t)^{\alpha-1}}ds + \sum_{k=0}^{n-1}(-1)^k\frac{h^{(k)}(L).(L-t)^{k-\alpha}}{\Gamma(k-\alpha+1)}, \tag{20}$$

therefore,

$$\langle D^{\alpha}_{0x} f, h\rangle = \frac{1}{\Gamma(2-\alpha)}\int_0^L \frac{d^2}{d\xi^2}\left\{\int_\xi^L (x-\xi)^{1-\alpha} h(x)dx\right\} f(\xi)d\xi = \langle f, D^{\alpha}_{xL} h\rangle, \tag{21}$$

hence,

$$D_{xL}^{\alpha}h = \frac{1}{\Gamma(2-\alpha)}\frac{d^2}{d\xi^2}\int_{\xi}^{L}(x-\xi)^{1-\alpha}h(x)dx. \tag{22}$$

Consequently, the adjoint operator

$$A^*(h) = h'' + c_1 D_{xL}^{\alpha}h,$$

therefore, we consider the following conjugated problem associated with problem (6)

$$\begin{aligned}&\widetilde{\chi}''(x) + c_1 D_{0x}^{\alpha}\widetilde{\chi}(x) + \lambda\widetilde{\chi}(x) = 0,\\ &\widetilde{\chi}(0) = \widetilde{\chi}(L) = 0; \qquad \alpha \in (1,2),\end{aligned} \tag{23}$$

where, $\widetilde{\chi}_j(x)$ be the eigenfunctions of problem (23), then

$$\widetilde{\chi}_j(x) = \chi_j(1-x). \tag{24}$$

Hence,

$$\widetilde{\chi}_j(x) = \sum_{n=0}^{\infty}(-1)^n\sum_{k=0}^{n}\frac{\binom{n}{k}c_1^k\lambda_j^{n-k}}{\Gamma(2n+2-k\alpha)}(1-x)^{2n+1-k\alpha}; j = 1,2,...,5. \tag{25}$$

The graphs of the first four eigenfunctions of $\chi_j(x)$ and $\widetilde{\chi}_j(x)$ are presented in figure 3.

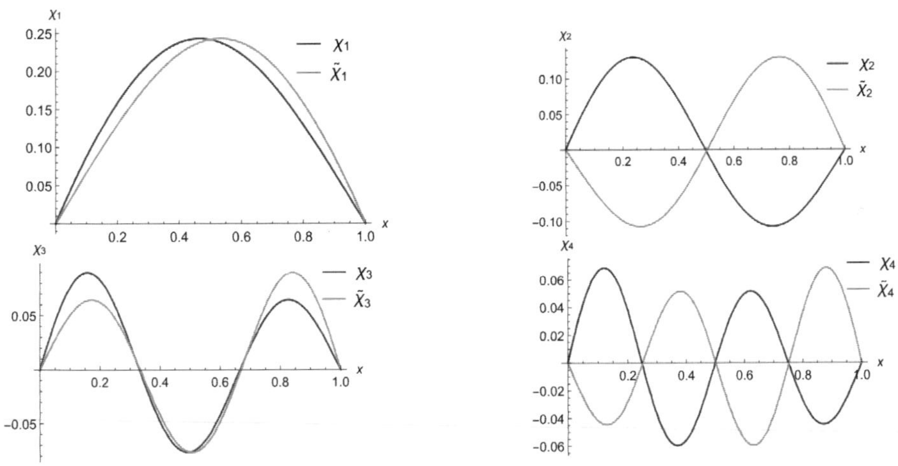

Figure 3. The graphs of the eigenfunctions $\chi_j(x), \widetilde{\chi}_j(x)$;$j = 1,...,4$ at $c_1 = 0.5$, $\alpha = 1.5$.

Utilizing the high-level technical computing language Wolfram Mathematica, we calculate the values of the scalar products $\langle\chi_m(x), \widetilde{\chi}_k(x)\rangle$, which are written in table 2.

Also, for $T(t)$ we obtain on the following equation of Caputo fractional derivative:

$$aT''(t) - c_0 D_{0t}^{\beta}T(t) + \lambda T(t) = 0. \tag{26}$$

Now, in order to seek a solution of equation (26), let us take

$$T''(t) = Y(t), \tag{27}$$

Table 2. The scalar products of the eigenfunctions $\langle \chi_m, \widetilde{\chi}_k \rangle$ at $c_1 = 0.5$ and $\alpha = 1.5$

$\langle \chi_m, \widetilde{\chi}_k \rangle$	χ_1	χ_2	χ_3	χ_4	χ_5
$\widetilde{\chi}_1$	0.02948	0	0	0	0
$\widetilde{\chi}_2$	0	−0.00695	0	0	0
$\widetilde{\chi}_3$	0	0	0.00289	0	0
$\widetilde{\chi}_4$	0	0	0	−0.00153	0
$\widetilde{\chi}_5$	0	0	0	0	0.00092

by integrating both sides from 0 to t

$$\begin{aligned} T'(t) &= \int_0^t Y(s)ds + T'(0); \\ T(t) &= \int_0^t (t-s)Y(s)ds + T'(0)t + T(0), \end{aligned} \tag{28}$$

substituting from equations (4), (27) and (28) in (26) and put $T'(0) = A, T(0) = B$, so

$$Y(t) + \int_0^t \left[\frac{-c_0}{a\Gamma(2-\beta)}(t-\zeta)^{1-\beta} + \frac{\lambda}{a}(t-\zeta) \right] Y(\zeta)d\zeta = -\frac{\lambda}{a}At - \frac{\lambda}{a}B, \tag{29}$$

therefore, the solution is

$$\begin{aligned} Y(t) = \frac{\lambda}{a}A &\left[-t + \sum_{m=1}^{\infty}(-1)^{m+1}\sum_{k=0}^{m} \frac{\binom{m}{k}(-c_0)^k\lambda^{m-k}}{a^m\Gamma(2m+2-k\beta)} t^{2m+1-k\beta} \right] \\ + \frac{\lambda}{a}B &\left[-1 + \sum_{m=1}^{\infty}(-1)^{m+1}\sum_{k=0}^{m} \frac{\binom{m}{k}(-c_0)^k\lambda^{m-k}}{a^m\Gamma(2m+1-k\beta)} t^{2m-k\beta} \right], \end{aligned} \tag{30}$$

hence,

$$\begin{aligned} T'(t) = A &\left[1 - \frac{\lambda t^2}{2a} + \sum_{m=1}^{\infty}(-1)^{m+1}\sum_{k=0}^{m} \frac{\binom{m}{k}(-c_0)^k\lambda^{m+1-k}}{a^{m+1}\Gamma(2m+3-k\beta)} t^{2m+2-k\beta} \right] \\ + B &\left[-\frac{\lambda t}{a} + \sum_{m=1}^{\infty}(-1)^{m+1}\sum_{k=0}^{m} \frac{\binom{m}{k}(-c_0)^k\lambda^{m+1-k}}{a^{m+1}\Gamma(2m+2-k\beta)} t^{2m+1-k\beta} \right], \end{aligned} \tag{31}$$

and

$$\begin{aligned} T(t) = A &\left[t - \frac{\lambda t^3}{6a} + \sum_{m=1}^{\infty}(-1)^{m+1}\sum_{k=0}^{m} \frac{\binom{m}{k}(-c_0)^k\lambda^{m+1-k}}{a^{m+1}\Gamma(2m+4-k\beta)} t^{2m+3-k\beta} \right]; \\ + B &\left[1 - \frac{\lambda t^2}{2a} + \sum_{m=1}^{\infty}(-1)^{m+1}\sum_{k=0}^{m} \frac{\binom{m}{k}(-c_0)^k\lambda^{m+1-k}}{a^{m+1}\Gamma(2m+3-k\beta)} t^{2m+2-k\beta} \right], \end{aligned} \tag{32}$$

put,

$$Z(t) = t - \frac{\lambda t^3}{6a} + \sum_{m=1}^{\infty}(-1)^{m+1}\sum_{k=0}^{m} \frac{\binom{m}{k}(-c_0)^k\lambda^{m+1-k}}{a^{m+1}\Gamma(2m+4-k\beta)} t^{2m+3-k\beta}, \tag{33}$$

$$\widetilde{Z}(t) = 1 - \frac{\lambda t^2}{2a} + \sum_{m=1}^{\infty}(-1)^{m+1}\sum_{k=0}^{m}\frac{\binom{m}{k}(-c_0)^k\lambda^{m+1-k}}{a^{m+1}\Gamma(2m+3-k\beta)}t^{2m+2-k\beta}, \tag{34}$$

which, deduce that the solution of equation (26) can be written as

$$T(t) = A.Z(t) + B.\widetilde{Z}(t). \tag{35}$$

Then, from equation (13) and equation (35) yields that the solution of the problem (1)-(3) has the following standard form as

$$V(x,t) = \sum_{m=1}^{\infty}\chi_m(x)T_m(t) = \sum_{m=1}^{\infty}\chi_m(x)\left[A_m.Z_m(t) + B_m.\widetilde{Z}_m(t)\right]. \tag{36}$$

Using the initial condition (2), that is lead to substituting in equation.(36) about $t = 0$

$$\begin{aligned} V(x,0) &= \sum_{m=1}^{\infty}\chi_m(x)\left[A_m.Z_m(0) + B_m.\widetilde{Z}_m(0)\right] = \varphi(x); \\ V_t(x,0) &= \sum_{n=1}^{\infty}\chi_m(x)\left[A_m.Z'_m(0) + B_m.\widetilde{Z}'_m(0)\right] = \psi(x), \end{aligned} \tag{37}$$

multiply both sides of the equation (37) with a scalar system $\widetilde{\chi}_m(x)$, so

$$\begin{aligned} &\sum_{m=1}^{\infty}\langle\chi_m(x),\widetilde{\chi}_m(x)\rangle\left[A_m.Z_m(0) + B_m.\widetilde{Z}_m(0)\right] = \langle\varphi(x),\widetilde{\chi}_m(x)\rangle; \\ &\sum_{m=1}^{\infty}\langle\chi_m(x),\widetilde{\chi}_m(x)\rangle\left[A_m.Z'_m(0) + B_m.\widetilde{Z}'_m(0)\right] = \langle\psi(x),\widetilde{\chi}_m(x)\rangle, \end{aligned} \tag{38}$$

since, $\{\chi_m(x)\}_{m=1,2,..}$ and $\{\widetilde{\chi}_m(x)\}_{m=1,2,..}$ are biorthogonal systems, then equation (38) can be written as

$$\begin{aligned} A_m.Z_m(0) + B_m.\widetilde{Z}_m(0) &= \frac{\langle\varphi(x),\widetilde{\chi}_m(x)\rangle}{\langle\chi_m(x),\widetilde{\chi}_m(x)\rangle}; \\ A_m.Z'_m(0) + B_m.\widetilde{Z}'_m(0) &= \frac{\langle\psi(x),\widetilde{\chi}_m(x)\rangle}{\langle\chi_m(x),\widetilde{\chi}_m(x)\rangle}, \end{aligned} \tag{39}$$

then, the solution of the problem (1)-(3) has the following form

$$V(x,t) = \sum_{m=1}^{\infty}\frac{\chi_m(x)}{\langle\chi_m(x),\widetilde{\chi}_m(x)\rangle}\left[Z_m(t)\langle\psi(x),\widetilde{\chi}_m(x)\rangle + \widetilde{Z}_m(t)\langle\varphi(x),\widetilde{\chi}_m(x)\rangle\right]. \tag{40}$$

3. Numerical Construction of the Solution

The numerical solution to the problem (1)-(3) can be written as:

$$V(x,t) \approx V_{(5)}(x,t) = \sum_{m=1}^{100}\chi_m(x)\left[A_m.Z_m(t) + B_m.\widetilde{Z}_m(t)\right]. \tag{41}$$

Consider the following example for illustrating the efficiency of the method. ($a = 1, c_1 = 0.5, c_0 = -1.8$ at $\alpha = 1.5, \beta = 1.47$),

$$a\frac{\partial^2 V(x,t)}{\partial t^2} = \frac{\partial^2 V(x,t)}{\partial x^2} + c_0 D^{\beta}_{0t} V(x,t) + c_1 D^{\alpha}_{0x} V(x,t), \quad 0 < x, t < 1, \tag{42}$$

constrained by the conditions,

$$\begin{aligned} V(0,t) &= V(1,t) = 0, \\ V(x,0) &= 0.1x(1-x), \\ V_t(x,0) &= -x^3. \end{aligned} \tag{43}$$

We could construct graphs for a solution by formula (41) numerically by using the high-level technical calculation language Wolfram Mathematica, where figure 4, and figure 5 show the solution at $a = 1$, $c_1 = 0.5$, $c_0 = -1.8$ when $\alpha = 1.5$, $\beta = 1.47$.

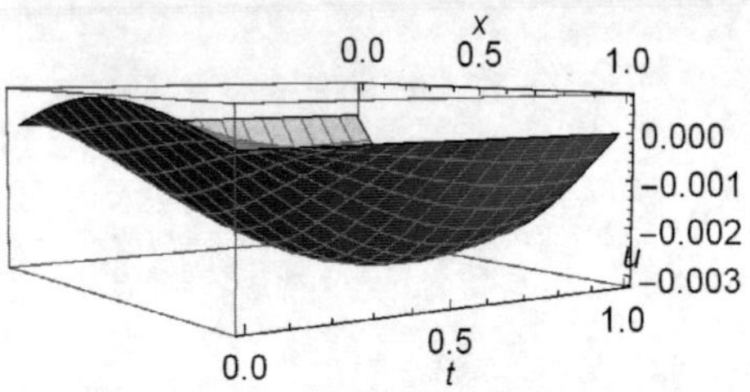

Figure 4. Graph the solution $V_{(5)}(x, t)$ of equation (41) at c_1=0.5, c_0=−1.8 when α=1.5, β=1.47.

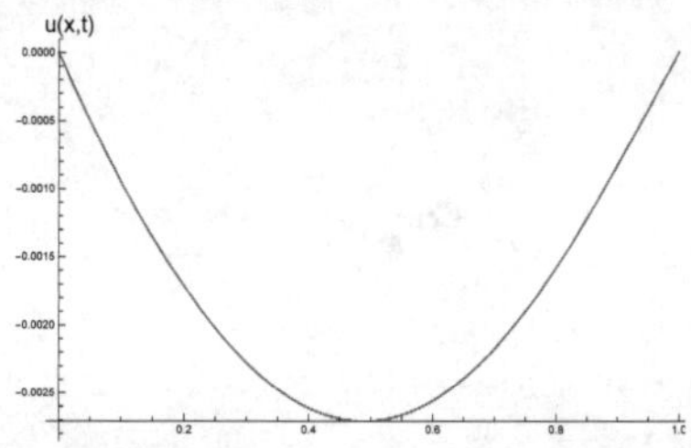

Figure 5. Graph the solution at $t = 1, c_1 = 0.5, c_0 = -1.8$ when $\alpha = 1.5, \beta = 1.47$.

4. Conclusion

In this paper, the vibration string equation of a fractional derivative with respect to time and space has been characterized and displayed. We obtained the solution of the vibration string equation by Fourier method. The example shows some advantages of the suggested approach to solve the fractional vibration string equation.

References

[1] Herrmann R 2014 Fractional Calculus: An Introduction for Physicists (2nd Edition) *World Scientific, Singapore*
[2] Sandev T and Tomovski Z 2019 Fractional Equations and Models: Theory and Applications *Springer, Berlin*
[3] Luchko Y 2019 Subordination principles for the multi-dimensional space-time fractional diffusion wave equations *Theor. Probab. Math. Stat.* **98** pp 127–47
[4] Aleroev T and Elsayed A 2020 Analytical and approximate solution for solving the vibration string equation with a fractional derivative *Mathematics* **8(7)** 1154
[5] Elsayed A and Orlov V 2020 Numerical scheme for solving time–space vibration string equation of fractional derivative *Mathematics* **8(7)** 1069
[6] Chen H, Xu D and Peng Y 2017 A second order BDF alternating direction implicit difference scheme for the two-dimensional fractional evolution equation *Appl. Math. Model* **41** pp 54–67
[7] Podlubny I 1999 Fractional Differential Equations *Academic Press New York USA*
[8] Aleroev T 2020 The analysis of the polymer concrete characteristics by fractional calculus *IOP Conf. Series: J. of Physics* 1425
[9] Agibalova A 2010 On the completeness of the systems of root functions of a fractional-order differential operator with matrix coefficients *Mat.Zametki* **88** pp 317–20

IOP Conf. Series: Materials Science and Engineering 1129 (2021) 012031 doi:10.1088/1757-899X/1129/1/012031

Mathematical model of the polymer concrete by fractional calculus with respect to a spatial variable

Temirkhan S Aleroev [1], Elsayed I Mahmoud [2,3] and Asmaa M Elsayed [1,2]

[1] Moscow State University of Civil Engineering, 129337 Moscow, Russia
[2] Zagazig University, Zagazig 44519, Egypt
[3] Nikolskii Mathematical Institute, Peoples Friendship University of Russia, Moscow 117198, Russia

asmaa_mahdy1991@yahoo.com

Abstract. The article presents a solution of the vibration string equation containing the fractional derivative concerning the spatial variable, in which the fractional derivative is defined by Caputo. This model is utilized for characterizing the oscillation mechanical process in a viscoelastic medium. The theoretical solution is presented, and the fractional derivative parameter is determined. We compared the theoretical solution and the experimental data for polymer concrete samples. A sample of the problem of structural mechanics has been considered. This sample allows the demonstration of some advantages of the application of the suggested approach to solve the fractional vibration equation.

1. Introduction

The partial differential equations (PDEs) for fractional order arise in many different areas of engineering and science. The use of fractional calculus has much attention to developing the improved mathematical modeling of several real-world problems. The theoretical development and application of the fractional calculus have been discussed by many authors in the previous works. Fractional differential equations (FDEs) are more realistic to present natural phenomena, and they have been used in several areas of applied mathematics. In engineering applications, fractional derivatives are advantageous for characterizing the happening of vibrations. In numerous researches, studies concerning fractional calculus and its applications to mechanical problems show widely [1].

Firstly, we notice that the fractional derivatives concerning time could be used to model some processes with a “memory". Particular attention must be paid to the equation of the form:

$$\frac{\partial^2 u}{\partial t^2} = \frac{\partial^2 u}{\partial x^2} + c_1 \cdot D_{ox}^{\alpha} u + c_0 \cdot D_{ot}^{\beta} u + F$$

That is used to characterize the vibration of string in a medium with fractal geometry considering friction, where this equation is used in this work to model changes in polymer concrete's deformation-strength characteristics under loading. This model is intrinsically a form of multi term space-time fractional wave equations. So far, there exist many works on analytical and numerical methods for multi term space -time fractional vibration string equation and fractional wave equations, see [2, 3]. The remnants of the article are arranged as follows. In section 2, we apply the separation of variables scheme based on the Fourier method for solving equation (1). In section 3, Parameter identification and experimental Studies are devoted.

2. Vibration string equation and Fourier method

In the region $\boldsymbol{D} = \{\boldsymbol{0} < \boldsymbol{x} < \boldsymbol{L}, \boldsymbol{t} > 0\}$ assume the vibration string equation with respect to fractional derivative of order α and partial variable

$$\frac{\partial^2 u}{\partial t^2} = \frac{\partial^2 u}{\partial x^2} + c_1 \cdot D_{ox}^{\alpha} u, \tag{1}$$

with boundary conditions

$$u(0,t) = u(1,t) = 0, \tag{2}$$

and initial conditions

$$u(x,0) = \varphi(x); \tag{3}$$

$$u_t'(x,0) = \psi(x), \tag{4}$$

where, $\boldsymbol{0} < \boldsymbol{\alpha} < 1$, is the viscoelasticity parameters of the medium, $\boldsymbol{c_1}$ is the viscosity modulus of resin, $\boldsymbol{D_{ox}^{\alpha}}$ is the Caputo fractional derivative of order $\boldsymbol{\alpha}$ with respect to $\boldsymbol{x}$. $\boldsymbol{u(x,t)}$ is the displacement.

The Caputo fractional derivative of the function $\boldsymbol{f(t)} \in \boldsymbol{C_{-1}^{m}}$ with order $\boldsymbol{\alpha} > 0, \boldsymbol{m} \in \boldsymbol{N} \cup \{\boldsymbol{0}\}$, is defined by the formula

$$D_{ox}^{\alpha} f(x) = \frac{1}{\Gamma(m-\alpha)} \int_0^x \frac{f^{(m)}(t)dt}{(x-t)^{\alpha+1-m}}, m-1 < \alpha < m. \tag{5}$$

To solve the vibration string equations (1)-(4) by using the method of separating variables, assume

$$u(x,t) = \chi(x)T(t), \tag{6}$$

and substituting from equation (6) in equation (1), we get on an ordinary linear differential equation for $\boldsymbol{T(t)}$:

$$T''(t) = \lambda T(t), \tag{7}$$

where λ is a positive constant parameter (eigenvalues), then the solution of the equation (7) is written as

$$T(t) = a\cos\left(t\sqrt{-\lambda}\right) + b\sin\left(t\sqrt{-\lambda}\right). \tag{8}$$

Also, for $\boldsymbol{\chi(x)}$ we obtain on a fractional ordinary linear differential equation with the Caputo derivative:

$$\begin{gathered} \chi''(x) + c_1 \cdot D_{ox}^{\alpha}\chi(x) = \lambda\chi(x); \\ \chi(0) = 0;\ \chi(1) = 0. \end{gathered} \tag{9}$$

Then the solution of the problem (1)-(4) can be written in the standard way:

$$u(x,t) = \sum_{m=1}^{\infty} \left(a_m\, cos\left(t\sqrt{-\lambda_m}\right) + b_m\, sin\left(t\sqrt{-\lambda_m}\right)\right)\chi_m(x). \tag{10}$$

The number λ is an eigenvalue of problem (9) - (10), if and only if $\boldsymbol{\lambda}$ is the zero of the function

$$\omega(\lambda) = \sum_{n=0}^{\infty}\sum_{k=0}^{n+1} \frac{\binom{n+1}{k}(-c_1)^k \lambda^{n-k}}{\Gamma(2n+4-k\alpha)}. \tag{11}$$

The next figure shows a graph of the function $\boldsymbol{\omega(\lambda)}$ for the case $\mathbf{0} < \boldsymbol{\alpha} < 1$ and $\boldsymbol{c_1} = \mathbf{1.8}$ on the interval $\boldsymbol{\lambda} \in (\mathbf{1}; \mathbf{610})$.

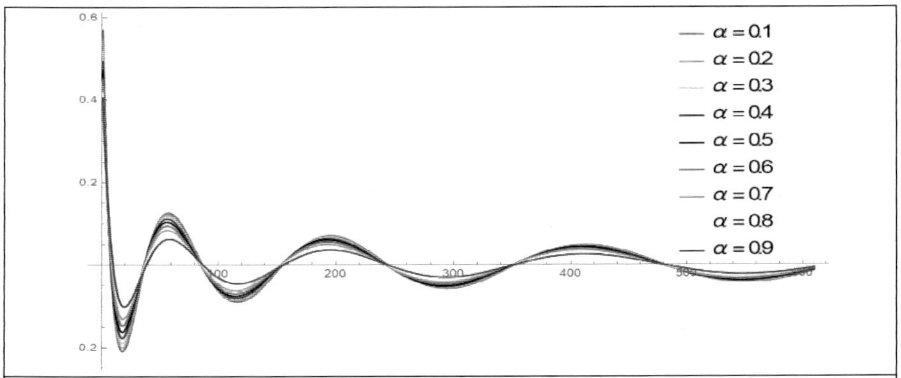

Figure 1. The graph of the function (λ) for $0 < \alpha < 1$ and $c_1 = 1.8$ on the interval $\lambda \in$ (1;610).

Now to seek a solution of equation (8), let us take

$$\chi''(x) = Y(x), \tag{12}$$

integrate both sides from 0 to x

$$\chi'(x) = \int_0^x Y(t)\, dt + \chi'(0), \tag{13}$$

here we used the Cauchy formula for repeated integration:

$$\int_{x_0}^{x}\int_{x_0}^{x} \dots\dots\dots (n - times) \dots \int_{x_0}^{x} f(x)\, dx = \frac{1}{(n-1)!}\int_{x_0}^{x} (x-z)^{n-1} f(z) dz\,, \tag{14}$$

then we integrate both sides of equation (13) from 0 to x we obtain

$$\chi(x) = \int_0^x (x-t) Y(t) dt + \chi(0) + x.\chi'(0), \tag{15}$$

substituting from equation (5), equation (12) and equation (15) in equation (8)

$$Y(x) + \frac{c_1}{\Gamma(1-\alpha)}\int_0^x (x-t)^{-\alpha}\left[\int_0^t Y(\xi)\, d\xi + \chi'(0)\right] dt = \lambda\left[\int_0^x (x-t)Y(t)dt + \chi(0) + x.\chi'(0)\right], \tag{16}$$

let's set $\chi'(0) = A,\ \chi(0) = B$. So

$$Y(x) + \int_0^x \left[\frac{c_1}{\Gamma(2-\alpha)}(x-t)^{1-\alpha} - \lambda(x-t)\right] Y(t) dt = \lambda.B + \lambda\, A.x - A.\frac{c_1}{\Gamma(2-\alpha)} x^{1-\alpha}. \tag{17}$$

The solution of equation (17) is

$$Y(x) = y_0(x) + \dots + y_n(x), \tag{18}$$

where:

$$y_0(x) = \lambda.B + \lambda A.x - A.\frac{c_1}{\Gamma(2-\alpha)}x^{1-\alpha}, \tag{19}$$

and

$$y_n(x) = \int_0^x [\frac{(-c_1)}{\Gamma(2-\alpha)}(x-t)^{1-\alpha} + \lambda(x-t)]y_{n-1}(t)dt. \tag{20}$$

By integrating equation (20), and consider equation (10), so we get on

$$y_0(x) = \lambda A.x - A.\frac{c_1}{\Gamma(2-\alpha)}x^{1-\alpha}, \tag{21}$$

$$y_n(x) = A\left[\sum_{k=0}^{n+1} \frac{\binom{n}{k}(-c_1)^k \lambda^{n-k}}{\Gamma(2n+2-k\alpha)} \chi^{2n+1-k\alpha}\right]. \tag{22}$$

then,

$$Y(x) = A\left[\sum_{n=0}^{\infty}\sum_{k=0}^{n+1} \frac{\binom{n+1}{k}(-c_1)^k \lambda^{n+1-k}}{\Gamma(2n+2-k\alpha)} \chi^{2n+1-k\alpha}\right], \tag{23}$$

$$\chi'^{(x)} = A\left[1 + \sum_{n=0}^{\infty}\sum_{k=0}^{n+1} \frac{\binom{n+1}{k}(-c_1)^k \lambda^{n+1-k}}{\Gamma(2n+3-k\alpha)} \chi^{2n+2-k\alpha}\right], \tag{24}$$

and

$$\chi(x) = A\left[x + \sum_{n=0}^{\infty}\sum_{k=0}^{n+1} \frac{\binom{n+1}{k}(-c_1)^k \lambda^{n+1-k}}{\Gamma(2n+4-k\alpha)} \chi^{2n+3-k\alpha}\right], \tag{25}$$

which is a solution of equation (8), then substituting from equation (26) into equation (11), deduce the solution of equations (1)-(4). We find the first several eigenfunctions of the boundary value problem (8) - (9) numerically using the high-level technical calculation language Wolfram Mathematica, and plot their graphs, taking $0 < \alpha < 1$ and $c_1 = 1.8$, where the first five eigenvalues [4] are written below in table1:

Table 1. The first five eigenvalues.

λ1	**λ2**	**λ3**	**λ4**	**λ5**
-16,51	-59,49	-125,13	-213,33	-323,27

The eigenfunctions of problem (8) - (9) can be written as

$$\chi_m(x) = x + \sum_{n=0}^{\infty}\sum_{k=0}^{n+1} \frac{\binom{n+1}{k}(-c_1)^k \lambda_m{}^{n+1-k}}{\Gamma(2n+4-k\alpha)} \chi^{2n+3-k\alpha}\ ;\ m = 1,2,\ldots \tag{26}$$

The system of eigenfunctions (26) is complete [5], but not orthogonal. Thus, we construct a system that will be biorthogonal to system (26); therefore, we consider the following problem associated with problem (9) - (10).

$$\chi''(x) + c_1 \cdot D_{x1}^{\alpha}\chi(x) = \lambda\chi(x), \tag{27}$$

IOP Conf. Series: Materials Science and Engineering 1129 (2021) 012031 doi:10.1088/1757-899X/1129/1/012031

where: D_{x1}^{α} is the fractional differentiation operator which is a start from x to 1, i.e. D_{x1}^{α} is conjugated with operator of fractional differentiation D_{0x}^{α} [6]

$$D_{x1}^{\alpha} f(x) = \frac{1}{\Gamma(1-\alpha)} \int_x^1 \frac{f'(t)dt}{(t-x)^{\alpha}}. \tag{28}$$

The solution of problem (28) leads to the following relation

$$\tilde{\chi}_m(x) = \chi_m(1-x), \qquad m = 1; 2; .... \tag{29}$$

The graphs of the first five functions $\chi_m(x)$ and the biorthogonal $\tilde{\chi}_m(x)$ functions are presented in the following figures.

Figure 2 show the graph of the functions $\chi_m(x), \tilde{\chi}_m(x)$; $m = 1,2,\ldots$; $0 < \alpha < 1$, c_1=1.8.

α = **0.1**

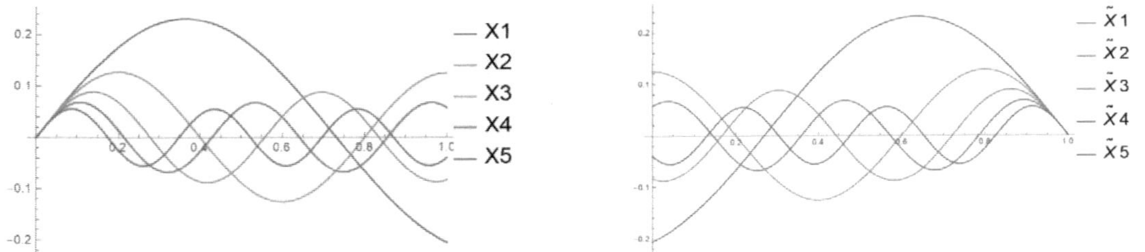

α = **0.2**

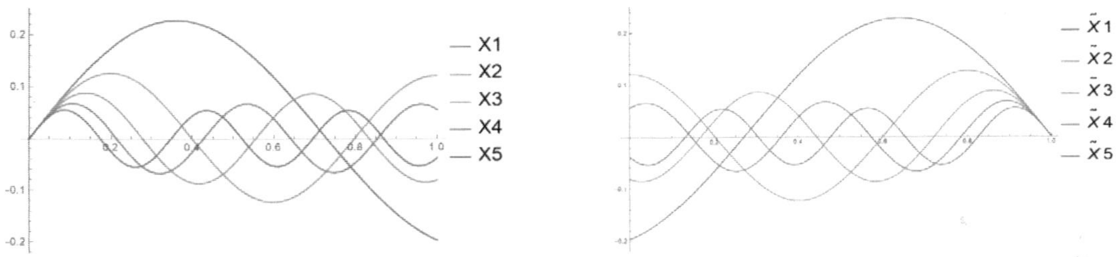

α = **0.3**

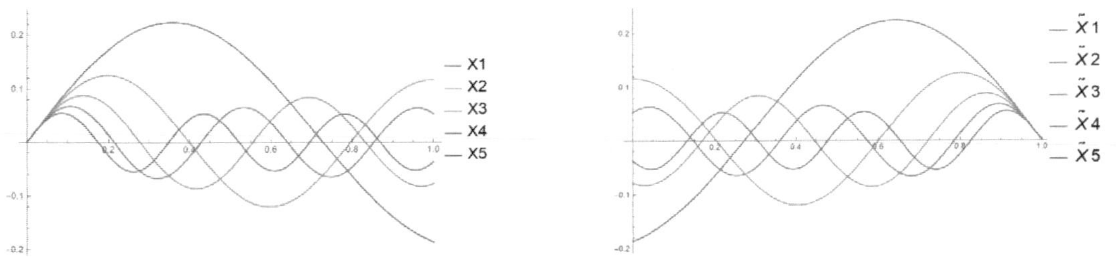

α = **0.4**

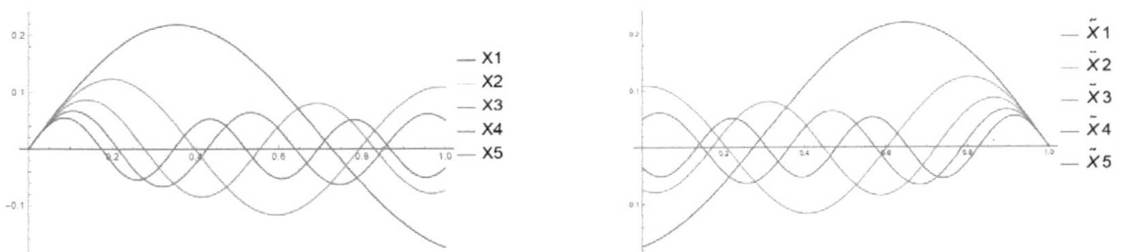

IOP Conf. Series: Materials Science and Engineering 1129 (2021) 012031 doi:10.1088/1757-899X/1129/1/012031

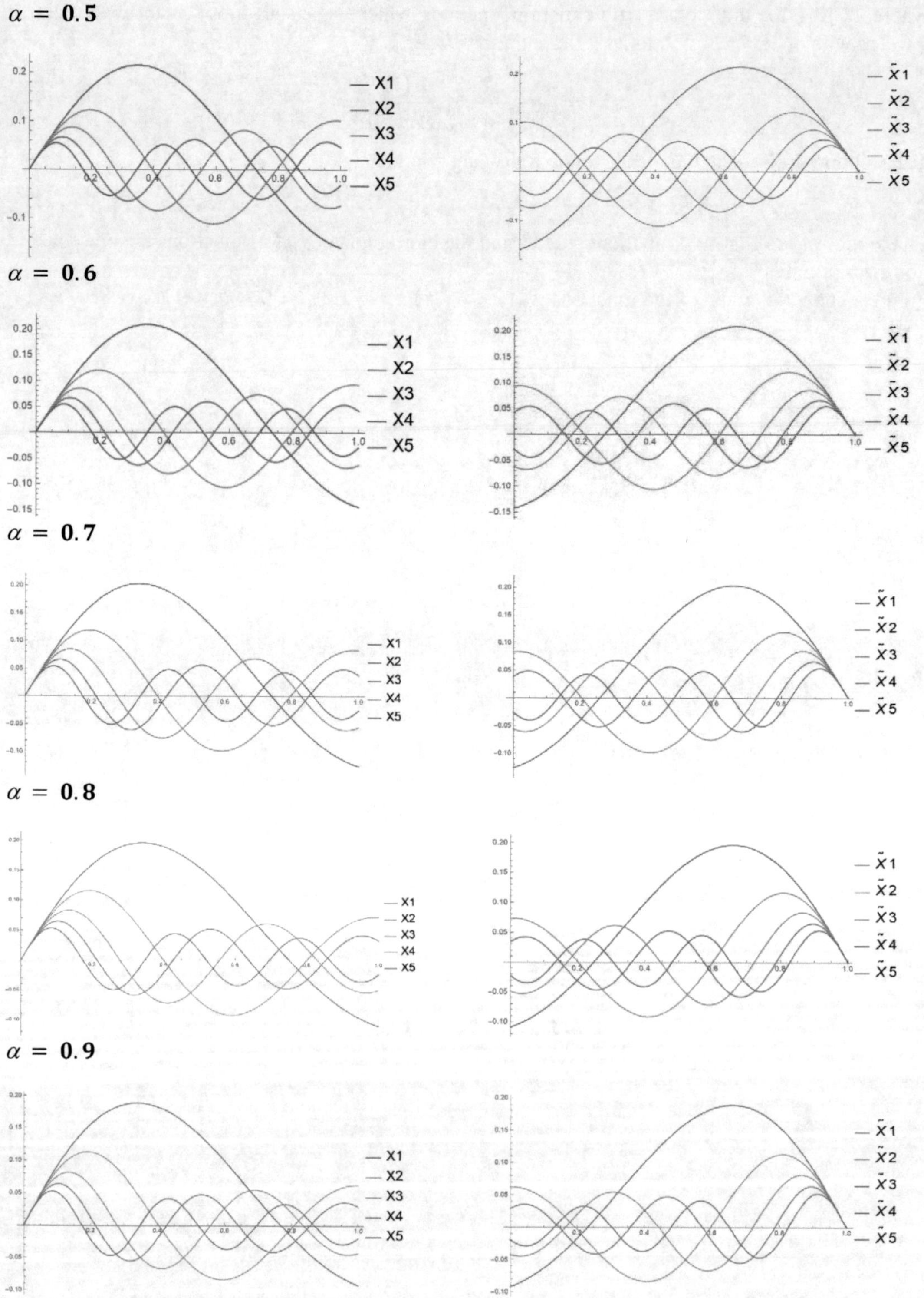

Figure 2. The graph of the functions $\chi_m(x), \tilde{\chi}_m(x)$; $m = 1; 2; \ldots, 0 < \alpha < 1$; $c_1 = 1.8$.

Now we determine the constants of the equation (10), as following,

$$\begin{cases} A_m = \dfrac{1}{\langle \chi_m(x), \tilde{\chi}_m(x) \rangle} \cdot \langle \varphi(x), \tilde{\chi}_m(x) \rangle, \\ B_m = \dfrac{1}{\langle \chi_m(x), \tilde{\chi}_m(x) \rangle} \cdot \dfrac{\langle \psi(x), \tilde{\chi}_m(x) \rangle}{\sqrt{-\lambda_m}}. \end{cases} \tag{30}$$

So, the solution of the problem (1)- (4) takes the form:

$$u(x,t) \approx u_{(5)}(x,t) = \sum_{m=1}^{5} \chi_m(x)\left[a_m cos\left(t\sqrt{-\lambda_m}\right) + b_m \cdot sin\left(t\sqrt{-\lambda_m}\right)\right]. \tag{31}$$

3. Parameter identification and experimental Studies

We use the experimental data obtained in [6], to validate the technique. The stress-strain state of these samples was well defined by models (9) and (10), and their parameter was also found to be $c_1 = 1.8$ [7], The values for polymer concrete samples based on polyester resin (Dian- and dihloangidride-1,1-dichloro-2,2-diethylene) are shown below in Table 2. The advantage of this technique of parametric identification compared with different descriptive techniques [8] consists of the quantitative accurate estimation of selecting the search parameter. The relation between the theoretical solution (31) and the experimental data is shown below in figure 3, when

$$\varphi(x) = 0; \ \psi(x) = 1; \ t \in [0,5].$$

Table 2. Experimental data for polymer concrete samples.

$x_i(c)$	0.25	0.5	0.75	1
U_i	0.05	-0.04	-0.01	0.02

4. Conclusion

This comparison between the experimental data and the theoretical solution at the various value of the parameter $0 < \alpha < 1$, allows us to deduce that the order of the fractional derivative in the model (1) is $\alpha = 0.9$, where we see from the comparison that the theoretical solution is in very good agreement with the experimental data at this value, that allows us to conclude that it is a suitable value. Knowing the parameter of the model lets us, predict the stress-strain characteristics of the material (asphaltic concrete, polymer concrete, etc.) when to undergo loading.

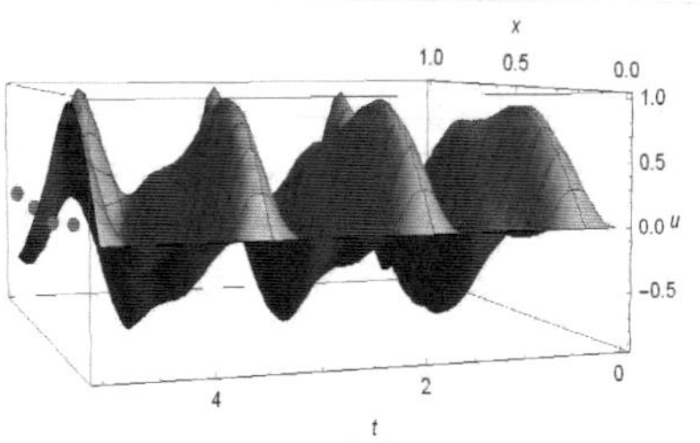

$\alpha = 0.1$

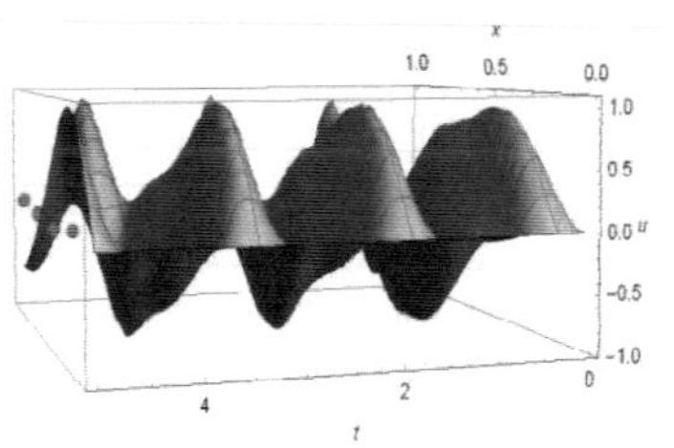

$\alpha = 0.2$

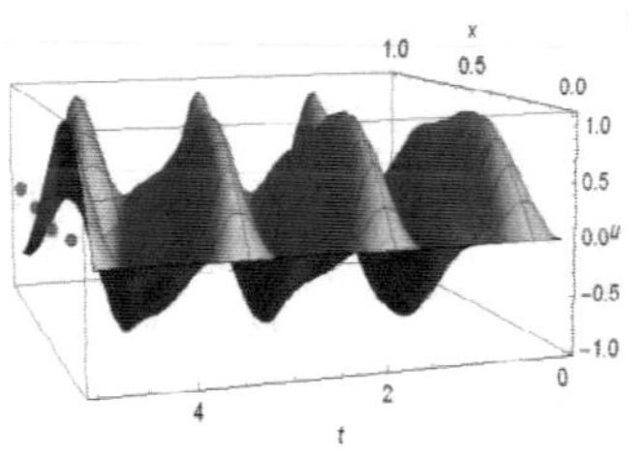

$\alpha = 0.3$

IOP Conf. Series: Materials Science and Engineering 1129 (2021) 012031 doi:10.1088/1757-899X/1129/1/012031

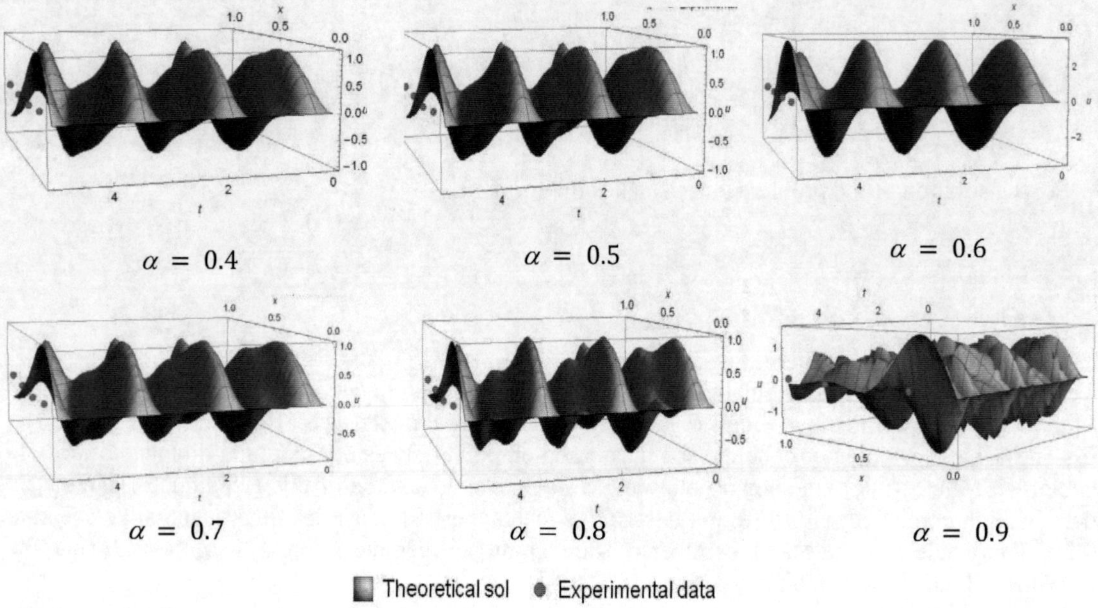

Figure 3. The comparison of the experimental data and the theoretical solution at various values of the parameter $0 < \alpha < 1$.

References

[1] Sandev T and Tomovski Z 2019 Fractional Equations and Models: Theory and Applications *Springer* (Berlin)

[2] Aleroev T and Elsayed A 2020 Analytical and approximate solution for solving the vibration string equation with a fractional derivative *Mathematics* **8(7)** 1154

[3] Elsayed A and Orlov V 2020 Numerical scheme for solving time-space vibration string equation of fractional derivative *Mathematics* **8(7)** 1069

[4] Aleroev T and L Kirianova 2018 Presence of basic oscillatory properties in the Bagley-Torvik model *MATEC Web of Conferences* **251** 04022

[5] Livshits M 1954 On spectral decomposition of linear non self-adjoint operators *Mat. Sb. (N.S.)* **34(76)** pp 145-99

[6] Dzhrbashian M 1970 The boundary-value problem for a differential fractional-order operator of the Sturm-Liouville type *Izv. Akad. Nauk ArmSSR Ser. Mat.* **5(2)** pp 71-96

[7] Aleroev T and Erokhin S 2019 Parametric Identification of the Fractional-Derivative Order in the Bagley-Torvik Model *Math. Models and Comp. Simul* **2(11)** pp 219–25

[8] Kekharsaeva E and Aleroev T 2001 Model of deformation-strength characteristics of chlorine-containing polyesters based on fractional derivatives *Plast. Massy* **3** p 35

Influence of technological load on the stability of synchronous anti-phase motion of the jaws of a vibratory crusher

G Altshul, A Gouskov and G Panovko

Mechanical Engineering Research Institute of the Russian Academy of Sciences, Moscow, Russia

grigoryaltshyl@gmail.com

Abstract. Vibratory jaw crushers are based on the effect of self-synchronization of jaws and body, when its elements vibrate with frequencies that are in a rational relationship due to the influence of links already existing in the system. The technological load acting on the jaws of a vibratory crusher during its operation can have a significant effect on which operating mode of the crusher (synchronous antiphase or synchronous inphase) is implemented. This paper investigates the influence of the proposed crushing model on the dynamics of a vibratory jaw crusher. The model takes into account the elastoplastic properties of the rock being crushed and its one-way connection with the jaws. The destruction of the material, its movement along the crushing chamber and the change in the properties of the processed rock as it is crushing has been implemented. Conclusions are made about the influence of the processed rock on the implementation of the crusher operating modes

1. Introduction

Vibratory jaw crushers (VJC) are common machines for processing and crushing rock. The main feature that sets them apart from other machines for crushing material is the use of self-synchronization effect in design. This effect consists in the fact that the elements of the system begin to vibrate with the same or multiple frequencies due to the connections already existing in the system. This makes it possible to exclude from the design of the machine rigid and unreliable kinematic connections that ensure the mutual movement of the crusher jaws [1]. However, practice has shown that, depending on the physical and mechanical properties of the treated rock and the conditions of impact with the jaws, the type of self-synchronization (phase of mutual oscillations) of vibration exciters can change, which affects the efficiency of the crushing process. The working mode is the synchronous antiphase mode of the crusher vibrations.

There are studies in which the force interaction between the VJC jaws and the processed material is modeled by introducing an equivalent dissipation coefficient into a dynamic system [2-3]. In these works, the conditions for the stability of the in-phase and anti-phase modes of operation of the VJC were obtained, but many properties of the crushed rock itself and the effects occurring in the crushing chamber (movement of the rock along it, changes in the properties of the material and its chipping during processing) were not taken into account. The article [4] proposes a model of material accounting in the VJC, based on the laws of crushing and distribution of the processed rock in the crushing chamber. In this case, the movement of the jaw is set by the kinematic functions of time, i.e. the rock being crushed does not affect this movement. Another approach that allows to simulate crushing of a single piece and crushing "under the blockage" is the use of the discrete element method (DEM) [5, 6]. This approach allows us to consider the crushing of each piece in the rock flow, however, the movement of the jaw is similarly set in a kinematic manner and does not experience the opposite effect of the processed material. Thus, the interaction between the jaws of the VJC and the rock on which they act is not fully understood. The influence of this interaction on the stability of the operating modes of the VSC is the subject of this article. The further development of the model created by the authors of [7] and its influence on the movement of the air pressure motor, taking into account the imperfection of electric motors, is proposed.

2. Vibratory jaw crusher model

The VJC (figure 1) consists of an absolutely rigid body with mass *m* and a moment of inertia *J* relative to its centre of mass *O*. At points D_1, D_2, the body is suspended on a package of soft springs with transverse and longitudinal stiffnesses and linear-viscous damping coefficients k_1, k_2, b_1, b_2. The position of the body is defined by offsets of the center of mass *x* and *y*, as well as by rotation relative to the center of mass φ. At points A_1 and A_2, at the distance of $2a$ from each other on elastic torsion bars, jaws with centres of mass C_1 and C_2 are symmetrically suspended. Torsion bars have the same linear characteristics with torsional stiffness k_3 and linear-viscous damping coefficient b_3. The jaws have the same masses μ and moments of inertia J_0 relative to their centres of mass. Jaw length $2l$. The position of the jaws is defined by the relative (relative to the body) angles φ_1 and φ_2. At points B_1 and B_2 at an angle β to the surfaces of the jaws at a distance *b*, two unbalanced vibrators are attached to the jaws. The mass of each unbalance is m_e and is concentrated at points E_1 and E_2, the eccentricity of both unbalances is *e*. The position of the unbalances is set by the angles θ_1 and θ_2 relative to the horizontal. Unbalances rotation speed ω_1 and ω_2. The characteristics of asynchronous motors driving the unbalances: critical moment M_c, critical slip s_c and coefficient of friction in their bearings *f*. Considering the limited engine power makes it possible to study the stability of synchronous modes.

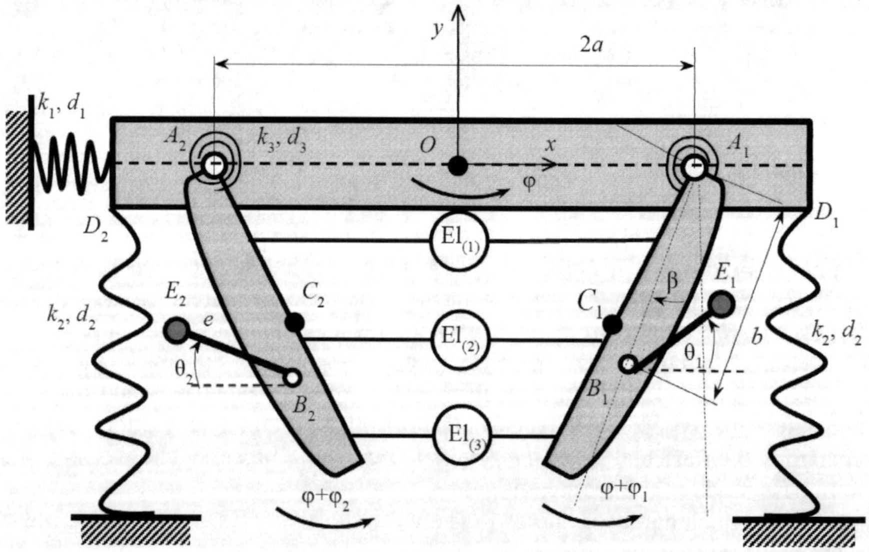

Figure 1. Vibratory jaw crusher scheme.

The case of uniform supply of crushed material to the crushing chamber and uniform distribution of the processed rock along the height of this chamber is simulated. Thus, the chamber is completely filled with rock at any given time. To describe the mechanical properties of the rock and its movement along the crushing chamber, the following phenomenological model is used: the entire crushing chamber is divided into three zones (figure 1), to each of which the element $El_{(i)}$ (figure 2) is added at a distance $l_{F(i)}$ from torsion bars. Each element is an ideally elastic-plastic body and consists of a linear elastic element with stiffness $k_{e(i)}$ and a dry friction element with the ultimate force $F_{lim(i)}$. During crushing, these elements do not move vertically. The cycle of movement of the jaws consists of two phases - the compression phase, when the jaws approach each other and the reverse phase of the jaws movement, when they move away from each other. During the compression phase, the elastic element is compressed until the ultimate force of the dry friction element is reached. After that, until the end of compression, the element's response stops increasing. After the elastic element has straightened to an unstressed state during the reverse movement phase, before the start of a new compression phase, the element does not exert forces on the jaws. Each such element can only withstand a certain number of cycles of interaction with the jaw, after which it collapses. The rock is loaded in the low-cycle fatigue mode. During compression, a complex stress-strain state is realized in the pieces, in the most weakened places of the piece, where pores and tiny cracks are present, damage begins to accumulate, which then develop into macro-cracks and the piece collapses. To simulate the accumulation of damage, each element is assigned a function of damage $\chi_{(i)}$, which at the initial moment has a value of 0, and at the moment of destruction turns into 1. Accumulation of damage proceeds in accordance with the hypothesis of linear damage accumulation, the kinetic equation is as follows

$$\frac{d\chi_{(i)}}{dt} = \frac{1}{t_{cr}\left[F_{(i)}(t)\right]} \qquad (1)$$

Here, t_{cr} is the time of rock destruction under loading by a symmetric sinusoidal load with an amplitude F. In this model, the simplest approximation $F=C(t_{cr})^n$ is used, where C and n are the experimental constants of the material; they are selected so that each element collapses in 10 vibrations. During the oscillations, function (1) is integrated over time. As soon as the damage of all elements reaches $\chi_{(i)} = 1$, the element is destroyed.

During the reverse phase of the jaws movement immediately following the fracture, there is no technological load. This reflects the movement of rock down the crushing chamber. Then the whole cycle is repeated.

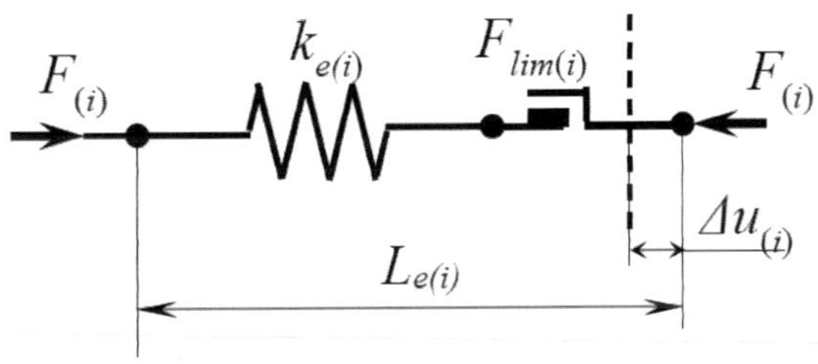

Figure 2. Scheme of the element that reproduces the interaction of the rock and jaw.

Consider the behavior of a part of an element consisting of a series-connected spring and a dry friction damper when compressed by a force $F_{(i)}$ in the compression phase. Relative displacement $\Delta u_{(i)}$ of points touching of the element and jaws.

$$\Delta u_{(i)} = \begin{cases} F_{(i)}/k_{e(i)}, & F_{(i)} < F_{\lim(i)} \\ \infty, & F_{(i)} = F_{\lim(i)} \end{cases} \qquad (2)$$

The dependence of the compressive force $F_{(i)}$ on the displacement $\Delta u_{(i)}$ is written in the form:

$$F_{(i)} = \begin{cases} k_{e(i)}\Delta u_{(i)}, & \Delta u_{(i)} < \dfrac{F_{\lim (i)}}{k_{e(i)}} \\ F_{\lim (i)}, & \Delta u_{(i)} \geq \dfrac{F_{\lim (i)}}{k_{e(i)}} \end{cases} \tag{3}$$

For subsequent numerical integration, it is expedient to replace the piecewise linear function (3) with a smooth function. As an approximation of the piecewise linear function (3), we will use a smooth function of the hyperbolic tangent.

$$\begin{aligned} F_{(i)} = &-\mathrm{H}(\dot{\varphi}_1 - \dot{\varphi}_2)\frac{F_{\lim(i)}}{2}\left(1 + \tanh\left(s\left(\Delta u_{(i)} - \Delta u_{\max(i)} - F_{\lim(i)}/k_{e(i)}\right)/L_{e(i)}\right)\right) + \\ &-\mathrm{H}(-(\dot{\varphi}_1 - \dot{\varphi}_2))\mathrm{H}\left(F_{\lim(i)} - k_{e(i)}\left(\Delta u_{(i)} - \Delta u_{\min(i)}\right)\right)\mathrm{H}\left(1-\chi_{(i)}\right)\left(F_{\lim(i)} - k_{e(i)}\left(\Delta u_{(i)} - \Delta u_{\min(i)}\right)\right) \end{aligned} \tag{4}$$

Here $\Delta u_{\max(i)}$ is the maximum distance between the touch points of the element and the jaws (end of the reverse phase); $\Delta u_{\min(i)}$ – the minimum distance between the points of contact of the element and the jaws (end of the compression phase); $\mathrm{H}(\dot{\varphi}_1 - \dot{\varphi}_2)$ – Heaviside step function, indicates that this part of the power characteristic works only in the compression phase; $\mathrm{H}(-(\dot{\varphi}_1 - \dot{\varphi}_2))$ is the Heaviside step function, indicates that this part of the power characteristic works only during the reverse phase of the jaws; $\mathrm{H}(F_{\lim(i)} - k_{e(i)}(\Delta u_{(i)} - \Delta u_{min(i)}))$ is the Heaviside function, indicates that only straightening occurs during the reverse phase of the jaws springs, without further stretching it; $\mathrm{H}(1 - \chi_{(i)})$ is the Heaviside function indicates the destruction of an element and the absence of a reaction after this destruction; $L_{e(i)}$ is the initial length of the element.

3. Differential equations of system movement

Crusher vibrations are described by the following differential equation:

$$\mathbf{M}\ddot{\bar{q}} = \bar{F}(\dot{\bar{q}}, \bar{q}) \tag{5}$$

Here $\bar{q} = \{x, y, \varphi, \varphi_1, \varphi_2, \theta_1, \theta_2\}^T$ is the vector of the degrees of freedom of the system; **M** - symmetric mass matrix of the system with dimension 7x7; $\bar{F}$ is the vector of the right parts, which includes elastic and dissipative forces in the package of soft springs and torsion bars, the interaction forces of the unbalanced vibrator and the jaw, the force of the electric motor, the friction force in the bearings of the unbalanced vibrator and the force of interaction of the jaws and elements.

$$\begin{aligned} &M_{11} = m + 2(\mu + m_e);\ M_{12} = 0; M_{13} = 2(\mu l\cos(\varphi_{cm}) + m_e b\cos(\beta));\ M_{14} = \mu l\cos(\varphi_{cm}) + m_e b\cos(\beta); \\ &M_{15} = M_{14}; M_{16} = -m_e e\sin(\theta_1); M_{17} = m_e e\sin(\theta_2); M_{22} = m + 2(\mu + m_e); M_{23} = 0; \\ &M_{24} = \mu l\sin(\varphi_{cm}) + m_e b\sin(\beta); M_{25} = -M_{24}; M_{26} = m_e e\cos(\theta_1); M_{27} = m_e e\cos(\theta_2); \\ &M_{33} = J + 2J_0 + 2\mu(a^2 + l^2) + 2m_e(a^2 + b^2) + 4\mu al\sin(\varphi_{cm}) + 4m_e ab\sin(\beta); \\ &M_{34} = J_0 + \mu l^2 + m_e b^2 + \mu al\sin(\varphi_{cm}) + m_e ab\sin(\beta); M_{35} = M_{34}; M_{36} = m_e e(a\cos(\theta_1) + b\sin(\beta - \theta_1)); \\ &M_{37} = -m_e e(a\cos(\theta_2) + b\cos(\beta - \theta_2)); M_{44} = J_0 + \mu l^2 + m_e b^2; M_{45} = 0; M_{46} = m_e eb\sin(\beta - \theta_1); \\ &M_{47} = 0; M_{55} = J_0 + \mu l^2 + m_e b^2; M_{56} = 0; M_{57} = -m_e eb\sin(\beta - \theta_2); M_{66} = m_e e^2; M_{67} = 0; M_{77} = m_e e^2 \end{aligned} \tag{6}$$

The load vector components are given by expressions (7) and (8):

$$
\begin{aligned}
&F_1 = -k_1 x - b_1 \dot{x} + m_e e\cos(\theta_1)\dot{\theta}_1^2 - m_e e\cos(\theta_2)\dot{\theta}_2^2;\\
&F_2 = -2k_2 y - b_2 \dot{y} + m_e e\sin(\theta_1)\dot{\theta}_1^2 + m_e e\sin(\theta_2)\dot{\theta}_2^2;\\
&F_3 = -2a^2 k_2 \varphi - b_3\dot{\varphi} + m_e e(a\sin(\theta_1) + b\cos(\beta-\theta_1))\dot{\theta}_1^2 - m_e e(a\sin(\theta_2) + b\cos(\beta-\theta_2))\dot{\theta}_2^2 +\\
&+\sum_{i=1}^{3} F_{(i)} l_{f(i)} \left\{\cos(-\varphi_{cm} + \varphi + \varphi_1) - \cos(\varphi_{cm} + \varphi + \varphi_2)\right\};
\end{aligned}
\tag{7}
$$

$$
\begin{aligned}
&F_4 = -k_3\varphi_1 - b_3\dot{\varphi}_1 + m_e eb\cos(\beta-\theta_1)\dot{\theta}_1^2 + \sum_{i=1}^{3} F_{(i)} l_{f(i)} \cos(-\varphi_{cm} + \varphi + \varphi_1);\\
&F_5 = -k_3\varphi_2 - b_3\dot{\varphi}_2 - m_e eb\cos(\beta-\theta_2)\dot{\theta}_2^2 - \sum_{i=1}^{3} F_{(i)} l_{f(i)} \cos(\varphi_{cm} + \varphi + \varphi_2);\\
&F_6 = 2M_{c1}\cdot\left(\left(\omega_1-\dot{\theta}_1\right)\cdot\left(\omega_1\cdot s_{c1}\right)^{-1} + \omega_1\cdot s_{c1}\cdot\left(\omega_1-\dot{\theta}_1\right)^{-1}\right)^{-1} - m_e e f_1\cdot\dot{\theta}_1^{\,2} sign(\dot{\theta}_1) - m_e eg\cos(\theta_1);\\
&F_7 = 2M_{c2}\cdot\left(\left(\omega_2-\dot{\theta}_2\right)\cdot\left(\omega_2\cdot s_{c2}\right)^{-1} + \omega_2\cdot s_{c2}\cdot\left(\omega_2-\dot{\theta}_2\right)^{-1}\right)^{-1} - m_e e f_2\cdot\dot{\theta}_2^{\,2} sign(\dot{\theta}_2) - m_e eg\cos(\theta_2)
\end{aligned}
\tag{8}
$$

4. Results and discussions

This system was solved numerically for different speeds of rotation of unbalances $\omega=\omega_1=\omega_2$ with the following parameters: m=4738 kg; J=2575 kg·m^2; μ=3253 kg; J_0=450 kg·m^2; m_e=172 kg; l=0.8 m; b=1.01 m; e=0.1 m; a=0.5 m; k_1=62000 N·m^{-1}; k_2=620000 N·m^{-1}; k_3=45.4·105 N·m; φ_{st}=15°; b_1=380 N·s·m^{-1}; b_2=380 N·s·m^{-1}; b_3=11198 N·m·s·rad^{-1}; β=-27°; M_c=600 N·m^{-1}; s_c=0.12; f=0.001; $F_{\lim(1)}$=25600 N; $F_{\lim(2)}$=32000 N; $F_{\lim(3)}$=38400 N; $k_{e(1)}=k_{e(2)}=k_{e(3)}=k_e$=530000 N·m^{-1}.

It is known that machines of this class have two characteristic modal frequencies - p_{sym}, corresponding to a symmetric natural shape (when the jaws vibrate in antiphase) and p_{asym}, corresponding to a skew-symmetric natural shape of the system (when the jaws vibrate in phase). For the case of absence of technological load, these frequencies set three possible section by the frequency of rotation of the unbalances ω [1], depending on which one or another mode of operation of the machine is stable. If $\omega<p_{asym}$, then the synchronous antiphase mode is stable, if $p_{sym}<\omega<p_{asym}$, then the synchronous in-phase mode is realized when $\omega > p_{asym}$, then the synchronous antiphase regime becomes stable again. For this crusher p_{sym}=52 rad·s^{-1}, p_{asym}=92 rad·s^{-1}.

Figure 3 shows a graph of changes in the phase difference of the unbalance rotation $\Delta\theta = \theta_1$-θ_2 depending on the unbalance rotation speed. For vibrating crushers, the phase of oscillation of the jaw correlates with the phase of rotation of its driving unbalance, therefore the value of $\Delta\theta$ shows which mode of movement of the crusher has been established. For the used directions of θ_1, θ_2, the synchronous antiphase regime corresponds to $\Delta\theta$=0, and the synchronous in-phase regime corresponds to $\Delta\theta=\pi$. It can be seen in this figure that the use of this model shifts the boundaries of the stability segments of oscillation modes towards decreasing frequencies by about 10 rad·s^{-1}. It can be noted that in the interresonance mode, corresponding to the in-phase movement of the jaws, communition is impossible, which is displayed by a strong spread of the frequency difference in the middle of the graph.

The plots in figure 4 ((a), (b)) oscillations of the jaws φ_1 and φ_2 for two frequencies of external excitation ω=22 rad·s^{-1} (a) and ω=100 rad·s^{-1} (b). I.e. demonstrate the behavior of crushing organs in pre-resonance and over-resonance modes. It can be seen that, for these frequency ranges, antiphase oscillations of the VJC are realized, corresponding to the operating modes of crushing. On these graphs, a different influence of the model is noticeable - the vibrations of the jaws are no longer harmonic, they are complex. Oscillatory motion is superimposed on a non-zero mean value of the angle of rotation, which indicates that the jaws are moving apart relative to each other. This phenomenon has been confirmed experimentally [1]. The change in the amplitude of vibrations is associated with taking into account the destruction of the rock. At the moment when the element collapses, the resistance force falls on the jaws, and the amplitude of the oscillations increases. On the next compression cycle, the element begins to resist again, which reflects the supply of new material to the crushing chamber, and the vibration amplitude begins to decrease.

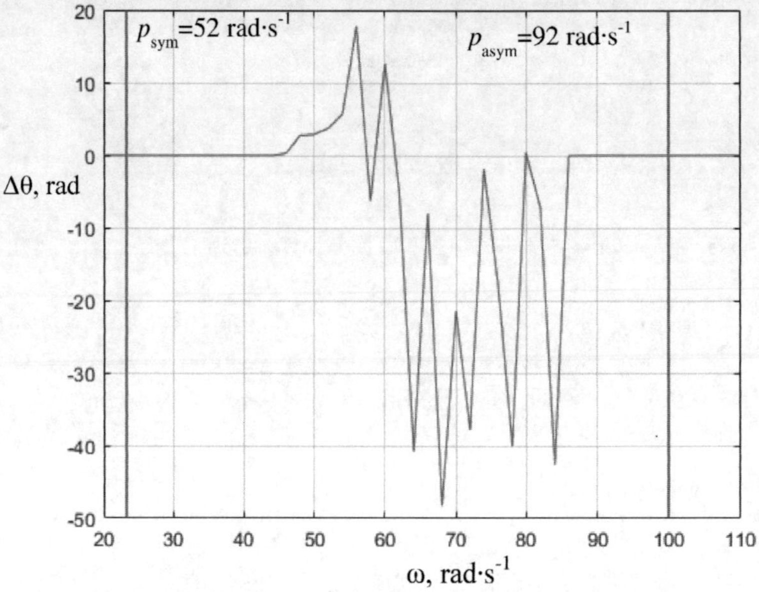

Figure 3. Plot of changes in the phase difference of unbalances

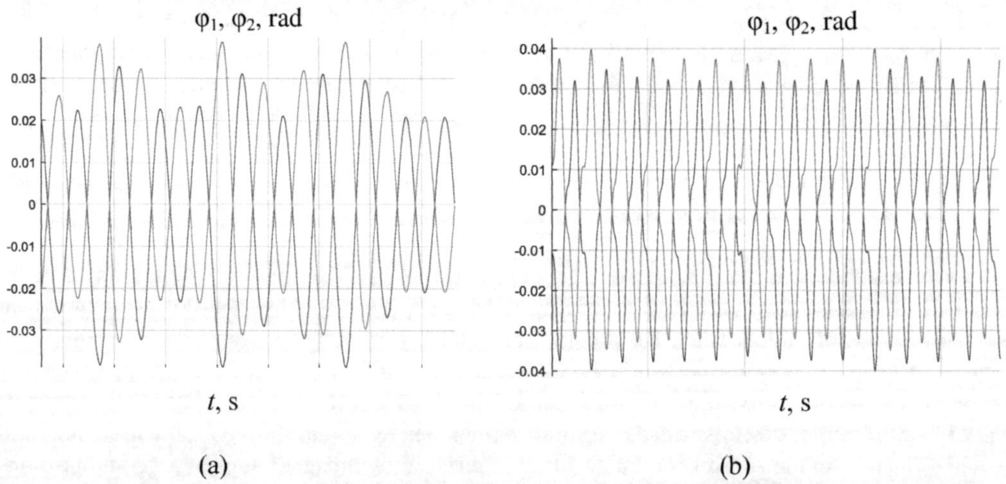

Figure 4. (a) Plot of the jaw vibrations for the case ω=22 rad·s^{-1}; (b) Plot of the jaw vibration for the case ω = 100 rad·s^{-1}

5. Conclusions

This article discusses the issues of modeling the behavior of the VJC and the treated rock. The VJC is modeled as an absolutely rigid body with 3 degrees of freedom and two absolutely rigid jaws, each of them has 2 degrees of freedom. The system is in motion with two unbalanced vibrators of limited power, fixed on the jaws. The rock being crushed is presented in the form of a phenomenological model: the entire crushing chamber is divided into 3 zones, in each zone there is an elastoplastic element. This model simulates the elastic and plastic properties of the crushed rock, the packing of the rock to the bottom of the crushing chamber, its destruction and chipping, and the change in these properties during the movement of the rock through the crushing chamber. The strength properties of the rock are taken into account using the hypothesis of linear damage accumulation. Damage to the elements accumulates with each interaction with the jaws. The results show the influence of such models of rock crushing on the movement of the VJC. Accounting for the material in this way leads to a change in the synchronous rotation frequency of the stability unbalances of the synchronous in-phase mode and synchronous anti-phase mode of the crusher operation. Also, due to the interaction with the cultivated rock, the nature of the machine vibrations changes.

Acknowledgements
This work was supported by the Russian Science Foundation (project no. 18-19-00708)

References

[1] Vaisberg L, Zarogatskiy L and Turkin V 2004 *Vibratory crushers. Basic of calculation, design and technological using* (Saint-Petersburg: VSEGEI)
[2] Shishkin E and Safronov A 2016 Dynamics of vibratory jaw crusher taking into account technological load *Obogashenie Rud Mineral Processing J.* **6** 39-43
[3] Shishkin E V, Pugina O A and Kazakov S V 2016 Shock-free vibrations of bent chamber vibratory jaw crusher *Obogashenie Rud Mineral Processing J.* **6** 55-9
[4] Johansson M, Bengtsson M, Evertsson M and Hulthén H 2017 A fundamental model of an industrial-scale jaw crusher *Minerals Engineering* **105** 69-78
[5] Beloglazov I, Stepanyan A, Feoktistov A and Usupov G 2018 Modelling the disintegration process in a jaw crusher with complex jaw oscillations *Obogashenie Rud Mineral Processing J.* **2** 3-8
[6] Refahi A, Aghazadeh Mohandesi J and Rezai B 2010 Discrete element modeling for predicting breakage behavior and fracture energy of a single particle in a jaw crusher *Int. J. of Mineral Processing* **94** 83-91
[7] Altshul G, Gouskov A, Panovko G and Shokhin A Interaction model of one jaw of a vibrating jaw crusher with the processed rock, taking into account the properties of the electric motor *IOP Conference Series: Materials Science and Engineering* **747**

Programming wave propagation problem in cable of wirerope barrier systems with mathematica

I V Demyanushko, I A Karpov, O V Titov and B T Tavshavadze

Moscow Automobile and Road Construction State Technical University (MADI), 64, Leningradsky Prospect, Moscow, 125319, Russia

ilya.karpov@outlook.com

Abstract. The article discusses the issues of modeling of longitudinal and shear waves propagation in the cables of a road cable barrier. The problem was solved using explicit methods with finite deformation. In the first part of the article, the procedure considered, that allows to solve the problem. In the second part, the procedures for P and S waves verified, using analytical solutions and the commercial LS-DYNA code.

1. Introduction

Despite the frequent occurrence of the wire rope safety barriers, there is still a number of understudied aspects, in particular, the relation between the structural specifics (of wire ropes, legs) and the operating properties (deflections, accelerations at car points).

The finite-element complexes [1] and simplified physical models [2] are mainly used to describe the behavior of the wire rope safety barrier when hit by a car that is an impact. The latter do not include the dynamics of the wire rope itself. Each approach does not allow conducting detailed parametric studies of the structure parameters. For instance, it is difficult to set the geometry parameters in the software complexes, while the analytical model without the wire rope dynamics does not allow studying the wire rope vibrations and their behavior under the load. This work represents the first stage to unify these two approaches. A simplified model of the structure behavior is planned to be developed as part of this work for the parametric study of the safety barrier with consideration of longitudinal and transversal wire rope vibrations.

2. Setting and Solving the Task

The interaction process between the car and the wire rope safety barrier, when the latter is hit, represents one of the frequent diagonal impact cases (impact at an angle to the wire rope and wire rope safety barrier axial line) resulting in longitudinal and transversal wire rope vibrations. In the present work the wire rope was considered to be in the simplest position as a rod. The main equation used to describe the dynamics was the virtual power equation [2]. After discretizing the equation by space and choosing the form functions, the equation takes the form of the equation of dynamics:

$$[M]\{\dot{v}\} + \{f^{int}\} - \{f^{ext}\} = 0 \quad (1)$$

$\{\dot{v}\}$, $\{f^{int}\}$, $\{f^{ext}\}$ – vectors *of* acceleration, external and internal forces, respectively. $[M]$ – lumped matrix.

Equation (1) does not require the calculation of the stiffness matrix or the damping matrix compare with implicit methods. Damping and stiffness is included in the value of the internal force, which is obtained by integrating the stresses. Stresses can be found by strain-displacements relations and constitutive equations.

According to J. Hallquist, it is the lumped mass matrix that shall be used in the calculation, as non-physical vibration modes occur if a distributed mass matrix is applied. Another advantage of the concentrated mass is its small width and the inverse matrix is determined faster.

The lumped mass matrix of elements is calculated:

$$M_{ijIJ} = \delta_{ij} \int_{\Omega} \rho N_I N_J d\Omega \tag{2}$$

δ is the Kronecker delta, ρ is the density, and N is the element shape functions.

To solve equations (1), they must be integrated over time. At the first stage of solution equation (3) should be solved:

$$\{\dot{v}\} = [M]^{-1}\{f^{int} - f^{ext}\} = [M]^{-1}\{f^n\} \tag{3}$$

f^n – forces vector at time n.

The initial conditions in the form of forces and velocities are imposed on equation (3) at the time t = 0. To find the displacements, a central-difference time integration scheme is used, so the velocities are found for the time n + 0.5 from the velocity values at the time n-0.5. If the velocity at time n are known, then the velocity at time n + 0.5 are determined by the following equation:

$$\{v^{n+0.5}\} = \alpha\Delta t[M]^{-1}\{f^n\} + \{v^{n+0.5-\alpha}\} \tag{4}$$

Δt – time step. α – parameter, for t=0 is α =0.5 and for other time steps $\alpha = 1$.

After finding the velocities, essential boundary conditions are imposed on the system and displacements are determined in the system for the next time step:

$$\{u^{n+1}\} = \{u^n\} + \Delta t\{v^{n+0.5}\} \tag{5}$$

u - nodes displacements.

Then nodal forces effective in the system were calculated. Forces can be found either through stress integration or by the use of the resulting elements directly connecting the displacements and forces. The resulting definitions were used in the said algorithm.

3. Internal Force Calculation

Elongation of for each element (figure 1 (a)) were calculated using equation (6). The finite length of the element (L) was determined based on the known displacements in the longitudinal (u_1, u_2) and lateral (w_1, w_2) directions, as well as the initial node coordinates (X_1, Y_1) and (X_2, Y_2):

$$L = (((X_2 + u_2) - (X_1 + u_1))^2 + ((Y_2 - w_2) - (Y_1 - w_1))^2)^{0.5} \tag{6}$$

When solving the ask of free wave propagation along the wire rope, a numerical instability was identified in the course of the calculation under formula 2 (figure 1 (b)) in case of large wire rope displacements and the use of single precision numbers.

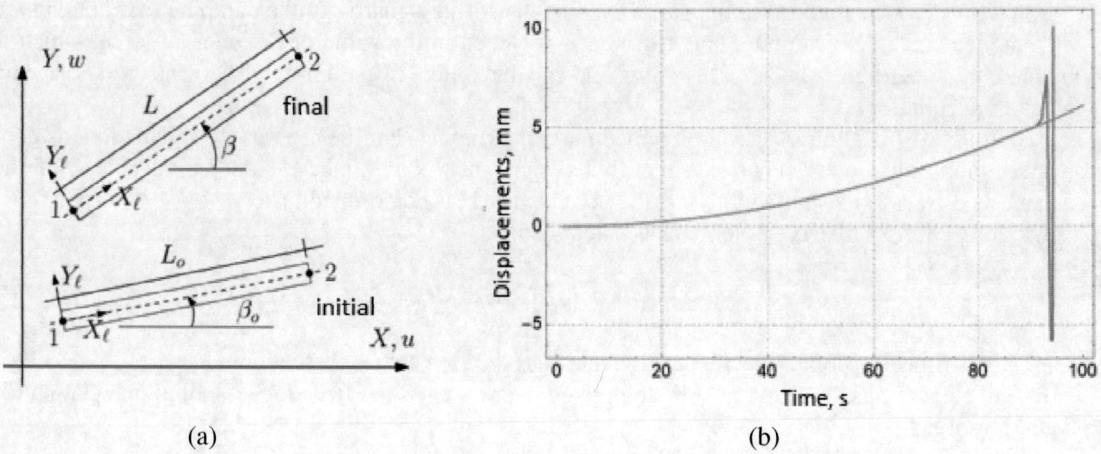

Figure 1. (a) Initial and final element positions in space, (b) point displacement under the impulse: blue under formula 2, brown under formula 3.

Thus, based on the recommendations of T. Belytschko [4], the element elongations were calculated under formula 3:

$$u_l = \frac{L^2 - L_0^2}{L - L_0} \tag{7}$$

L is the finite element length, L_0 is the initial element length, and u_l is the change in the element length.

In case of small displacements (the task of lateral vibrations of the string fixed at both edges with the amplitude of 4.5 mm at the cable length of 1000 mm), as well as the applied machine accuracy of the calculations, no difference was found between the solutions under the formulas (2) and (3).

Then, based on the known displacements, inner element forces in the local coordinate system were calculated and displayed in the global coordinate system:

$$f = \frac{EAu_l}{L}\{-cos\theta, -sin\theta, cos\theta, sin\theta\}^T \tag{8}$$

Equations (7) and (8) simplify the calculation of the forces f^n for equation (3), since forces are found directly from displacements.

4. Model Verification

Two calculations were made according to the said technique to determine the longitudinal and transversal wire rope vibrations under the above-stated conditions. In case of longitudinal vibrations, an impulse load of 10 kN was applied at one end, while the other end was recorded by displacements. In case of transversal waves, the load was applied to the cable center. The technique stated above was written in Wolfram Mathematica programming language for the wave process modelling in the wire rope. The operation of this code was compared to that of the commercial code for the wave process study (LS-DYNA). The results for the longitudinal impact are shown on figure 2. Local element equations were united into global ones under the congruent transformation method. A model of 50 mm long 10 rod elements with the time integration step of $\Delta t=10^{-5}$ s, $E=1.55\cdot10^5$ MPa and the density of $ro=7.850\cdot10^{-9}$ $t\cdot mm^{-3}$ was used to verify the solution.

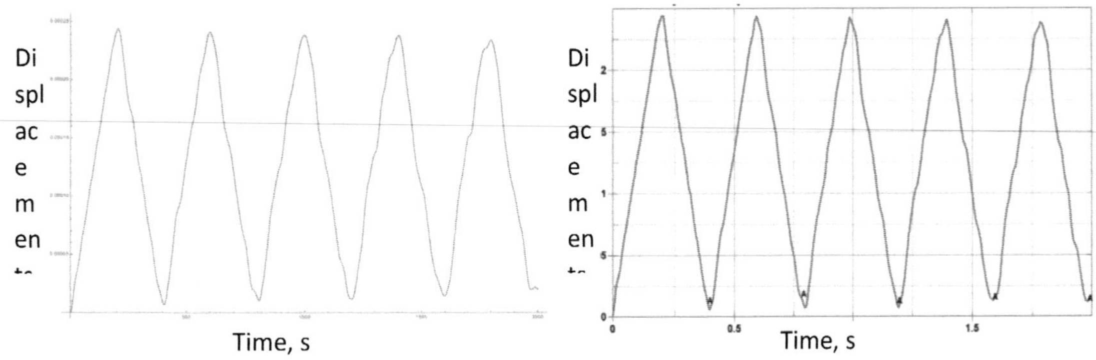

(a) (b)

Figure 2. Modelling results for the longitudinal impact: (a) author code, (b) LS-DYNA.

The comparison gave close results for transversal wire rope vibrations with the author code and LS-DYNA code. A remarkable thing was noted when analyzing the code operation. If the physical wave velocity is v=5000 m·s^{-1}, the impulse at the element length of 100 mm shall be transmitted to the next solution region only in t=0.1/5000=20 ms. At that, the information in the code is distributed in the element in 1 time step, i.e. with the velocity of $v_m = l_e/\Delta t = 0.1/10^{-6} = 10^5$ m·s^{-1}. It turns out that the information distribution velocity in the model is always higher than in the physical environment and is equal to the latter only if the time increment and the element size have the relation of $v_{ph} = \sqrt{E/\rho} = l_e/\Delta t$ (the lower velocity will result in the solution instability as the Courant-Friedrichs-Levi condition is not met [4]). This circumstance is planned to be studied in further works. In addition, the impulse shape loss shall be noted, which reduces with the increase in the number of elements.

Comparison of the transversal wire rope vibrations is shown on figure 3. Both models demonstrated close solution results.

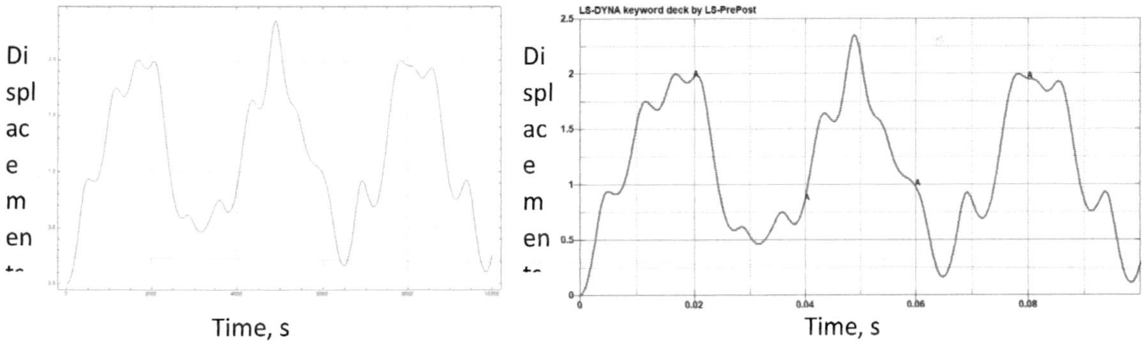

(a) (b)

Figure 3. Modelling results for the transversal impact: (a) author code, (b) LS-DYNA.

When analyzing the results of transversal vibrations, it can be noted that the results obtained when using both models are close by their amplitude and frequency properties.

The wire ropes have a number of specific features complicating their physical descriptions: the wire rope elasticity modulus is a generalized structural indicator and is not related with the real wave propagation velocity in the material, the flexure-torsion and transversal rigidities, as well as the deformation properties are non-linear functions of the respective geometric parameters. It was also found [5, 6] that the deformation velocities influence the wire ropes to a great extent, especially the elasticity modulus. As a result of the experiments it has been determined that the modulus increases by

20-30% even at the velocities $\dot{e} = 100$ mm·(s·mm)$^{-1}$ being the operating local velocities for wire ropes when the car hits the safety barrier. Thus, a separate non-resulting definition of the finite elements is planned to be developed through the deformation velocities.

5. Conclusions

The article considers the procedure to solve the task of the longitudinal and transversal vibrations for the wire rope represented as a rod. The program was developed in Wolfram Mathematica to obtain an analytical solution. The model was verified through comparison with the solution obtained when using the LS-DYNA commercial code. Both solutions had close results. Compared to LS-DYNA, the author's program is easier to parameterize, which is supposed to be used to study the influence of the elastic modulus, total length of the barrier, initial tension, stiffness and type of posts on its characteristics. The program is the first step in the work of the authors to obtain a more accurate model of the barrier, which makes it possible to take into account the change in the characteristics of the cables depending on the strain rate.

References

[1] Mohan P, Marzougui D, Meczkowski L and Bedewi N 2004 *Int. J. of Crashworthiness* **10** pp 267-74
[2] Bateman M B, Howard I C, Johnson A R, and Walton J M 2001 *Int. J. of Impact Engineering* **25** pp 67-85
[3] Benson D J 1992 *Computational Methods in Applied Mech. and Engineering* **99** pp 235–394
[4] Belytschko T and Hsieh B J 1973 *Int. J. for Numerical Methods in Engineering* **7** pp 255-71
[5] Malinovsky V A 2001 *Steel Ropes* (Odessa: Astroprint) p 188
[6] Svetlitsky V A 1978 *Flexible Rod and Cable Mechanics* (Moscow: Mechanical Engineering) p 222

Influence of a thin interlayer and a distribution of microdefects on guided wave dispersion properties in an elastic laminate plate

A A Eremin, M V Golub, O V Doroshenko and S Yu Gurinovich

Institute for Mathematics, Mechanics and Informatics, Kuban State University, 350040 Krasnodar, Russian Federation

sveta.gurinovitch@yandex.ru

Abstract. Adequate mathematical and computational models for damaged interfaces and thin interlayers between two solids are necessary for the non-destructive ultrasonic evaluation of the reliability and safety of bonded structures in service. However, the complexity of real interfaces does not allow to describe their mechanical behaviour precisely. In this paper, the problem of elastic guided wave propagation in a multi-layered structure with thin polymeric plies with possible weakened adhesion is treated uniformly employing effective boundary conditions at sublayer interfaces. While for the pristine structure continuity of the displacements and stresses at the interfaces is assumed, for the damaged interfaces, an approach relying on spring-like boundary conditions is used. The effect of the internal layer and damaged interfaces between sub-layers on an elastodynamic behaviour of a laminate is analysed. Their influence on the dispersion properties of guided waves propagating in a laminated composite structure is illustrated both theoretically and experimentally.

1. Introduction

Ultrasonic guided waves are widely applied in the nondestructive evaluation and structural health monitoring for an inspection of laminate thin-walled composite structures. They are sensitive to possible faults such as cracks, pitting corrosion, voids etc. The accuracy of damage identification and detection might be improved employing relevant computational models of wave propagation, which take into account intrinsic mechanical properties of the inspected waveguide structure [1, 2, 3]. Among the latter, the models taking into account the presence and the influence of adhesive connections (which, in turn, could be either pristine or damaged) between the plies of a multi-layered composite are of particular interest [4, 5, 6, 7]. Typically, the complexity of interfaces does not allow for an explicit description of their mechanical behaviour. Thus, the derivation of boundary conditions that approximate the interfacial properties of the contact between two solid bodies has been the subject of extensive investigations [8, 9]. One of the basic approaches is to treat the interface and its vicinity as an additional thin elastic layer typically with specific, probably viscoelastic, properties and to assume its ideal contact with surrounding substrates. As an alternative, various phenomenological approaches that intend characterizing the overall strength of the joint are considered. For example, within the well-developed quasi-static approach [10, 11] imperfect interface due to the presence of cracks is treated as a set of distributed springs connecting one material to another. Parameters for this model in the case of isotropic materials are spring stiffnesses, which bind the normal and transverse displacement to the traction vector at the interface. Approximate boundary conditions derived from these "spring-like" ideas could be also used for describing guided wave propagation in a multi-layered medium with imperfect contact between substrates [12, 13].

In the present study, the influence of thin intermediate adhesive-like films between laminate sub-layers on elastodynamic properties of the whole waveguide structure is considered both experimentally and theoretically. To address this problem, laminate specimens have been manufactured from two isotropic aluminium plates glued together using two-sided thin epoxy tape. Since the adhesion quality could not be assured in advance and is influenced by various external factors (i.e., poor surface preparation, non-uniform pressure during attachment, etc.), along with the basic elastodynamic problem for a three-layered structure with an ideal contact between plies, a more sophisticated formulation is considered. The latter takes into account the presence of partially debonded interfaces or zones of imperfect contact between sub-layers in such a laminate employing the distributed spring model. For this purpose, analytic frequency-dependent expressions for spring stiffnesses derived in [14, 15, 16] using the ensemble average technique and the boundary integral equation method [17, 18]. The results of corresponding numerical evaluations have revealed the strong influence of polymeric interlayer and possible adhesion imperfections on the wave dynamic properties of the considered laminate structure. Employing frequency-wavenumber analysis applied to the out-of-plane velocities of piezo-induced wave motion acquired with the laser Doppler vibrometry, the wavenumbers of propagating guided waves in a wide frequency range have been evaluated as well. The strong influence of the adhesive sub-layer is confirmed manifesting itself in remarkable changes in dispersion curves of both the fundamental and the high-order guided waves being in a good coincidence with corresponding theoretically predicted wavenumbers.

2. Boundary conditions

Let us consider two isotropic and homogeneous elastic plates D_1 and D_2 of thicknesses h_1 and h_2 and a thin film D_0 of thickness h_0 between them as shown in figure 1. Materials of the

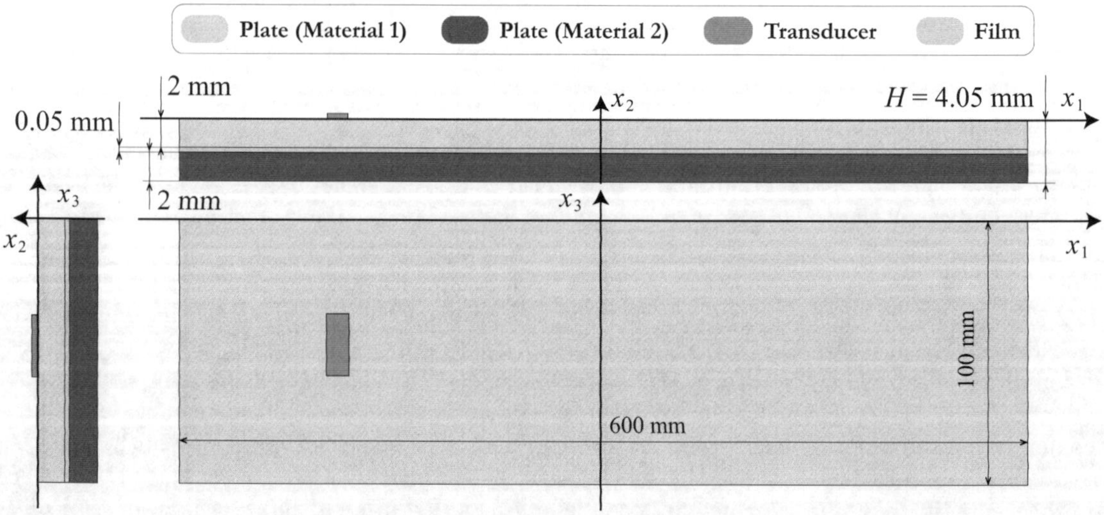

Figure 1. Geometry of the problem.

layers are characterized by the mass density ρ_j and Lame constants λ_j, μ_j. The latter can be also expressed in terms of wave velocities c_{js} (here subscript s corresponds to longitudinal L and transversal T waves correspondingly). For time-harmonic in-plane motion with the angular frequency ω, the displacement vector $\boldsymbol{u}^{(j)} = \{u_1^{(j)}, u_2^{(j)}\}$ obeys Lame equation

$$k_{j\mathrm{L}}^{-2}\nabla\nabla \cdot \boldsymbol{u}^{(j)} - k_{j\mathrm{T}}^{-2}\nabla \times \left(\nabla \times \boldsymbol{u}^{(j)}\right) + \boldsymbol{u}^{(j)} = 0 \tag{1}$$

written in terms of wave numbers $k_{js} = \omega/c_{js}$. The stress tensor components $\boldsymbol{\tau}^{(j)}$ by the Hooke's law are expressed through the components of the displacement vector.

2.1. Thin interlayers

As it is mentioned in the introduction, a thin elastic layer the interface between two different solids can be substituted by an interface with effective properties. For this case, the asymptotic approach was developed in [9] for a thin curved layer, and the numerical analysis of its application was provided for a thin spherical layer. As a result, the authors conclude that this approach is quite accurate if the layer thickness is small enough relatively to the wavelength.

Thus, for the calculation of the dispersion properties of a three-layered plate of thickness $H = h_1 + h_0 + h_2$, the continuity of the traction and displacement vectors at the interfaces is assumed:

$$\boldsymbol{\tau}^{(1)} - \boldsymbol{\tau}^{(0)} = \boldsymbol{u}^{(1)} - \boldsymbol{u}^{(0)} = 0, \qquad x_2 = -h_1, \tag{2}$$

$$\boldsymbol{\tau}^{(2)} - \boldsymbol{\tau}^{(0)} = \boldsymbol{u}^{(2)} - \boldsymbol{u}^{(0)} = 0, \qquad x_2 = -h_1 - h_0. \tag{3}$$

Here $\tau_k^{(j)}$ are components of the traction vector $\{\sigma_{12}^{(j)}, \sigma_{22}^{(j)}\}$. At the surfaces of the waveguide, stress-free boundary conditions

$$\boldsymbol{\tau}^{(1)} = 0, \quad x_2 = 0; \qquad \boldsymbol{\tau}^{(2)} = 0, \quad x_2 = -H. \tag{4}$$

are assumed.

The effective boundary conditions derived in [9] can be applied for the simulation of dynamic behaviour of a thin interlayer. In this case, (4) and total thickness $H = h_1 + h_0 + h_2$ of the waveguide should be kept, two thicker sub-layers should have thicknesses $h_1 + h_0/2$ and $h_2 + h_0/2$ and the thin interlayer is substituted by the effective boundary conditions, i.e. boundary conditions (2) and (3) are replaced by the relations:

$$\begin{cases} u_1^{(1)} - u_1^{(2)} = s_{13}^{(1)}\tau_1^{(1)} + s_{13}^{(2)}\tau_1^{(2)} \\ u_2^{(1)} - u_2^{(2)} = s_{21}^{(1)}u_1^{(1)} + s_{21}^{(2)}u_1^{(2)} + s_{24}^{(1)}\tau_2^{(1)} + s_{24}^{(2)}\tau_2^{(2)} \\ \tau_1^{(1)} - \tau_1^{(2)} = s_{31}^{(1)}u_1^{(1)} + s_{31}^{(2)}u_1^{(2)} + s_{34}^{(1)}\tau_2^{(1)} + s_{34}^{(2)}\tau_2^{(2)} \\ \tau_2^{(1)} - \tau_2^{(2)} = s_{42}^{(1)}u_2^{(1)} + s_{42}^{(2)}u_2^{(2)} \end{cases}, \qquad x_2 = -h_1 - h_0/2, \tag{5}$$

where

$$\begin{gathered} s_{13}^{(j)} = \frac{h_0}{2}\left(\frac{1}{\mu_0} - \frac{1}{\mu_j}\right), \qquad s_{34}^{(j)} = h_0(\lambda_j\mu_0 - \lambda_0\mu_j)b_0 b_j\frac{\partial}{\partial x_1}, \\ s_{21}^{(j)} = \frac{h_0}{2}(\lambda_j b_j - \lambda_0 b_0)\frac{\partial}{\partial x_1}, \qquad s_{24}^{(j)} = \frac{h_0}{2}(b_0 - b_j), \\ s_{31}^{(j)} = h_0(\mu_j - \mu_0 + \lambda_j\mu_j - \lambda_0\mu_0)\frac{\partial^2}{\partial x_1^2} + \frac{h_0}{2}(\rho_j - \rho_0)\omega^2, \\ s_{42}^{(j)} = \frac{h_0}{2}(\mu_0 - \mu_j)\frac{\partial^2}{\partial x_1^2} + \frac{h_0}{2}(\rho_j - \rho_0)\omega^2, \qquad b_j = \frac{1}{\lambda_j + 2\mu_j}. \end{gathered} \tag{6}$$

2.2. Concentration of micro-defects or damaged interface

On the other hand, a damaged interface can be described by a distributed spring model [10], where stiffness characterizes the weakness of adhesion. The effective elastic moduli of elastic solids containing randomly distributed micro-cracks can be estimated in terms of crack density. Consider a random distribution of cracks at the interface, where the distribution is assumed translationally invariant, and all the cracks are of the same size. The wave interaction between the cracks is neglected by assuming small cracks compared to the wavelength. Interface damage is defined through crack density $C = \dfrac{S_{dam}}{S_{total}}$. The ensemble average technique is employing for the build of the scattered field, which is represented in the form of plane waves in a far-field zone from the interface [14, 19] and finally, the total transmission coefficient for distribution of cracks is derived. The random distribution of cracks is then compared with the distributed spring model, where the damaged interface (for instance, between media D_0 and D_1) is described via spring boundary conditions

$$\boldsymbol{\tau}^{(1)} = \boldsymbol{\tau}^{(0)} = \begin{pmatrix} \kappa_{\mathrm{T}} & 0 \\ 0 & \kappa_{\mathrm{N}} \end{pmatrix} \cdot \left(\boldsymbol{u}^{(1)} - \boldsymbol{u}^{(0)}\right) \tag{7}$$

written in terms of tangential κ_{T} and normal κ_{N} spring stiffnesses. The spring boundary conditions demand that the stresses are continuous, while the displacement jump is proportional to the stresses. According to [14], normal ($s = N$) and tangential ($s = T$) spring stiffnesses, which define the severity of damage, can be represented as follows

$$\kappa_s = \frac{f_s}{C(\boldsymbol{p}_s \cdot \overline{\triangle \boldsymbol{u}_s})} - \frac{f_s}{2}, \quad f_s = \frac{\mathrm{i} c_{1s} k_{1s} c_{0s} k_{0s}}{c_{1s} k_{1s} + c_{0s} k_{0s}}. \tag{8}$$

Here $\boldsymbol{p}_s$ is a unit wave vector, and $\overline{\triangle \boldsymbol{u}_s}$ is the average of the crack opening displacement (COD).

Relying on the asymptotic analysis in the frequency domain and assuming a certain shape of micro-cracks, approximate analytical representations for SBCs could be derived [16]. The substitution of asymptotic average COD into (8) results in a closed-form relations for tangential κ_{T} and normal κ_{N} spring stiffnesses. In the case of a distribution of identical strip-like cracks of width a, the spring stiffnesses can be expressed in terms of the circular frequency ω as follows

$$\kappa_{\mathrm{T}} = \kappa_{\mathrm{N}} = \frac{8}{C\pi a \beta_1}, \tag{9}$$

where

$$\beta_1 = \frac{\lambda_1 + \mu_1}{(\lambda_1 + \mu_1)\mu_1} + \frac{\lambda_0 + 2\mu_0}{(\lambda_0 + \mu_0)\mu_0}.$$

Thus, if a damaged interface is assumed, then dispersion relations for the considered three-layered waveguide can be composed of (4) and (7) at interfaces $x_2 = -h_1$ and $x_2 = -h_1 - h_0$ instead of (2) and (3).

3. Results of numerical and experimental analysis

Further on, the dispersion properties of GWs propagating in a three-layered plate composed of two 2-mm thickness plates made of aluminium ($\lambda_1 = \lambda_2 = 50.5$ GPa, $\mu_1 = \mu_2 = 26.3$ GPa, $\rho_1 = \rho_2 = 2700$ kg m^{-3}) and two-sided tape between them ($\lambda_0 = 1.39$ GPa, $\mu_0 = 0.227$ GPa, $\rho_0 = 930$ kg m^{-3}) of 0.05 mm thickness are investigated. Possible contact imperfections between the layers are simulated employing SBCs according to the theory presented above. Spring stiffnesses in normal and tangential directions are assumed to be equal and frequency-independent. In the case under consideration, the 3D dispersion equation for modes of an infinite plate allows decomposition into two independent ones.

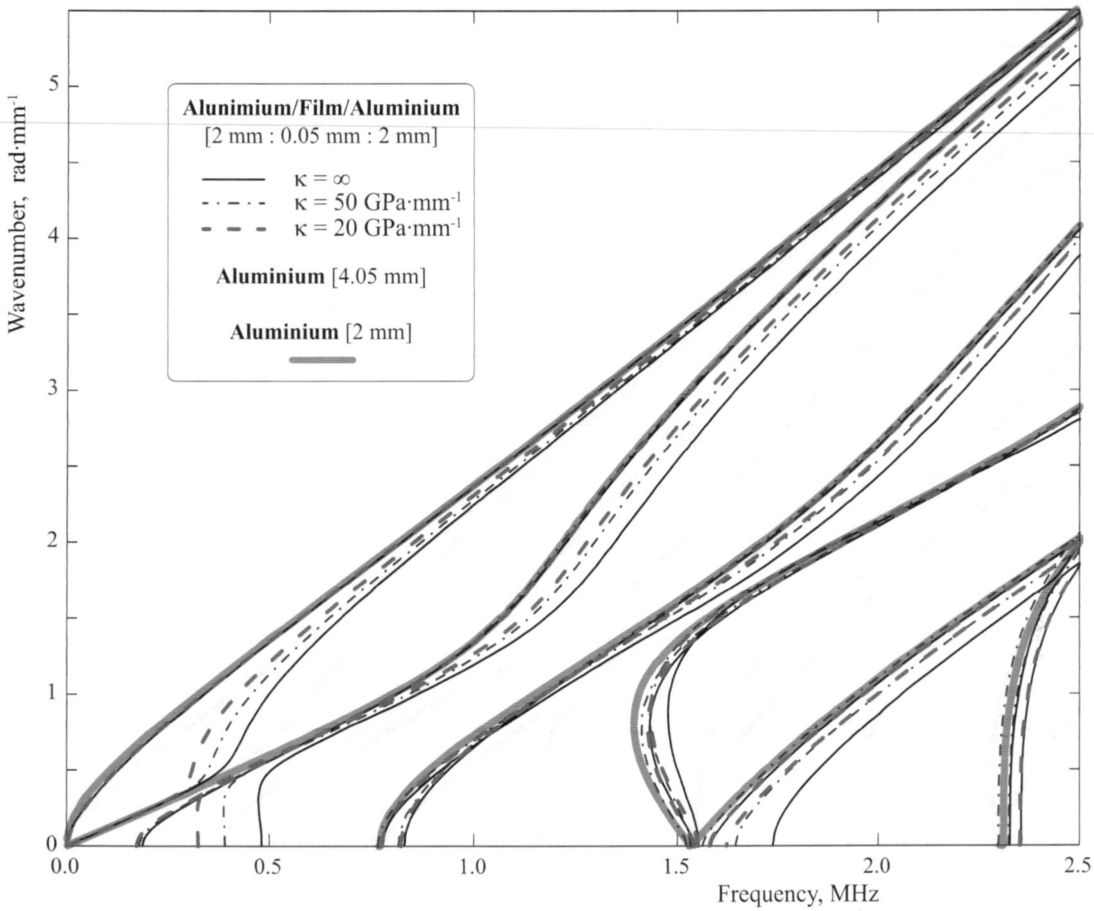

Figure 2. Wavenumbers of GWs propagating in a three-layered plates with various contact at the interfaces simulated via the SBC and in a homogeneous aluminium plates.

The wavenumbers of propagating Lamb waves (LWs) are shown in figure 2. First of all, this plot demonstrates that the presence of a thin interlayer sufficiently changes the dispersion properties of the waveguide. Namely, there is almost no coincidence between dispersion curves for a three-layered structure and a monolithic aluminium layer of the same thickness. On the contrary, the behaviour of certain normal modes for a laminate structure coincides with the dispersion curves for a single 2-mm aluminium plate, which could be explained by a softness of the polymeric interlayer, i.e., aluminium substrates though being attached remain "quasi-decoupled" for these particular guided waves. Taking possible contact imperfections at interfaces into account provides additional changes into dispersion curves, and the sensitivity to adhesion disturbances varies depending on the frequency and the number of the mode. The fundamental modes in the low-frequency range (much lower than the first cut-off frequency) are practically insensitive to the weakened adhesion of considered severity, whereas its effect is pronounced at $f \in (0.25, 1)$ MHz, especially for the fundamental symmetric Lamb wave S_0. Two next LWs also show good sensitivity in this region.

For the experimental observation, a corresponding laminate sample was made from two aluminium plates with a thickness of 2 mm attached together with a thin 0.05 mm double-sided epoxy tape. The out-of-plane velocities of piezo-induced wave packages were measured on the specimen surface by laser Doppler vibrometry and were post-processed via the frequency-wavenumber analysis (FWA) [20]. To minimize parasitic side-wall reflections and to improve the FWA quality the specimen edges were extensively covered by plasticine. In ffigure 3, the FWA results (i.e., dark-coloured areas indicate experimental dispersion curves of LWs) and corresponding numerical data for three-layered waveguide as well as for a single 2 mm aluminium layer are presented. A detailed analysis shows that it is essential to consider a thin interlayer in simulations, otherwise, the experimental frequency dependencies of LW wavenumbers are not captured adequately. Moreover, assuming contact imperfections at sublayer interfaces via boundary conditions (6) improves the coincidence between predicted and experimental dispersion curves of higher LWs. Surprisingly, however, is that in figure 3 at low frequencies experimental wavenumbers of the fundamental antisymmetric Lamb wave A_0 agree with the A_0 dispersion curve for a single-layered 2 mm aluminium waveguide rather than with the results for the three-layered structure. An adequate explanation of this phenomena would be a subject of further research.

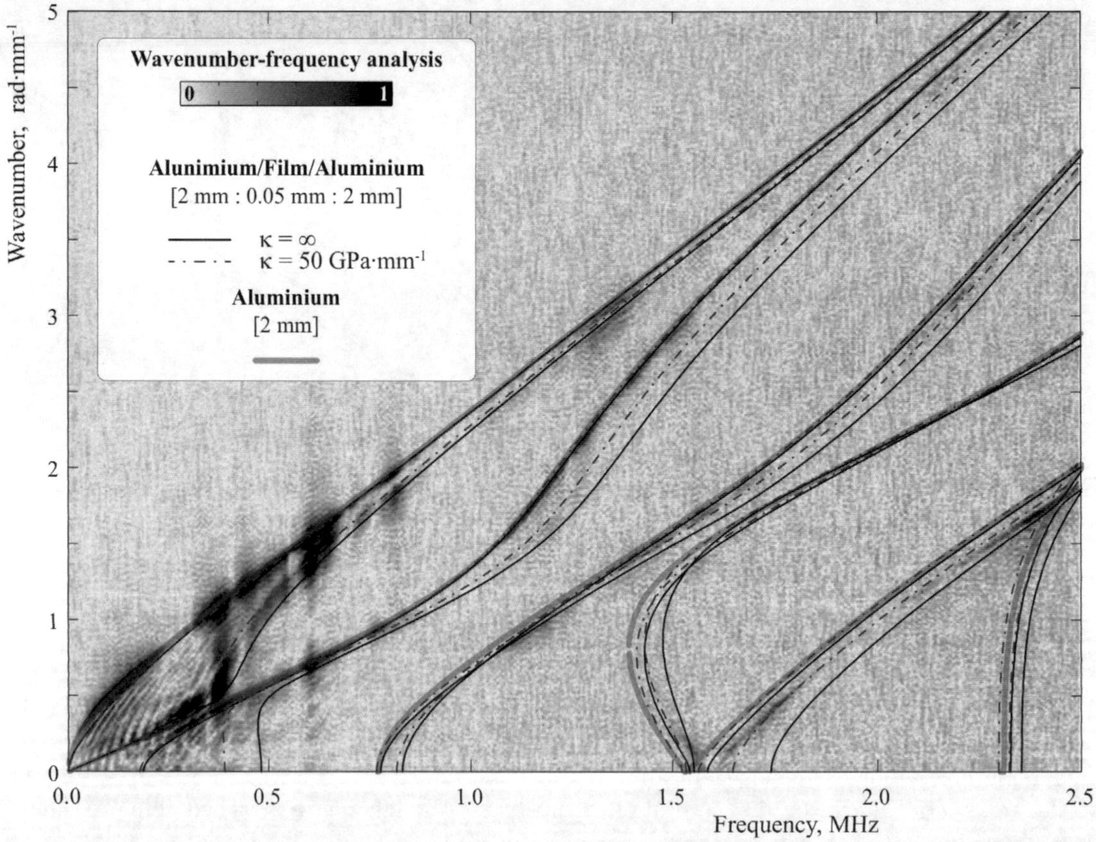

Figure 3. Results of the FWA and wavenumbers of GWs propagating in a three-layered pristine (thin solid line) and damaged simulated via the SBC (dashed line) plates made from aluminium and epoxy tape and in 2 mm thickness aluminium plate (thick solid line).

4. Conclusions

Computational models aiming at describing the influence of thin interlayers between two solids with possible contact imperfections are proposed. They are essential for predictive simulations in ultrasonic based non-destructive and structural health monitoring of adhesively bonded constructions in service to ensure their reliability and safety. Within these models, the effect of the internal thin polymeric layer and imperfect interfaces between sub-layers in a laminate composite structure is analysed revealing their non-neglectable influence on dispersion properties of propagating Lamb waves, which, in turn, is verified experimentally with laser Doppler vibrometry measurements of piezoelectric induced LWs.

Acknowledgments
The authors are grateful to the support by the Russian Science Foundation (project 17-11-01191).

References

[1] Tamborrino R, Palumbo D, Galietti U, Aversa P, Chiozzi S and Luprano V 2016 *Compos. Part B-Eng.* **91** 337–45
[2] Attar L, Leduc D, Ech Cherif El Kettani M, Predoi M, Galy J and Pareige P 2020 *NDT&E Int.* **111** 102213
[3] Yilmaz B and Jasiuniene E 2020 *Int. J. Adhes. Adhes.* **102** 102675
[4] Jiao D and Rose J L 1991 *J. Adhes. Sci. Tech.* **5** 631–46
[5] Siryabe E, Renier M, Meziane A, Galy J and Castaings M 2017 *Ultrasonics* **79** 34–51
[6] Rucka M, Wojtczak E and Lachowicz J 2018 *Appl. Sci.* **8** 522
[7] Dahmen S 2019 *Ultrasonics* **94** 37–49
[8] Rokhlin S I and Wang Y J 1991 *J. Acoust. Soc. Am.* **89** 503–15
[9] Boström A, Bövik P and Olsson P 1992 *J. Nondestruct. Eval.* **11** 175–84
[10] Baik J M and Thompson R B 1984 *J. Nondestruct. Eval.* **4** 177–96
[11] Margetan F, Thompson R, Rose J and Gray T 1992 *J. Nondestruct. Eval.* **11** 109–26
[12] Mal A 1988 *Int. J. Eng. Sci.* **26** 873–81
[13] Balvantin A, Baltazar A and Aranda-Sanchez J I 2012 *Int. J. Mech. Sci.* **63** 66–73
[14] Golub M V and Boström A 2011 *Wave Motion* **48(2)** 105–15
[15] Golub M V and Doroshenko O V 2019 *Int. J. Sol. Struct.* **165** 115–26
[16] Golub M V and Doroshenko O V 2020 *Eur. J. Mech. A-Sol.* **81** 103894
[17] Krenk S and Schmidt H 1982 *Philos. Trans. R. Soc.* A **308** 167–98
[18] Glushkov Y V and Glushkova N V 1996 *J. Appl. Math. Mech.* **60** 277–83
[19] Glushkov E V and Glushkova N V 2001 *J. Comp. Acoust.* **9(3)** 889–98
[20] Wilde M V, Golub M V and Eremin A A 2019 *Ultrasonics* **98** 88–93

Determination of trapped mode eigenfrequencies of an elastic plate with a strip-like delamination

Mikhail V Golub[1], Artem A Eremin[1], Jochen Moll[2], Diana A Sverkunova[1] and Rolf Lammering[3]

[1] Institute for Mathematics, Mechanics and Informatics, Kuban State University, 350040 Krasnodar, Russian Federation
[2] Department of Physics, Terahertz Photonics, Goethe University of Frankfurt am Main, Frankfurt am Main, Germany
[3] Institute of Mechanics, Helmut-Schmidt-University, Hamburg, Germany

golub@inbox.ru

Abstract. Resonance scattering of Lamb waves by strip-like interface delamination is investigated. Trapped mode resonance frequencies of the plate with interface delamination are experimentally studied via the application of the Fourier transform to point-wise measurements along a line obtained using laser Doppler vibrometry. Eigenfrequencies are also calculated employing semi-analytical and mesh-based computational methods. A good agreement between experimentally extracted and numerically predicted trapped mode eigenfrequencies is demonstrated.

1. Introduction

Since composite structures are widely employed nowadays, effective methods for damage detection in them are necessary [1, 2]. Some techniques used in non-destructive testing and structural health monitoring are based on elastic guided waves because their reflection allows detecting hidden defects [3, 4, 5]. Therefore, damage detection methods can be improved by employing information on the scattering of guided waves by various obstacle types.

It has been shown in [6] that Lamb wave scattering by delamination-like obstacles when featured by resonance interaction can be used for damage identification. Resonance scattering by crack-like defect manifests itself in a localized wave motion at the defect's vicinity at its eigenfrequencies [7, 8, 9, 10]. This phenomenon is also known as trapped modes [11], scattering resonances [12] or local defect resonances [10, 13, 14]. The focus of the current study, considered as the further development of the ideas from ref. [6], is on the determination of trapped mode resonance frequencies of an interface strip-like delamination in an elastic laminate fabricated from different materials.

In this paper, delamination resonance scattering phenomena are investigated experimentally and numerically. For this purpose, several specimens made from aluminium and glass with two-sided epoxy tape as a couplant and introduced strip-like artificial delaminations are considered. The spectral analysis is applied to experimental transient B-scans and point-wise measurements obtained with scanning laser Doppler vibrometry (LDV) for resonance frequency extraction from the data (so called, S-scans). Some postprocessing relying on moving windowed Fourier transform and averaging is proposed to increase the reliability of the proposed evaluation procedure. Developed semi-analytical and mesh-based computational models are used to compute eigenfrequencies and eigenforms. A good agreement between experimentally extracted and numerically predicted trapped mode eigenfrequencies is demonstrated.

2. Experimental setup

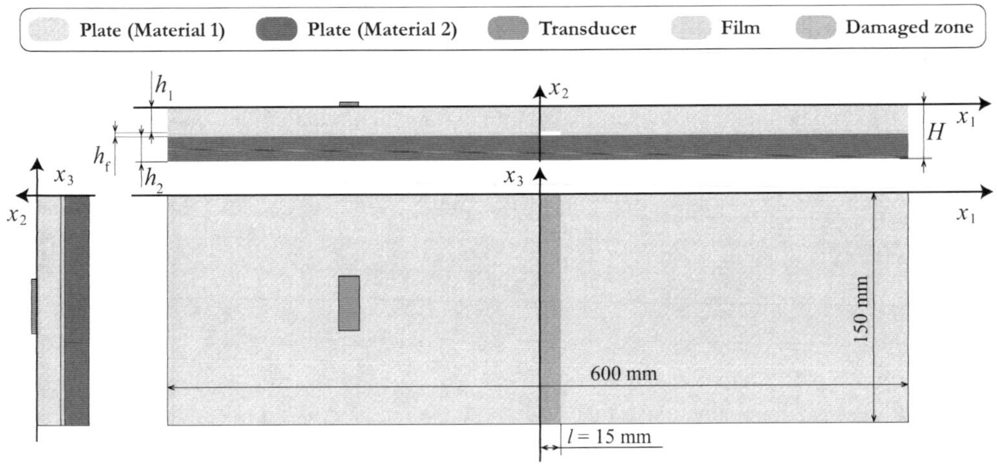

Figure 1. Geometry of the problem.

In order to investigate resonance scattering by a strip-like delamination several specimens were manufactured composing two plates of thicknesses h_1 and h_2 and of equal in-plane dimensions 600 × 150 mm^2. The plates were glued together using a two-sided epoxy tape of 0.05 mm thickness, whereas a delamination was introduced as its absence in the rectangular area of $l = 15$ mm width. The geometry of the specimen is shown in figure 1.

Wave motion in the sample is excited by a rectangular piezoelectric transducer ($10 \times 30 \times 0.25$ mm^3), which is glued on the surface of the plate parallel to its edges. The distance between two closest edges of the piezoelectric transducer and the delamination is 150 mm. The actuator is driven with a transient voltage $p(t)$ applied at its electroded surface. Two pulse configurations $p(t)$ are employed, i.e., in the form of either a broad-band rectangular pulse of short duration or a narrow-band Hann-windowed five-cycle sine burst, which spectrum is strongly concentrated near its central frequency f_0. The latter are generated by a Tektronix AFG 3022B arbitrary signal generator and are pre-amplified to the range 70 V-pp by a NF HSA4101 external high-frequency power amplifier before being applied to the PWAS. The out-of-plane velocity field of propagating GWs is measured on the plate surfaces by means of a Polytec PSV-500 one-dimensional scanning LDV.

3. The boundary value problem

The time-harmonic motion with the angular frequency $\omega = 2\pi f$ of each layer $j = \overline{1,3}$ in a composite laminate plate $-H \leq x_2 \leq 0$ is governed by the Lamé equations

$$(\lambda^{(j)} + 2\mu^{(j)})\nabla\left(\nabla \cdot \mathbf{u}^{(j)}\right) - \mu^{(j)}\nabla \times (\nabla \times \mathbf{u}^{(j)}) + \rho^{(j)}\omega^2\mathbf{u}^{(j)} = 0,$$

which can be written in terms of the displacement vector $\mathbf{u}^{(j)}(\mathbf{x})$, Lamé elastic constants $\lambda^{(j)}$, $\mu^{(j)}$ and mass density $\rho^{(j)}$. The surfaces of the plate $x_2 = -H$ and $x_2 = 0$ as well as the surfaces of the delamination Ω are assumed stress-free

$$\sigma_{j2}(x_1, x_2, 0) = \sigma_{j2}(x_1, x_2, -H) = 0$$

except the domain of the contact between the plate and the piezoelectric transducer. Here σ_{ij} is the stress tensor related with displacements u_i by the Hooke's law. The material properties of elastic materials considered in the computations are given in table 1.

Table 1. The material properties.

Material	Poisson's ratio ν	Shear modulus μ [GPa]	Young's modulus E [GPa]	λ [GPa]	Density ρ [kg×m^{-3}]	SV-wave c_T [m×s^{-1}]	P-wave c_L [m×s^{-1}]
Aluminium	0.33	26.3	70	50.5	2700	3098.15	6816.15
Two-sided tape	0.43	0.227273	0.65	1.3961	930	494.348	1801.87
Glass	0.25	27.66	69.15	27.66	2770	3159.99	5473.27

To determine eigenfrequencies of the boundary-value problem, two methods have been used. The boundary integral equation method [15, 16] relies on integral representations for the wave-field scattered by the crack in the form of the convolution of the corresponding Green matrices with unknown crack opening displacement $\Delta\mathbf{u}$. In this case, the complex-valued resonance poles in the frequency plane or eigenfrequencies f_n are spectral points of the integral operator $\mathcal{A}$ describing stresses induced by the delamination. Numerically, they are approximated by the zeros of the determinant of the linear algebraic system to which the boundary integral equation is reduced via Galerkin method.

The finite element method (implemented with COMSOL Multiphysics 5.5) can be also applied employing the perfectly matched layers (PML) to simulate an open or unbounded waveguide. The comparison of the results computed using both methods shows a good agreement.

4. Resonance frequencies extraction from experimental data

To extract the values of trapped mode eigenfrequencies from a line scan made by laser vibrometer, the Laplace transform can be applied to the part of the signal measured on the plate surface. It is preferable to excite the piezoactuator with a broad-band voltage $p(t)$, e.g., in the form of 0.5 μs rectangular pulse, see [6] for more details. However, the direct application of the Laplace transform does not reveal localization phenomena. Therefore, the spectral analysis of the "tail" of the transient signal measured at the damage, i.e

$$\mathcal{L}[v](x_1, f) = \int_{t_1}^{t_2} v(t)\mathrm{e}^{\mathrm{i}2\pi ft}\mathrm{d}t, \quad (1)$$

should be performed. An example of the application of (1) to the out-of-plane velocities $v(x_1, t)$ measured by the LDV along the line $x_2 = 0$ for the specimen composed of 2 mm Aluminium and 3 mm Glass plates is demonstrated in figure 2 (a). Two high peaks near 80 kHz are clearly visible on the surface $|\mathcal{L}\ [v]\ (x_1,\ f\ )|$. This allows to determine the second eigenfrequency, whereas the third and fourth eigenfrequencies cannot be determined in such a straightforward manner. To meet this problem, an averaging technique is proposed. This approach takes into account that information about different eigenfrequencies are distributed in different parts of the signal.

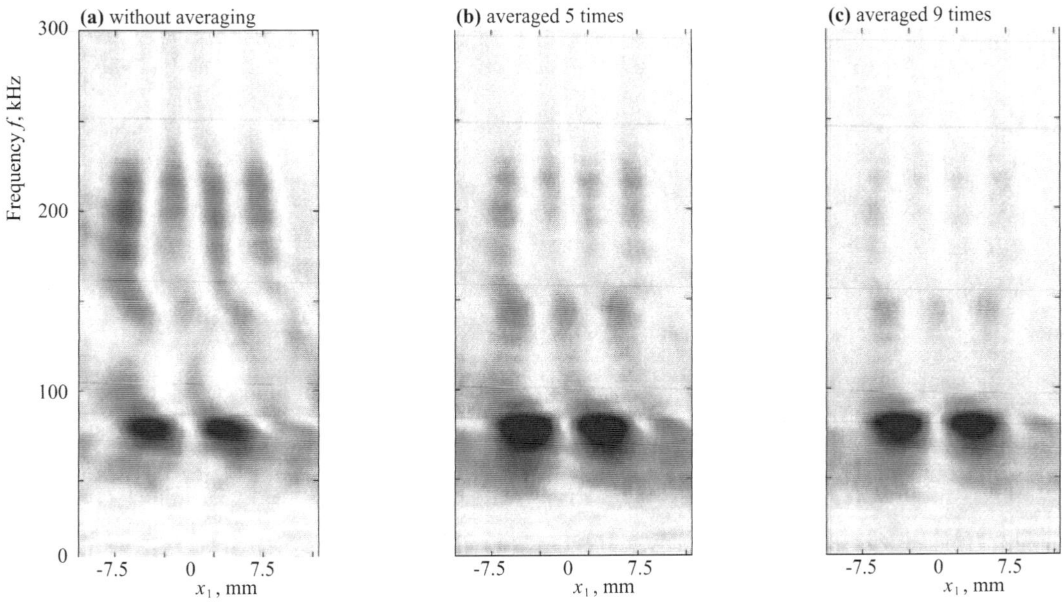

Figure 2. The absolute value of $|\mathcal{L}[v](x_1, f)|$ for different number of averages: $M = 1$ (a), $M = 5$ (b), $M = 9$ (c). From the B-scan at the surface of aluminium of the specimen 2 mm Aluminium — 3 mm Glass; $a = 0.64\mu$s.

Therefore, (1) is substituted by

$$\mathcal{L}\left[v\right](x_1, f) = \sum_{k=1}^{M} \int_{t_1+ka}^{t_2+ka} v(x_1, t)\mathrm{e}^{\mathrm{i}2\pi ft}\mathrm{d}t, \tag{2}$$

where a is the shift of the windows in time, M is the number of averages. The improvement of the procedure of the determination of peaks corresponding to eigenfrequencies of trapped modes via the application of (2) is shown in figure 2 (b), (c).

5. Numerical analysis

Figure 3 demonstrates surfaces results of the application of (2) to B-scans measured at both surfaces of two specimens: 2 mm Aluminium / 3 mm Glass (figure 3 (a), (b)) and 3 mm Aluminium / 2 mm Glass (figure 3 (c), (d)). One can see periodically situated peaks localized at the same frequencies. Centers of narrow frequency ranges, where wave motion localization is clearly visible, can be identified as resonance frequencies. It should be noted that resonance frequencies are more accurately identified at the glass-side of the specimen surface. The results of the determination of resonance frequencies f_n numerically and experimentally are summarized in table 2.

To demonstrate that trapped mode phenomena can be clearly seen at the determined frequencies a B-scan has been made at the glass surface of the specimen 2 mm Aluminium / 3 mm Glass, see figure 4. In the measurement, Hann-windowed five-cycle sine burst with the central frequency $f_0 = f_2 = 82$ kHz has been used for actuator excitation. One can see a strong localization above the delamination, which becomes weaker if the central frequency is not equal to one of eigenfrequencies.

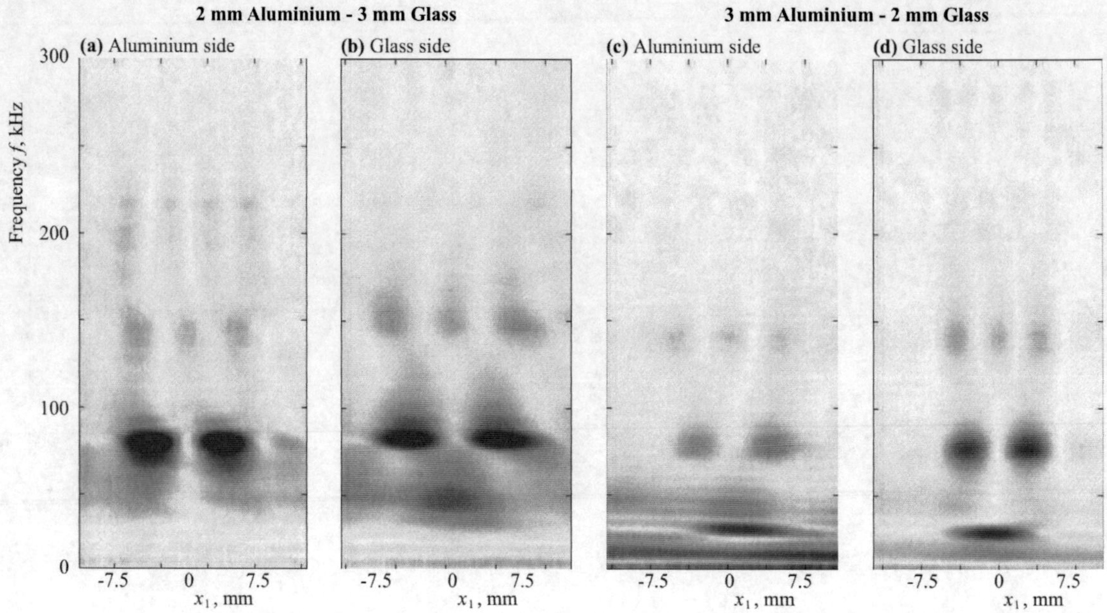

Figure 3. The absolute value of $|\mathcal{L}[v](x_1, f)|$ obtained from B-scans at the surface of aluminium ((a), (c)) and glass ((b), (d)) of the specimen 2 mm Aluminium / 3 mm Glass ((a), (b)) and 3 mm Aluminium / 2 mm Glass ((c), (d)); $a = 0.64$ μs, $M = 9$.

Eigenfrequency		2 mm Aluminium / 3 mm Glass		2 mm Aluminium / 3 mm Glass	
		Aluminium side	Glass side	Aluminium side	Glass side
f_1	Experiment		45.6	31	30
	Simulation	47		33	
f_2	Experiment	80	82.5	76	76
	Simulation	78		76	
f_3	Experiment	145	147	140	140
	Simulation	150		150	
f_4	Experiment	217	217	217	219
	Simulation	240		242	
f_5	Experiment		295		279.5
	Simulation	292		283	

Table 2. Eigenfrequencies for two specimens considered.

6. Conclusions

Resonance scattering of Lamb waves by strip-like interface delamination is investigated. Trapped mode resonance frequencies of the plate with interface delamination are experimentally studied via the application of the Fourier transform to point-wise measurements along a line obtained using LDV. Proposed data postprocessing increases the reliability of the experimental results. Eigenfrequencies and eigenforms are also calculated employing semi-analytical and mesh-based computational methods. A good agreement between experimentally extracted and numerically predicted trapped mode eigenfrequencies is demonstrated. Since these frequencies strongly depend on the obstacle geometry and location within the waveguide thickness, this technique could have a potential for damage identification in a combination with other methods.

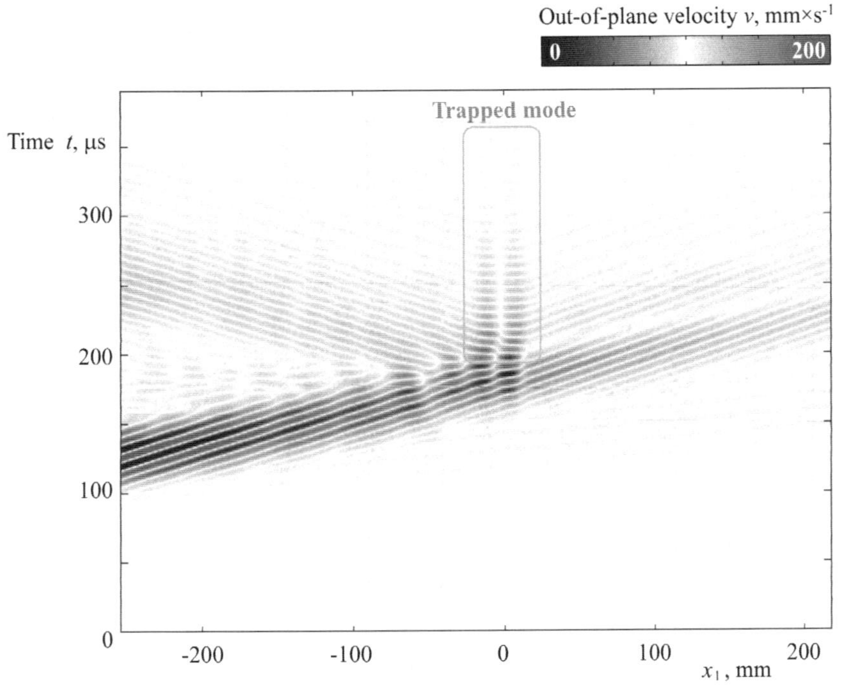

Figure 4. The absolute value of $|\mathcal{L}[v](x_1, f)|$ obtained from B-scans at the surface of aluminium ((a), (c)) and glass ((b), (d)) of the specimen 2 mm Aluminium / 3 mm Glass ((a), (b)) and 3 mm Aluminium / 2 mm Glass ((c), (d)); $a = 0.64\mu$s, $M = 9$.

Acknowledgments

The authors are grateful to the support by the Russian Science Foundation (project 17-11-01191).

References

[1] Zhao G, Wang B, Hao W, Luo Y and Chen H 2020 *Compos. Struct.* **238** 111972
[2] Xu W, Su Z, Liu J, Cao M and Ostachowicz W 2020 *Smart Mater. Struct.* **29** 045001
[3] Wojtczak E and Rucka M 2019 *Materials* **12**
[4] Moll J, Kexel C, Kathol J, Fritzen C P, Moix-Bonet M, Willberg C, Rennoch M, Koerdt M and Herrmann A 2020 *Appl. Sci.* **10** 3068
[5] Spada A, Capriotti M and Lanza di Scalea F 2020 *Int. J. Sol. Struct.* **182-183** 267–80
[6] Eremin A A, Golub M V, Glushkov E V and Glushkova N V 2019 *NDT & E Int.* **103** 145–53
[7] Glushkov E, Glushkova N, Golub M, Moll J and Fritzen C P 2012 *Smart Mater. Struct.* **21** 125001
[8] Solodov I and Kreutzbruck M 2019 *NDT & E Int.* **107** 102146
[9] Hanuman N, Roy S and Bose T 2019 *Int. J. Mech. Sci.* **163** 105092
[10] Segers J, Hedayatrasa S, Verboven E, Poelman G, Van Paepegem V and Kersemans M 2019 *NDT & E Int.* **102** 218–25
[11] Linton C M and Evans D V 1991 *Quart. J. Mech. Appl. Math.* **44(3)** 487–506
[12] Glushkov Y V and Glushkova N V 1996 *J. Appl. Math. Mech.* **60** 277–83
[13] Solodov I 2017 *Res. Nondestr. Eval.* **28** 28–44
[14] Solodov I and Kreutzbruck M 2020 *Ultrasonics* **108** 106221
[15] Glushkov E V and Glushkova N V 2001 *J. Comp. Acoust.* **9(3)** 889–898
[16] Glushkov E V, Glushkova N V and Ekhlakov A V 2002 *J. Appl. Math. Mech.* **66** 141–49

Investigation of the stability of a cantilevered pipeline with a fluid flow compressed at the free end by the tracking force

D V Kapitanov, O V Kapitanova and A G Bobylkov

National Research Lobachevsky State University of Nizhny Novgorod, Nizhny Novgorod, 603950, Russia

dis-kdv@mail.ru

Abstract. The paper presents a numerical and analytical investigation of the stability of a cantilevered pipeline transporting an incompressible fluid flow at a constant speed compressed by tracking force at the free end. To solve the problem, a dynamic approach based on the analysis of the behavior of the system's own characteristics is used. In contrast to the classical research approaches, which are based on the representation of the solution of the problem using a finite number of basic functions, an algorithm that does not require such a restriction is proposed.

1. Introduction

Consideration of problems related to the interaction of elements of mechanical structures and their streamlined fluid flow is relevant for the modern development of the technical industry, including mechanical engineering. Quite often, structural elements are represented by a system of pipes for which the liquid is the internal medium. The main feature of the interaction of a structure with a liquid can be considered a significant change in the dynamic properties of the structure and the occurrence of vibrations for its individual components, which can lead to its destruction or decommissioning [1, 2]. Quite often, the pipeline is modeled as hollow rods, streamlined by the flow of liquid. Under this assumption, the hydrodynamic component of the system can be eliminated by adding hydrodynamic forces acting from the fluid on the structural elements to the mathematical model of the system. In this case, everything is reduced to solving problems related to the study of dynamics and stability of distributed mechanical systems using the theory of rods.

2. Mathematical model

Let us consider small flat low-frequency bending vibrations of a cantilevered pipeline loaded at the free end by a tracking force, which transports an incompressible fluid flow at a constant speed. In [2, 3], the pipeline oscillation equation is derived. In [4-5], the results of stability studies in the absence of compressive load are presented.

The equation describing the movement of the pipeline taking into account internal friction in the material according to the Kelvin-Voigt hypothesis and the corresponding boundary conditions have the form:

$$\mu EI\frac{\partial^5 y}{\partial x^4 \partial t} + EI\frac{\partial^4 y}{\partial x^4} + (Mv^2 + P)\frac{\partial^2 y}{\partial x^2} + 2Mv\frac{\partial^2 y}{\partial x \partial t} + (m + M)\frac{\partial^2 y}{\partial t^2} + \xi\frac{\partial y}{\partial t} = 0, \quad (1)$$

$$y|_{x=0} = \frac{\partial y}{\partial x}\bigg|_{x=0} = EI\frac{\partial^2 y}{\partial x^2}\bigg|_{x=l} = EI\frac{\partial^3 y}{\partial x^3}\bigg|_{x=l} = 0, \tag{2}$$

where $y(x,t)$ – transverse displacement of the pipeline, m and M – per unit length weight of pipeline and liquid, respectively, EI – is the exural rigidity of thc pipeline, P – is the longitudinal compressive load, l – is the length of the pipeline, μ and ξ – are the coefficients of internal and external friction, respectively, v – the speed of fluid flow.

To write the equation of motion (1) and boundary conditions (2) in dimensionless parameters, we replace the variables:

$$x = l\varphi\ (0 \le \varphi \le 1), \qquad t = \left(\frac{m+M}{EI}\right)^{1/2} l^2\tau.$$

As a result, we get:

$$\gamma\frac{\partial^5 y}{\partial\varphi^4\partial\tau} + \frac{\partial^4 y}{\partial\varphi^4} + (b+c)\frac{\partial^2 y}{\partial\varphi^2} + 2a\frac{\partial^2 y}{\partial\varphi\partial\tau} + \frac{\partial^2 y}{\partial t^2} + \delta\frac{\partial y}{\partial t} = 0, \tag{3}$$

$$y|_{\varphi=0} = \frac{\partial y}{\partial\varphi}\bigg|_{\varphi=0} = \frac{\partial^2 y}{\partial\varphi^2}\bigg|_{\varphi=1} = \frac{\partial^3 y}{\partial\varphi^3}\bigg|_{\varphi=1} = 0, \tag{4}$$

where:

$$a = \frac{Mvl}{EI}\left(\frac{EI}{m+M}\right)^{1/2}, \qquad b = \frac{Mv^2l^2}{EI} = a^2\frac{m+M}{M}, \qquad c = \frac{Pl^2}{EI},$$

$$\delta = \frac{\xi l^2}{EI}\left(\frac{EI}{m+M}\right)^{1/2}, \qquad \gamma = \frac{\mu}{l^2}\left(\frac{EI}{m+M}\right)^{1/2}.$$

3. Stability study of a compressed cantilevered pipeline with a liquid flow

We will search for the solution of problem (3, 4) using the variable separation method:

$$y(\varphi,\tau) = W(\varphi)e^{\lambda\tau}. \tag{5}$$

After substituting the expression (5) into equation (3) and boundary conditions (4), a boundary value problem for the eigenvalue problem is obtained:

$$(1+\gamma\lambda)\frac{d^4W(\varphi)}{d\varphi^4} + (b+c)\frac{d^2W(\varphi)}{d\varphi^2} + 2a\lambda\frac{dW(\varphi)}{d\varphi} + (\delta\lambda+\lambda^2)W(\varphi) = 0 \tag{6}$$

$$W(\varphi)|_{\varphi=0} = \frac{dW(\varphi)}{d\varphi}\bigg|_{\varphi=0} = \frac{d^2W(\varphi)}{d\varphi^2}\bigg|_{\varphi=1} = \frac{d^3W(\varphi)}{d\varphi^3}\bigg|_{\varphi=1} = 0 \tag{7}$$

The eigenvalue problem consists of an ordinary fourth-order differential equation (6) and homogeneous boundary conditions (7).

The general solution of the problem (6) has the form:

$$W(\varphi) = \sum_{i=1}^{4} A_i e^{W_i\varphi},$$

where $W_i = W_i(\lambda, a, b, c, \gamma, \delta)$, $i = \overline{1,4}$, – roots of the characteristic equation

$$F(\lambda, W) = (1+\gamma\lambda)W^4 + (b+c)W^2 + 2a\lambda W + (\delta\lambda+\lambda^2) = 0 \tag{8}$$

The coefficients A_i ($i = \overline{1,4}$) are determined from the boundary conditions of problem (7) as a result of solving a system of linear homogeneous algebraic equations

$$\begin{cases} A_1 + A_2 + A_3 + A_4 = 0, \\ W_1A_1 + W_2A_2 + W_3A_3 + W_4A_4 = 0, \\ W_1^2e^{W_1}A_1 + W_2^2e^{W_2}A_2 + W_3^2e^{W_3}A_3 + W_4^2e^{W_4}A_4 = 0, \\ W_1^3e^{W_1}A_1 + W_2^3e^{W_2}A_2 + W_3^3e^{W_3}A_3 + W_4^3e^{W_4}A_4 = 0. \end{cases} \quad (9)$$

For the existence of a non-trivial solution of system (9), it is necessary and sufficient that its determinant $D(W_1, W_2, W_3, W_4)$ vanishes:

$$D(W_1, W_2, W_3, W_4) = 0 \quad (10)$$

From equations (8) and (10), we can write a system of five equations with five unknowns considered over a field of complex numbers $W_1, W_2, W_3, W_4, \lambda$:

$$\begin{cases} F(\lambda, W_i) = 0, \ i = \overline{1,4}, \\ D(W_1, W_2, W_3, W_4) = 0. \end{cases} \quad (11)$$

Starting from zero values of the dimensionless parameter of the flow velocity of the fluid (parameter a) and gradually increasing it, and by varying a dimensionless parameter of compressive load (option c) from zero to some value when the current value of the dimensionless speed parameter, it is possible to determine critical values of parameters a and c, which will cause the loss of stability in the system (figure 1). This happens when the real part of one of the eigenvalues of λ in the system (11) becomes positive. All parameters describing the mechanical system, with the exception of the fluid flow rate and compressive load, are fixed. The system (11) is solved by the iterative Newton method, and the initial values are taken as the values in the absence of compressive load and friction losses at zero fluid flow velocity [6], and later, when the parameters a and c are changed, the solutions of the system obtained at the previous step of changing these parameters are taken.

4. The results of the research

The conducted studies allow us to divide the plane of parameters of the dimensionless fluid flow rate and dimensionless load (a;c) into stable and unstable regions of the system (figure 1). Four cases are considered in this paper: no friction losses ($\gamma = 0, \delta = 0$, curve 1), the presence of only internal friction ($\gamma = 0.0015, \delta = 0$, curve 2), the presence of only external friction ($\gamma = 0, \delta = 2.52$, curve 3) and the presence of both types of friction ($\gamma = 0.0015, \delta = 2.52$, curve 4). The results are presented for a system corresponding to the flow of water in a steel pipe. The pipe has a length of $l = 1\ m$, its external and internal radiuses are $R = 5 \cdot 10^{-3}\ m$ and $r = 4 \cdot 10^{-3}\ m$, respectively.

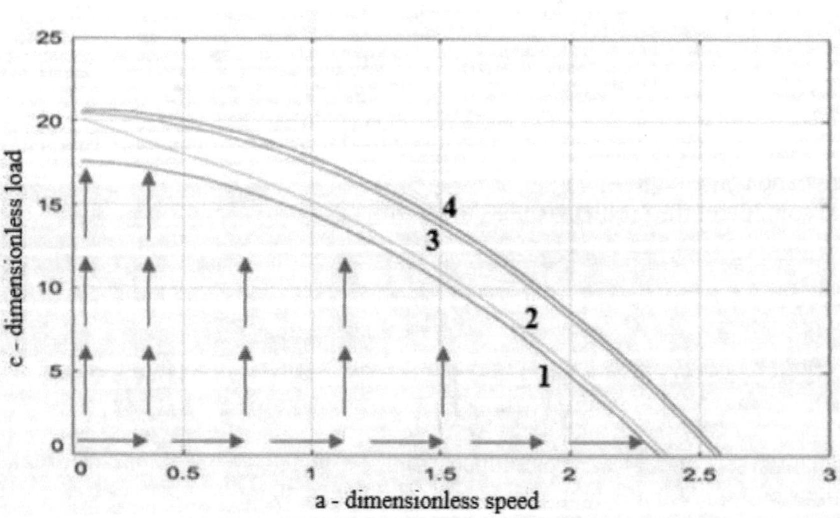

Figure 1. Stability region of the system.

Studies have shown that the loss of stability is of the flutter type and occurs in the second form of vibrations. Figure 2-5 provides a qualitative representation of the hodographs of the first three eigenvalues of the system on the complex plane as the fluid flow rate increases from zero at a constant dimensionless load c for four different cases presented earlier. In all cases, the eigenvalues start moving to the left. Then the eigenvalue corresponding to the second oscillation mode changes its direction and passes into the positive real half-plane (there is a loss of stability). If there is no friction, the eigenvalues begin their movement from the imaginary axis, and if there is friction from the negative real half-plane. If there is only external friction, the movement begins with an axis parallel to the imaginary one, and if there is only internal friction, the real parts of the eigenvalues in the absence of a fluid flow have different values. It can be noted that internal friction has a stronger effect on the older modes of vibrations compared to the younger ones. In the presence of both types of friction, the critical velocity of the fluid flow at which the loss of stability occurs is of the greatest importance compared to the other cases.

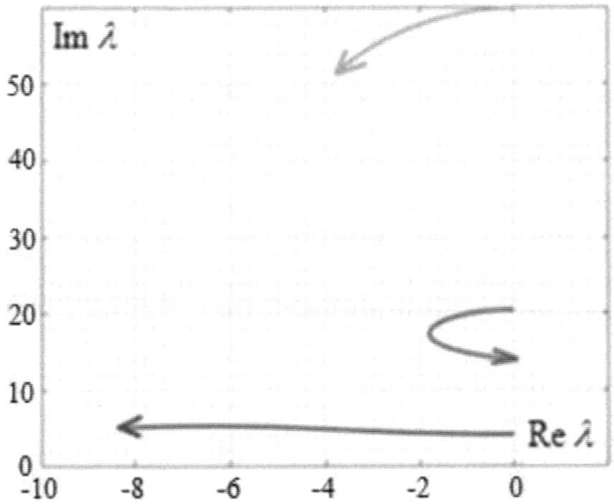

Figure 2. No friction in the system.

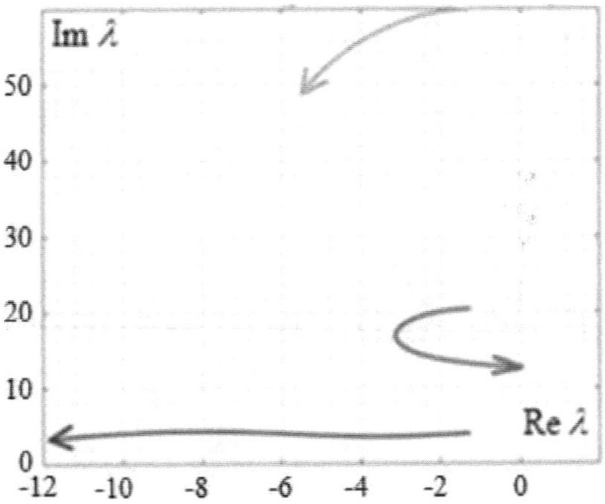

Figure 3. The presence of only external friction.

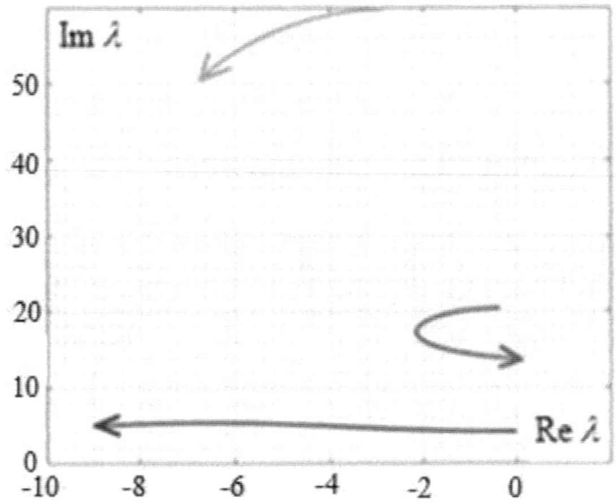

Figure 4. The presence of only internal friction.

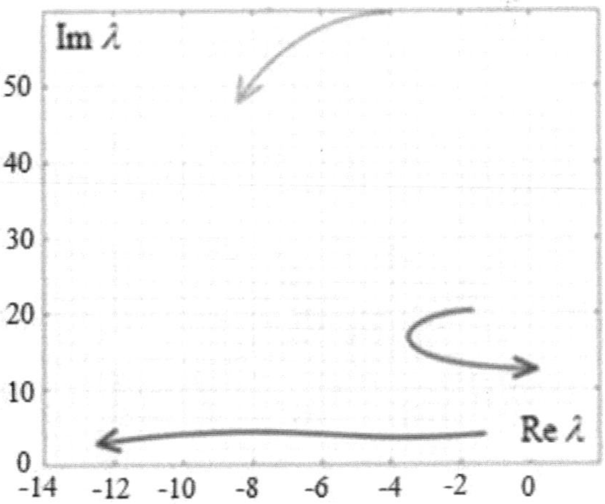

Figure 5. Presence of both types of friction.

Table 1 shows the numerical values of the critical dimensionless load for some values of the dimensionless fluid flow velocity, depending on the parameters of internal and external friction.

Table 1. Dependence of the dimensionless load on the dimensionless fluid flow rate.

a	$\begin{pmatrix}\gamma = 0, \\ \delta = 0\end{pmatrix}$	$\begin{pmatrix}\gamma = 0, \\ \delta = 2.52\end{pmatrix}$	$\begin{pmatrix}\gamma = 0.0015, \\ \delta = 0\end{pmatrix}$	$\begin{pmatrix}\gamma = 0.0015, \\ \delta = 2.52\end{pmatrix}$
0.10	17.578	20.469	20.078	20.625
0.50	16.797	19.688	18.047	20.078
1.00	14.453	17.422	15.234	17.891
1.50	10.469	13.750	11.094	14.141
2.00	4.844	8.281	5.391	8.672
2.30	0.703	4.219	1.250	4.609

Figure 6-8 shows, respectively, graphs of approximations of the dependence of the real parts of the first three eigenvalues on the dimensionless parameters of the fluid flow velocity a and the compressive load c. According to the corresponding graphs of the level lines, it can be seen that the loss of stability also occurs in the second form of system oscillations.

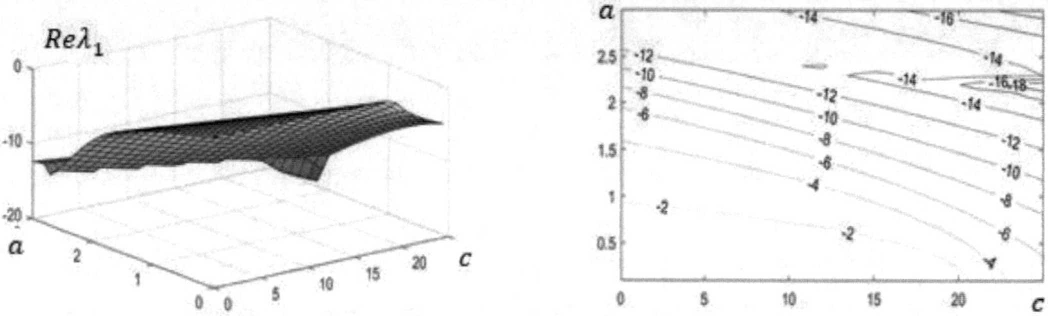

Figure 6. Dependence of the real part of the first eigenvalue on the system parameters.

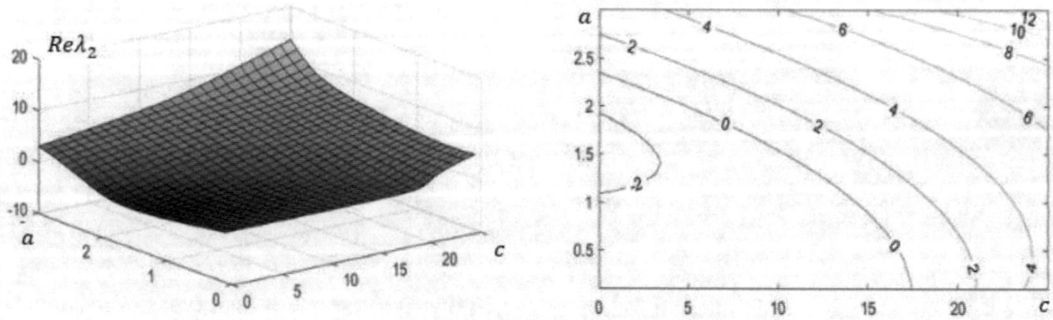

Figure 7. Dependence of the real part of the second eigenvalue on the system parameters.

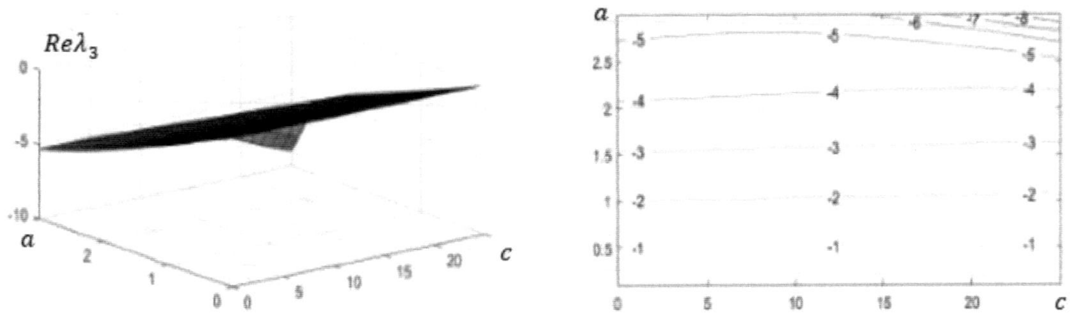

Figure 8. Dependence of the real part of the third eigenvalue on the system parameters.

Acknowledgments
This work is supported by the grant of Russian Foundation for Basic Research (Project № 19-08-01092 А)

References

[1] Volmir A S 1979 *Shells in the flow of liquid and gas. The problems of hydroelasticity* (Moscow: science) p 320

[2] Frolov K V, Mahutov N A, Kaplunov S M, Smirnov L V *et al* 2002 *Dynamics of structures of hydroaeroelastic* (Moscow: science) p 397

[3] Paidoussis M P and Issid N T 1974 Dynamic stability of pipes conveying fluid *J. of Sound and Vibration* **30-3** 267–94

[4] Kapitanov D V, Ovchinnikov V F and Smirnov L V 2010 Nonconservative stability of a pipeline and cantilever rod *J. of Machinery Manufacture and Reliability* **39-2** 197–202

[5] Kapitanov D V 2011 Investigation of non-conservative stability of a pipeline with a fluid flow as a hydroelastic system *Bulletin of the Lobachevsky University of Nizhny Novgorod* **4-2** 169–71

[6] Korenev B G and Rabinovich I M 1972 *Handbook of structure dynamics* (Moscow: Stroyizdat) p 511

Simulation of wave propagation in a periodic layered elastomer using spectral element method

A D Khanazaryan, M V Golub, O V Doroshenko, A N Shpak and S I Fomenko

Institute for Mathematics, Mechanics and Informatics, Kuban State University, 350040 Krasnodar, Russian Federation

E-mail: artur97.10@mail.ru

Abstract. The problem of wave propagation in a periodic elastomer is investigated. Dispersion equation for a periodic layered composite made of elastomers is formulated and solved employing the spectral element method. The problem is formulated in a variational form taking into account that stiffness tensor has no symmetry when an elastomer is considered. The results are compared with the same calculations made using the transfer matrix method. The method can be naturally extended to dielectric elastomers and to the three-dimensional case.

1. Introduction

In recent years, acoustic metamaterials and phononic crystals have received special attention since they provide unique properties such as negative refraction [1], band-gaps [2, 3], tunable properties [4, 5] etc. Employment of periodic composite structures has a great potential in development of sensors/actuators along with the systems of active noise reduction, adaptive optics, energy harvesting etc. Introduction of dielectric elastomers, as components of these structures, allows more efficient manipulating of the wave energy flow. However, wave propagation and diffraction in acoustic metamaterials with elastomers have been studied for a relatively low number of problems [6, 7]. The standard finite element software, e.g. COMSOL Multiphysics, does not include tools for the simulation of arbitrary stiffness, piezoelectric and dielectric constants tensors. The main objective of this study is to extend the spectral element method [8] to the simulation of periodic elastomer composite, also called phononic crystal, and provide investigation results of the propagating waves properties.

2. Statement of the problem

2.1. Finite layered phononic crystal between two half-planes

Let us consider a layered phononic crystal (PnC) situated between two isotropic elastic half-planes, see Fig. 1. In the context of this work we consider a 2D problem. The PnC is made of M elastic unit-cells of thickness H composed of L sublayers. Therefore, Cartesian coordinate system $\mathbf{x} = (x_1, x_2)$ is introduced so that the lower boundary of the PnC is parallel to Ox_1 axis. Index $n = 1, ..., ML$ is used to number sub-layers $V^{(n)} = \{|x_1| \leq \infty,\, z_{n-1} \leq x_2 \leq z_n\}$ of finite thickness $h_n = z_n - z_{n-1}$. The lower $\{x_2 \leq z_0\}$ and upper $\{x_2 \geq z_{ML}\}$ half-planes are denoted as $V^{(0)}$ and $V^{(ML+1)}$ respectively. Material properties of the n-th layer ($n = 1, \ldots, ML$) and

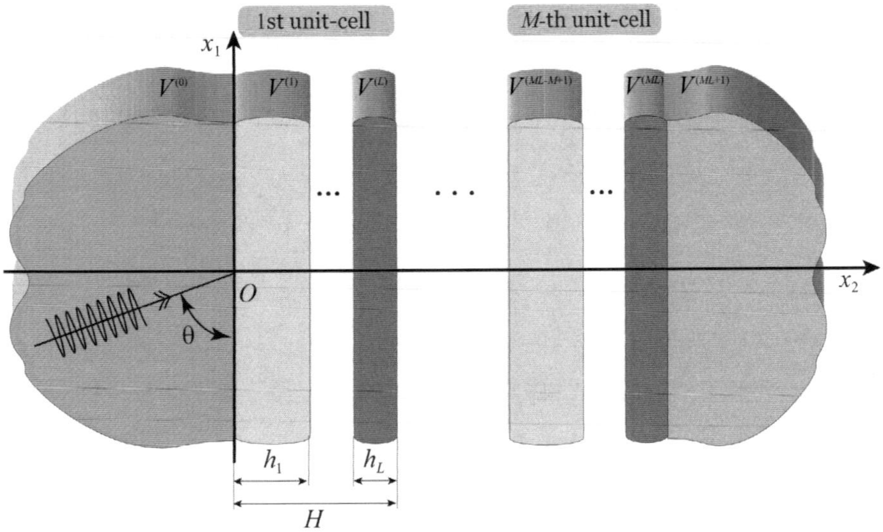

Figure 1. The geometry of the problem for a finite phononic crystal between two half-planes.

half-planes ($n = 0, ML+1$) are given by tensor $C^{(n)}_{ijkl}$, which in general case for the elastomer has no symmetry, and the mass density $\rho^{(n)}$.

The governing equations for the time-harmonic wave motion with the angular frequency ω in n-th layer $V^{(n)}$ are written in terms of the components of stress tensor σ_{ij} and the displacement vector with two non-zero components $\mathbf{u} = \{u_1, u_2\}$ as follows

$$\sum_{j=1}^{2} \frac{\partial \sigma^{(n)}_{ij}(\mathbf{x})}{\partial x_j} + \rho^{(n)}\omega^2 u^{(n)}_i(\mathbf{x}) = 0. \tag{1}$$

Constitutive equations

$$\sigma^{(n)}_{ij} = C^{(n)}_{ijkl} \frac{\partial u^{(n)}_k}{\partial x_l} \tag{2}$$

relate components of the stress tensor σ_{ij} and the displacement vector u_k. For in-plane motion, the substitution of constitutive equations (2) into (1) gives governing equations in terms of the displacement vector $\mathbf{u}$ for n-th layer as follows:

$$\sum_{j=1}^{2}\sum_{k=1}^{2}\sum_{l=1}^{2} C^{(n)}_{ijkl} \frac{\partial^2 u^{(n)}_k(\mathbf{x})}{\partial x_l \partial x_j} + \rho^{(n)}\omega^2 u^{(n)}_i(\mathbf{x}) = 0. \tag{3}$$

Hereafter, upper index n is omitted where possible for easier perception of the equations.

2.2. Infinite layered phononic crystal

An infinite phononic crystal with a two-layered unit-cell is considered as shown in figure 2, where governing equations (1) are valid. To describe wave propagating at angle θ with respect to Ox_2, wavenumber ζ is introduced following the Floquet-Bloch theory for periodic structures. Therefore, the following periodic boundary conditions are assumed for an infinite periodic structure:

$$u_k(x_1, H) = u_k(x_1, 0)\mathrm{e}^{-\mathrm{i}\zeta\cos\theta H}, \qquad 0 \le x_1 \le d, \tag{4}$$

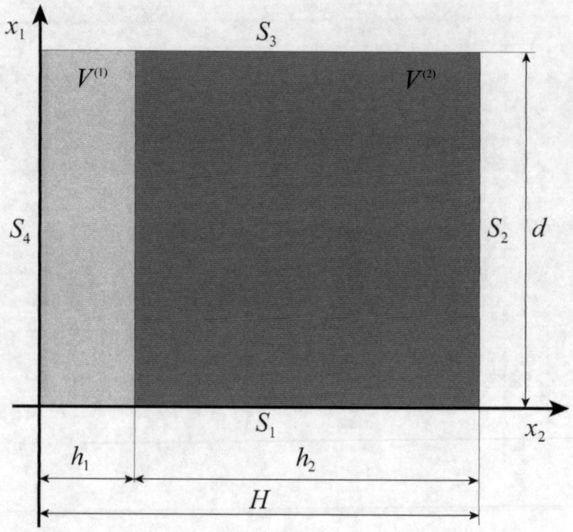

Figure 2. The geometry of the problem for an infinite phononic crystal.

$$u_k(d, x_2) = u_k(0, x_2)\mathrm{e}^{-\mathrm{i}\zeta \sin\theta d}, \qquad 0 \le x_2 \le H, \tag{5}$$

$$\sigma_{2j}(x_1, H) = \sigma_{2j}(x_1, 0)\mathrm{e}^{-\mathrm{i}\zeta \cos\theta H}, \qquad 0 \le x_1 \le d, \tag{6}$$

$$\sigma_{1j}(d, x_2) = \sigma_{1j}(0, x_2)\mathrm{e}^{-\mathrm{i}\zeta \sin\theta d}, \qquad 0 \le x_2 \le H. \tag{7}$$

The value of d is usually chosen smaller than the thickness of unit-cell H (in this study $d = 0.2H$).

3. The spectral element method

The problem of a finite phononic crystal situated between two half-spaces can be solved using the extended transfer matrix method [9]. At the same time, infinite PnC can be simulated with the mesh-based numerical methods, for example finite element method. However, the standard software, e.g. COMSOL Multiphysics, usually does not include modules, which allow to consider arbitrary stiffness and piezoelectric tensor. Therefore, the spectral element method is applied to solve the boundary value problem (3)–(7). The variational form

$$\sum_{i=1}^{2}\sum_{j=1}^{2}\oint_{\partial V} \sigma_{ij}(\mathbf{x})v_i^{k'}(\mathbf{x})n_j \mathrm{d}\ell - \sum_{i=1}^{2}\sum_{j=1}^{2}\int_V \sigma_{ij}(\mathbf{x})\frac{\partial v_i^{k'}(\mathbf{x})}{\partial x_j}\mathrm{d}V + \rho\omega^2 \sum_{i=1}^{2}\int_V u_i(\mathbf{x})v_i^{k'}(\mathbf{x})\mathrm{d}V = 0, \tag{8}$$

of the governing equations for a given domain V with boundary ∂V is used in the spectral element method.

Interpolation polynomials based on Gauss-Legendre-Lobatto grid $C^{i_k}(\xi_k^{l_k})$, where $k = 1, 2$, are employed as basis functions

$$u_k(\mathbf{x}) = \sum_{l_1=1}^{M_1}\sum_{l_2=1}^{M_2}\sum_{i_1=1}^{N}\sum_{i_2=1}^{N} u_k^{l_1 l_2 i_1 i_2} C^{i_1}\left(\xi_1^{l_1}\right) C^{i_2}\left(\xi_2^{l_2}\right) = \sum_{I} u_k^I C^{i_1}\left(\xi_1^{l_1}\right) C^{i_2}\left(\xi_2^{l_2}\right). \tag{9}$$

The global coordinates x_i are transformed into local coordinates ξ_i^l at the each element l:

$$\xi_i^l = \frac{2x_i - x_i^{l+1} - x_i^l}{x_i^{l+1} - x_i^l}, \quad x_i = \frac{x_i^{l+1} - x_i^l}{2}\xi_i^l + \frac{x_i^{l+1} + x_i^l}{2}, \quad i = 1, 2. \tag{10}$$

$$\frac{\mathrm{d}}{\mathrm{d}x_i} = r_i^l \frac{\mathrm{d}}{\mathrm{d}\xi_i^l}, \quad r_i^l = \frac{2}{x_i^{l+1} - x_i^l}, \quad i = 1, 2.$$

Here index $I(l_1, l_2, i_1, i_2)$ is introduced for simplicity, and I is in the range $1, \ldots, G = M_1 M_2 N^2$, more details can be found in [10]. According to the Bubnov-Galerkin method, projection functions coincide with basis functions

$$v_i^{k'}(\mathbf{x}) = \delta_{k'i} C^{i'_1}\left(\xi_1^{l'_1}\right) C^{i'_2}\left(\xi_2^{l'_2}\right). \tag{11}$$

Here index $I'(l'_1, l'_2, i'_1, i'_2)$ is used to address projection functions in the same way as index I.

Then, the relations for stresses can be also written via (9) and (11):

$$\sigma_{ij}(\mathbf{x}) = \sum_{k=1}^{2}\sum_{l=1}^{2} C_{ijkl} \frac{\partial}{\partial x_l}\left(\sum_{I}^{G} u_k^I C^{i_1}\left(\xi_1^{l_1}\right) C^{i_2}\left(\xi_2^{l_2}\right)\right) =$$

$$= \sum_{k=1}^{2}\sum_{I}^{G} u_k^I \left(C_{ijk1}\frac{\mathrm{d}C^{i_1}\left(\xi_1^{l_1}\right)}{\mathrm{d}x_1} C^{i_2}\left(\xi_2^{l_2}\right) + C_{ijk2} C^{i_1}\left(\xi_1^{l_1}\right)\frac{\mathrm{d}C^{i_2}\left(\xi_2^{l_2}\right)}{\mathrm{d}x_2}\right).$$

Employing periodic Floquet boundary conditions (6) and (7), contour integrals can be rewritten as follows:

$$\oint_{\partial\Omega} \sum_{i=1}^{2}\sum_{j=1}^{2} \sigma_{ij}(\mathbf{x}) v_i^{k'}(\mathbf{x}) n_j(\mathbf{x}) \mathrm{d}\ell =$$

$$\sum_{i=1}^{2}\int_0^d \left(-\sigma_{i2}(x_1, 0) v_i^{k'}(x_1, 0) + \sigma_{i2}(x_1, H) v_i^{k'}(x_1, H)\right) \mathrm{d}x_1 +$$

$$\sum_{i=1}^{2}\int_0^H \left(-\sigma_{i1}(0, x_2) v_i^{k'}(0, x_2) + \sigma_{i1}(d, x_2) v_i^{k'}(d, x_2)\right) \mathrm{d}x_2.$$

These integrals can be evaluated using quadrature formula based on N Gauss-Legendre-Lobatto grid nodal points χ_k:

$$\int_{-1}^{1} f(x) dx = \sum_{k=1}^{N} \mathrm{w}_k f(\chi_k). \tag{12}$$

Let us consider contour integrals

$$P_1 = -\sum_{i=1}^{2}\int_0^d \sigma_{i2}(x_1, 0) v_i^{k'}(x_1, 0) \mathrm{d}x_1.$$

Using (2), (9) and (11), one can derive

$$P_1 = -\sum_{i=1}^{2}\sum_{k=1}^{2}\sum_{I} u_k^I \left(\int_0^d C_{i2k1}\frac{\mathrm{d}C^{i_1}\left(\xi_1^{l_1}\right)}{\mathrm{d}x_1} C^{i_2}\left(\xi_2^{l_2}\right) \delta_{k'i} C^{i'_1}\left(\xi_1^{l'_1}\right) C^{i'_2}\left(\xi_2^{l'_2}\right) \mathrm{d}x_1 + \right.$$

$$\left. + \int_0^d C_{i2k2} C^{i_1}\left(\xi_1^{l_1}\right)\frac{\mathrm{d}C^{i_2}\left(\xi_2^{l_2}\right)}{\mathrm{d}x_2} \delta_{k'i} C^{i'_1}\left(\xi_1^{l'_1}\right) C^{i'_2}\left(\xi_2^{l'_2}\right) \mathrm{d}x_1\right).$$

Gauss-Legendre-Lobatto polynomials $C^i(\chi_j)$ are non-zero only if $i = j$: $C^i(\chi_j) = \delta_{ij}$. The values of derivative at the nodal points χ_k are designated as

$$\frac{\mathrm{d}C^i(\chi_k)}{\mathrm{d}\xi} = D^i_k.$$

Therefore, the following expressions are valid taking into account (12):

$$C^{i_2}\left(\xi_2^{l_2}\right) C^{i'_2}\left(\xi_2^{l'_2}\right) = \delta_{l_2 1}\delta_{l'_2 1}\delta_{i_2 1}\delta_{i'_2 1}, \qquad x_2 = 0,$$

$$\frac{\mathrm{d}C^{i_2}\left(\xi_2^{l_2}\right)}{\mathrm{d}x_2} C^{i'_2}\left(\xi_2^{l'_2}\right) = \delta_{l_2 1}\delta_{l'_2 1} D_1^{i_2} r_2^{l_2} \delta_{i'_2 1}, \qquad x_2 = 0,$$

$$\int\limits_0^d \frac{\mathrm{d}C^{i_1}\left(\xi_1^{l_1}\right)}{\mathrm{d}x_1} C^{i'_1}\left(\xi_1^{l'_1}\right) \mathrm{d}x_1 = \delta_{l'_1 l_1} \mathrm{w}_{i'_1} D^{i_1}_{i'_1},$$

$$\int\limits_0^d C^{i_1}\left(\xi_1^{l_1}\right) C^{i'_1}\left(\xi_1^{l'_1}\right) \mathrm{d}x_1 = \frac{1}{r_1^{l_1}} \mathrm{w}_{i'_1} \delta_{l'_1 l_1} \delta_{i'_1 i_1}.$$

Then, integral P_1 is calculated as:

$$P_1 = -\sum_{k=1}^{2}\sum_{I} u_k^I \delta_{l_1 l'_1}\delta_{l_2 1}\delta_{l'_2 1}\delta_{i'_2 1}\mathrm{w}_{i'_1}\sum_{i=1}^{2}\delta_{k' i}\left(C_{i2k1}\delta_{i_2 1}D^{i_1}_{i'_1} + C_{i2k2}D_1^{i_2}\delta_{i'_1 i_1}\frac{r_2^{l_2}}{r_1^{l_1}}\right).$$

The remaining are calculated in the same manner:

$$P_2 = \mathrm{e}^{-\mathrm{i}\zeta\cos\theta H}\cdot\sum_{i=1}^{2}\int\limits_0^d \sigma_{i2}(x_1,0) v_i^{k'}(x_1,H)\mathrm{d}x_1 =$$

$$\mathrm{e}^{-\mathrm{i}\zeta\cos\theta H}\sum_{k=1}^{2}\sum_{I} u_k^I \delta_{l_1 l'_1}\delta_{l_2 1}\delta_{l'_2 M_2}\delta_{i'_2 N}\mathrm{w}_{i'_1}\sum_{i=1}^{2}\delta_{k' i}\left(C_{i2k1}\delta_{i_2 1}D^{i_1}_{i'_1} + C_{i2k2}D_1^{i_2}\delta_{i'_1 i_1}\frac{r_2^{l_2}}{r_1^{l_1}}\right).$$

$$P_3 = -\sum_{i=1}^{2}\int\limits_0^H \sigma_{i1}(0,x_2) v_i^{k'}(0,x_2)\mathrm{d}x_2 =$$

$$-\sum_{k=1}^{2}\sum_{I} u_k^I \delta_{l_2 l'_2}\delta_{l_1 1}\delta_{l'_1 1}\delta_{i'_1 1}\mathrm{w}_{i'_2}\sum_{i=1}^{2}\delta_{k' i}\left(C_{i1k1}D_1^{i_1}\delta_{i'_2 i_2}\frac{r_1^{l_1}}{r_2^{l_2}} + C_{i1k2}\delta_{i_1 1}D^{i_2}_{i'_2}\right).$$

$$P_4 = \sum_{i=1}^{2}\int\limits_0^H \sigma_{i1}(0,x_2)\mathrm{e}^{-\mathrm{i}\zeta\sin\theta d} v_i^{k'}(d,x_2)\mathrm{d}x_2 =$$

$$\mathrm{e}^{-\mathrm{i}\zeta\sin\theta d}\sum_{k=1}^{2}\sum_{I} u_k^I \delta_{l_2 l'_2}\delta_{l_1 1}\delta_{l'_1 M_1}\delta_{i'_1 N}\mathrm{w}_{i'_2}\sum_{i=1}^{2}\delta_{k' i}\left(C_{i1k1}D_1^{i_1}\delta_{i'_2 i_2}\frac{r_1^{l_1}}{r_2^{l_2}} + C_{i1k2}\delta_{i_1 1}D^{i_2}_{i'_2}\right).$$

To satisfy periodic Floquet boundary conditions (4) and (5), the following equalities must be satisfied:

$$u_k^{l_1 M_2 i_1 N} = u_k^{l_1 1 i_1 1}\mathrm{e}^{-\mathrm{i}\zeta\cos\theta H}, \tag{13}$$

$$u_k^{M_1 l_2 N i_2} = u_k^{1 l_2 1 i_2} \mathrm{e}^{-\mathrm{i}\zeta \sin\theta d}. \tag{14}$$

Therefore, the following system can be composed if a given surface load is applied (the components $f_{I'}$ in the right-hand side describe this load):

$$\sum_I \sum_{k=1}^{2} \left(A_{I'I}^{k'k} + \widehat{A}_{I'I}^{k'k}(k) - \omega^2 \breve{A}_{I'I}^{k'k} \right) u_k^{l_1 l_2 i_1 i_2} = f_{I'}.$$

Three summands in the variational form (8) are evaluated separately. Thus, the contour integral gives

$$\widehat{A}_{I'I}^{k'k} = \delta_{l_1 l_1'} \delta_{l_2 1} \mathrm{w}_{i_1'} \sum_{i=1}^{2} \delta_{k'i} \left(C_{i2k1} \delta_{i_2 1} D_{i_1'}^{i_1} + C_{i2k2} D_1^{i_2} \delta_{i_1' i_1} \frac{r_2^{l_2}}{r_1^{l_1}} \right) \left(\delta_{l_2' M_2} \delta_{i_2' N} \mathrm{e}^{-\mathrm{i}\zeta \cos\theta H} - \delta_{l_2' 1} \delta_{i_2' 1} \right) +$$
$$\delta_{l_2 l_2'} \delta_{l_1 1} \mathrm{w}_{i_2'} \sum_{i=1}^{2} \delta_{k'i} \left(C_{i1k1} D_1^{i_1} \delta_{i_2' i_2} \frac{r_1^{l_1}}{r_2^{l_2}} + C_{i1k2} \delta_{i_1 1} D_{i_2'}^{i_2} \right) \left(\delta_{l_1' M_1} \delta_{i_1' N} \mathrm{e}^{-\mathrm{i}\zeta \sin\theta d} - \delta_{l_1' 1} \delta_{i_1' 1} \right).$$

The volume integrals can be separated into diagonal matrix

$$\breve{A}_{I'I}^{k'k} = -\frac{\rho}{r_1^{l_1} r_2^{l_2}} \delta_{l_1' l_1} \delta_{l_2' l_2} \delta_{kk'} \mathrm{w}_{i_1} \delta_{i_1' i_1} \mathrm{w}_{i_2} \delta_{i_2' i_2},$$

which is frequency dependent, and matrix

$$A_{I'I}^{k'k} = \delta_{l_1' l_1} \delta_{l_2' l_2} \Bigg(-\frac{r_1^{l_1}}{r_2^{l_2}} \mathrm{w}_{i_2} \delta_{i_2' i_2} C_{k'1k1} \sum_{n=1}^{N} \mathrm{w}_n D_n^{i_1'} D_n^{i_1} - \frac{r_2^{l_2}}{r_1^{l_1}} \mathrm{w}_{i_1} \delta_{i_1 i_1'} C_{k'2k2} \sum_{n=1}^{N} \mathrm{w}_n D_n^{i_2'} D_n^{i_2} -$$
$$- C_{k'1k2} D_{i_1}^{i_1'} D_{i_2'}^{i_2} \mathrm{w}_{i_1} \mathrm{w}_{i_2'} - C_{k'2k1} D_{i_1'}^{i_1} D_{i_2}^{i_2'} \mathrm{w}_{i_1'} \mathrm{w}_{i_2} \Bigg).$$

Boundary conditions (13) and (14) are also taken into account so that

$$\breve{A}_{I'(l_1', M_2, i_1', N)I}^{k'k} = \widehat{A}_{I'(l_1', M_2, i_1', N)I}^{k'k} = A_{I'(l_1', M_2, i_1', N)I}^{k'k} = 0,$$

$$\breve{A}_{I'(M_1, l_2', N, i_2')I}^{k'k} = \widehat{A}_{I'(M_1, l_2', N, i_2')I}^{k'k} = A_{I'(M_1, l_2', N, i_2')I}^{k'k} = 0,$$

except for

$$\breve{A}_{I'(l_1', M_2, i_1', N)I(l_1', M_2, i_1', N)}^{kk} = 1, \qquad \breve{A}_{I'(l_1', M_2, i_1', N)I(l_1', 1, i_1', 1)}^{kk} = -\mathrm{e}^{-\mathrm{i}\zeta \cos\theta H},$$

$$\breve{A}_{I'(M_1, l_2', N, i_2')I(M_1, l_2', N, i_2')}^{kk} = 1, \qquad \breve{A}_{I'(1, l_2', 1, i_2')I(1, l_2', 1, i_2')}^{kk} = -\mathrm{e}^{-\mathrm{i}\zeta \sin\theta d},$$

where $I = \overline{1, G}$, $l_1 = \overline{1, M_1}$, $l_2 = \overline{1, M_2}$, $i_1 = \overline{1, N}$, $i_2 = \overline{1, N}$.

Finally, the dispersion equation

$$\det \left(\mathbf{A}_{I'I}^{k'k} + \widehat{\mathbf{A}}_{I'I}^{k'k}(k) - \omega^2 \breve{\mathbf{A}}_{I'I}^{k'k} \right) = 0$$

can be written with respect to the wavenumber ζ. The latter can be reduced to the eigenvalue problem:

$$\mathbf{B} - \lambda_n \mathbf{E} = 0, \qquad \mathbf{B} = \breve{\mathbf{A}}^{-1} \left(\mathbf{A} + \widehat{\mathbf{A}}(\zeta) \right), \quad \omega_n = \sqrt{\lambda_n},$$

here $\mathbf{E}$ is the identity matrix.

4. Conclusions

In this study, the spectral element method [8] was extended to the simulation of periodic elastomer composite. The obtained mathematical model is convenient for the investigation of the properties of the waves propagating in phononic crystals and acoustic metamaterials including elastomers. The method can be naturally extended to dielectric elastomers and to the three-dimensional case.

Acknowledgments

The research was carried out with the financial support of the Russian Foundation for Basic Research and the Administration of Krasnodar Region (project 19-41-230012).

References

[1] Lan J, Zhang X, Wang L, Lai Y and Liu X 2020 *Sci. Rep.* **10** 10794
[2] An X, Fan H and Zhang C 2020 *J. Sound Vib.* **475** 115292
[3] Ishii Y and Biwa S 2018 *J. Appl. Phys.* **113** 101906
[4] Ren T, Liu C, Li F and Zhang C 2020 *J. Intell. Mater. Syst. Struct.* **31** 843–59
[5] Grinberg I and Matlack K H 2020 *Wave Motion* **93** 102466
[6] Galich P I and Rudykh S 2017 *J. Appl. Mech.* **84** 9 091002
[7] Li J, Slesarenko V, Galich P I and Rudykh S 2018 *Mech. Res. Commun.* **87** 21–8
[8] Komatitsch D, Vilotte J P, Vai R, Castillo-Covarrubias J M and Sanchez-Sesma F J 1999 *Int. J. Numer. Meth. Eng.* **45** 1139–64
[9] Fomenko S I, Golub M V, Chen A, Wang Y and Zhang C 2019 *J. Sound Vib.* **439** 219–40
[10] Golub M V and Shpak A N 2019 *Appl. Math. Model.* **65** 234–55

Modeling of a vibrating machine with spatial oscillations of the working body

I P Lyan, G Y Panovko and A E Shokhin

Mechanical Engineering Research Institute of the Russian Academy of Sciences (IMASH RAN), Moscow, Russia

LyanIlyaIMASH@yandex.ru

Abstract. In modern industry, vibration technological machines with unbalanced vibration exciters (vibrating conveyors, vibrating screens, vibrating crushers, etc.) are widely used. The variety of vibration technological machines typical schemes leads to a need to take into account the dynamic system features in the mathematical models that are being created. The aim of this work is to create a spatial mathematical model and determine the dynamic system unknown parameters of a vibrating screen experimental sample with two self-synchronizing unbalanced vibration exciters that create the working body spatial motion. The mathematical model motion equations are derived using the Lagrange equations of the second kind. Using the obtained experimental data (natural frequencies and logarithmic damping decrement), the mathematical model mass-geometric parameters and the damping parameters values were calculated. The investigation result is a verified mathematical model of a vibrating screen sample with two self-synchronizing unbalanced vibration exciters.

1. Introduction

In modern industry, vibrating technological machines (vibrating conveyors, vibrating screens, vibrating crushers, etc.) are widely used [1-3]. Nowadays, self-synchronizing unbalanced vibration exciters based on asynchronous electric motors with a squirrel-cage rotor [1, 3, 4] are widely used as drives of vibration technological machines. In the induction motors with the elastic system of a vibrating machine interaction, various nonlinear effects are often observed, manifested in the form of jumps in the working body oscillations frequencies and amplitudes [2, 3, 5, 6]. When using several unbalanced vibration exciters, the possibility of self-synchronization phenomenon usage is crucial [1, 3, 7].

In most cases, modern vibrating technological machines with unbalanced vibration exciters operate in a resonant mode, when the frequency of forced vibrations exceeds the natural frequency of the working body [1, 4]. This mode allows to ensure the machine vibration stability in a wide range of load parameters. However, due to the need to overcome resonant frequencies, it is necessary to use electric motors with excess power [8, 9]. This leads to the fact that in the operating mode the drive motor is significantly underloaded, as a result of which energy consumption increases and its service life decreases. In addition, the excess electric drive power narrows the frequencies range of unbalances synchronous rotation when using several self-synchronizing vibration exciters [5, 7].

One of the creating energy-efficient vibrating machines principles is based on the use of working body resonant modes of vibration. At the same time, the required masses of unbalances and electric motor power are significantly reduced, which leads to an increase in the efficiency and service life of vibration exciters [3, 10, 11]. However, maintaining such an operation mode requires the use of automated systems for collecting, processing and analyzing information about the current state of the machine dynamic characteristics with the control actions simultaneous formation [3, 10-12]. The creation of rational systems for automatic control for vibration machines is based on a mathematical model of a machine, which should take into account the essential features of its dynamic properties (dynamic system with an engine interaction, self-synchronization various types stability areas, dynamic system to regulation response, etc.).

The purpose of this study is to develop a design scheme and a mathematical model of a vibrating screen with two self-synchronizing unbalanced vibration exciters that can create a working body spatial motion. To achieve this goal, the following main tasks are solved in the work: the machine design scheme formation and its modeling based on the motion differential equations, the parameters identification of the machine design scheme based on the experimental studies results.

2. Calculation model

The existing designs of screens, basically, use the layout of two unbalanced vibration exciters, the axes of rotation of which are perpendicular to the vertical plane passing through the technological axis of the machine (the main movement of the processed material direction) [1, 4, 13]. A feature of the machine design considered in this work is that the axes of both unbalanced vibration exciters rotation are located in vertical planes parallel to the technological axis of the machine. Figure 1 shows a model of ULS2010.12W vibrating screen manufactured by Kroosh Technologies Ltd, where it is indicated: 1 - vibration exciters; 2 - working body; 3 - springs of elastic suspension, 4 - movable support of the bed; *X* - the direction of the technological axis of the machine [14].

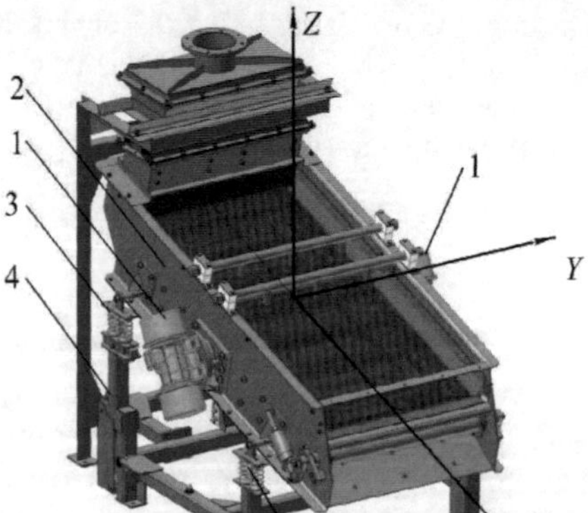

Figure 1. Vibration screen ULS2010.12W.

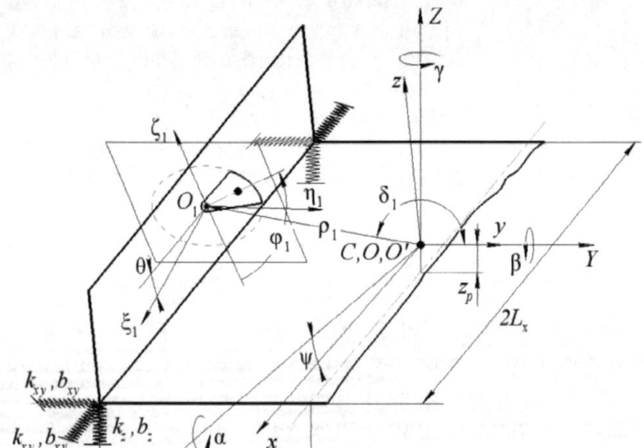

Figure 2. Calculation scheme of mathematical model.

Due to the screen symmetry relative to the vertical plane passing through the technological axis of the machine, in figure 2 only one (right) half of the calculation scheme is presented. The working body movement is described relative to the global coordinate system *OXYZ*, the beginning of which at rest coincides with the position of the center of mass of the system *C*. The working body is modeled by an absolutely rigid body of mass *M* of length $2L_x$ and width $2L_y$. The position of the unbalances and the spring fixing points are set in the local system *O′xyz*, rigidly connected with the working body and the reference point coinciding with the origin of the system *OXYZ*, and the axes *OY* and *O′y* coincide, and the axes *OX* and *O′x* form an angle ψ - the angle of the working body inclination. The inertia moments of the working body around the axes *O'x*, *O'y*, *O'z* are designated J_x, J_y, J_z, respectively. Angular vibrations are described using the Krylov-Bulgakov angles [15, 16], which in figure 2 to simplify the figure are presented as the angles of working body rotation α, β, γ around the axes *OX*, *OY* and *OZ*. As a result, the working body position relative to the global coordinate system can be specified using six coordinates: three displacements *x*, *y*, *z* and three angles α, β, γ.

The fastening points of the springs are located in the working body corners, and are displaced by a distance z_p along the *O′z* axis, which determines the level of oscillations along the *OX* and *OY* axes with rotations β and α, respectively connectedness. It is believed that each of the elastic suspension springs has linear characteristics of stiffness in three mutually perpendicular directions with the coefficients k_z, k_x, k_y, which in the design scheme is represented by three elastic elements whose axes are parallel to the global coordinate system, and $k_x = k_y$ (in figure 2 and hereinafter these coefficients are denoted as k_{xy}). Damping in the system is only due to the energy dissipation in the springs and is described by a linear viscous friction model with coefficients b_z, b_x, b_y, and $b_x = b_y = b_{xy}$.

In the design scheme, each vibration exciter has one unbalance of mass m_i, eccentricity r_i and inertia moment J_{di} ($i = 1{,}2$ is the vibration exciter number) fixed in the *zO'y* plane. Each unbalance position in the local coordinate system *O′xyz* is described using an additional local coordinate system $O_i\xi_i\eta_i\zeta_i$. The position of the local coordinate system $O_i\xi_i\eta_i\zeta_i$ is specified by the radius vector $\boldsymbol{\rho}_i = (0,\ \ \rho_i\cos\delta_i\ \ \ \rho_i\sin\delta_i)$, where ρ_i and δ_i are the modulus and the inclination angle of the radius vector to the positive direction of the *O′y* axis, and the inclination angle θ_i of the $O_i\xi_i$ axis to the *O′x* axis, measured from the *O′x* axis positive direction counterclockwise. The unbalance position angles φ_i are counted from the $O_i\zeta_i$ axis negative direction counterclockwise.

The vibration exciters asynchronous electric motors have torque characteristics L_i. To determine the direction of the motors rotation, the parameter $\sigma_i = \pm 1$ is used, where the positive value corresponds to the rotation direction of the *i*-th unbalance counterclockwise, and the negative value corresponds to the clockwise direction.

To connect the local coordinate systems $O_i\xi_i\eta_i\zeta_i$ and *OXYZ*, the rotation matrix **T** is used:

$$\mathbf{T} = \begin{pmatrix} \cos(\theta+\psi)-\beta\sin(\theta+\psi) & -\cos(\theta+\psi)(\alpha\sin\theta+\gamma\cos\theta) & \sin(\theta+\psi)+\beta\cos(\theta+\psi) \\ \alpha\sin\theta+\gamma\cos\theta & 1 & 0 \\ -\sin(\theta+\psi)-\beta\cos(\theta+\psi) & \sin(\theta+\psi)(\alpha\sin\theta+\gamma\cos\theta) & \cos(\theta+\psi)-\beta\sin(\theta+\psi) \end{pmatrix}. \quad (1)$$

The equations of the system motion, which are obtained using the Lagrange equation of the second kind and geometric rotation matrices, have the form:

$$\mathbf{M}\ddot{\mathbf{q}} + \mathbf{B}\dot{\mathbf{q}} + \mathbf{K}\mathbf{q} = \mathbf{F}(\mathbf{q},\dot{\mathbf{q}}) \quad (2)$$

where $\mathbf{q}^T = (x \ \ y \ \ z \ \ \alpha \ \ \beta \ \ \gamma \ \ \varphi_1 \ \ \varphi_2)$ is the column vector of displacements, **M**, **K** and **B** are symmetric matrices (8x8) of masses, stiffness rates and damping, respectively, $\mathbf{F}(\mathbf{q},\dot{\mathbf{q}})$ – is the column vector of nonlinear functions describing the disturbing effect arising from the unbalances rotation.

The components of the mass matrix **M** are as follows:

$$M_{21} = M_{31} = M_{32} = M_{52} = M_{62} = M_{87} = 0; \quad M_{11} = M_{22} = M_{33} = m_1 + m_2 + M;$$

$$M_{41} = \sum_{i=1,2} m_i(\rho_i\cos\delta_i\sin\psi - r_i\cos(\theta+\psi)\sin\theta\sin\varphi_i); \quad M_{51} = \sum_{i=1,2} m_i(\rho_i\cos\delta_i\sin\psi - r_i\cos(\theta+\psi)\sin\varphi_i);$$

$$M_{61} = \sum_{i=1,2} m_i(-\rho_i\cos\delta_i\cos\psi - r_i\cos(\theta+\psi)\cos\theta\sin\varphi_i);$$

$$M_{(i+6)1} = m_i r_i\{\sin\varphi_i[\sin(\theta+\psi)+\beta\cos(\theta+\psi)] - \cos(\theta+\psi)\cos\varphi_i(\alpha\sin\theta+\gamma\cos\theta)\};$$

$$M_{42} = -\sum_{i=1,2} m_i\rho_i\sin\delta_i; \quad M_{(i+6)2} = m_i r_i\cos\varphi_i; \quad M_{43} = \sum_{i=1,2} m_i(\rho_i\cos\delta_i\cos\psi + r_i\sin(\theta+\psi)\sin\theta\sin\varphi_i);$$

$$M_{53} = \sum_{i=1,2} m_i (r_i \sin(\theta+\psi)\cos\varphi_i - \rho_i \sin\delta_i \sin\psi);$$
$$M_{63} = \sum_{i=1,2} m_i (\rho_i \cos\delta_i \sin\psi + r_i \sin(\theta+\psi)\cos\theta \sin\varphi_i);$$
$$M_{(i+6)3} = m_i r_i \{\sin\varphi_i [\cos(\theta+\psi) - \beta\sin(\theta+\psi)] + \sin(\theta+\psi)\cos\varphi_i(\alpha\sin\theta + \gamma\cos\theta)\};$$
$$M_{44} = J_x + \sum_{i=1,2} m_i (\rho_i^2 + r_i \sin^2\theta \sin\varphi_i (2\rho_i \cos\delta_i + r_i \sin\varphi_i]);$$
$$M_{54} = \sum_{i=1,2} m_i r_i \sin\theta(\rho_i \cos\delta_i \cos\varphi_i + [r_i \cos\varphi_i - \rho_i \cos\theta \sin\delta_i]\sin\varphi_i);$$
$$M_{64} = \sum_{i=1,2} m_i r_i \cos\theta \sin\theta \sin\varphi_i (2\rho_i \cos\delta_i + r_i \sin\varphi_i);$$
$$M_{(i+6)4} = m_i r_i (\rho_i \cos\delta_i \cos\theta \sin\varphi_i - \rho_i \cos\varphi_i \sin\delta_i + \sin\theta[\rho_i \cos\delta_i + r_i \sin\varphi_i] \times$$
$$\times [\cos\varphi_i \{\alpha\sin\theta + \gamma\cos\theta\} - \beta\sin\varphi_i]); \quad (3)$$
$$M_{55} = J_y + \sum_{i=1,2} m_i ([\rho_i \cos\psi \sin\delta_i - r_i \cos(\theta+\psi)\cos\varphi_i]^2 + [r_i \sin(\theta+\psi)\cos\varphi_i - \rho_i \sin\psi \sin\delta_i]^2);$$
$$M_{65} = \sum_{i=1,2} m_i \rho_i \cos\delta_i (r_i \cos\theta \cos\varphi_i - \rho_i \sin\delta_i) + m_i r_i \sin\varphi_i \cos\theta (r_i \cos\varphi_i - \rho_i \cos\theta \sin\delta_i);$$
$$M_{(i+6)5} = m_i r_i (\rho_i \sin\delta_i \sin\theta \sin\varphi_i + [r_i \cos\varphi_i - \rho_i \sin\delta_i \cos\theta][\cos\varphi_i \{\alpha\sin\theta + \gamma\cos\theta\} - \beta\sin\varphi_i]);$$
$$M_{66} = J_z + \sum_{i=1,2} m_i ([\rho_i \cos\delta_i \cos\psi + r_i \cos\theta \cos(\theta+\psi)\sin\varphi_i]^2 +$$
$$+ [\rho_i \cos\delta_i \sin\psi + r_i \cos\theta \sin(\theta+\psi)\sin\varphi_i]^2);$$
$$M_{(i+6)6} = m_i r_i (2\cos\theta[\rho_i \cos\delta_i + r_i \sin\varphi_i][\cos\varphi_i \{\alpha\sin\theta + \gamma\cos\theta\} - \beta\sin\varphi_i] - \rho_i \cos\delta_i \sin\theta \sin\varphi_i;$$
$$M_{(i+6)(i+6)} = J_{di} + m_i r_i^2 (1 + [\cos\varphi_i \{\alpha\sin\theta + \gamma\cos\theta\} - \beta\sin\varphi_i]^2).$$

The components of the mass matrix **K** are as follows:

$$K_{21} = K_{31} = K_{41} = K_{61} = K_{32} = K_{52} = K_{62} = K_{43} = K_{63} = K_{54} = K_{56} = K_{7j} = K_{8j} = 0, \ j = 1...8;$$
$$K_{11} = K_{22} = 4k_{xy}; K_{33} = 4k_z; K_{44} = 4k_z L_y^2 \cos^2\psi + 4k_{xy}(z_p^2 + L_y^2 \sin^2\psi);$$
$$K_{55} = 4k_z (z_p^2 \sin^2\psi + L_x^2 \cos^2\psi) + 4k_{xy}(z_p^2 \cos^2\psi + L_x^2 \sin^2\psi); \quad (4)$$
$$K_{66} = 4k_z L_y^2 \sin^2\psi + 4k_{xy}(L_x^2 + L_y^2 \cos^2\psi); K_{51} = -4k_{xy} z_p \cos\psi;$$
$$K_{42} = 4k_{xy} z_p; K_{53} = 4k_{xy} z_p \sin\psi; K_{64} = 2L_y^2 (k_z - k_{xy})\sin 2\psi.$$

The components of the mass matrix **B** are as follows:

$$B_{21} = B_{31} = B_{41} = B_{61} = B_{32} = B_{52} = B_{62} = B_{43} = B_{63} = B_{54} = B_{56} = B_{7j} = B_{8j} = 0, \ j = 1...8;$$
$$B_{11} = B_{22} = 4b_{xy}; B_{33} = 4b_z; B_{44} = 4b_z L_y^2 \cos^2\psi + 4b_{xy}(z_p^2 + L_y^2 \sin^2\psi);$$
$$B_{55} = 4b_z (z_p^2 \sin^2\psi + L_x^2 \cos^2\psi) + 4b_{xy}(z_p^2 \cos^2\psi + L_x^2 \sin^2\psi); \quad (5)$$
$$B_{66} = 4b_z L_y^2 \sin^2\psi + 4b_{xy}(L_x^2 + L_y^2 \cos^2\psi); B_{51} = -4b_{xy} z_p \cos\psi;$$
$$B_{42} = 4b_{xy} z_p; B_{53} = 4b_{xy} z_p \sin\psi; B_{64} = 2L_y^2 (b_z - b_{xy})\sin 2\psi.$$

The $\mathbf{F}(\mathbf{q}, \dot{\mathbf{q}})$ vector components of nonlinear functions are as follows:

$$F_1 = \sum_{i=1,2} m_i r_i \dot{\varphi}_i (2\cos(\theta+\psi)[\dot{\beta}\sin\varphi_i - \cos\varphi_i\{\dot{\alpha}\sin\theta + \dot{\gamma}\cos\theta\}] +$$
$$+ \dot{\varphi}_i[\cos\varphi_i\{\sin(\theta+\psi) + \beta\cos(\theta+\psi)\} + \cos(\theta+\psi)\sin\varphi_i\{\alpha\sin\theta + \gamma\cos\theta\}]);$$
$$F_2 = -\sum_{i=1,2} m_i r_i \dot{\varphi}_i^2 \sin\varphi_i;$$
$$F_3 = g(m_1 + m_2 + M) + \sum_{i=1,2} m_i r_i \dot{\varphi}_i (2\sin(\theta+\psi)[\cos\varphi_i\{\dot{\alpha}\sin\theta + \dot{\gamma}\cos\theta\} - \dot{\beta}\sin\varphi_i] +$$
$$+ \dot{\varphi}_i[\cos(\theta+\psi)\cos\varphi_i - \sin(\theta+\psi)\{\alpha\sin\varphi_i\sin\theta + \gamma\sin\varphi_i\cos\theta + \beta\cos\varphi_i\}]);$$
$$F_4 = \sum_{i=1,2} m_i r_i \dot{\varphi}_i ([\rho_i\cos\delta_i + r_i\sin\varphi_i][2\dot{\alpha}\cos\varphi_i\sin^2\theta - 2\dot{\beta}\sin\theta\sin\varphi_i + \dot{\gamma}\cos\varphi_i\sin 2\theta] +$$
$$+ \dot{\varphi}_i[\rho_i\{\cos\delta_i\cos\varphi_i\cos\theta + \sin\delta_i\sin\varphi_i\} - \sin\theta\{\rho_i\cos\delta_i + r_i\sin\varphi_i\}\times \tag{6}$$
$$\times\{\alpha\sin\varphi_i\sin\theta + \gamma\sin\varphi_i\cos\theta + \beta\cos\varphi_i\}]);$$
$$F_5 = \sum_{i=1,2} m_i r_i \dot{\varphi}_i (2[r_i\cos\varphi_i - \rho_i\cos\theta\sin\delta_i][\cos\varphi_i\{\dot{\alpha}\sin\theta + \dot{\gamma}\cos\theta\} - \dot{\beta}\sin\varphi_i] +$$
$$+ \dot{\varphi}_i[\rho_i\cos\varphi_i\sin\delta_i\sin\theta - \{r_i\cos\varphi_i - \rho_i\sin\delta_i\cos\theta\}\{\alpha\sin\varphi_i\sin\theta + \gamma\sin\varphi_i\cos\theta + \beta\cos\varphi_i\}]);$$
$$F_6 = \sum_{i=1,2} m_i r_i \dot{\varphi}_i^2 (2\cos\theta[\rho_i\cos\delta_i + r\sin\varphi_i][\cos\varphi_i\{\dot{\alpha}\sin\theta + \dot{\gamma}\cos\theta\} - \dot{\beta}\sin\varphi_i] -$$
$$- \dot{\varphi}_i[\rho_i\cos\varphi_i\cos\delta_i\sin\theta + \cos\theta\{\rho_i\cos\delta_i + r\sin\varphi_i\}\{\alpha\sin\varphi_i\sin\theta + \gamma\sin\varphi_i\cos\theta + \beta\cos\varphi_i\}]);$$
$$F_{i+6} = \sigma_i M_{Si} - \sigma_i L_i + 2m_i r_i^2(\cos\varphi_i[\alpha\sin\theta + \gamma\cos\theta] - \beta\sin\varphi_i)\times$$
$$\times(\dot{\varphi}_i[\cos\varphi_i\{\dot{\alpha}\sin\theta + \dot{\gamma}\cos\theta\} - \dot{\beta}\sin\varphi_i),$$

where L_i is the i-th electric drive driving moment, M_{Si} is the rotation resistance moment of the i-th unbalance, the index $(i + 6)$ is equal to 7 or 8 depending on $i = 1.2$.

3. System parameters identification

The system parameters J_x, J_y, J_z, k_{xy}, k_z, z_p were determined from the equation

$$\det\left(\mathbf{K}_0 - \mathbf{M}_0\mathbf{p}^2\right) = 0, \tag{7}$$

where $\mathbf{p}^2 = \begin{pmatrix} p_1^2 & p_2^2 & p_3^2 & p_4^2 & p_5^2 & p_6^2 \end{pmatrix}^T$ is the column vector of squares of natural frequencies, the number of which corresponds to the working body degrees of freedom number, $\mathbf{M}_0$ and $\mathbf{K}_0$ are the matrices of masses and stiffness, which are obtained from the matrices $\mathbf{M}$ and $\mathbf{K}$ included in the equations of motion of the working body (2), at $\varphi_i = \rho_i = 0$, $\mathbf{B} = 0$ and $\mathbf{F}(\mathbf{q}, \dot{\mathbf{q}}) = 0$.

The (7) is an equation of the sixth degree relative to the square of the natural frequency. To determine the parameters J_x, J_y, J_z, k_{xy}, k_z, z_p, the inverse problem of dynamics is solved, when the components of the matrices $\mathbf{M}_0$ and $\mathbf{K}_0$ are calculated from the values of natural frequencies known from the experiment. From the solutions obtained, those in which the values of k_z differ significantly from the known one are discarded, as well as those that do not satisfy physical and geometric representations. The damping parameters were estimated using the experimentally obtained envelope of damped oscillations by varying the logarithmic damping coefficient ς in the equation of the form $A(t) = A_0 e^{-\varsigma t/T}$. The viscous friction coefficient was calculated by the formula $b = 2\varsigma Mp$, where p is the corresponding natural vibration frequency.

The required system parameters were determined from specially set experiments performed on an experimental sample of a vibrating screen manufactured by REC «Mekhanobr-Tekhnika» at an angle of the working body inclination $\psi = 0$ (figure 3).

Figure 3. Photographs of experimental sample of vibrating screen.

Oscillations of the working body were excited in the direction of one of the global coordinate system axes of the vibrating machine by hitting the working body with a dynamometric hammer. The vibrations of the working body were measured using piezoaccelerometers of the AP2038P-50 type. The data obtained were processed in the Wolfram Mathematica 10 software package, the envelopes of damped oscillations were constructed, and the center of mass vibration acceleration spectra along the global coordinate system axes were calculated using the fast Fourier transform. As a result of processing the experimental data, 6 vibrations natural frequencies of the vibrating machine working body were calculated $p_1 = 6.5$ Hz, $p_2 = 6.35$ Hz, $p_3 = 7.5$ Hz, $p_4 = =10.2$ Hz, $p_5 = 12.6$ Hz, $p_6 = 9.3$ Hz, according to the values of which the following values were obtained system parameters: $J_x = 0.456$ kg·m^2, $J_y = 0.478$ kg·m^2, $J_z = 1.036$ kg·m^2, $k_{xy} = 10475$ N·m^{-1}, $k_z = =13141$ N·m^{-1}, $z_p = 0.054$ m, as well as the coefficients damping system $b_z = 35.5$ N·s·m^{-1}, $b_{xy} = 38.8$ N·s·m^{-1}. The model parameters $\rho_i = 0.23$ m and $\delta_1 = -0.07$ rad, $\delta_2 = \pi + 0.07$ rad were calculated from the known screen size and distance z_p.

4. Mathematical model verification

The verification obtained mathematical model was carried out by comparing the calculated and experimental dependences of the steady oscillations acceleration amplitudes along the axes *OX*, *OY*, *OZ* on the unbalances rotation frequency. To conduct the experiment, a three-axis accelerometer (AP2038P-10) and two encoders (E40H8-2500-6-L5) on the axes of vibration exciters were fixed on a laboratory sample of a vibrating screen. The motors were recorded from a frequency converter with a current with a frequency *f* ranging from 10 to 60 Hz, varied discretely with a variable step after the establishment of system oscillations. As a result of processing the experimental data, the maximum amplitudes of accelerations and the averaged frequency of rotation of the unbalances $\dot{\varphi}$ of steady-state oscillations were obtained at each frequency of the supply voltage, which is shown by dots in figure 4.

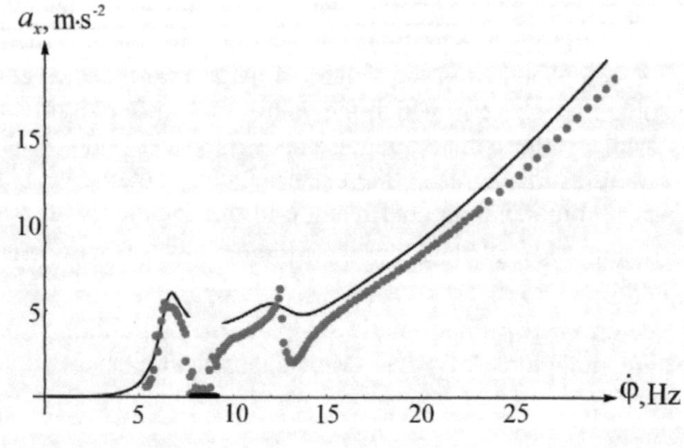

Figure 4. The amplitudes of the acceleration of steady-state oscillations along OX axis.

When calculating the dependence of the acceleration amplitude on the unbalance rotation frequency, the supply voltage frequency change by the converter was taken into account using the modified Kloss formula [10]:

$$L(\dot{\varphi}_i) = \frac{2M_{Cr}\left(\omega_s - \omega_{Cr}\right)\left(\Delta\omega - \dot{\varphi}_i\right)}{\left(\Delta\omega - \dot{\varphi}_i\right)^2 + \left(\omega_s - \omega_{Cr}\right)^2}, \tag{8}$$

where L_{Cr} = 1.232 N·m is the critical moment of the electric motor, ω_{Cr} = 115 rad·s^{-1} is the frequency at which the critical moment is reached, $\omega_s = 2\pi f_0 K^{-1}$ = 157 rad·s^{-1} is the no-load frequency, where f_0 = 50 Hz is the supply voltage frequency, K = 2 is the electric motor number of poles, $\Delta\omega$ = $=2\pi f\mathrm{K}^{-1}$ is the torque characteristic displacement of the electric motor caused by a change in the frequency of the current f. In the range $\Delta\omega$ = 0 ... 188 rad·s^{-1} (which corresponds to a change in the supply frequency in the range f = 0 ... 60 Hz). The calculation results are shown by lines in figure 4-6.

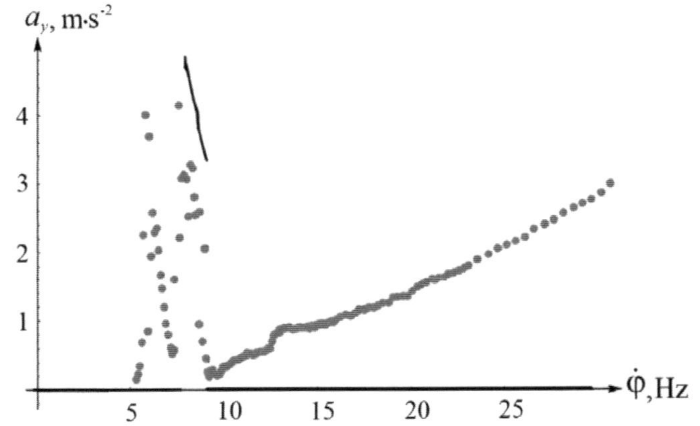

Figure 5. The amplitudes of the acceleration of steady-state oscillations along OY axis.

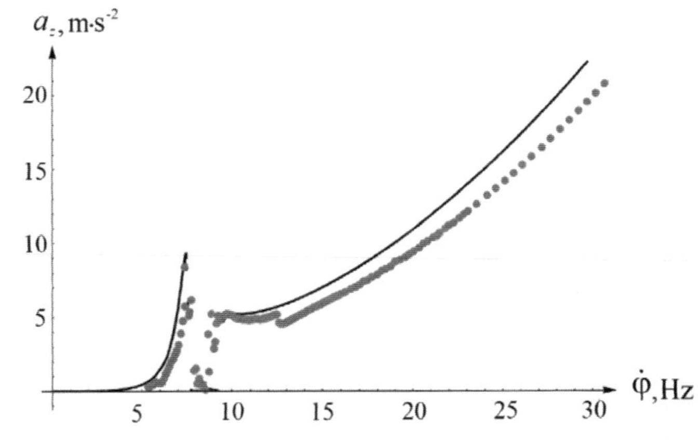

Figure 6. The amplitudes of the acceleration of steady-state oscillations along OZ axis.

5. Results and conclusions

As a result of the study, a vibrating screen design scheme and a mathematical model with the spatial motion of the working body were developed, taking into account the torque characteristics of vibration exciters asynchronous electric motors and their rotation speed frequency regulation. The model describes such features of the dynamic screening system as the variable inclination angle of vibration exciters axes of rotation, the possibility of changing the inclination angle of the working body, the connectivity of angular and longitudinal vibrations. It should be noted that the calculated and experimental data match is quite accurate. The existing discrepancies between the model and the experiment can be explained by the natural difference in the parameters of the drive electric motors and unbalances, as a result of which one unbalance lags behind the other during synchronous rotation, which leads to the appearance of a small disturbing force in the direction of the OY axis. In the mathematical model, the motors and unbalances are exactly the same and oscillations along the OY axis occur only in case of a change in the synchronous rotation type of the unbalances.

The obtained mathematical model verification was carried out using comparison of the steady oscillations acceleration amplitudes dependences along the main axes of the installation.

Solutions comparison of the obtained motion differential equations system showed satisfactory coincidence of the calculated and experimental characteristics of the system, which, among other things, indicates the correct parameters determination of the vibrating screen laboratory sample.

The results obtained in carrying out numerical experiments correspond to the physical concept of the processes occurring in the dynamic system of a vibrating technological machine.

The proposed mathematical model and universal algorithm for determining the parameters of a vibrating machine can be used to study the behavior of vibrating technological machines with a similar design scheme.

Acknowledgments
The reported study was funded by RSF, project number 18-19-00708.

References

[1] Blekhman I I 2018 *Vibration Mechanics and Vibration Rheology (theory and applications)* (Moscow: Fismatlit) p 752
[2] Vulfson I 2015 *Dynamics of cyclic machines* (Heidelberg, New York, Dordrecht, London: Springer) p 390
[3] Fedorenko I Ya 2016 *Vibration processes and devices in agro-industrial complex: monograph* (Barnaul: EPD of Altai State Agrarian University) p 289
[4] Vaisberg L A 1986 *Design and Calculation of Vibrating Screens* (Moscow: Nedra) p 144
[5] Blekhman I I 2000 *Vibrational mechanics: nonlinear dynamic effects, general approach, applications* (World Scientific) p 536
[6] Rand R H, Kinsey R J and Mingori D L 1992 Dynamics of spin up through resonance *Int. J. of Non-Linear Mechanics* **27(3)** pp 489-502
[7] Blekhman I I 1973 *Synchronization of dynamic systems* (Moscow: Nauka) p 896
[8] Blekhman I I, Vasilkov V.B. and Yaroshevich N P 2013 On some possibilities of improving vibration machines with self-synchronizing inertial vibration exciters *Problems of mechanical engineering and reliability of machines* **3** pp 18-22
[9] Blekhman I I, Blekhman L I, Vaisberg L A and Vasilkov V B 2019 Energy Consumption in Vibrational Transportation and Process *Obogashchenie Rud* **1(379)** pp 18-27
[10] Lyan I P , Panovko G Y and Shokhin A E 2019 Comparative analysis of energy efficiency in the use of vibration-type process machines in resonant and superresonant operating modes *Obogashchenie Rud* **6** pp 40-6
[11] Dentsov N N 2015 Prospects for the development of resonant vibration technology *Modern trends in the development of science and technology* **2(2)** pp 66-8
[12] Astashev V K and Pichugin K A 2019 Resonant tuning and vibration parameters of the bar system with a piezoelectric exciter *J. of Machinery Manufacture and Reliability* **48(6)** pp 535-40
[13] *Handbook: Vibration in Technics Vol 4. Vibration processes and machines* 1981 ed E E Levendel (Moscow: Mashinostroenie) p 509
[14] Available at: http://kroosh.com/products/uls201012w-screen
[15] Zhuravlev V F 2008 *Fundamentals of Theoretical Mechanics* (Moscow: Fizmatlit) p 304
[16] Ganiev R F and Kononenko V O 1976 *Oscillations of solids* (Moscow: Nauka) p 432

Analytical modelling of the influence of temperature and capillary diameter on the sonocapillary effect for liquids with different density

N Mikhailova and I Smirnov

Saint Petersburg University, Universitetskaya nab 7/9, Saint Petersburg, 199034, Russia

n.v.mikhailova@spbu.ru

Abstract. The paper examines an approach to calculating the sonocapillary effect. The proposed model for calculating the initial sonocapillary pressure considers the temperature and density of a liquid, the capillary diameter, and the ultrasound frequency. The calculations are based on the incubation time criterion of cavitation. This approach allows us to find the threshold cavitation amplitude of ultrasound for a given frequency and to estimate the corresponding sonocapillary pressure. The calculation results show that the sonocapillary pressure is directly related to the liquid density, increasing the capillary diameter reduces the intensity of the sonocapillary effect, and a rising temperature increases the sonocapillary pressure. However, minimal experimental data are available at present to verify the obtained theoretical results. Therefore, a wider range of new experimental results should be obtained for further studies of the sonocapillary effect based on the presented results.

1. Introduction

Ultrasonic vibration of the volume of a liquid near narrow channels, capillaries, pores and cracks can cause the liquid to rise to abnormally high levels in these voids compared to the usual capillary effect. This phenomenon is called the sonocapillary effect [1, 2]. This effect finds broad application – for example, in the impregnation of fabrics and insulation materials for coiled products, in soldering and in ultrasonic cleaning. Despite the widespread use of the sonocapillary effect, many opportunities for its experimental and theoretical study remain to be exploited.

In a number of experiments, it was shown [2-4] that a necessary condition for the sonocapillary effect to occur is cavitation near the capillary entrance (figure 1). Studies of ultrasound-induced cavitation in liquid are discussed in detail in [5, 6]. These studies have shown that acoustic cavitation strongly depends on the amplitude-frequency characteristics of ultrasound and on the temperature and density of the liquid. However, the available literature lacks a detailed presentation of these relationships for the case of the sonocapillary effect. At the same time, some experimental studies [7, 8] note the influence of capillary size and liquid density.

In this paper, we calculate the influence of capillary size and the density and temperature of the liquid on the sonocapillary effect at various ultrasound frequencies. To calculate this effect, we use the incubation time criterion of cavitation [9, 10], which enables us to introduce an integral characteristic of the complex cavitation process, significantly simplifying the calculation. Based on this approach, a model was presented in [11] for determination of the threshold sonocapillary pressure for various ultrasound frequencies. In addition, it was shown in [12] that the incubation time criterion allows us to consider the effect of temperature on the threshold vibration amplitudes leading to cavitation. Thus, the available results lead us to propose a model for assessing the effect of temperature on the sonocapillary pressure for liquids with different densities, for different ultrasound frequencies and capillary diameters.

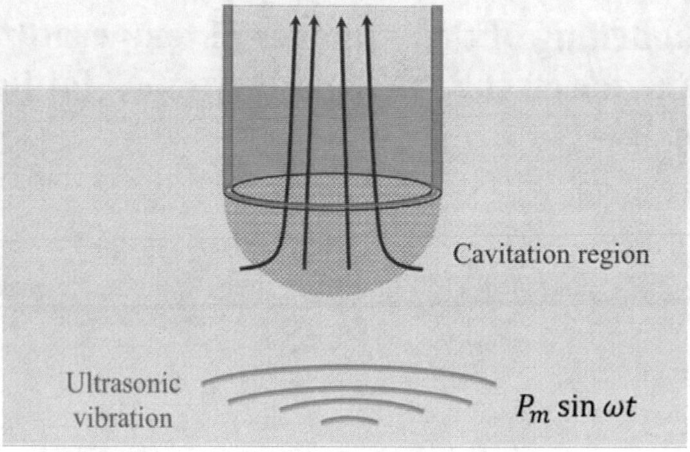

Figure 1. Schematic illustration of the sonocapillary effect with cavitation processes under the capillary edge.

2. Modelling of sonocapillary effect

2.1. Calculation of threshold loading

In the model, the sonocapillary pressure is calculated for the threshold modes of ultrasonic impact; in other words, for a given frequency, we consider the minimum vibration amplitude at which the sonocapillary effect (SCE) is realized. Experimental studies associate the occurrence of SCE with the initiation of cavitation under the edge of the capillary [2, 3, 7, 13] (figure 1). Therefore, the model uses the starting conditions for cavitation in the liquid to determine the threshold loads of the SCE.

Due to high-frequency ultrasonic vibrations, the effect on the liquid is dynamic. Thus, if we consider the formation of cavitation bubbles as a fracture or disturbance in liquid continuity, then to determine the conditions under which cavitation starts, we propose to use the incubation time criterion, which takes into account the time and energy parameters of the process. Under this criterion, cavitation can begin only when the impulse reaches a critical value. This is determined by the incubation time τ and the pressure P_c, which under quasi-static (low-frequency or constant) loading will lead to cavitation. The criterion has an integral form [9, 10]:

$$\frac{1}{\tau}\int_{t^*-\tau}^{t^*}\left|\frac{P(t)}{P_c}\right|^{\alpha} dt \geq 1 \tag{1}$$

where $P(t)$ is the tensile load in the fluid, t^* is the time of the cavitation generation, and α is a dimensionless parameter characterizing the fluid sensitivity to the loading history. The parameter α depends on the liquid's viscosity, which is derived from the bubble dynamics equation [10]. The liquids considered in this work are non-viscous; therefore, α is taken to be 0.5. The pressure P_c determines the strength of the liquid under quasi-static loading; it is calculated using the phase diagram of the liquid [12].

Cavitation occurs as a result of a tensile load, which in the case of ultrasonic action is represented in the form $P(t)=P_m \sin \omega t$, $t \in (0, \pi/\omega)$ where $\omega = 2\pi f$ is the angular frequency. By substituting the load $P(t)$ into the criterion (1), the threshold amplitude of ultrasonic vibrations can be determined for a given frequency:

$$P_m=\begin{cases} P_c\cdot\left|\frac{1}{\lambda}\max_{t_z\in(0;\pi)}\int_{t_z-\lambda}^{t_z}|\sin z|^{\alpha}\,dz\right|^{\frac{1}{\alpha}}, & \lambda\leq\pi \\ P_c\cdot\left|\frac{1}{\lambda}\int_0^{\pi}|\sin z|^{\alpha}\,dz\right|^{\frac{1}{\alpha}}, & \lambda>\pi \end{cases} \tag{2}$$

where, as a result of the change $z=\omega t$, the following parameters were introduced: $\lambda=\omega\tau$ and $t_z=\omega t^*$.

2.2. Calculation of sonocapillary pressure

In the model it is assumed that in the case of SCE, liquid rises through the capillary due to the collapse of bubbles formed in the cavitation region under it. Therefore, to calculate the sonocapillary pressure we consider the energy of the cavitating region to be the surface energy of N bubbles at the moment of their maximum expansion:

$$A_\sigma=4\pi\sigma R_m^2 N \tag{3}$$

Here σ is the surface tension coefficient and R_m is the maximum radius of cavitating bubbles, which can be estimated for a liquid with density ρ using the amplitude P_m and frequency $f=\frac{1}{T}$ of the ultrasonic field as follows [5]:

$$R_m\approx\frac{T}{4}\left(\frac{2}{3}\frac{P_m}{\rho}\right)^{0.5} \tag{4}$$

The number of bubbles in the cavitation region is estimated using the solution to the problem of packing equal-diameter circles into a larger circle. For packing density D_b and cavitation-region radius R_{cav}, the number of bubbles is calculated as [11]:

$$N=D_b\left(\frac{R_{cav}}{3R_m}\right)^3 \tag{5}$$

The derived energy of the collapsed bubbles of the cavitation cloud lifts the liquid up the capillary. The liquid is carried through the capillary by the acoustic flow at the moment of the compression phase [2]. Flow velocity depends on the speed of sound and density of the liquid; it is determined by the amplitude of the ultrasonic field $u=\frac{P_m}{\rho c}$. From this, the height the liquid rises in the capillary can be calculated by the law of conservation of total mechanical energy as $h=\frac{u^2}{2g}$.

The sonocapillary pressure can then be expressed as the work A_σ done by the collapsed bubbles as they raise a liquid column of height h in one cycle of oscillations along a capillary of radius R_c:

$$P_{sc}=\frac{A_\sigma}{\pi R_c^2 h} \tag{6}$$

3. Results and discussion

In [11] the modelling of the SCE for water showed good agreement with the experimental data. Therefore, this paper considers the effect of fluid parameters on the SCE; to this end, SCE modeling was carried out for ethanol, heptane and heavy water. In the calculations for each liquid, the threshold parameters of ultrasonic loading were determined. The properties of the liquids are presented in table 1. As can be seen, ethanol and heptane are lighter than water, while deuterium oxide has a higher density.

A comparison of the sonocapillary pressure for the studied liquids is shown in figure 2. The pressure P_{sc} was normalized to the maximum value of the sonocapillary pressure in water. Calculations show that the SCE grows with liquid's density increasing.

Table 1. Properties of liquids.

	Water	Heavy water (D_2O)	Ethanol	Heptane
Density ($kg \cdot m^{-3}$)	1000	1105	789	680
Sound speed ($m \cdot s^{-1}$)	1485	1400	1180	1130
Surface tension coefficient ($mN \cdot m^{-1}$)	72.5	71.87	22.39	20.86

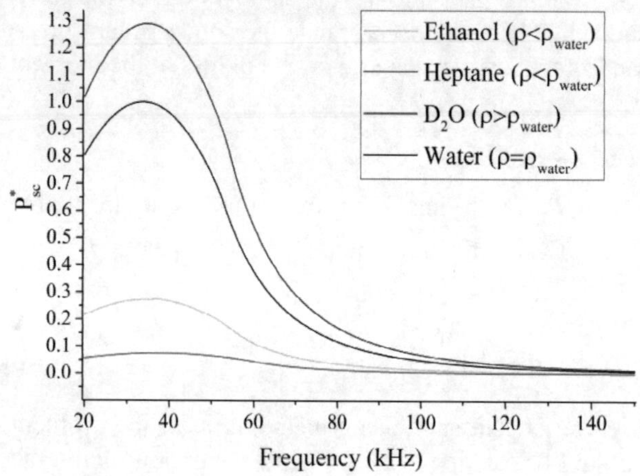

Figure 2. Frequency dependence of normalized sonocapillary pressure for liquids with different densities.

The influence of fluid temperature on the SCE was also considered. It is shown in [12] that the incubation time changes with increasing temperature as $\tau(T)=\tau_0\exp\left(\frac{G}{kT}\right)$, where G is the energy required for the onset of cavitation on an elementary volume, and τ_0 is a parameter characterizing the strength of the liquid at a given scale level. The strength parameters of the liquids considered in this work were calculated and verified by experimental data in [12].

Figure 3 depicts the results of modeling SCE for temperatures ranging from 10˚C to 90˚C. The sonocapillary pressure is also normalized to the sonocapillary pressure in water at a temperature of 20˚C. Modelling results show that the rise in temperature contributes to an increase in the SCE. The most rapid increase in sonocapillary pressure is observed as the liquid approaches its boiling point, which is clearly apparent in the case of ethanol. Its boiling point at atmospheric pressure is 78.37 ˚C, and its SCE value at that temperature is the highest among the four liquids.

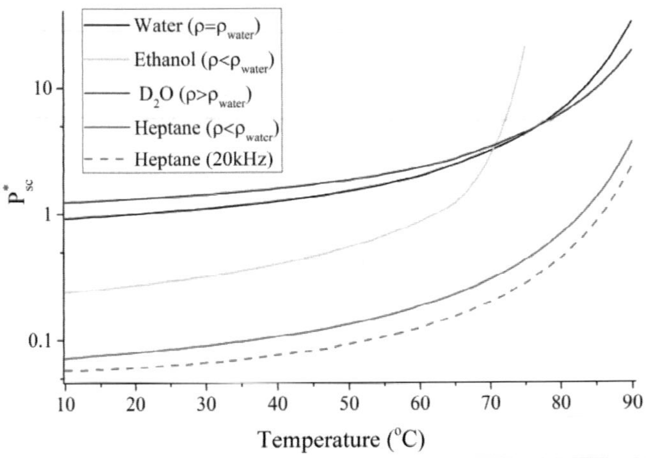

Figure 3. Dependence of normalized sonocapillary pressure on temperature for liquids with different densities.

In addition, figure 3 shows the influence of frequency on the temperature dependence of the SCE. Solid lines are calculated for a frequency of 41.9 kHz and the corresponding threshold amplitude. For heptane, an additional calculation was performed for an ultrasound frequency of 20 kHz, which is indicated in figure 3 with a dotted line. The modelling shows that lowering the frequency decreases the temperature dependence of the sonocapillary pressure.

The results of the SCE testing show a reduction in the effect's intensity with an increase in the capillary diameter [2, 13], which was also confirmed by modelling the acoustic capillary pressure for water. Figure 4 reveals the modelling results for two ultrasound frequencies, 18.5 kHz and 41.9 kHz, with experimental data [2, 7, 13]. Experiments conducted at the frequency of 41.9 kHz [7, 13] had a slightly higher amplitude of ultrasound than threshold value corresponding to this frequency. This leads to higher sonocapillary pressure compared to the theoretical calculation. For the SCE at 18.5 kHz, there is good agreement between the modelling result and the experimental data.

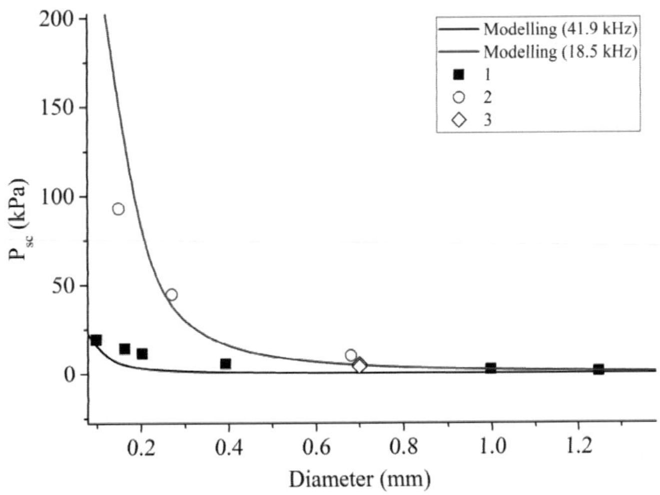

Figure 4. Sonocapillary pressure versus capillary diameter. Solid lines are theoretical calculations; points 1, 2, 3 are experimental data [2], [13], [7], respectively.

4. Conclusions

The model presented in this study allows us to calculate the influence of temperature, fluid density and capillary diameter on the sonocapillary effect (SCE). Modelling results show that the rise in liquid's density contributes to an increase in the SCE at low ultrasound frequencies. However, as frequency rises the sonocapillary pressure declines and the influence of liquid density on SCE levels out.

The temperature analysis of the SCE shows that near the boiling point the sonocapillary effect soars. Additionally, we compared temperature dependencies of sonocapillary pressure at two different frequencies. Modelling the dependence of the SCE on the capillary diameter revealed a good agreement with the experimental data for both of the ultrasonic frequencies considered.

Acknowledgments

This research was supported by the Grant of the President of the Russian Federation for young scientists (MK-269.2019.8).

References

[1] Konovalov G E and Germanovich I N 1962 Sonocapillary effect *Doklady AS BSSR* **6** pp 492–3
[2] Prokhorenko P P, Dezhkunov N V and Konovalov G E 1981 *Sonocapillary Effect* (Minsk: Nauka i Tekhnika) p 136
[3] Rosina E Yu 2003 Cavitation regime of sonocapillary effect *Akustichniy Vistnik* **6** pp 48–59
[4] Malykh N V and Sankin G N 2010 Stabilization and acoustic spectra of a cavitation cluster in an ultrasonic spherical cavity *Tech. Phys.* **55** pp 92–7
[5] Rosenberg L D 1968 *Physics and Technology of Powerful Ultrasound* (Moscow: Nauka) p 269
[6] Brennen C E 2013 *Cavitation and Bubble Dynamics* (Cambridge: Cambridge University Press) p 249
[7] Dezhkunov N and Francescutto A 2003 Ultrasonic capillary effect and sonoluminescence *Proc. of the World Congress on Ultrasonics* (Paris) pp 597–600
[8] Malykh N V, Petrov V and Sankin G N 2003 On sonocapillary effect *Proc. of the World Congress on Ultrasonics* (Paris) pp 1343–6
[9] Volkov G A, Gruzdkov A A and Petrov Y V 2007 The incubation time criterion and the acoustic strength of sea water *Acoust. Phys.* **53** pp 119–22
[10] Gruzdkov A A and Petrov Y V 2008 Cavitation breakup of low-and high-viscosity liquids *Tech. Phys.* **53** pp 291–5
[11] Mikhailova N V, Smirnov I V, Sharipova A and Slesarenko V 2020 Calculation of the dependence of the sonocapillary effect on the frequency of ultrasound based on the criterion of threshold cavitation *Problemy Prochnosti i Plastichnosti* **82** pp 64–74
[12] Mikhailova N, Smirnov I and Volkov G 2019 Modelling of pressure-temperature conditions for cavitation initiation in different liquids *Vibroengineering Proc.* vol 27 ed M Ragulskis (Leipzig: JVE International Ltd) pp 121–5
[13] Rosina E Yu 2003 On the nature of the force acting on the cavitating medium at the capillary edge *Akustichniy Vistnik* **6** pp 60–8

IOP Conf. Series: Materials Science and Engineering 1129 (2021) 012040 doi:10.1088/1757-899X/1129/1/012040

Wake-induced vibration of two-phase conductors connected by spacers

R A Shavnya and A N Danilin

Institute of Applied Mechanics of Russian Academy of Sciences, Moscow, Russia

shavnya.ruslan@yandex.ru

Abstract. Considering finite element approach applied to modelling of wind-induced split-phase overhead power line conductor oscillations caused by aerodynamic wake. Model represents an aeroelastic system including twin bundled conductor with spacers. Modified Simpson's theory is used for conductor wake coupling simulation. Assuming small oscillating conductor deflection from element ends connection line compared to element length, thus the element tension and element tensile strain can be considered as constant values. Strain, as function of transverse displacements, defines as quadratic approximation. Element absolute displacements and twisting angles, together with Ritz's method coefficients is taken as generalized coordinates, in which conductor move equations are compiled with non-linear elastic, inertial and aerodynamic forces. Appropriate linearized small oscillation equations near the static equilibrium are derived – those can be used for defining system flutter critical wind velocity. An example with split-phase twin bundled conductor, three-sub-span system is presented.

1. Introduction

Wind-induced overhead power line conductor oscillation phenomena can be subdivided on three groups depending of frequencies and amplitudes.

First one – high frequency low amplitude vibrations, also called "Aeolian vibrations", occurring when wind are weak and stable and conductor itself is strained. Those vertical plane oscillations in form of steady-waves have frequencies about 5 to 100 Hz with initial wind speed of 0.5 to 8 $m\cdot s^{-1}$. The cause of vibration excitement is the air flow interruption behind the conductors in forms of Karman vortex streets [1, 2].

Next group contains low frequency (0.2 to 3 Hz) high amplitude oscillations, so called "Conductor galloping".This phenomenon occurs with combination of quasi-stable wind (speeds of 5-20 $m\cdot s^{-1}$) and conductor ice accretion [3, 4]: due to asymmetrical iced conductor cross-section various aerodynamic forces can arise and change, leading to oscillation. Conductor galloping is a variety of flutter oscillations, "peak-to-peak" type, amplitude of which can reach several meters vertically.

Last group contains split-phase power line bundled conductors oscillations, usually called sub-span oscillations (figure 1). In this case most common frequencies are 0.7 to 5 Hz. Necessary initial conditions for this phenomenon to occur consists approximately close location of two-bundled conductors in one horizontal plane and a wind velocity of 6-15 $m\cdot s^{-1}$; thus, one conductor (leeward) is in the wake of another (windward), which leads to aerodynamic lift and drag changes, causing windward conductor and, due to spacers connections, leeward, to oscillate [5-7]. Sub-span oscillations (in combination with constant vibrations) can lead to various power line structural elements (spacer clamps, dampers, etc.) damages and wear-outs.

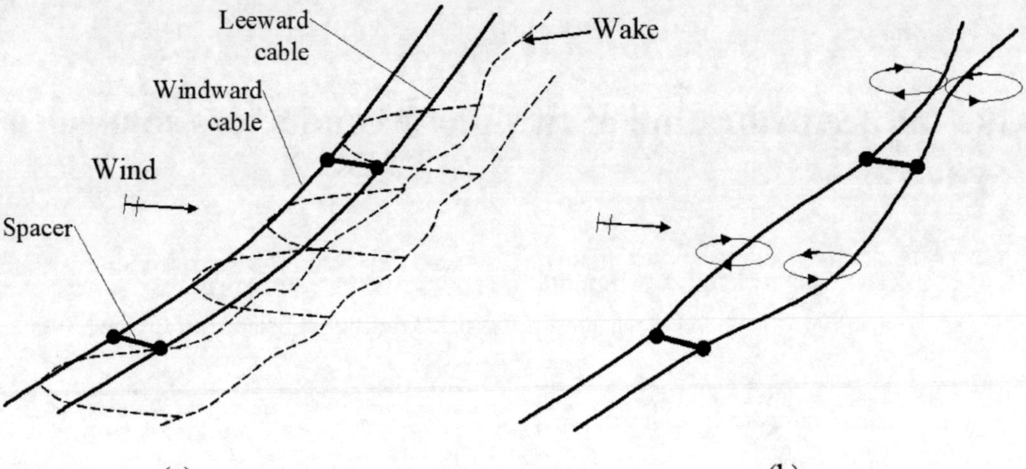

Figure 1. Sub-span oscillations: (a) wake-induced oscillations scheme; (b) case of oscillations.

2. Basic hypotheses, defining relations and equations

A strained conductor in the span of a power line is considered, assuming conductor having only tensile elasticity with rigidity of *EF*, where *E* – is a Young's modulus, *F* – effective cross-section area.

Initial surplus of the conductor length:

$$\Delta^{\circ} = L^{\circ}(1 + \alpha T) - L, \tag{1}$$

where L^0 – initial conductor length in normal temperature conditions, L – span length, T – temperature increment, α– linear temperature expansion coefficient.

In purpose of dimensional coordinates discretisation finite element approach is using. With each *i* element a local coordinate system $\tilde{O}^{(i)}x^{(i)}y^{(i)}z^{(i)}$, which moving relatively to global (inertial) system OXYZ, is connected (next we will neglect upper *(i)* index for the case of convenience). Local axes are parallel to global axe OX and in model only rotations around OX axe are considered (figure 2).

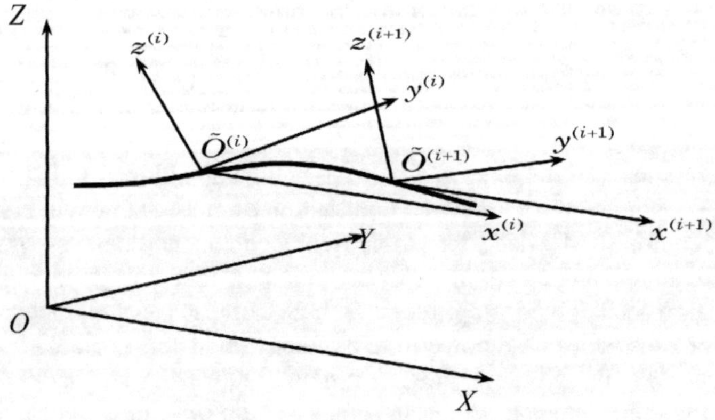

Figure 2. Global and local coordinate systems.

By the influence of gravitation and wind conductor is oscillating near static position. Nodes displacements along OX, OY, OZ axes and twisting angle around OX axe denote as u, v, w and φ.

Then, local displacements and conductor cross-section twisting angle within finite element length of l can be described using Ritz method as:

$$\tilde{u} = (u_1 - u_0)\frac{x}{l},$$
$$\tilde{v} = (v_1 - v_0)\frac{x}{l} + \sum_k q_k \sin\left(\frac{k\pi x}{l}\right),$$
$$\tilde{w} = (w_1 - w_0)\frac{x}{l} + \sum_k r_k \sin\left(\frac{k\pi x}{l}\right),$$
$$\tilde{\varphi} = (\varphi_1 - \varphi_0)\frac{x}{l} + \textstyle\sum_k p_k \sin\left(\frac{k\pi x}{l}\right), \quad (2)$$

where $u_k(t), v_k(t), w_k(t), \phi_k(t)(k = 0,1)$ are boundary values of displacements and twisting angle given that $x = 0$ $(k = 0)$ and $x = l$ $(k = 1)$, that is a generalized coordinates. Besides, problem generalized coordinates are coefficients $p_k(t)$, $q_m(t)$, $r_n(t)$, that defines finite element inner freedom degrees relative to sine forms.

According to Hook's law, given that $\varepsilon > \frac{\Delta^o}{l}$ and considering (1), tensile force$N$ denotes:

$$N = EF\left(\varepsilon - \frac{\Delta^\circ}{L}\right), \quad (3)$$

where ε – longitudinal strain, which will be denoted in the form of quadratic dependence as:

$$\varepsilon = \frac{d\tilde{u}}{dx} + 0.5\left[\left(\frac{d\tilde{v}}{dx}\right)^2 + \left(\frac{d\tilde{w}}{dx}\right)^2\right]. \quad (4)$$

Torque of the arbitrary conductor cross-section is $M_t = GJ\frac{d\varphi}{dx}$, where GJ is torsional rigidity.

EF and GJ rigidities are assuming to be constant along the span.

Because of longitude inertial forces neglection tensile force will be constant along the conductor length, $N(x,t) \approx N(t)$. Thus by using (3) same can be applied to the strains, $\varepsilon(x,t) \approx \varepsilon(t)$. By taking integral of (4) considering (2) we will get:

$$\varepsilon = \frac{1}{l}\left\{u_1 - u_0 + \frac{1}{2l}\left[(v_1 - v_0)^2 + (w_1 - w_0)^2 + \frac{1}{2}\textstyle\sum_k(\pi k)^2(q_k^2 + r_k^2)\right]\right\}, \quad (5)$$

then potential energy will be denoted as:

$$U = \frac{1}{2}\int_0^l \left[EF\left(\varepsilon - \frac{\Delta^\circ}{L}\right)^2 + GJ_t\varphi'^2\right]dx = \frac{lEF}{2}\left(\varepsilon - \frac{\Delta^\circ}{l}\right)^2 + \frac{GJ}{2l}\left[(\varphi_1 - \varphi_0)^2 + \frac{1}{2}\textstyle\sum_k(\pi k)^2 p_k^2\right]. \quad (6)$$

Virtual work principle is using in form of $\delta U = \delta A_g + \delta A_a + \delta A_{int}$ to get the nonlinear move equations of the system. Potential energy variation:

$$\delta U = \textstyle\sum_{i=0}^1\left(\frac{\partial U}{\partial u_i}\delta u_i + \frac{\partial U}{\partial v_i}\delta v_i + \frac{\partial U}{\partial w_i}\delta w_i\right) + \textstyle\sum_k\left(\frac{\partial U}{\partial q_k}\delta q_k + \frac{\partial U}{\partial r_k}\delta r_k + \frac{\partial U}{\partial p_k}\delta p_k\right), \quad (7)$$

and the gravity, aerodynamic and inertial forces work variations are:

$$\delta A_g = -\mu g\int_0^l \delta(w + \eta_C \sin\varphi + \vartheta_C \cos\varphi)\,dx, \delta A_a = \int_0^l (Y\delta v + Z\delta w)\,dx,$$
$$\delta A_i = \int_0^l \mu[\ddot{u}\delta u + (\ddot{v} - \eta_C\ddot{\varphi} - \vartheta_C\dot{\varphi}^2)\delta v + (\ddot{w} + \vartheta_C\ddot{\varphi} - \eta_C\dot{\varphi}^2)\delta w + (-\eta_C\ddot{v} + \vartheta_C\ddot{w} + \rho_C^2\ddot{\varphi})\delta\varphi]dx, \quad (8)$$

Where μ – linear mass, ϑ_c and η_c – cross-section center-of-mass coordinates (according to cross-section coordinate system), $\mu\rho_c^2$ – polar moment of inertia; Y and Z– aerodynamic load components.

Integrals in (8) are taking by using (2), (3) and (5).

3. Aerodynamics

While inside the aerodynamic wake, average flow speed is slower compared to free-stream. This phenomenon can be described by using empirical functions [8]. Implying d as diameter of conductor, we denote coordinates of $\xi = y/d$, $\eta = z/d$ and also free-stream drag coefficient C_{DMax} which for smooth cylinder is $C_{DMax} \approx 1.2$. Ratio between the free-stream velocity and a local velocity inside the wake (correspondingly V and $V_l(\xi, \eta)$) is:

$$b(\xi,\eta) = \frac{V_l(\xi,\eta)}{V} = 1 - \psi(\xi,\eta)\sqrt{\frac{\psi(\xi,\eta)}{\eta+6}},$$

$$\psi(\xi,\eta) = C_{Dmax}\exp\left(-\frac{\xi^2}{0.23C_{Dmax}(\eta+6)}\right) \tag{9}$$

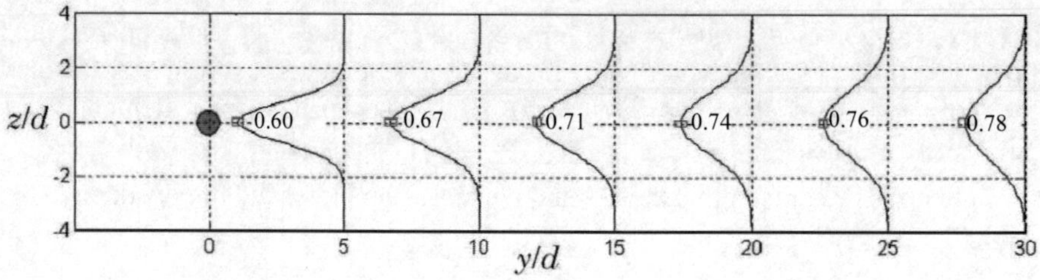

Figure 3. Distribution of $b(\xi, \eta)$ function for various profiles along the aerodynamic wake.

References [9-13] contain coefficients for a wide range of various intervals. In this particular paper we use smooth cylinder coefficients of Price. Also by using analytical expressions of drag and lift coefficients (C_D and C_L) derivative taking problem are significantlysimplified.

Our polynomial expressions are similar to [14] except the higher precision and covering of the wider aerodynamic wake area:

$$\begin{gathered} C_D(\eta,\xi) = \sum_{i=0}^{5} a_{2i}\,\xi^{2i}, \\ a_i = b_{i0} + b_{i1}\eta + b_{i2}\eta^2, \\ C_L(\eta,\xi) = \sum_{i=0}^{4} p_{2i+1}\,\xi^{2i+1}, \\ p_i = q_{i0} + q_{i1}\eta + q_{i2}\eta^2, \\ |\xi| \le 4;\ 5 \le \eta \le 20.3. \end{gathered} \tag{10}$$

Outside the designated intervals of ξ,η drag and lift coefficients are$C_D = 1.2$, $C_L = 0$; values for b_{i0}, b_{i1}, b_{i2}, q_{i0}, q_{i1}, q_{i2} are given in [15].

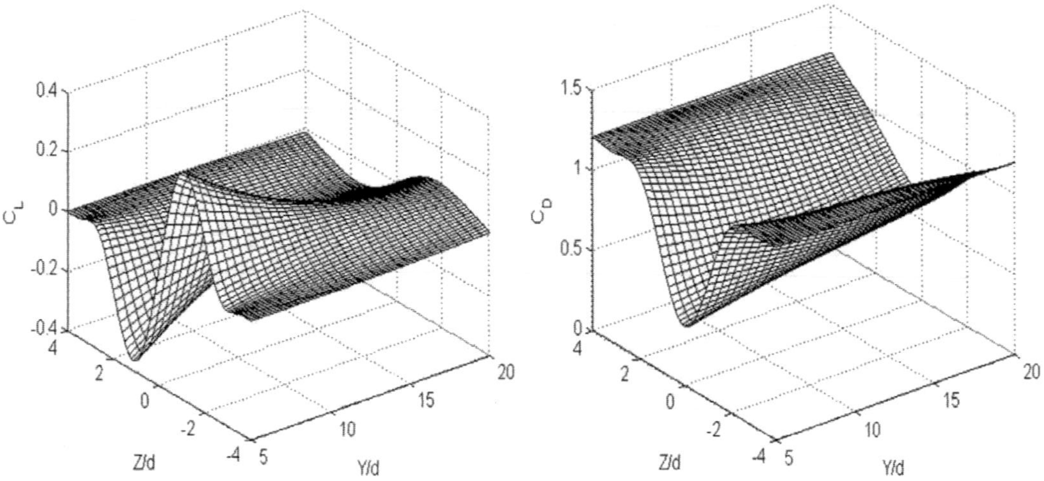

Figure 4. Aerodynamic coefficients surfaces (10) inside the designated intervals.

4. Example. Split-phase twin bundle conductor, three sub-spans

Considering the case where the line supposed to be strongly prone to sub-span oscillations: large and equal sub-spans, tilted bundle plane. Wind velocity is 15 m·s^{-1}. Model scheme are in figure 5 and detailed model parameters are in table 1. Results illustrated in figures 6 and 7.

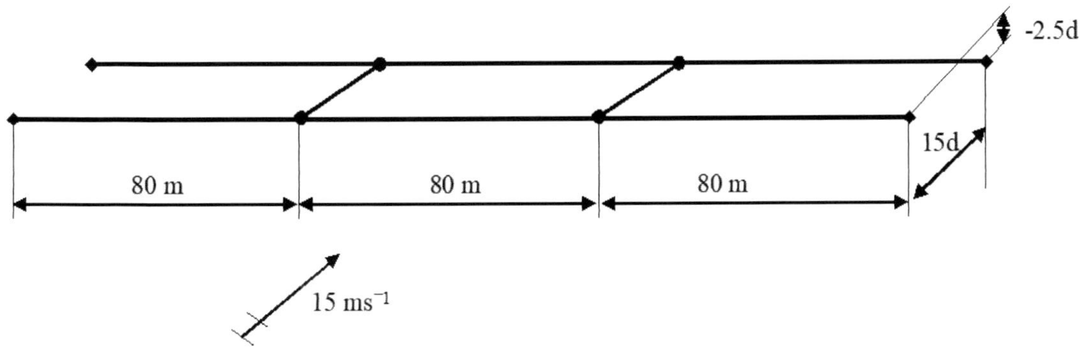

Figure 5. Example model scheme.

Table 1. Model parameters.

Span length	240 m
Sub-spans	3 x 80 m
Conductor diameter	2.54 cm
Conductor horizontalspacing	15 conductor diameters
Conductor verticalspacing	2.5 conductor diameters
Conductor tension	19 kN Windward; 19.1 kN Leeward
Conductor linear mass	1.25 kg·m^{-1}

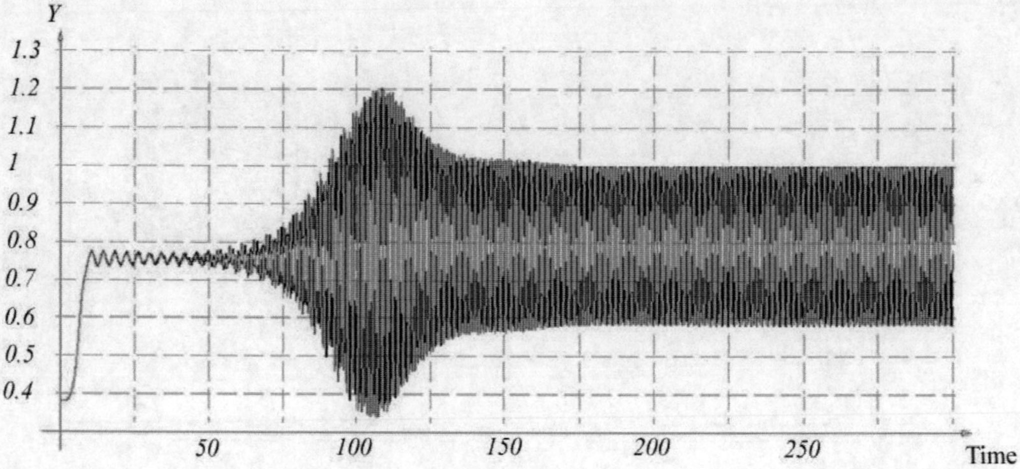

Figure 6. Red and black curves are representing correspondingly horizontal displacement of leeward sub-conductors in the middle of sub-spans 1 and 3.

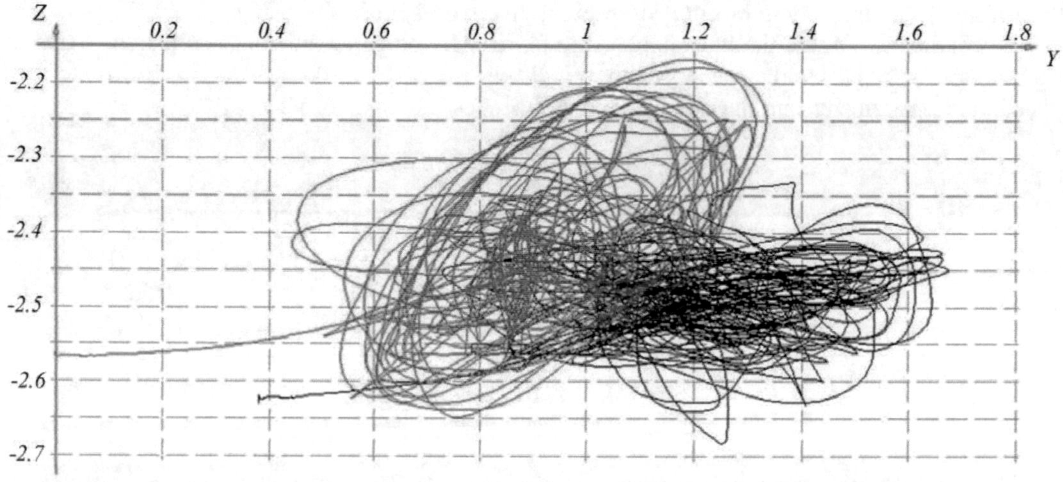

Figure 7. Windward (left, red curve) and leeward (right, black curve) sub-conductor orbits due to sub-oscillations, middle of end sub-span.

Acknowledgments

This investigation was performed under the State Task for Basic Researches (register number AAAA-A19-119012290118-3).

References

[1] Chan J, Havard D, Rawlins C, Diana G, Cloutier L, Lilien J L, Hardy C, Wang J and Goel A 2009 *EPRI Transmission line reference book. Wind-induced Conductor motion. Second edition* (Palo Alto, CA: Electric Power Research Institute)

[2] Van Dyke M 1982 *An Album of Fluid Motion* (Standford, CA: The Parabolic Press)

[3] Lilien J L *et al* 2007 *State of the art of conductor galloping. A complementary document to "Transmissionline reference book – Wind induced conduction motion Chapter 4: Conductorgalloping", Based on EPRI Research project 792* (CIGRE)

[4] Keutgen R 1999 *Galloping Phenomena: A Finite Element ApproachPh.D. thesis* (Liège:

Appliuées de l'Université de Liège)
[5] Lilien J L and Snegovski D 2004 Wake-Induced vibration of power transmission line: parametric study *Proc. of the 8th Int. Conf. on Flow-Induced Vibration6-9 July 2004 (Paris)*
[6] Mair W A and Maull D J 1971 Aerodynamic behaviour of bodies in the wake of other bodies*Phil. Trans. Roy. Soc. Lond.* **268** pp 425–37
[7] Rawlins C B 1976 Fundamental concepts in the analysis of wake-induced oscillation of bundled conductors *IEEE Transactions on Power Apparatus and Systems* **95** pp 1377–93
[8] Blevins B D 1990 *Flow-Induced Vibration* (New York: Van Nostrand Reinhold)
[9] Diana G and Giordana F 1972 Sulle vibrazioni di un cilindro in scia di un altro – Analisi del fenomeno con modello analitico *L'Energia Elettrica* **7** pp 448–57
[10] Price S J 1975 Wake induced flutter of power transmission conductors *J. of Sound and Vibration* **38** pp 125–47
[11] Price S J and Maciel Y 1990 Solution of the nonlinear equations for wake-induced flutter via the Krylov and Bogoliubov method of averaging *J. of Fluids and Structures* **4** pp 519–40
[12] Price S J and Païdoussis M P 1984 An improved mathematical model for the stability of cylinder rows subject to cross-flow *J. of Sound and Vibration* **97** pp 615–40
[13] Price S J and Païdoussis M P 1984 The aerodynamic forces acting on groups of two and three circular cylinders when subject to a cross-flow *J. of Wind Engineering and Industrial Aerodynamics* **17** pp 329–47
[14] Bokaian A 1989 Galloping of a circular cylinder in the wake of another *J. of Sound and Vibration* **128** pp 71–85
[15] Snegovski D 2010 *Wake-Induced Oscillations in Cable Structures: Finite Element Approach Ph.D. thesis* (Liège: University of Liège, Montefiore Institute of Electricity)

IOP Conf. Series: Materials Science and Engineering 1129 (2021) 012041 doi:10.1088/1757-899X/1129/1/012041

On self-synchronization of inertial vibration exciters in a vibroimpact three-mass system

A E Shokhin, K V Krestnikovskii and A N Nikiforov

Mechanical Engineering Research Institute of the Russian Academy of Sciences, Moscow, Russia

shohinsn@mail.ru

Abstract. The dynamics of a vibration machine's model in the form of a vibroimpact three-mass system with one shock pair, which oscillations are excited by two non-ideal inertial vibration exciters, is considered. The collision of the shock pair elements is modeled on the basis of the classical impact theory. A numerical simulation of the model dynamics at varying the excitation frequency, as well as the parameters of the initial gap and the coefficient of restitution, characterizing the conditions of contact interaction of the machine's working bodies with the processed material, has been carried out. The influence of these parameters values on the frequency range of stable synchronization of vibration exciters and stable periodic vibroimpact modes of the system motion has been established. It is shown that at the same excitation frequency, a change in these parameters can lead to a change in the periodicity of the vibroimpact mode and the working bodies' oscillations mode from synchronous-antiphase to practically synchronous-inphase.

1. Introduction

To excite oscillations of vibrating machines and mechanisms for various technological purposes, systems of several self-synchronizing inertial (unbalance) vibration exciters are widely used [1-3]. One of the main problems of their effective application is to ensure synchronous rotation of vibration exciters with given ratio of speeds and mutual phase, which is required for efficient implementation of the technological process, when interacting with the processed material.

The general theory of synchronization of dynamic systems and its application to the solution of some problems of self-synchronization of inertial vibration exciters in linear mechanical systems and in the systems with weak nonlinearity are considered in [4]. In [5-8], the issues of self-synchronization of vibration exciters in mechanical systems with nonlinear characteristics of elastic elements that simulate interaction with the processed material are considered for various types of nonlinearity. In this works the vibration exciters are attached to a single carrying body. In [1, 9-11], the dynamics of vibration machines with self-synchronizing vibration exciters located on various carrying bodies is analyzed. In these cases, the interaction with the processed material is taken into account on the basis of simplified representations, for example, in the form of an additionally attached mass, or a linear visco-elastic element. The interaction with the material in the form of a non-instantaneous impact of the supporting bodies is taken into account in [12, 13]. It is shown that taking into account the vibroimpact nature of the interaction has a significant effect on the frequency range of vibration exciter's self-synchronization and the modes of system's oscillation. At the same time, practical application of such a model is complicated by the need to conduct relatively complex experiments to determine the dynamic characteristics of contact interaction.

In this paper, the dynamics of a model of a vibrating machine with two self-synchronizing vibration exciters located on different carrying bodies, which are its working bodies, is consider. Interaction with the processed material is taken into account in the form of an instant impact of the working bodies against each other on the basis of the classical theory of impact [14].

2. Mathematical model

The design scheme of the vibrating machine model is shown in figure 1. The model consists of a solid body of mass m_1, which simulates the frame of the machine, attached to a fixed foundation by means of a linear spring with the coefficients of stiffness and viscosity c_1 and b_1, respectively; and two rigid bodies with the same mass m_2, simulating the working bodies of the machine, symmetrically fixed to the frame using identical linear springs with the stiffness and viscosity coefficients c and b, respectively. Motions of each of the bodies are described by the displacements x_1, x_2, x_3 of their centers of mass along the horizontal axis Ox, measured relative to their equilibrium positions.

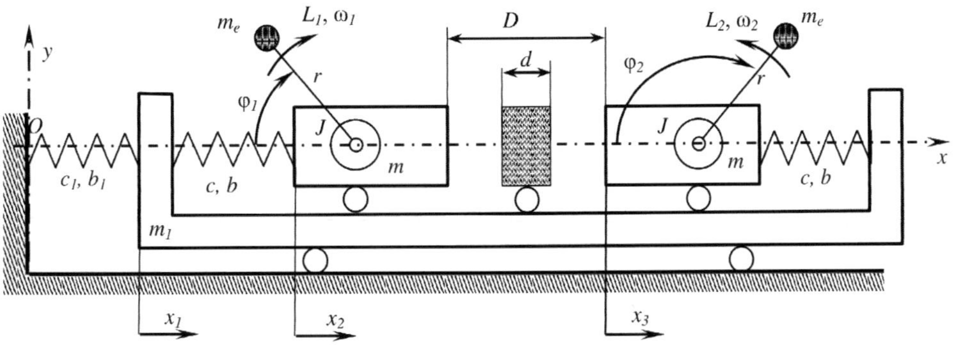

Figure 1. The design scheme.

Oscillations of the system are excited by two identical unbalance vibration exciters with an imbalance mass m_e and an eccentricity r, rigidly fixed to each working body. The angular positions of the imbalances are described by the angles φ_i ($i = 1,2$ is the vibration exciter number) measured from the negative direction of the Ox axis. Each vibration exciter is driven by an asynchronous electric motor with the moment of inertia J_0. The driving moment L_i of the i-th motor is taken into account by its static characteristic $L_i = 2\sigma_i M_{ci}/(s_{ci}/s_i - s_i/s_{ci})$ [4], where M_{ci} and s_{ci} are critical torque and critical slip of the motor respectively, $s_{ci} = (\omega_{0i} - \sigma_i\dot{\varphi}_i)/\omega_{0i}$, $\omega_{0i} = \omega_e/p_i$ – synchronous speed of the motor, p_i – quantity of the motor's pole pair, ω_e – power supply frequency, $\sigma_i = \pm 1$ – coefficient taking into account the direction of the motor torque action. Friction in the bearings supporting the exciter's shaft is considered as the moment of Coulomb friction forces $L_{Ri} = 0.5 m_e r d_s f_i \dot{\varphi}_i^2 \mathrm{sign}(\dot{\varphi}_i)$, f_j – coefficient of Coulomb friction, d_s – diameter of vibration exciter's shaft.

When the machine is operating, the processed material is fed into the space between the working bodies (working chamber). The material's mass is significantly less than the mass of the working bodies and, therefore, is not taken into account. The filling of the working chamber with the material is characterized by the value Δ (hereinafter "the initial gap"), which is the difference between the distance D between the working bodies at undeformed elastic elements of the system and the characteristic transverse dimension d of the material's volume in the working chamber. A decrease in the value of Δ corresponds to an increase in the fullness of the working chamber's volume. The interaction of the working bodies with the material occurs when $x_2 - x_3 = \Delta$, and is modeled by the instant impact of the working bodies against each other. Energy losses spent on material processing at impact are characterized by the coefficient of restitution R. Velocities $\dot{x}_{2+}, \dot{x}_{3+}, \dot{\varphi}_{1+}, \dot{\varphi}_{2+}$ of the shock pair elements after impact are defined by the expressions:

$$\begin{aligned} \dot{x}_{2+} &= ((1+R)A_2\dot{x}_{3-} + (A_1 - RA_2)\dot{x}_{2-})/(A_1 + A_2), \\ \dot{x}_{3+} &= ((1+R)A_1\dot{x}_{2-} + (A_2 - RA_1)\dot{x}_{3-})/(A_1 + A_2), \\ \dot{\varphi}_{1+} &= \dot{\varphi}_{1-} - m_e r \sin(\varphi_1)(\dot{x}_{2+} - \dot{x}_{2-})/J, \\ \dot{\varphi}_{2+} &= \dot{\varphi}_{2-} - m_e r \sin(\varphi_2)(\dot{x}_{3+} - \dot{x}_{3-})/J, \end{aligned} \tag{1}$$

where $A_1 = m_2 + m_e - (m_e r \sin(\varphi_1))^2/J$, $A_2 = m_2 + m_e - (m_e r \sin(\varphi_2))^2/J$, $J = J_0 + m_e r^2$, $\dot{x}_{2-}, \dot{x}_{3-}, \dot{\varphi}_{1-}, \dot{\varphi}_{2-}$ – velocities of the shock pair elements before impact.

Motion of the system between impacts is described by a system of equations (presented in dimensionless form):

$$\begin{cases} \ddot{q}_1 + 2\lambda\mu_1(\beta_1\dot{q}_1 + \beta_2\dot{q}_3) + n_1^2 q_1 = \mu_1(\sum_{i=1}^{2}(\ddot{\varphi}_i \sin\varphi_i + \dot{\varphi}_i^2 \cos\varphi_i)) \\ \ddot{q}_2 + 2\lambda\dot{q}_2 + q_2 = (\sum_{i=1}^{2}(-1)^i(\ddot{\varphi}_i \sin\varphi_i + \dot{\varphi}_i^2 \cos\varphi_i)) \\ \ddot{q}_3 + 2\lambda\mu_2(\beta_2\dot{q}_1 + \beta_3\dot{q}_3) + n_3^2 q_3 = \mu_2(\sum_{i=1}^{2}(\ddot{\varphi}_i \sin\varphi_i + \dot{\varphi}_i^2 \cos\varphi_i)), \\ \ddot{\varphi}_1 = N_1(\dot{\varphi}_1) + J_*(\ddot{q}_1 - \ddot{q}_2 + \ddot{q}_3)\sin\varphi_1 \\ \ddot{\varphi}_2 = N_2(\dot{\varphi}_2) + J_*(\ddot{q}_1 + \ddot{q}_2 + \ddot{q}_3)\sin\varphi_2 \end{cases} \tag{2}$$

where $\mathbf{q} = \frac{\mathbf{U^{-1}x}}{X_*}$, $\mathbf{q} = [q_1, q_2, q_3]^T, \mathbf{x} = [x_1, x_2, x_3]^T$, $\mathbf{U} = \begin{bmatrix} u_{11} & 0 & u_{31} \\ 1 & -1 & 1 \\ 1 & 1 & 1 \end{bmatrix}$,

$$u_{11} = -\frac{2 + \frac{c_1}{c} - \frac{m_1}{m} - \sqrt{D}}{2\frac{m_1}{m}}, \quad u_{31} = -\frac{2 + \frac{c_1}{c} - \frac{m_1}{m} + \sqrt{D}}{2\frac{m_1}{m}}, \quad D = 4(1 + \frac{c_1}{c} + \frac{m_1}{m}) + \left(\frac{c_1}{c} - \frac{m_1}{m}\right)^2,$$

$$n_1^2 = \frac{2 + \frac{c_1}{c} + \frac{m_1}{m} - \sqrt{D}}{2\frac{m_1}{m}}, \quad n_3^2 = \frac{2 + \frac{c_1}{c} + \frac{m_1}{m} + \sqrt{D}}{2\frac{m_1}{m}}, \quad n_2^2 = 1,$$

$$T_* = \sqrt{m/c}, \quad \mu_1 = m_{22}^*/m_{11}^*, \quad \mu_2 = m_{22}^*/m_{33}^*,$$

$$\beta_1 = b_{11}^*/b_{22}^*, \quad \beta_2 = b_{13}^*/b_{22}^* = b_{31}^*/b_{22}^*, \quad \beta_3 = b_{33}^*/b_{22}^*,$$

$$\lambda = b_{22}^* T_*/2m_{22}^*, \quad N_i(\dot{\varphi}_i) = (L_i(\dot{\varphi}_i) - L_{Ri}(\dot{\varphi}_i))/L_*, \quad L_* = J/T_*^2, \quad J_* = m_{22}^* X_*^2/J,$$

$$\Delta^* = \Delta/X_*, \quad X_* = m_e r/m_{22}^*,$$

where m_{ij}^*, b_{ij}^* - corresponding elements of inertia matrix $\mathbf{M}^*$ and damping matrix $\mathbf{B}^*$, determined as follows: $\mathbf{M}^* = \mathbf{U}^\mathrm{T}\mathbf{M}\mathbf{U}$, $\mathbf{B}^* = \mathbf{U}^\mathrm{T}\mathbf{B}\mathbf{U}$, $\mathbf{M} = \begin{bmatrix} m_1 & 0 & 0 \\ 0 & m & 0 \\ 0 & 0 & m \end{bmatrix}$, $\mathbf{B} = \begin{bmatrix} 2b + b_1 & -b & -b \\ -b & b & 0 \\ -b & 0 & b \end{bmatrix}$, $m = m_2 + m_e$, dots denote differentiation with respect to dimensionless time $\tau = t/T_*$.

3. Simulation results and their analysis

The simulation of the model dynamics was carried out numerically in Matlab. Integration of the system of differential equations (2) was performed by standard integration functions by the Runge-Kutta method of the 4th order with an additional option to accurately determine the moments of working bodies' collision. At the moments of collision, the integration was interrupted, and the velocities were recalculated according to (1), after which the integration continued.

To study the dynamic characteristics of the system, the dimensionless frequency of the supply voltage $\tilde{\omega}_e = \omega_e/n_1$ was set, which gradually changed in a stepwise manner in the range of values $0.1 \le \tilde{\omega}_e \le 3.6$ with a frequency step of $\Delta\tilde{\omega}_e = 0.1$. The maximum torque of the motors was considered constant in the frequency range. The calculations were performed both with increasing and decreasing $\tilde{\omega}_e$. At each value of the frequency, the calculations were carried out during a period of time sufficient for the transient processes decay and the steady oscillations establishment. After the steady oscillations were achieved, the averaged frequency $\tilde{\omega}$ and the mutual phase $\Delta\varphi$ of the exciters rotation, the peak-to-peak values and period of the working bodies' oscillations, as well as the speed of their collisions (impact rate), were determined. To establish the influence of Δ^* and R on the dynamic characteristics of the system, the calculations were repeated at their different values in the ranges $-5 \leq \Delta^* \leq 15$ and $0.1 \leq R \leq 0.9$. The simulation was carried out for the following values of the system parameters: $n_1 = 0{,}166$, $n_3 = 1{,}524$, $\mu_1 = 0.576$, $\mu_2 = 0.423$, $\beta_1 = 0{,}473$, $\beta_2 = -0.579$, $\beta_3 = 6{,}279$, $\lambda = 0{,}03$, $J_* = 0{,}013$, $M_{c1}/L_* = M_{c2}/L_* = 0.248$, $s_{c1} = s_{c2} = 0.15$, $0.5m_e r d_s f_1 / L_* = 0.5m_e r d_s f_2/L_* = 1 \cdot 10^{-4}$.

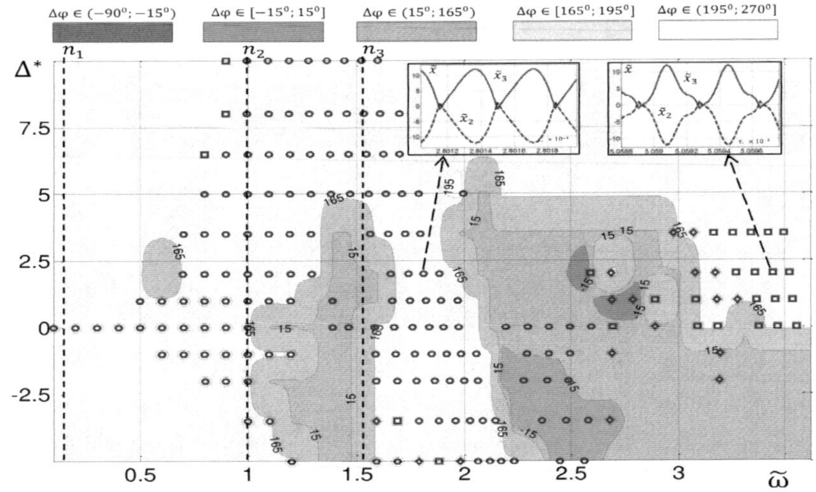

(a)

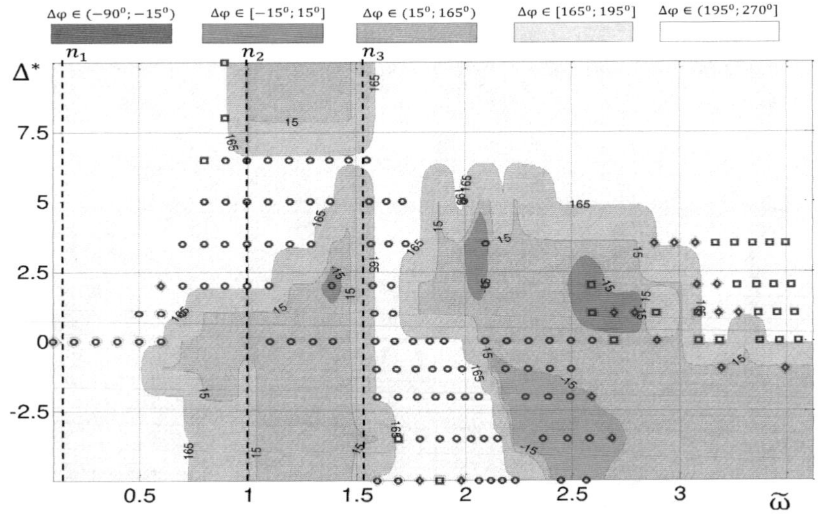

(b)

Figure 2. Motion modes for *R*=0.5.

Figures 2-4 show stable periodic vibroimpact modes of the model motion and areas of synchronous rotation of vibration exciters with different values of the mutual phase $\Delta\varphi$ depending on the frequency of the exciters rotation $\tilde{\omega}$ and the value of Δ^*, calculated for $R = \{0.1,\ 0.3,\ 0.5\}$ at an increase in the frequency $\tilde{\omega}_e$ (figures 2-4 (a)) and with its decrease (figures 2-4 (b)). Single markers denote vibroimpact modes of motion with different periods relative to the period of the exciters rotation

$T = 1/\tilde{\omega}$ and one collision during the period found: markers "O" denote modes with a period equal to the main excitation period T, markers "□" denote modes with period $2T$, markers "◊" denote modes with a period greater than $2T$. Vibroimpact modes of the corresponding periodicity with several impacts found during the period (multi-impact modes) are marked with similar, but double markers. To display changes in the steady-state mutual phase $\Delta\varphi$, the entire range of possible values was divided into intervals$\{(-90^{\circ};-15^{\circ}),\ [-15^{\circ};15^{\circ}],\ (15^{\circ};165^{\circ}),\ [165^{\circ};195^{\circ}],\ (195^{\circ};270^{\circ}]\}$. Regions of the varied parameters values at which $\Delta\varphi$ is set within the corresponding interval are shown in different colors. Note that the values of $\Delta\varphi \in [-15^{\circ};15^{\circ}]$ correspond to the rotation of vibration exciters close to synchronous-in-phase, and $\Delta\varphi \in [165^{\circ};195^{\circ}]$ - close to synchronous-antiphase, required for normal operation of the machine.

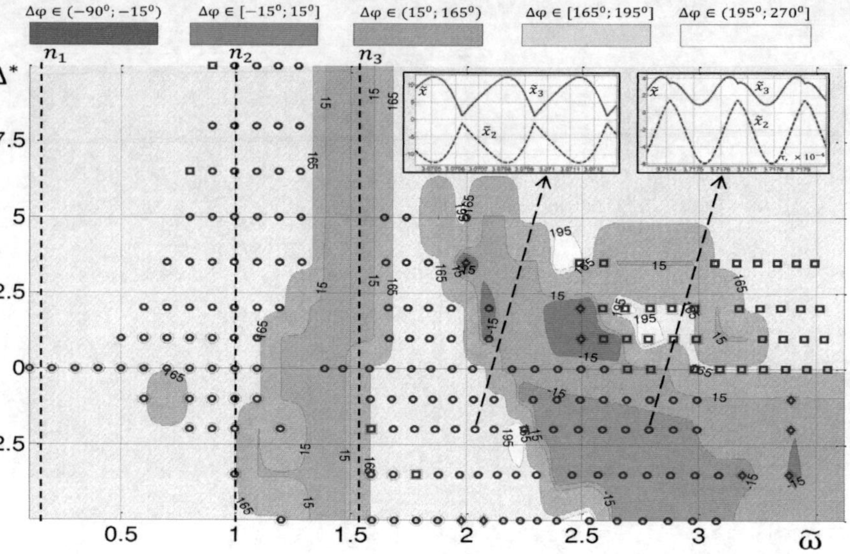

(a)

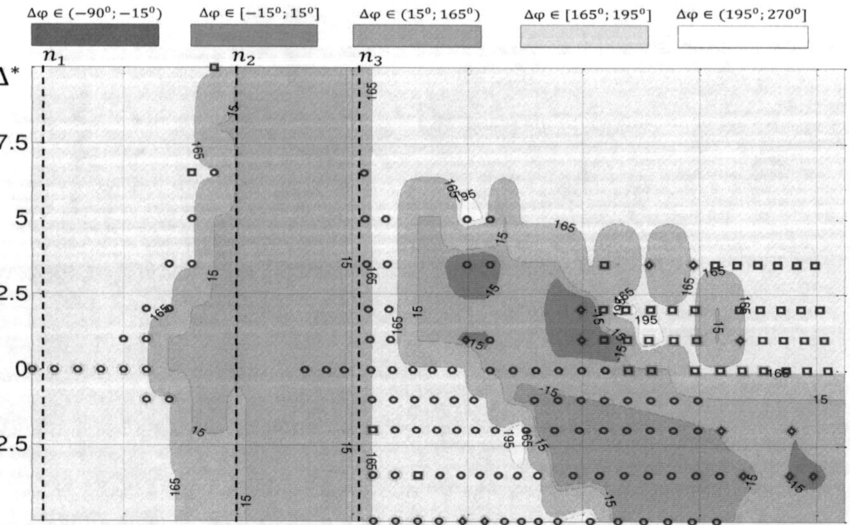

(b)

Figure 3. Motion modes for R=0.3.

The presented results show that in the frequency range $n_1 < \tilde{\omega} < 1$ (pre-resonant frequency range), T-periodic vibroimpact modes are usually excited, characterized by a synchronous-antiphase mode of the exciters rotation and working bodies oscillations, low impact rates, as well as weak sensitivity to changes of R. The vibroimpact mode characteristic, determined by the number of collisions per period and their impact rates, significantly depends on the value of the initial gap Δ^*. A decrease in Δ^* leads to excitation of multi-impact modes. The frequency range of vibroimpact modes excitation practically coincides with the frequency range of synchronous-antiphase rotation of vibration exciters, the boundaries of which differ significantly depending on the method of reaching a given excitation mode - with an increase or decrease in the excitation frequency.

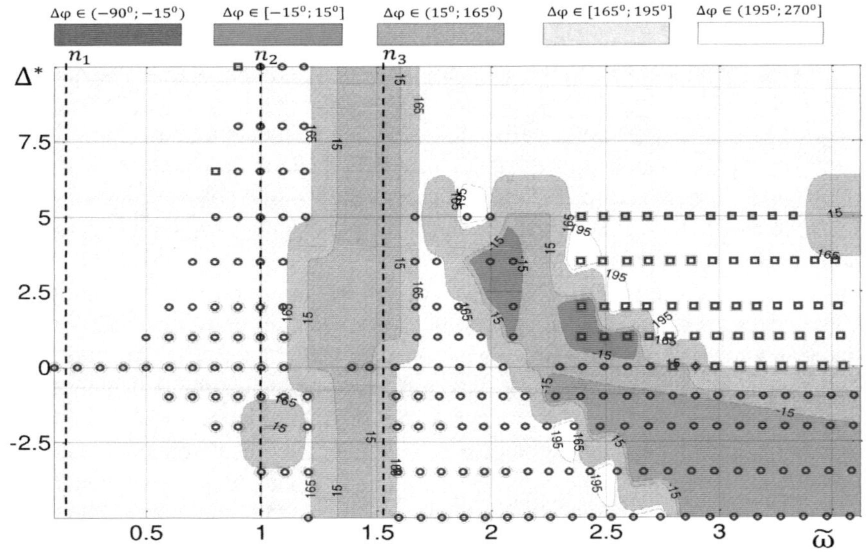

(a)

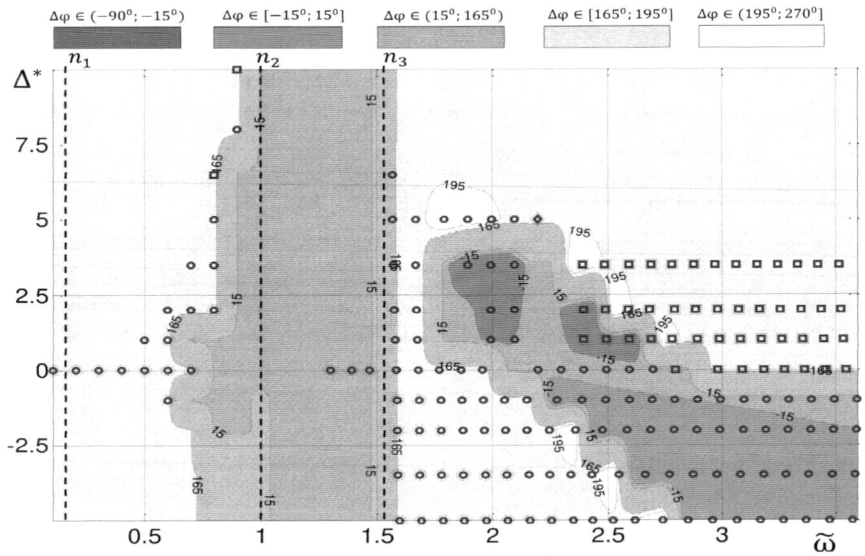

(b)

Figure 4. Motion modes for R=0.1.

In the frequency range $1 < \tilde{\omega} < n_3$ (inter-resonant frequency range), it is possible to excite T-periodic vibroimpact modes, characterized by a synchronous-antiphase rotation of vibration exciters and oscillations of working bodies, as well as relatively high impact rates (see figures 2-4 (a)). The implementation of these modes turns out to be possible due to the change in the resonance characteristics of the system because of the impacts occurrence. However, since in the absence of impacts in this frequency range a synchronous-in-phase mode of vibration exciters rotation is realized (see figures 2-4 (b)), to excite these vibro-shock modes, certain conditions must be created, for example, due to initial excitation vibroimpact oscillations in the adjacent frequency ranges, followed by adjustment of the excitation frequency to the specified range. With a decrease in the values of both Δ^* and R, the frequency range of excitation of stable vibroimpact modes narrows, due to the shift to the left of its upper boundary.

In the frequency range $\tilde{\omega} > n_3$ (above-resonant frequency range), near its lower boundary, there is a frequency range in which T-periodic vibroimpact modes are excited, characterized by a synchronous-antiphase mode of vibration exciters rotation and oscillations of the working bodies, as well as high impact rates. A change in the parameters Δ^* and R leads to a change in the width of this frequency range due to a shift in its upper boundary. As Δ^* decreases, this frequency range expands. At positive values of Δ^*, a decrease in R leads to a narrowing of this frequency range, and at negative values of Δ^* - to expansion. Regardless of the sign of Δ^* at the same excitation frequency, a decrease in R leads to a decrease in the impact rates.

In addition, at $\Delta^* \geq 0$ and far from n_3, a frequency range in which it is possible to excite stable vibroimpact $2T$-periodic modes, characterized by a synchronous-antiphase mode of vibration exciters rotation and working bodies oscillations, as well as high impact rates, is observed. Note that for excitation of these modes, it is necessary that the amplitudes of the working bodies relative oscillations, which in the resonant frequency range in the absence of impacts decrease with an increase in the excitation frequency, exceed the value of Δ^*. As R decreases, this frequency range expands due to the leftward shift of its lower boundary. In this case, at the same excitation frequency, the impact rates practically do not change. In the remaining frequency range of the above-resonant range, predominantly chaotic vibroimpact modes are observed.

For $\Delta^* < 0$ at a distance from n_3, there is a frequency range in which T-periodic vibroimpact modes are excited, characterized by a synchronous mode of vibration exciters rotation with stable values of $\Delta\varphi$ in the range $-90° < \Delta\varphi < 0°$, tending to 0° as the excitation frequency increases. In these modes, the oscillations of the working bodies occur almost in-phase, but with different amplitudes. As a result, their collisions occur at low impact rates. With a decrease in R, the frequency range of stable excitation of these modes expands due to the shift to the right of its upper boundary.

4. Conclusion

As a result of the model dynamics analysis, the influence of the parameters Δ^* and R, characterizing the conditions of the working bodies' interaction with the processed material, on the frequency range of self-synchronization of vibration exciters and periodic vibroimpact modes of the system motion is established. In the pre-resonant and inter-resonant frequency ranges, periodic vibroimpact modes with period equal to the period of excitation (T-periodic modes) are predominantly excited. An increase in the excitation frequency leads to an increase in the impact rates of the working bodies in these modes, while a decrease in both Δ^* and R leads to a narrowing of the frequency range of their excitation.

In the above-resonant frequency range at $\Delta^* \geq 0$, there are two frequency ranges with T-periodic vibroimpact modes, near the third natural frequency, and $2T$-periodic vibroimpact modes, at a distance from it, with synchronous antiphase rotation of vibration exciters and oscillations of working bodies and high impact rates. With a decrease in R, the frequency range of excitation of T-periodic modes narrows, and of $2T$-periodic modes - expands. At the same time, the impact rate of the working bodies' collision in T-periodic modes decreases, and in $2T$-periodic modes it practically does not change. When $\Delta^* < 0$, near the third natural frequency, there is a frequency range with the required T-periodic vibroimpact modes, which expands with a decrease in R. In the remaining part of this frequency range,

T-periodic vibroimpact modes arise when the vibration exciters rotate close to synchronous-in-phase mode. In this case, the oscillations of the working bodies occur almost in phase, but with different amplitudes, as a result of which their collisions occur at low impact rates. The upper boundary of the frequency range of stable excitation of these modes shifts significantly to the right with decreasing *R*.

It is shown that at the same excitation frequency, a change in the parameters of the machine's working bodies interaction with the material can lead to a change in both the periodicity of the vibroimpact mode and the form of working bodies oscillations from synchronous-antiphase to practically synchronous-inphase.

Acknowledgments
The reported study was funded by RFBR according to the research project №18-08-01491_a.

References

[1] Vaisberg L A, Zarogatsky L P and Turkin V Y 2004 *Vibrating crushers. Basics of calculation, design and technological application* (St.Peterburg: Izdatelstvo VSEGEI)
[2] Nagaev R F 2003 *Dynamics of synchronizing systems* (Berlin, Heidelberg: Spriger)
[3] Lavendelis E E 1981 *Vibration in the Technique, Handbook, Vibration Processes and Machines* vol 4 (Moscow: Mashinostroenie)
[4] Blekhman I I 1971 *Synchronization of dynamical systems* (Moscow: Nauka)
[5] Li Y, Li H, Wei X and Wen B 2014 Self-synchronization theory of a nonlinear vibration system driven by two exciters. Part 1: Theoretical analysis *J. of Vibroengineering* **16**(2) pp 725-34
[6] Zhang X, Wen B and Zhao C 2016 Theoretical study on synchronization of two exciters in a nonlinear vibrating system with multiple resonant types *Nonlinear Dynamics* **85**(1) pp 141–54
[7] Zhang N 2020 Self-synchronization characteristics of a class of nonlinear vibration system with asymmetrical hysteresis *J. of Low Freq. Noise, Vib. and Active Contr.* **39**(1) pp 114-28
[8] Bikhovskiy I I 1968 *Fundamentals of the theory of vibration technology* (Moscow: Mashinostroenie)
[9] Nagaev R and Karagulov R 2001 Dynamics of a vibration machine with allowance for the influence of treated material *Problemy Mashinostraeniya i Nadezhnos'ti Mashin* **1** pp 48-51
[10] Arkhipov M, Vetyukov M, Nagaev R and Utimishev M 2006 Dynamics of a vibrational crusher, taking account of the material being crushed *J. of Machinery Manufacture and Reliability* **1** pp 15-8
[11] Shishkin E and Safronov A 2016 Vibratory jaw crusher dynamics with consideration of working load effect *Obogashchenie Rud* **6** pp 39-43
[12] Panovko G, Shokhin A and Lyan I 2019 On a model of interaction of a vibratory jaw crusher's working bodies with the processed medium *Vibroengineering PROCEDIA* **25** pp 83-8
[13] Shokhin A E 2020 Self-synchronization of a vibrating jaw crusher with allowance for interaction with the medium processed *J.of Machinery Manufacture and Reliability* **49**(6) pp 500–10
[14] Panovko Y G 1977 *Introduction into the theory of mechanical shock* (Moscow: Nauka)

Nonlinear oscillation modes of double pendulum

A S Smirnov[1,2] **and B A Smolnikov**[1,2]

[1] Peter the Great St. Petersburg Polytechnic University, Saint Petersburg, Russia
[2] Institute for Problems in Mechanical Engineering of the Russian Academy of Sciences, Saint Petersburg, Russia

smirnov.alexey.1994@gmail.com

Abstract. The article investigates nonlinear oscillations of the double mathematical pendulum having identical parameters of its links and end loads. Approximate solutions in the nonlinear zone are constructed and analyzed on the basis of exact solution in the linear zone. These solutions clearly illustrate the drift of frequencies and modes of double pendulum with increasing oscillation amplitudes. The obtained analytical results are very useful for the synthesis of controlled movement modes of wide range of manipulators and various robotic structures, and they are also necessary for constructing optimal locomotion modes of walking machines and mechanisms. These solutions are also interesting as illustrative examples of nonlinear mechanics in the teaching and engineering practice.

1. Introduction

The study of movements of various pendulum structures has already been one of the main directions of research in analytical mechanics for a long time [1]. They are not only of fundamental theoretical importance, but also a large number of practical applications. Multi-link pendulums are of particular interest in this respect. They easily allow to increase the number of degrees of freedom, introducing and changing the inertial and dissipative parameters of individual links and also, if necessary, adding control actions in the articulated joints [2]. The double flat mathematical pendulum is the simplest of these pendulums [3]. This pendulum has found increasing application in recent decades in robotics as a manipulator and pedipulator, where it must perform working movements with both small and very large amplitudes [4].

The question arises, how such two-link link robot passes from the zone of small (i.e., linear) oscillations to the zone of weak nonlinearity. At the same time, regular modes with periodicity are the most valuable of all possible modes of motion of nonlinear systems, because they are advisable to use in practice. The study of such modes is closely related to the well-known problem of finding nonlinear vibration modes, which has recently attracted a large number of researchers [5]. We emphasize that double pendulum is a system with two degrees of freedom, for which the construction of analytical solutions in nonlinear zone turns out to be much more difficult than for systems with one degree of freedom. Therefore, nonlinear oscillations of the double pendulum are investigated only using numerical methods in most works on its dynamics. Nevertheless, the study of their qualitative nature on the basis of analytical dependences is also important in describing the regular movements of the system. These expressions would clearly demonstrate the drift of oscillations frequencies and modes with a gradual increasing the oscillation amplitudes of the links.

In this regard, the main attention in this article is focused precisely on the analytical study of nonlinear oscillations of double pendulum using specially developed asymptotic methods. The main purpose of this article is obtaining and interpreting formulas that make it possible to analyze in detail the nonlinear modes of oscillations of a double pendulum and establish their main qualitative features, and to accompany solutions with visual graphic illustrations as well.

2. Design scheme and selection of generalized coordinates

We consider the flat double mathematical pendulum with the same links of length l and identical end loads of mass m (figure 1). Such a system can be interpreted as the simplest nonlinear mechanical system with two degrees of freedom.

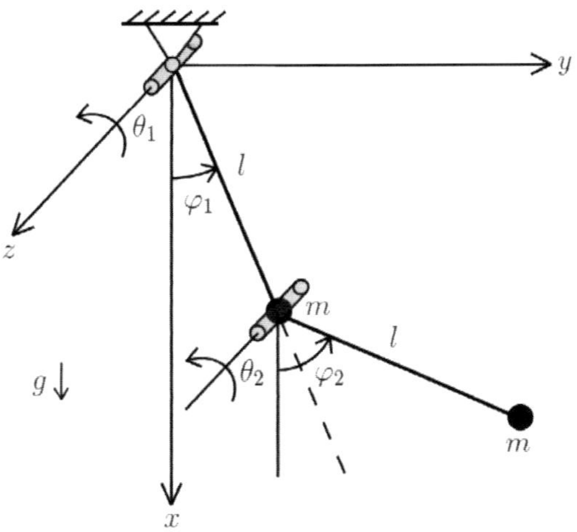

Figure 1. Double mathematical pendulum.

It is most convenient to choose the absolute angles φ_1 and φ_2 of deviations of the pendulum links from the vertical as generalized coordinates. We note that sometimes relative angles θ_1 and θ_2 of rotation in the joints of the pendulum are taken as generalized coordinates [6]. They are related to absolute angles by obvious relationships: $\theta_1 = \varphi_1$ and $\theta_2 = \varphi_2 - \varphi_1$. The angles θ_1 and θ_2 are often used in tasks of control motions because they are measurable. It should be emphasized that the choice of one or another generalized coordinates is conditioned by the specific research objectives. We will use absolute angles in this work, since in this case the equations of motion have the simplest form. Nevertheless, we will also demonstrate the behavior of the angles of rotation in the joints.

3. Derivation of nonlinear equations of motion

Proceeding to the derivation of motion equations of the double mathematical pendulum, we write its kinetic and potential energies [7]:

$$T = \frac{1}{2}ml^2\left[2\dot{\varphi}_1^2 + \dot{\varphi}_2^2 + 2\cos(\varphi_2 - \varphi_1)\dot{\varphi}_1\dot{\varphi}_2\right] = \frac{1}{2}\dot{\boldsymbol{\varphi}}^{\mathrm{T}}\mathbf{A}(\boldsymbol{\varphi})\dot{\boldsymbol{\varphi}}, \quad \Pi = mgl(3 - 2\cos\varphi_1 - \cos\varphi_2), \qquad (1)$$

where $\boldsymbol{\varphi} = [\varphi_1, \varphi_2]^{\mathrm{T}}$ is the column of generalized coordinates and matrix $\mathbf{A}(\boldsymbol{\varphi})$ of inertial coefficients has the form

$$\mathbf{A}(\boldsymbol{\varphi}) = ml^2\begin{bmatrix} 2 & \cos(\varphi_2 - \varphi_1) \\ \cos(\varphi_2 - \varphi_1) & 1 \end{bmatrix}. \qquad (2)$$

Substituting expressions (1) into Lagrange equations of the second kind [8], we obtain the nonlinear equation of system's motion in the well-known matrix form:

$$\mathbf{A}(\boldsymbol{\varphi})\ddot{\boldsymbol{\varphi}} + \mathbf{B}(\boldsymbol{\varphi},\dot{\boldsymbol{\varphi}}) + \mathbf{C}(\boldsymbol{\varphi}) = 0, \tag{3}$$

where columns $\mathbf{B}(\boldsymbol{\varphi},\dot{\boldsymbol{\varphi}})$ and $\mathbf{C}(\boldsymbol{\varphi})$ are determined by expressions

$$\mathbf{B}(\boldsymbol{\varphi},\dot{\boldsymbol{\varphi}}) = ml^2\sin(\varphi_2-\varphi_1)\begin{bmatrix}-\dot{\varphi}_2^2\\ \dot{\varphi}_1^2\end{bmatrix}, \quad \mathbf{C}(\boldsymbol{\varphi}) = mgl\begin{bmatrix}2\sin\varphi_1\\ \sin\varphi_2\end{bmatrix}. \tag{4}$$

The considered system is conservative, therefore there is the energy integral:

$$E = T + \Pi = \frac{1}{2}ml^2\left[2\dot{\varphi}_1^2 + \dot{\varphi}_2^2 + 2\cos(\varphi_2-\varphi_1)\dot{\varphi}_1\dot{\varphi}_2\right] + mgl(3-2\cos\varphi_1-\cos\varphi_2) = \text{const.} \tag{5}$$

Linearizing matrix equation (3), we obtain the equation that describes small oscillations:

$$\mathbf{A}_0\ddot{\boldsymbol{\varphi}} + \mathbf{C}_0\boldsymbol{\varphi} = 0, \tag{6}$$

where the constant matrices of inertial $\mathbf{A}_0$ and quasi-elastic $\mathbf{C}_0$ coefficients are [9]

$$\mathbf{A}_0 = ml^2\begin{bmatrix}2 & 1\\ 1 & 1\end{bmatrix}, \quad \mathbf{C}_0 = mgl\begin{bmatrix}2 & 0\\ 0 & 1\end{bmatrix}. \tag{7}$$

It is well-known, see [10], that frequencies and modes of linear model are defined by expressions

$$k_{s0} = \left(2\pm\sqrt{2}\right)^{1/2}\cdot(g/l)^{1/2} \quad \boldsymbol{\Phi}_{(s)} = \begin{bmatrix}1\\ \mp\sqrt{2}\end{bmatrix}, \quad s = 1,2. \tag{8}$$

Hereinafter, the lower sign corresponds to the first mode ($s=1$), and the upper sign corresponds to the second mode ($s=2$). We also need expressions for the norms of oscillations modes for subsequent constructions:

$$N_s = \boldsymbol{\Phi}_{(s)}^{\mathrm{T}}\mathbf{A}_0\boldsymbol{\Phi}_{(s)} = 2(2\mp\sqrt{2})ml^2. \tag{9}$$

4. Construction of asymptotic solutions

We now turn to the problem of finding nonlinear oscillation modes. For this we will use asymptotic methods of nonlinear mechanics [11]. Let's rewrite (3), so that on the left side there will be only linear terms corresponding to small oscillations (6), and all other terms will be on the right side. So we get:

$$\mathbf{A}_0\ddot{\boldsymbol{\varphi}} + \mathbf{C}_0\boldsymbol{\varphi} = \tilde{\mathbf{A}}(\boldsymbol{\varphi})\ddot{\boldsymbol{\varphi}} + \tilde{\mathbf{B}}(\boldsymbol{\varphi},\dot{\boldsymbol{\varphi}}) + \tilde{\mathbf{C}}(\boldsymbol{\varphi}) = \mathbf{Q}(\boldsymbol{\varphi},\dot{\boldsymbol{\varphi}},\ddot{\boldsymbol{\varphi}}), \tag{10}$$

where the matrices $\mathbf{A}_0$ and $\mathbf{C}_0$ are determined by expressions (7) and also new designations are made:

$$\tilde{\mathbf{A}}(\boldsymbol{\varphi}) = \mathbf{A}_0 - \mathbf{A}(\boldsymbol{\varphi}), \quad \tilde{\mathbf{B}}(\boldsymbol{\varphi},\dot{\boldsymbol{\varphi}}) = -\mathbf{B}(\boldsymbol{\varphi},\dot{\boldsymbol{\varphi}}), \quad \tilde{\mathbf{C}}(\boldsymbol{\varphi}) = \mathbf{C}_0\boldsymbol{\varphi} - \mathbf{C}(\boldsymbol{\varphi}). \tag{11}$$

Then column $\mathbf{Q}$, that is interpreted as column of disturbing forces acting in a linear system, takes the form:

$$\mathbf{Q} = ml^2\sin(\varphi_2-\varphi_1)\begin{bmatrix}\dot{\varphi}_2^2\\ -\dot{\varphi}_1^2\end{bmatrix} + mgl\begin{bmatrix}2(\varphi_1-\sin\varphi_1)\\ \varphi_2-\sin\varphi_2\end{bmatrix} + ml^2\left(1-\cos(\varphi_2-\varphi_1)\right)\begin{bmatrix}\ddot{\varphi}_2\\ \ddot{\varphi}_1\end{bmatrix}. \tag{12}$$

We will be interested in the construction and detailed analysis of the formulas for the first approximation, i.e., for a weakly nonlinear model. To do this, we expand the trigonometric functions in (12) in Taylor series so as to keep in the expression for $\mathbf{Q}$ the terms no higher than the third order of smallness in generalized coordinates, velocities and accelerations:

$$\mathbf{Q} = ml^2(\varphi_2 - \varphi_1)\begin{bmatrix} \dot{\varphi}_2^2 \\ -\dot{\varphi}_1^2 \end{bmatrix} + \frac{1}{6}mgl\begin{bmatrix} 2\varphi_1^3 \\ \varphi_2^3 \end{bmatrix} + \frac{1}{2}ml^2(\varphi_2 - \varphi_1)^2\begin{bmatrix} \ddot{\varphi}_2 \\ \ddot{\varphi}_1 \end{bmatrix}. \tag{13}$$

We will look for an approximate solution in the form of oscillations of constant amplitude according to one of the modes $\mathbf{\Phi}_{(s)}$ of the linear system, but with some correction to the expression for the frequency:

$$\boldsymbol{\varphi} = \mathbf{\Phi}_{(s)} a \cos\psi, \quad k_s = \dot{\psi} = k_{s0}(1 + \rho_s a^2), \quad a = \text{const}. \tag{14}$$

The correction factor ρ_s follows from the harmonic balance equation for oscillation mode $\mathbf{\Phi}_{(s)}$:

$$\int_0^{2\pi} \mathbf{\Phi}_{(s)}^{\mathrm{T}} \left(\mathbf{A}_0\ddot{\boldsymbol{\varphi}} + \mathbf{C}_0\boldsymbol{\varphi} - \mathbf{Q}\right)\cos\psi d\psi = 0. \tag{15}$$

Calculating the columns of generalized velocities $\dot{\boldsymbol{\varphi}}$ and accelerations $\ddot{\boldsymbol{\varphi}}$ accurate to third-order values

$$\dot{\boldsymbol{\varphi}} = -\mathbf{\Phi}_{(s)} a \sin\psi k_{s0}(1 + \rho_s a^2), \quad \ddot{\boldsymbol{\varphi}} = -\mathbf{\Phi}_{(s)} a \cos\psi k_{s0}^2(1 + 2\rho_s a^2), \tag{16}$$

we define the column $\mathbf{Q}$ with the same accuracy:

$$\mathbf{Q} = (\mathbf{q}_{1s}\cos\psi + \mathbf{q}_{3s}\cos 3\psi)mgla^3, \quad \mathbf{q}_{1s} = \begin{bmatrix} \dfrac{14 \pm 9\sqrt{2}}{4} \\ \dfrac{-22 \mp 17\sqrt{2}}{8} \end{bmatrix}, \quad \mathbf{q}_{3s} = \begin{bmatrix} \dfrac{46 \pm 33\sqrt{2}}{12} \\ \dfrac{-54 \mp 41\sqrt{2}}{24} \end{bmatrix}. \tag{17}$$

Now we substitute expressions (16) and (17) into the harmonic balance equation (15) and take into account that oscillation modes satisfy equation $(\mathbf{C}_0 - k_{s0}^2\mathbf{A}_0)\mathbf{\Phi}_{(s)} = 0$, and also use the notation (9) for the norms of these modes. Then we arrive at the following expressions for the correction factor ρ_s and the frequency of nonlinear oscillations k_s after transformations:

$$\rho_s = -\frac{mgl}{2N_s k_{s0}^2}\mathbf{\Phi}_{(s)}^{\mathrm{T}}\mathbf{q}_{1s} = -\frac{31 \pm 20\sqrt{2}}{32}, \quad k_s(a) = k_{s0}\left(1 - \frac{31 \pm 20\sqrt{2}}{32}a^2\right). \tag{18}$$

It is easy to see that for the first frequency the correction factor is (–0.0849), while for the second one it is (–1.8526), i.e., the second frequency changes more significantly with increasing of amplitude.

However, the most interesting is not the frequency, but the nonlinear oscillations mode. We obtain a solution up to the third order of smallness to describe it. It is known from the theory of oscillations that if the harmonic force $\mathbf{Q}_0\cos\omega t$ acts in a linear system with two degrees of freedom, i.e.

$$\mathbf{A}_0\ddot{\boldsymbol{\varphi}} + \mathbf{C}_0\boldsymbol{\varphi} = \mathbf{Q}_0\cos\omega t, \tag{19}$$

then it excites the forced oscillations in the system:

$$\boldsymbol{\varphi}_Q = \left[\mathbf{\Phi}_{(1)}\frac{\mathbf{\Phi}_{(1)}^{\mathrm{T}}\mathbf{Q}_0}{N_1(k_{10}^2 - \omega^2)} + \mathbf{\Phi}_{(2)}\frac{\mathbf{\Phi}_{(2)}^{\mathrm{T}}\mathbf{Q}_0}{N_2(k_{20}^2 - \omega^2)}\right]\cos\omega t. \tag{20}$$

It is easy to understand based on this expression what forced oscillations each of the harmonic components of expression (17) excites. When we construct the solution on the first mode, we should take into account only the second term for the first harmonic in expression (20), since this harmonic is already balanced on the first mode. On the contrary, only the first term should be taken into account when the solution is constructed on the second mode. Summing up the solutions corresponding to each harmonic of expression (17), we obtain the so-called "regularized oscillation" excited in the unperturbed system by generalized forces $\mathbf{Q}(\mathbf{\Phi}_{(s)} a\cos\psi,...)$

$$\boldsymbol{\varphi}_Q(a,\psi) = \mathbf{\Phi}_{(p)} \frac{\mathbf{\Phi}_{(p)}^{\mathrm{T}} \mathbf{q}_{1s} mgla^3}{N_p(k_{p0}^2 - k_{s0}^2)} \cos\psi + \left[\mathbf{\Phi}_{(1)} \frac{\mathbf{\Phi}_{(1)}^{\mathrm{T}} \mathbf{q}_{3s} mgla^3}{N_1(k_{10}^2 - 9k_{s0}^2)} + \mathbf{\Phi}_{(2)} \frac{\mathbf{\Phi}_{(2)}^{\mathrm{T}} \mathbf{q}_{3s} mgla^3}{N_2(k_{20}^2 - 9k_{s0}^2)} \right] \cos 3\psi, \tag{21}$$

where the additional index p is introduced, and $p=1$ at $s=2$ and vice versa $p=2$ at $s=1$. Adding (21) to the first expression (14), we obtain a solution up to the third order of smallness in the form

$$\boldsymbol{\varphi} = \mathbf{\Phi}_{(s)} a\cos\psi + \boldsymbol{\varphi}_Q(a,\psi) = (\mathbf{\Phi}_{(s)} a + \boldsymbol{\alpha}_{1s} a^3)\cos\psi + \boldsymbol{\alpha}_{3s} a^3 \cos 3\psi, \tag{22}$$

where columns $\boldsymbol{\alpha}_{1s}$ and $\boldsymbol{\alpha}_{3s}$ have the form

$$\boldsymbol{\alpha}_{1s} = \begin{bmatrix} \dfrac{\pm\sqrt{2}+1}{32} \\ \dfrac{2\pm\sqrt{2}}{32} \end{bmatrix}, \quad \boldsymbol{\alpha}_{3s} = \begin{bmatrix} \dfrac{\mp 552\sqrt{2} - 437}{2688} \\ \dfrac{\pm 781\sqrt{2} + 576}{2688} \end{bmatrix}. \tag{23}$$

We also can write the expression of the column of generalized velocities:

$$\dot{\boldsymbol{\varphi}} = -\left[(\mathbf{\Phi}_{(s)} a + \boldsymbol{\beta}_{1s} a^3)\sin\psi + \boldsymbol{\beta}_{3s} a^3 \sin 3\psi \right] k_{s0}, \tag{24}$$

where columns $\boldsymbol{\beta}_{1s}$ and $\boldsymbol{\beta}_{3s}$ have the form

$$\boldsymbol{\beta}_{1s} = \begin{bmatrix} \dfrac{\mp 19\sqrt{2} - 30}{32} \\ \dfrac{21\pm 16\sqrt{2}}{16} \end{bmatrix}, \quad \boldsymbol{\beta}_{3s} = \begin{bmatrix} \dfrac{\mp 552\sqrt{2} - 437}{896} \\ \dfrac{\pm 781\sqrt{2} + 576}{896} \end{bmatrix}. \tag{25}$$

It is possible to calculate the total mechanical energy of system (5) accurate to terms of the fourth smallness order as one of the checks of the obtained results and make sure that it is a constant for a given value of a, i.e., it doesn't depend on ψ. Expanding the trigonometric functions included in (5) into series, one can obtain

$$E = \frac{1}{2} ml^2 \left[2\dot{\varphi}_1^2 + \dot{\varphi}_2^2 + 2\dot{\varphi}_1\dot{\varphi}_2 \left(1 - \frac{(\varphi_2 - \varphi_1)^2}{2} \right) \right] + \frac{1}{2} mgl \left(2\varphi_1^2 + \varphi_2^2 - \frac{\varphi_1^4}{6} - \frac{\varphi_2^4}{12} \right). \tag{26}$$

Substituting solutions (22) and (24) into this formula, we find the expression for the dimensionless energy level with the required accuracy:

$$\gamma = \frac{E}{mgl} = 2a^2 + \frac{\mp 20\sqrt{2} - 37}{32} a^4. \tag{27}$$

It can be seen that there is no dependence on ψ in this formula which confirms the validity of the performed calculations.

5. Results and discussion

Now we turn to the interpretation of obtained results. We see, turning to the constructed solution (22) for oscillations modes in a weakly nonlinear model, that the motion on each mode is periodic, but no longer harmonic, and represents the sum of two harmonics. It is important to emphasize that if it is necessary to excite oscillations on one mode $\mathbf{\Phi}_{(s)}$ in linear model then the columns $\boldsymbol{\varphi}_0$ and $\dot{\boldsymbol{\varphi}}_0$ must be proportional to this mode, while in a nonlinear system the situation is much more complicated. Indeed, the ratio of the oscillation amplitudes of the deflection angles or angular velocities does not completely determine the nonlinear oscillations mode. Moreover, these ratios may be completely meaningless for sufficiently large oscillations, since generalized coordinates or velocities may have more than two extremes on the oscillations period.

We restrict ourselves to a detailed discussion of the first oscillation mode, i.e., the basic tone which is of main practical importance. Let's turn to the construction of the system's phase portrait in order to illustrate visually the process of the oscillation mode drift during the transition from the linear zone to the weakly nonlinear zone. It is clear that the phase space in our problem is four-dimensional. Therefore, it is convenient to construct the phase portrait for each generalized coordinate separately, that is on the planes $(\varphi_1, \dot{\varphi}_1 / k_{10})$ and $(\varphi_2, \dot{\varphi}_2 / k_{10})$, where the division of the generalized velocities by the value k_{10} is made from dimensional considerations. We will construct a phase portrait using the analytical dependences (22) and (24) and compare it with the results that are obtained using numerical integration. We note that the same dimensionless energy level γ must correspond to the approximate and exact phase trajectories for their adequate comparison. Let us assume for simplicity that $\dot{\boldsymbol{\varphi}}_0 = 0$, i.e., the movement begins without an initial velocity. Then we have an exact expression for γ according to (5):

$$\gamma = mgl(3 - 2\cos\varphi_{10} - \cos\varphi_{20}). \tag{28}$$

It is necessary to select the values φ_{10} and φ_{20} fixing a certain value γ and taking into account the expression (28), so that the movement of the system is periodic. Besides, the oscillation frequency of this movement should be close to the first frequency of the linear system. We use numerical methods for integrating the nonlinear matrix equation (3) to do this procedure.

Figures 2 and 3 demonstrate the phase portraits on the planes $(\varphi_1, \dot{\varphi}_1 / k_{10})$ and $(\varphi_2, \dot{\varphi}_2 / k_{10})$, where several phase trajectories corresponding to different values γ are shown. The solid lines show here the phase trajectories constructed using numerical integration and the circles indicate the values obtained using formulas (22) and (24). There is a good agreement between the numerical and analytical results, although some differences appear with increase of the energy level. In addition, it can be seen that for small values γ phase trajectories are concentric circles which corresponds to motion on the linear oscillations mode. These circles gradually begin to distort with increasing γ and turn into other phase trajectories that are more complex in character.

It is also of interest to construct a phase portrait for the interlink angle $\theta_2 = \varphi_2 - \varphi_1$, i.e., on the plane $(\theta_2, \dot{\theta}_2 / k_{10})$. It is shown in figure 4 and has a very unusual appearance. It is clearly seen here that the circles corresponding to analytical results taken at regular intervals are located on the phase trajectories more and more unevenly with increasing energy level.

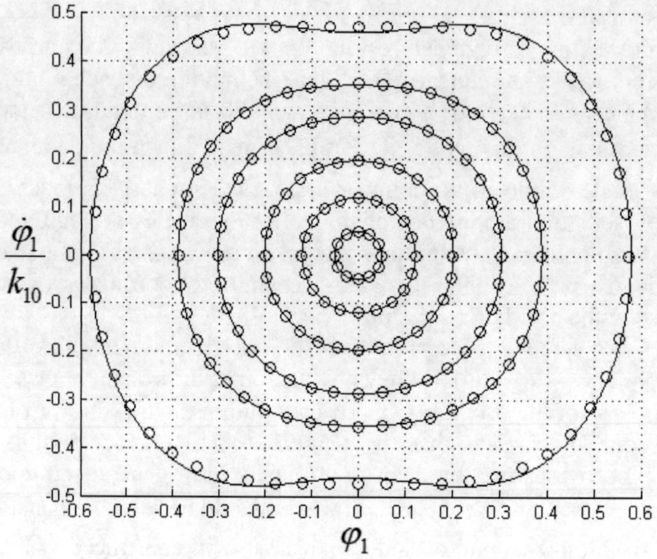

Figure 2. Phase portrait for coordinate φ_1.

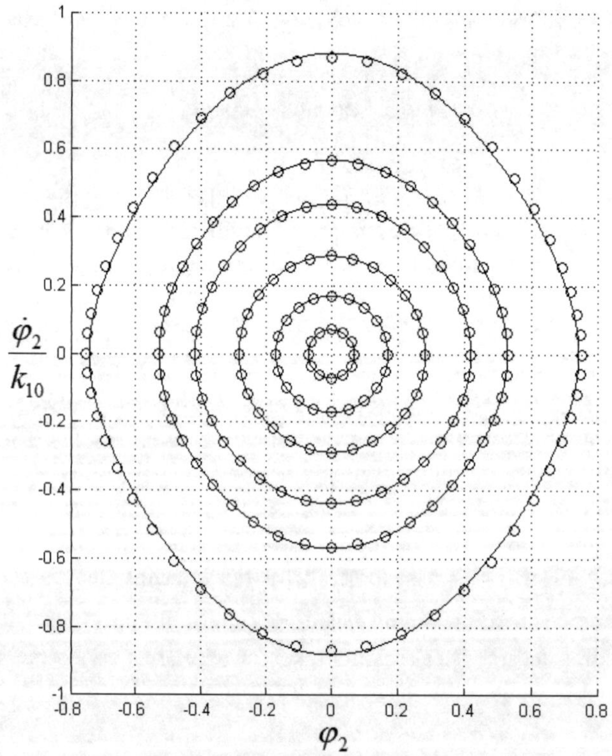

Figure 3. Phase portrait for coordinate φ_2.

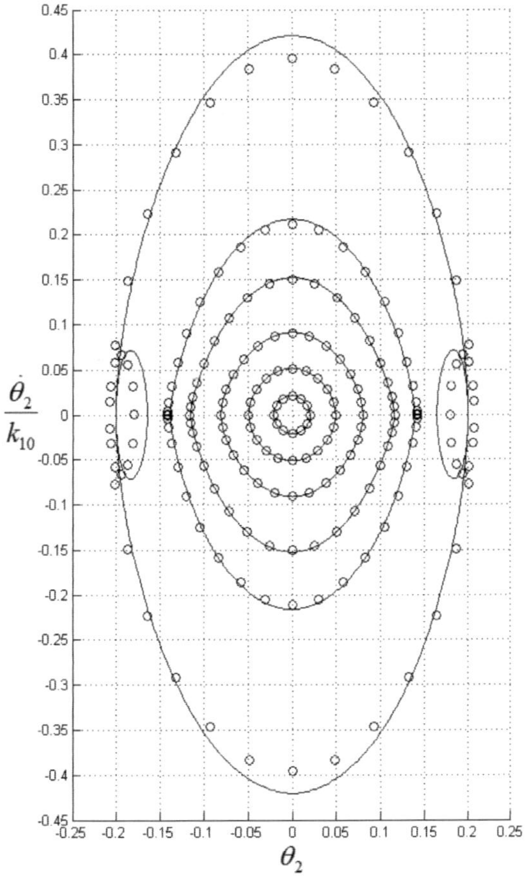

Figure 4. Phase portrait for coordinate θ_2.

We emphasize here, that it is possible to obtain analytical solutions in the second approximation using the harmonic balance equation (15). Of course, they will quantitatively give more accurate results. However, such solutions will turn out to be rather cumbersome, while solutions (22) and (24) are compact and give an adequate qualitative picture of the double pendulum movements, clearly reflecting the influence of nonlinear factors.

Figure 5 shows the graphs of the dependence of the absolute angles φ_1 and φ_2, and also the interlink angle θ_2 on time t at one oscillation period. We note that these graphs correspond to the phase trajectory with the highest energy level of those plotted in figures 2-4. The numerical dependences are shown here as before by the solid line and the analytical values are represented by the circles. These graphs clearly demonstrate the non-trivial character of the double pendulum movements and there is also a good agreement between the numerical and analytical results. Similar graphs for angular velocities are shown in figure 6. It should be noted that the angle θ_2, as well as the angular velocities $\dot{\varphi}_1$ and $\dot{\theta}_2$, for the given case differ significantly from the sinusoidal functions and have six extremes on the oscillation period, and not two as in the small oscillations case. At the same time, the angles φ_1 and φ_2, as well as the angular velocity $\dot{\varphi}_2$, have two extremes on the oscillation period as in the small oscillations case, but their form also differs significantly from sinusoidal functions.

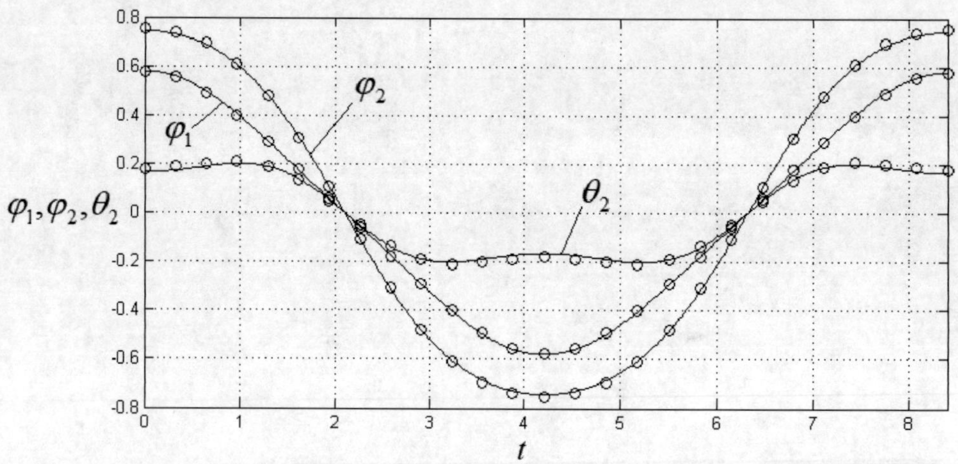

Figure 5. Graphs of the dependence of angles φ_1, φ_2 and θ_2 on time t.

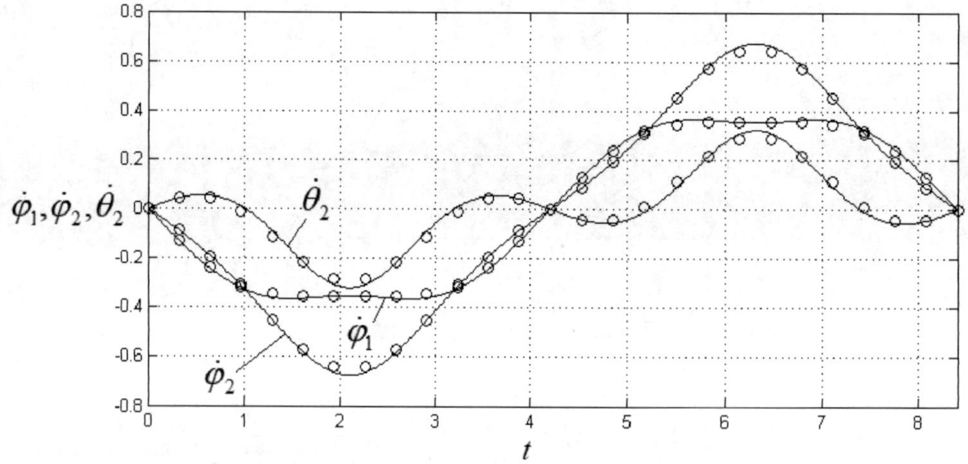

Figure 6. Graphs of the dependence of angle velocities $\dot{\varphi}_1$, $\dot{\varphi}_2$ and $\dot{\theta}_2$ on time t.

6. Conclusion

The approximating analytical solutions for the oscillations frequencies and modes of double pendulum, which were constructed in this work, allow us to consider a smooth transition from its linear model to a nonlinear one and vice versa from the unified standpoint. These results are very necessary in the development and design of various devices of modern robotics, primarily androids and other mobile robots, and exactly this area is currently attracting great attention of specialists.

The main value of the obtained results lies in the fact that the role of nonlinear terms in the general picture of behavior sharply increases with the transition from small oscillations of such manipulators to the large ones. Therefore, it is necessary to know the qualitative character of their movements in order to synthesize the adequate control. The undoubted advantage of the above expressions is expressed in their simplicity and adequate correspondence to the results of numerical integration. This also refers to their visual interpretation in the form of phase portraits and dependence graphs of the deviation angles of the links on time.

As final conclusion, we can note that described method of analytical search of frequencies and modes of nonlinear oscillations can be used not only for a two-link device, but also for three-link and four-link manipulators simulating the movements of the components of mobile robots.

References

[1] Loytsyansky L G and Lurie A I 1934 *Theoretical mechanics. Part 3. Dynamics of a non-free system and the theory of oscillations* (Leningrad, Moscow: ONTI GTTI)
[2] Formalsky A M 2012 *Motion control of unstable objects* (Moscow: Fizmatlit)
[3] Leontev V A, Smirnov A S and Smolnikov B A 2019 Collinear control of dissipative double pendulum *Robotics and technical cybernetics* **1** pp 65-70
[4] Leontev V A, Smirnov A S and Smolnikov B A 2018 Optimal damping of two-link manipulator oscillations *Robotics and technical cybernetics* **2** pp 52-59
[5] Klimenko A A and Mikhlin Yu V 2014 Normal oscillations modes in a nonlinear system containing a pendulum vibration damper *Problems of mechanical engineering* **3** pp 38-44
[6] MacMillan V D 1936 *Dynamics of rigid bodies* (New York: Dover Publications)
[7] Gantmakher F R 1966 *Lectures on analytical mechanics* (Moscow: Nauka, GRFML)
[8] Lurie A I 1961 *Analytical mechanics* (Moscow: GIFML)
[9] Vilke V G 2003 *Theoretical mechanics* (Saint Petersburg: Lan)
[10] Butenin N V, Lunts Ya L and Merkin D R 1979 *Theoretical Mechanics Course. Part 2. Dynamics* (Moscow: Nauka, GRFML)
[11] Bogolyubov N N and Mitropol'skiy Yu A 1958 *Asymptotic methods in the theory of nonlinear oscillations* (Moscow: GIFML)

IOP Conf. Series: Materials Science and Engineering 1129 (2021) 012043 doi:10.1088/1757-899X/1129/1/012043

The study of the motion of the truck vibration motors by the oscillating of a pendulum of variable length

Sergey V Strygin, Vladimir D Levin, Nikolai V Loschinin, Ilya N Kiryushin, Sergey A Parshukov and Vladimir V Metik

Ryazan Institute (branch) of Moscow Polytechnic University, Ryazan, Russia

strsw@mail.ru

Abstract. The article provides a solution to the problem, according to the condition of which there is a body – a cart, which must be brought into directional motion by means of a swinging elastic pendulum of variable length, pivotally mounted on it. A variant of creating the initial momentum of the trolley movement due to the movement of the pendulum, made in the form of a spring-loaded load with the possibility of movement along a straight guide when it rotates relative to the pendulum suspension point, is proposed and justified. The mode of movement of the cart-pendulum system is studied when the individual components of the system are gradually released from the connections to provide the initial impulse of the cart movement. The full-turn shape of the trajectory of a spring-loaded pendulum pivotally mounted on a cart is shown. We use numerical analysis methods (numerical integration of systems of nonlinear differential equations) with modeling by means of the Mathcad package.

The development of mechanical engineering, transport, and other branches of science and technology requires the use of new effective ways to move transport vehicles.

The goals of this work are:

- identification of the dependence of the speed of the pendulum vibration motor at the same initial angle of deviation of its pendulum from the value of the spring stiffness of the pendulum, made in the form of a spring-loaded load with the ability to move along a straight guide when it rotates relative to the point of suspension of the pendulum;
- identification of the dependence of the speed of the pendulum vibration motor on the value of the initial angle of deflection of its pendulum with the same stiffness of the pendulum spring;
- identification of the dependence of the speed of the pendulum vibration motor on the mass of the vibration motor trolley with different spring stiffness of the pendulum.

To achieve these goals, you need to solve the following tasks:

1. Generalization of theoretical foundations and practical experience in creating vibration motors.
2. Research and identification of the main characteristics of the pendulum vibration motor.

In [1], four geometric forms of periodic trajectories of movement of a load attached to a spring suspended from a movable carriage are considered. In this case, the swinging spring is used as a mover when moving the carriage along the horizontal support surface. Also shown is a method for determining the trajectory of a load attached to the end of the specified spring. In the formulation of the dynamic analysis described thus transport machines libredigital determine the initial length of the spring, which provides a periodic trajectory of movement of freight and translational movement of the carriage in the horizontal direction. In this case, the mass of the carriage, the weight of the load, the spring stiffness, the initial position of the spring with the load (other than vertical), and the initial impulse applied to the carriage are taken as known parameters. However, there are other geometric forms of periodic trajectories of movement of the load attached to a spring suspended from a movable carriage, in particular, this is a full-turn rotation of the spring with the load. Some drive systems of pendulum vibration motors are reduced to such a dynamic model, for example, shown in computer modeling on the Internet site [2]. Because of the potential structural simplification of the pendulum vibration motor, its dynamic model with a special operating mode is also of interest, in which the initial impulse applied to the carriage is provided only by the initial position of the spring with the load and the next two stages of starting work [3]. 1. The Load is released from the deflected position at an angle φ_0 at a fixed length of the spring (released from the connection that restrains its rotation) no initial speed. The cart position is also fixed. 2. At the moment when the spring with the load passes the vertical position, the load and the cart are released (released from ties) and acquire the ability to move with the spring deformation along the pendulum and translational horizontal movement, respectively.

Consider the movement of the trolley suspended by a spring, the weight of the load m_1 (figure 1).

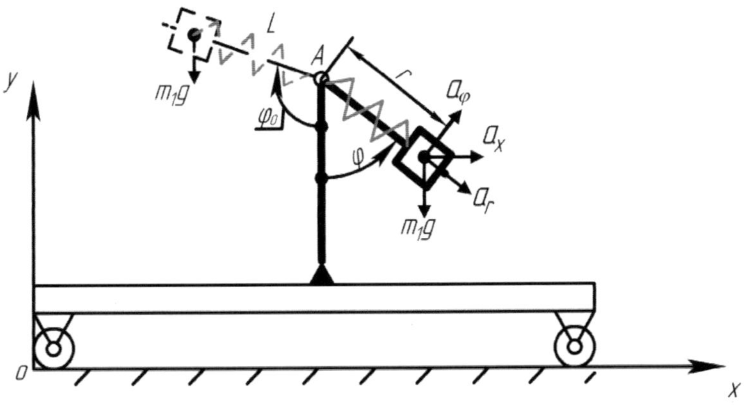

Figure 1. Diagram of a trolley with a swinging pendulum of variable length.

Figure 1 shows: φ_0 – the angle by which the spring with the load is deflected at the initial time; L – the length of the spring in the undeformed state; r – its length in the deformed state at a given time t; φ – the angle corresponding to a given time; a_r, a_φ - components of the acceleration vector in the polar coordinate system.

Since the movement of the load occurs along a flat trajectory different from the circle, the movement of the load in relative motion is convenient to describe in the polar coordinate system (r, φ).

In the polar coordinate system, the velocities along the coordinate lines $r = const$, $\varphi = const$, are determined by the formulas:

$$v_r = \dot{r}, \qquad v_\varphi = r\dot{\varphi}, \tag{1}$$

acceleration –

$$a_r = \ddot{r} - r\dot{\varphi}^2, \quad a_\varphi = \frac{1}{r}\frac{d}{dt}(r^2 \cdot \dot{\varphi}) \tag{2}$$

When the truck is moving, the load makes a plane-parallel movement: translational motion together with the truck and rotational with a variable radius r around the point A of suspension.

To derive the equations of motion, we apply the method of kinetostatics [4]. According to this method, the load will be in equilibrium if it is applied along with the specified forces (m_1g, m_2g – gravity of the load and the cart, $F_{\varphi n}$ – spring elastic force) apply inertia forces F_r, F_φ, F_{xt}, F_x

(figure 2). Their directions are opposite to the corresponding acceleration projections. We make the equilibrium equation in the projections on the coordinate lines (r, φ) for the forces acting on the load:

$$\Sigma r = -F_{\varphi n} - F_r + m_1\, g\, cos\ \varphi - F_x \cdot sin\ \varphi = 0, \tag{3}$$
$$\Sigma \varphi = -F_\varphi - m_1\, g\, sin\ \varphi - F_x \cdot cos\ \varphi = 0,$$

where according to (2)

$$F_r = m_1 \cdot (\ddot{r} - r \cdot \dot{\varphi}^2), \qquad F_\varphi = m_1 \cdot (2\dot{r}\dot{\varphi} + r \cdot \ddot{\varphi}), \tag{4}$$
$$F_{\varphi n} = c \cdot (r - L),$$

c – spring stiffness.

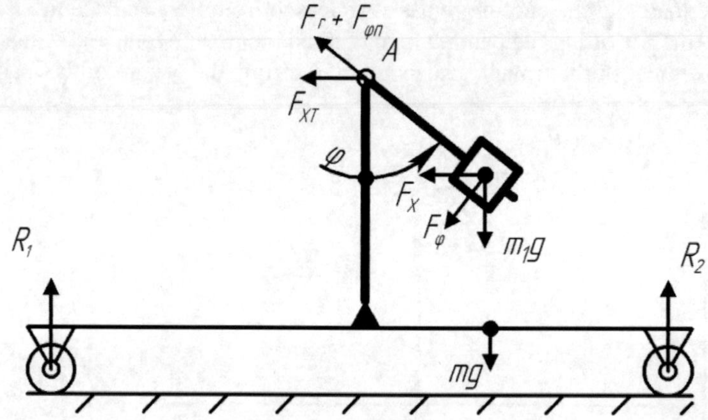

Figure 2. Dynamic model of a trolley with a swinging pendulum of variable length.

Since the cart makes a translational motion, the speed and acceleration of all its points are the same. Then the force of inertia of the portable motion of the load will be equal to:

$$F_x = m_1 \cdot \ddot{x}, \tag{5}$$

Substituting the expression of forces (4), (5) in equation (3), we obtain the differential equations of motion of the load relative to the fixed coordinate system *xoy*:

$$\ddot{r} + \frac{c}{m_1}(r - L) - r \cdot \dot{\varphi}^2 + \ddot{x} \cdot \sin\varphi - g \cdot \cos\varphi = 0 \tag{6}$$
$$\ddot{\varphi} + \frac{2\dot{r} \cdot \dot{\varphi}}{r} + \frac{\ddot{x}}{r} \cdot \cos\varphi + \frac{g}{r} \cdot \sin\varphi = 0.$$

The third equation is obtained by considering the motion of the cart. The impact of the truck load is carried out by means of the elastic force of the spring. The force of inertia acting on the cart will be associated with the acceleration of the cart in its forward motion.

$$F_{xt} = m \cdot \ddot{x} \tag{7}$$

Projecting the forces acting on the cart on the *ox* axis, we obtain the equation:

$$\ddot{x} = \frac{c}{m}(r - L) \cdot \sin\varphi \tag{8}$$

Substitute equality (8) in equations (6) and solve them with respect to the higher derivatives.

$$\ddot{r} = -\frac{c}{m_1}(1 + \frac{m_1}{m}\sin^2\varphi)(r - L) + r\,\dot{\varphi}^2 + g\cos\varphi\,,$$

$$\ddot{\varphi} = -\frac{2\dot{r}\dot{\varphi}}{r} - \frac{C}{m}\left(1 - \frac{L}{r}\right)\sin\varphi\, cos\varphi - \frac{g}{r}\sin\varphi\,, \tag{9}$$

$$\ddot{x} = \frac{C}{m}(r - L)\sin\varphi\,.$$

Using Lagrange equations of the 2nd kind, the authors also obtained three differential equations composed with respect to r, φ, x (the corresponding calculations are not given in this article due to their bulkiness).

$$m_1\,(\ddot{r} + \ddot{x}\sin\varphi - r\dot{\varphi}^2) = -c\,(r - L) + m_1\, g\cos\varphi\,, \tag{10}$$

$$2\dot{r}\dot{\varphi} + r\ddot{\varphi} + \ddot{x}\cos\varphi \;= -g\sin\varphi\,, \tag{11}$$

$$(m + m_1)\ddot{x} + m_1\,(\ddot{r}\sin\varphi + 2\dot{r}\dot{\varphi}\cos\varphi \;+ r\ddot{\varphi}\cos\varphi \;- r\dot{\varphi}^2\sin\varphi) = 0. \tag{12}$$

Expressing $\ddot{r}$ and $\ddot{\varphi}$ from equations (10), (11) and substituting them into equation (12), we obtain the equation

$$\ddot{x} = \frac{C}{m}(r - L)\sin\varphi\,.$$

Thus, the result obtained by the method of Lagrange equations of the 2nd kind coincides with the result obtained by the method of kinetostatics.

The initial conditions must be added to the system of differential equations (9).

Consider the movement of cargo, consisting of two stages:

1. The Load is released from the deflected position at an angle φ_0 clockwise with a fixed length of the spring without the initial speed. The cart position is also fixed.

2. At the time of passage by a spring with the load the vertical position, load and truck are released (freed from bonds). This gives the cart an impulse that drives it forward.

When the spring passes the vertical position ($\varphi = 0$) after releasing the load from the bonds, the spring receives an instantaneous elongation

$$\Delta r = \frac{m_1 \cdot \dot{\varphi}_k^2 * L}{C} = \frac{m_1 g \cdot 2(1 - \cos\varphi_0)}{C} \tag{13}$$

and has an angular velocity

$$\dot{\varphi}_{\text{к}} = \frac{[2gL(1 - \cos\varphi_0)]^{0.5}}{L} = \left[\frac{2g}{L}(1 - \cos\varphi_0)\right]^{0.5} \tag{14}$$

Thus, the initial conditions for the second stage of movement, on the basis of (13) and (14) are:

$$r(0) = L + \Delta r,\;\; \dot{r}(0) = 0,\;\; \varphi(0) = 0,\;\; \dot{\varphi}(0) = \dot{\varphi}_K. \tag{15}$$

The numerical solution of the system of equations (9) with initial conditions (12) can be obtained using the computational complex Mathcad.

An interesting case is when $\varphi_0 = \pi$. In this case, the load makes a continuous rotational movement around point A, and the cart moves on average at a constant speed.

It should be noted that perfectly smooth connections were considered in the problem.

Figures 3-8 show graphs of the movement of the trolley at the values: $L = 0.1$ *m*, $\varphi_0 = 0.5\pi$ *and* $\varphi_0 = \pi$, $m_1 = 1$ *kg*, $m = 2$ *kg and* $m = 4$ *kg*, $C = 490$ Nm^{-1} *and* $C = 49000$ Nm^{-1}.

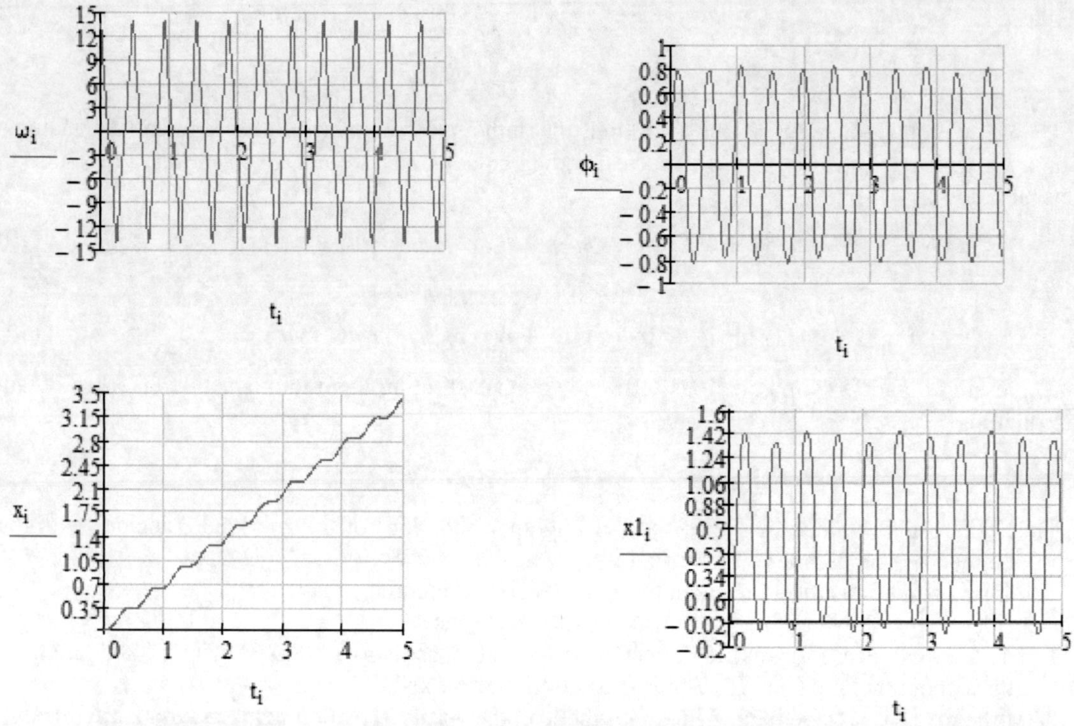

Figure 3. Numerical solution of the system of equations (9) with initial conditions (15) at values: $L = 0.1\ m$, $\varphi_0 = 0.5\pi$, $m_1 = 1\ kg$, $m = 2\ kg$, $C = 490\ Nm^{-1}$.

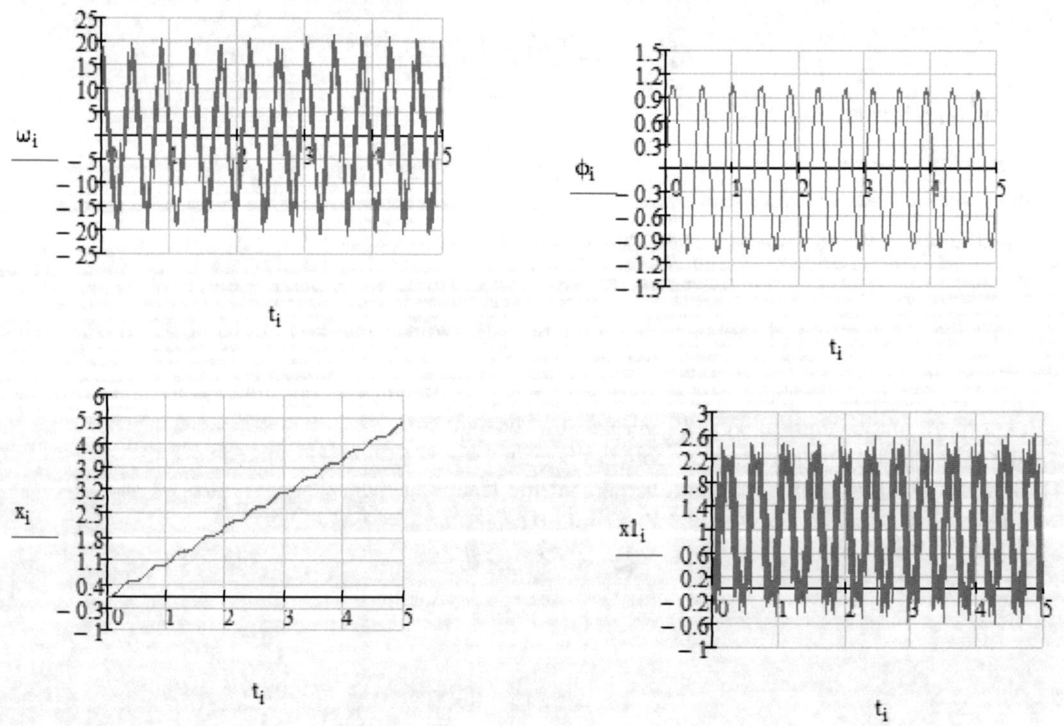

Figure 4. Numerical solution of the system of equations (9) with initial conditions (15) at values: $L = 0.1\ m$, $\varphi_0 = 0.5\pi$, $m_1 = 1\ kg$, $m = 2\ kg$, $C = 49000\ Nm^{-1}$.

The considered problem is devoted to the study and identification of individual characteristics of a pendulum vibration motor. Statement of the problem of translational directional movement of the cart of a pendulum vibration motor on a horizontal support surface due to the movement of a load attached to a spring suspended from the cart. For a fixed value of the initial length of the spring, determine the effect of the initial angle of deflection of the pendulum, spring stiffness, trolley mass, pendulum load on providing the initial impulse of the trolley for its directional movement in the horizontal plane with the gradual release of individual components of the system from connections. There are two stages. 1. the Load is released from the deflected position at an angle φ_0 at a fixed length of the spring (released from the connection that restrains its rotation) no initial speed. The cart position is also fixed. 2. At the moment when the spring with the load passes the vertical position, the load and the cart are released from the bonds and acquire the ability to move with the spring deformation along the pendulum and translational horizontal movement, respectively. The direction of movement of the cart coincides with the direction of the circumferential speed of the load in stage 2.

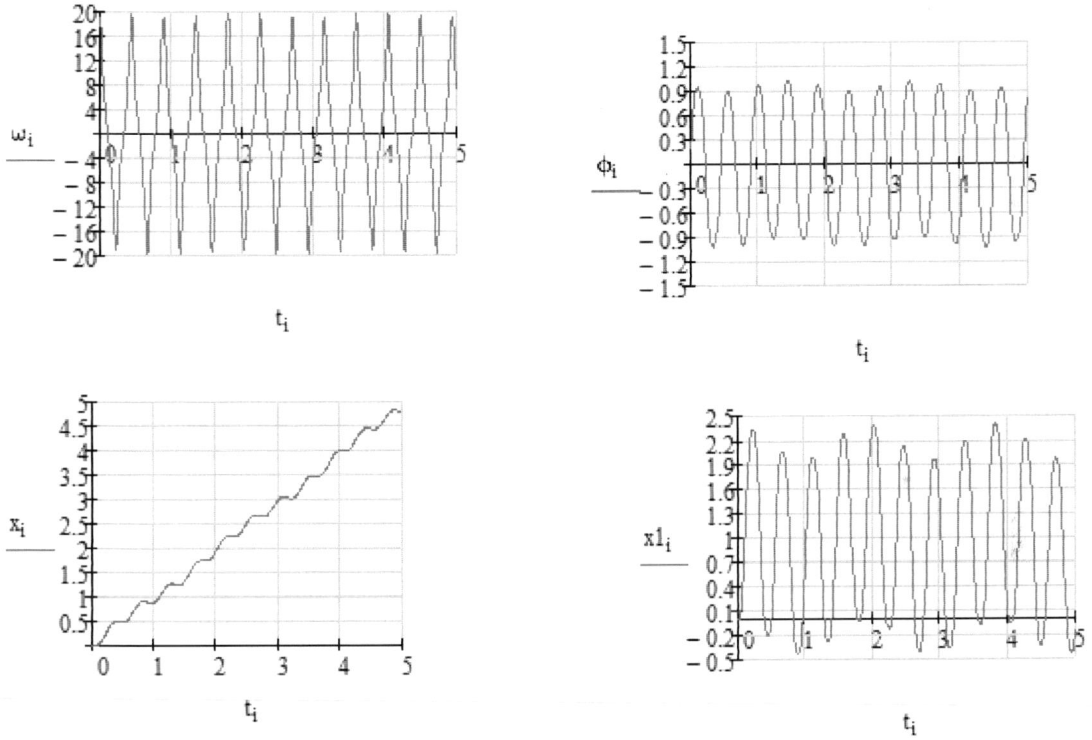

Figure 5. Numerical solution of the system of equations (9) with initial conditions (15) at values: $L = 0.1\ m$, $\varphi_0 = \pi$, $m_1 = 1\ kg$, $m = 2\ kg$, $C = 490\ Nm^{-1}$.

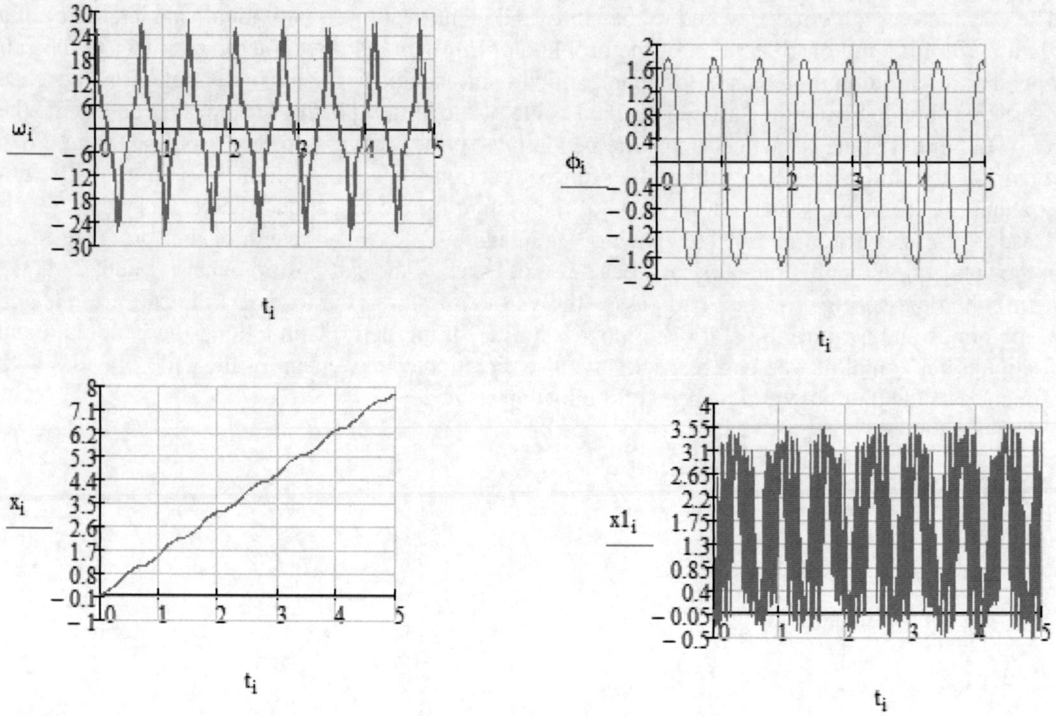

Figure 6. Numerical solution of the system of equations (9) with initial conditions (15) at values: $L = 0.1\ m$, $\varphi_0 = \pi$, $m_1 = 1\ kg$, $m = 2\ kg$, $C = 49000\ Nm^{-1}$.

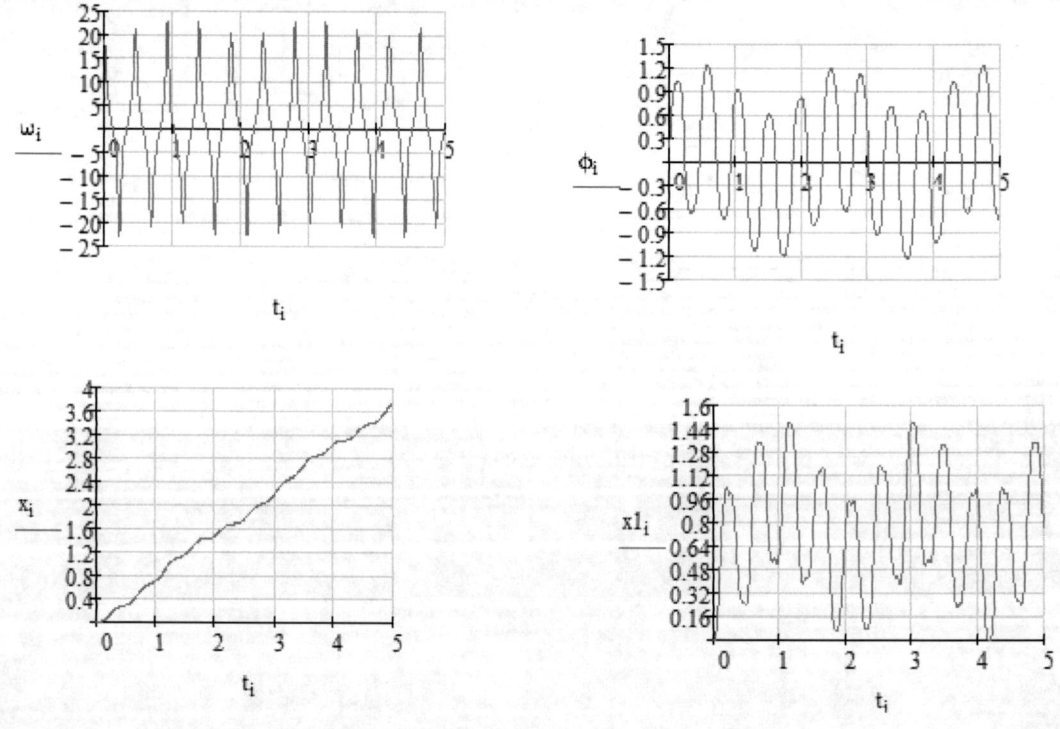

Figure 7. Numerical solution of the system of equations (9) with initial conditions (15) at values (e): $L = 0.1\ m$, $\varphi_0 = \pi$, $m_1 = 1\ kg$, $m = 4\ kg$, $C = 490\ Nm^{-1}$.

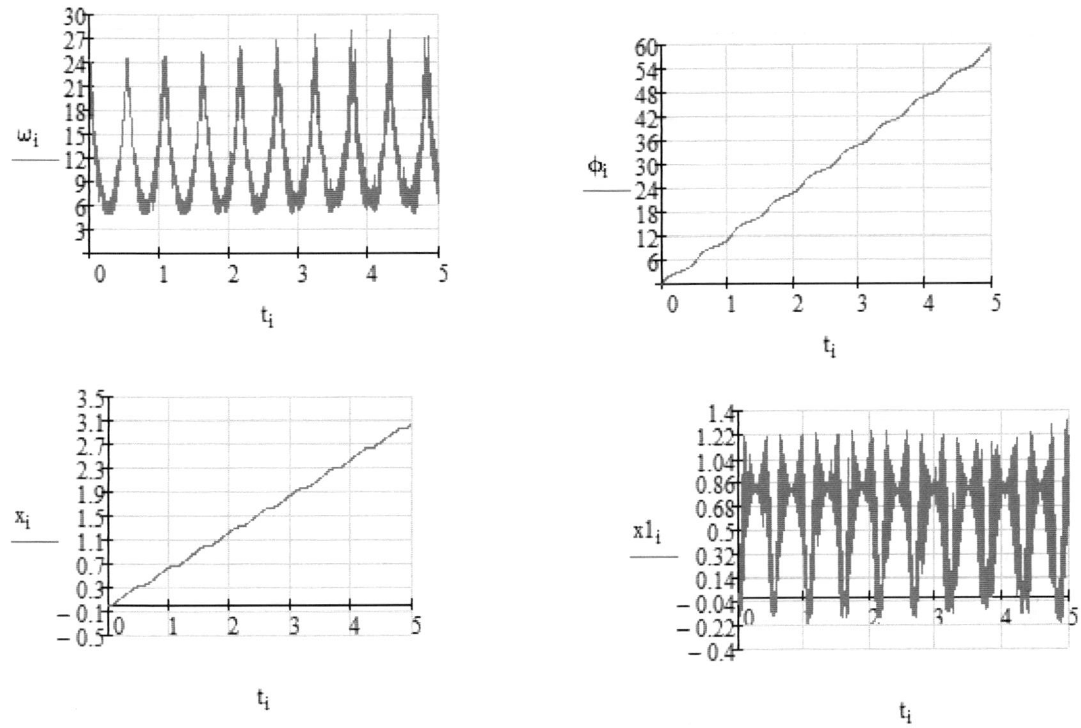

Figure 8. Numerical solution of the system of equations (9) with initial conditions (15) at values: $L = 0.1\ m$, $\varphi_0 = \pi$, $m_1 = 1\ kg$, $m = 4\ kg$, $C = 49000\ Nm^{-1}$.

As follows from the graphs of the obtained solutions, the rate of directional movement of the pendulum libredigital with the same initial deflection angle of the pendulum increases it by increasing the spring rate of the pendulum and decreases with decrease in the initial deflection angle of the pendulum and/or the increase of the weight of the truck libredigital with the same rigidity of a spring of pendulum. However, the identified mode, wherein the increase in the rate of libredigital occurs with a decrease in spring rate, excluding full-circle rotation of the pendulum, when the initial deflection angle of the pendulum $\varphi_0 = \pi$. In this case, the load without a spring makes a continuous rotational movement around the suspension point of the pendulum of the vibration motor, and the cart moves on average at a constant speed. If the load is connected to the pendulum through a spring, the speed of the cart increases in comparison with the operation of the pendulum without a spring, all other things being equal, and the full-rotation movement of the pendulum is excluded.

The paper describes the theoretical regularities of mechanical movements in the design of mechanisms for reproducing movements, including vibration. It is shown that there is a full-turn shape of the trajectory of a spring-loaded pendulum pivotally mounted on a cart. The substantiation of the main characteristics of the pendulum vibration motor, which allows the design of transport and technological means for various applications, is carried out.

References

[1] Kutsenko L, Vanin V, Zapolskiy L, Yablonskyi P, Vasyliev S, Danylenko V, Sukharkova O, Rudenko S and Zhuravskij M 2019 Synthesis and classification of periodic motion trajectories of the swinging spring load *Eastern-European J. of Ent. Technol* vol 2 Issue **7 (98)** pp 26–37

[2] Strygin S 2018 *Pendulum motor with self-oscillations* (https://grabcad.com/library/pendulum-motor-with-self-oscillations-1)

[3] Kalinkin D S and Strygin S V 2018 Pendulum three-movable vibration motor *A collection of High school. New technologies of science, technology, and pedagogy: Materials of the all-Russian scientific and practical conference* (Moscow: Moscow Polytechnic University) pp 487-90

[4] Loitsyansky L G and Lurie A I 1983 *Course of theoretical mechanics: in 2 volumes. Vol. II. Dynamics* (Moscow: Nauka, Main edition of physical and mathematical literature)

Bearings stiffness parameter estimation by sensitivity method

V N Vlasevskiy[1,2], **A E Belkin**[2] **and I I Ivanov**[1,2]

[1] Central Institute of Aviation Motors, Aviamotornaya Street 2, Moscow, 111116, Russia
[2] Bauman Moscow State Technical University, 2nd Baumanskaya Street 5, Moscow, 105005, Russia

vnvlasevskiy@ciam.ru

Abstract. The sensitivity method is applied for bearings stiffness parameter estimation of an experimental rotor system. For model updating, critical shaft speeds are used. The reliability of the sensitivity method is tested on a simple example. The results show that this method can be successfully applied to adjust the finite element model to the experimental data.

1. Introduction

In complex engineering problems, the agreement between analytical and experimental results is not always good. The discrepancy can be explained by the assumptions used in the development of the mathematical model. Often, more detailed modelling of a physical system does not yield the desired agreement of results and also is limited by the computing capacity of the hardware used. Because of this, methods that allow changing the model parameters to achieve better agreement between analytical and measured data are developed.

Description of the procedures for updating the model parameters and their application for simple test examples can be found in [1, 2], where mainly direct and iterative methods are presented. Direct methods allow obtaining the values of the updated parameters without iterative calculations. However, these values can go beyond the physically possible ranges after updating. So, contrary to iterative methods, direct methods are not widely used in practice.

Among iterative methods, the sensitivity method is actively used. It allows using different parameters and experimental data for adjusting the finite element model (FEM) to the test results. One can get acquainted with the application examples of the sensitivity method, as well as its description in [3, 4].

Other methods of updating the FEM parameters have been developing lately. The method of reducing the parameter identification problem to an optimization problem has gained huge popularity. The method allows imposing different constraints on the updated parameters, whereas many other methods cannot do it because of the method of solution they are based on. In [5], approaches to updating FEM using an optimization method were considered. The successful application of these algorithms is presented in [6-9].

Over the last decades, the neural network-based approach has been actively developing [10, 11].

For estimating dynamic characteristics of rotor system bearings, most of the above-mentioned methods have to be modified, but special identification methods are developed sometimes. For example, a method of estimation of non-linear stiffness parameters of rolling element bearings from

IOP Conf. Series: Materials Science and Engineering 1129 (2021) 012044 doi:10.1088/1757-899X/1129/1/012044

random response of rotor-bearing systems was developed in [12]. Paper [13] presented an algorithm of identification of bearings parameters for multi-degree-of-freedom rotor-bearing systems where the shaft was considered flexible, and the parameters of supports depended on the rotation speed. Many approaches to identification of bearing parameters are presented in [14].

In this study, the sensitivity method was used to estimate the bearings stiffness parameter of an experimental rotor system. For model updating, critical shaft speeds are used. The method can be applied to adjust a rotor model based on the test results of a more complicated dynamic system. Calculations were done in MATLAB.

2. Theory

Let us introduce a vector of updated model parameters $\boldsymbol{p}$, a vector of analytical z and experimental $z_{\mathbf{m}}$ critical speeds

$$\boldsymbol{p} = \begin{Bmatrix} p_1 \\ p_2 \\ p_3 \\ \vdots \\ p_k \end{Bmatrix}, \quad z = \begin{Bmatrix} \omega_1 \\ \omega_2 \\ \omega_3 \\ \vdots \\ \omega_n \end{Bmatrix}, \quad z_{\mathbf{m}} = \begin{Bmatrix} \omega_{\mathrm{m}_1} \\ \omega_{\mathrm{m}_2} \\ \omega_{\mathrm{m}_3} \\ \vdots \\ \omega_{\mathrm{m}_n} \end{Bmatrix}. \tag{1}$$

The updating equations for this method are based on a linear approximation of the experimental data using Taylor's series expansion

$$z_{\mathbf{m}} = z_j + \boldsymbol{S}_j \left(\boldsymbol{p} - \boldsymbol{p}_j\right), \tag{2}$$

where j is the iteration number, $\boldsymbol{S}_j$ is the sensitivity matrix (rectangular in a general case). Its components are the first derivatives of the critical speeds with respect to the updated parameters.

From (2), the following equation for updating model parameters can be derived:

$$\boldsymbol{p}_{j+1} = \boldsymbol{p}_j + \left[\boldsymbol{S}_j^{\mathrm{T}} \boldsymbol{S}_j\right]^{-1} \boldsymbol{S}_j^{\mathrm{T}} \left(z_{\mathbf{m}} - z_j\right). \tag{3}$$

After the initial parameter values are set, the calculations are conducted according to equation (3) until the change in the parameters between successful iterations becomes small.

3. Simple example

Consider an example of rotor supports stiffness updating using the sensitivity method. This example demonstrates the reliability of the applied parameter estimation approach.

The rotor model is comprised of a shaft, a disk, and supports on the left and right shaft ends. The shaft model consists of ten beam elements with a length of 50 mm and a circular solid cross section with a diameter of 10 mm. The elasticity modulus and the density of the material are $2 \cdot 10^5$ MPa and 7800 $\mathrm{kg \cdot m^{-3}}$, accordingly.

The rotor supports are modelled as isotropic constant stiffness springs. The stiffness of both springs is 100000 $\mathrm{N \cdot m^{-1}}$.

The disk is located at the shaft center. It is modelled as a lumped inertia element having the mass of 1 kg, the polar moment of inertia of $5 \cdot 10^{-4}$ $\mathrm{kg \cdot m^2}$ and the diameter moments of inertia of $2.5 \cdot 10^{-4}$ $\mathrm{kg \cdot m^2}$.

In this example, the simulated data is obtained from a similar model with other stiffness of springs equal to 150000 $\mathrm{N \cdot m^{-1}}$. For updating the parameter, the first critical shaft speed is used. Tables 1 and 2 show the results of applying the sensitivity method.

Table 1. Simulated and analytical data.

Critical speed, rpm			Error, %	
Initial FEM	Updated FEM	Simulated	Initial FEM	Updated FEM
1608	1658	1658	3	0

Table 2. Updated parameter.

Parameter	Value			Error, %
	Initial FEM	Updated FEM	Simulated	
Springs stiffness, $N \cdot m^{-1}$	100000	150000	150000	0

Figure 1 shows the plot of the parameter value vs. the iteration number.

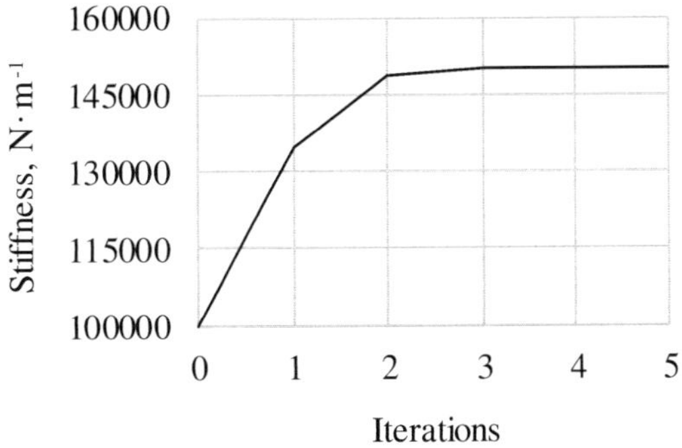

Figure 1. Convergence of the parameter for simple example.

It is seen that stiffness of the springs can be determined quite accurately. The method works correctly, and it can be applied for the estimation bearings stiffness parameter of an experimental rotor system.

4. Experimental example

This section presents the application of the sensitivity method to the problem of estimation bearings stiffness parameter of an experimental rotor system *Bently Nevada RK4 Rotor Kit*, figure 2. The kit comprises a frame, a shaft, a disk, supports, an electric motor with a coupling, proximity probes, a motor control unit, and proximity probes' data acquisition unit. In each rotor support, a bearing with a rubber ring seal is installed, figure 3.

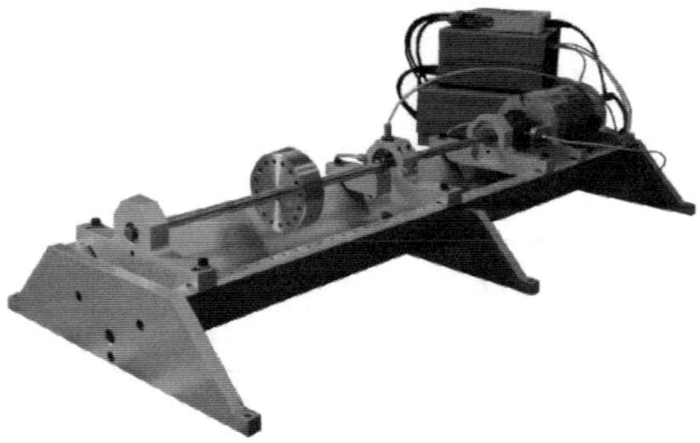

Figure 2. Experimental setup.

Figure 3. Bearing.

Table 3 shows the parameters of the shaft and disk. The disk was located closer to the left support. The distance between them was 170 mm.

Table 3. Parameters of the shaft and disk.

Parameter	Value	Unit
Shaft length	552	mm
Shaft diameter	10	mm
Disk mass	0.812	kg
Disk diameter	75	mm
Disk thickness	25	mm
Shaft material density	7796	$kg \cdot m^{-3}$
Shaft material Young modulus	$2.1 \cdot 10^5$	MPa

On the experimental system, the relationships of vibration amplitude vs. shaft speed for run-up and run-down were obtained, figure 4. Based on the plots, critical rotor speeds were obtained. The data was acquired near the support located closer to the electric motor. It is known [15] that the maximum displacement amplitude during the run-up and run-down is observed at differed speeds, so an average between the run-up and run-down critical speed was taken (1874 rpm).

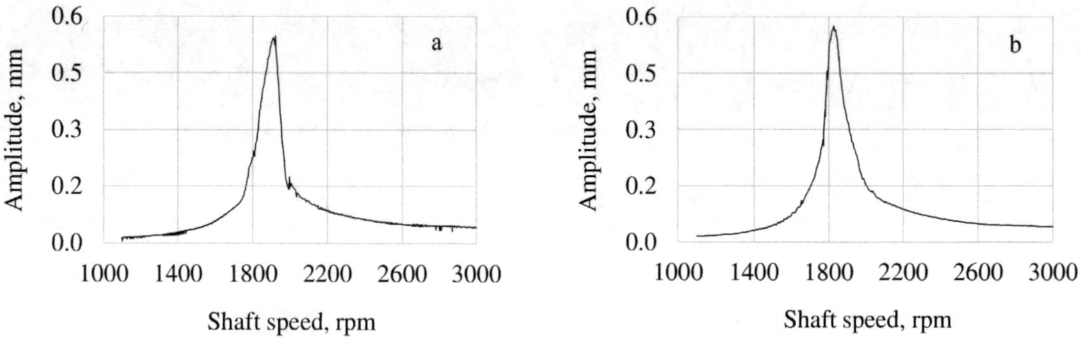

Figure 4. Vibration amplitude: (a) run-up; (b) run-down.

Figure 5 shows the rotor model.

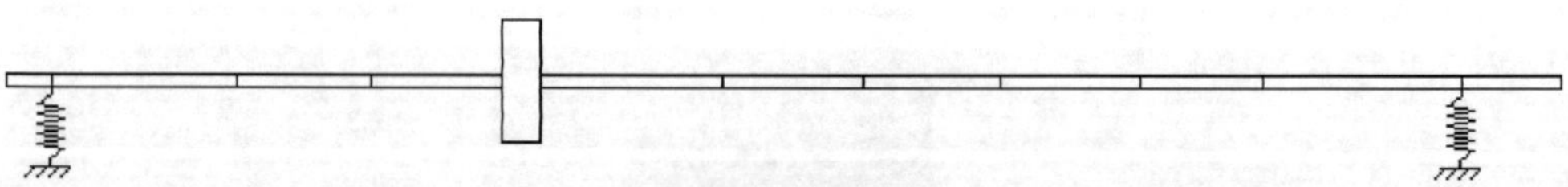

Figure 5. Rotor model.

The shaft is modelled with 11 beam elements. The disk is modelled as a lumped inertial element with a mass, a polar moment of inertia and the equal diameter moments of inertia. The rotor supports were modelled as isotropic constant stiffness springs.

Shaft model parameters correspond to the data given in table 3. Table 4 gives the parameters of the inertia element. The distance between the inertia element and the supports correspond to the disk location in the experimental system. The value of springs stiffness is not known, and its initial value was set to 50000 $N \cdot m^{-1}$.

Table 4. Inertia element parameters.

Parameter	Value	Unit
Mass	0.812	kg
Polar moment of inertia	$5.75 \cdot 10^{-4}$	$kg \cdot m^2$
Diameter moment of inertia	$3.30 \cdot 10^{-4}$	$kg \cdot m^2$

The analytical first critical rotor shaft speed was 1668 rpm. The error between analytical and experimental critical speed was 11%. The inertia rotor characteristics were modelled accurately enough. As the influence of damping on the critical speeds is small, springs stiffness (bearings stiffness parameter) is selected to update. Its initial value corresponds to the data presented in the model description. Tables 5 and 6 show the results of applying the sensitivity method.

Table 5. Analytical and experimental data.

Critical speed, rpm			Error, %	
Initial FEM	Updated FEM	Experimental	Initial FEM	Updated FEM
1668	1874	1874	11	0

Table 6. Updated parameter.

Parameter	Value	
	Initial FEM	Updated FEM
Springs stiffness, $N \cdot m^{-1}$	50000	119731

Figure 6 shows the plot of the parameter value vs. the iteration number.

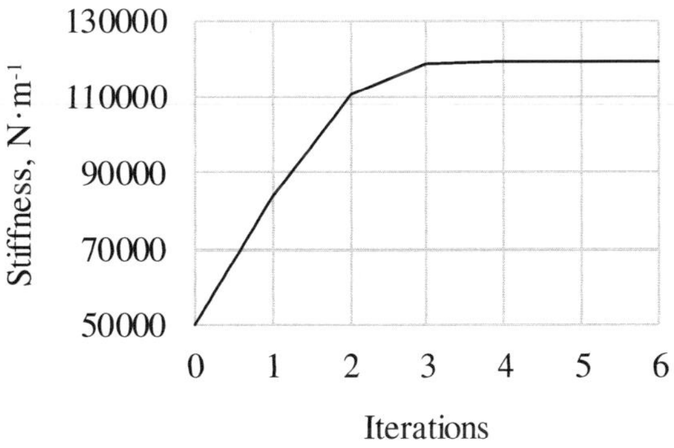

Figure 6. Convergence of the parameter for experimental example.

It is seen from figure 6 that the iterative process of bearings stiffness parameter estimation converged successfully. After updating, the error between the analytical and experimental critical speed is small.

IOP Conf. Series: Materials Science and Engineering 1129 (2021) 012044 doi:10.1088/1757-899X/1129/1/012044

5. Conclusion

In this study, the sensitivity method was applied to updating the rotor dynamic model. The reliability of the method was tested on a simple example. The bearings stiffness parameter of an experimental rotor system was estimated using the sensitivity method. The considered method allows adjusting the rotor model to replicate the experimentally observed critical speed.

References

[1] Friswell M I and Mottershead J E 1995 *Finite element model updating in structural dynamics* (Dordrecht: Kluwer Academic Publishers)

[2] Friswell M I and Mottershead J E 1993 Model updating in structural dynamics: a survey *J. of Sound and Vibration* **167** pp 347-75

[3] Mottershead J E, Link M and Friswell M I 2011 The sensitivity method in finite element model updating: a tutorial *Mechanical Systems and Signal Processing* **25** pp 2275-96

[4] Chouksey M, Dutt J K and Modak S V 2013 Model updating of rotors supported on ball bearings and its application in response prediction and balancing *Measurement* **46** pp 4261-73

[5] Marwala T 2010 *Finite element model updating using computational intelligence techniques: applications to structural dynamics* (Heidelberg: Springer)

[6] Levin R I and Lieven N A J 1998 Dynamic finite element model updating using simulated annealing and genetic algorithms *Mechanical Systems and Signal Processing* **12** pp 91–120

[7] Modak S V, Kundra T K and Nakra B C 2000 Model updating using constrained optimization *Mechanics Research Communications* **27** pp 543-51

[8] Yao J, Liu L, Yang F, Scarpa F and Gao J 2018 Identification and optimization of unbalance parameters in rotor-bearing systems *J. of Sound and Vibration* **431** pp 54-69

[9] Xia Z, Li A, Li J, Shi H, Duan M and Zhou G 2020 Model updating of an existing bridge with high-dimensional variables using modified particle swarm optimization and ambient excitation data *Measurement* **159** 107754

[10] Levin R I and Lieven N A J 1998 Dynamic finite element model updating using neural networks *J. of Sound and Vibration* **210** pp 593-607

[11] Atalla M J and Inman D J 1998 On model updating using neural networks *Mechanical Systems and Signal Processing (MSSP)* **12** pp 135-61

[12] Tiwari R and Vyas N S 1995 Estimation of non-linear stiffness parameters of rolling element bearings from random response of rotor-bearing systems *J. of Sound and Vibration* **187** pp 229-39

[13] Tiwari R, Lees A W and Friswell M I 2002 Identification of speed-dependent bearing parameters *J. of Sound and Vibration* **254** pp 967-86

[14] Tiwari R, Lees A W and Friswell M I 2004 Identification of dynamic bearing parameters: a review *The Shock and Vibration Digest* **36** pp 99-124

[15] Biderman V L 1980 *Theory of mechanical vibration* (Moscow: High school)

Contact-communicative approach to research and development the smart prosthesis for the transradial amputees

N S Kovalev, A A Solovyev and A M Sergeev

Mechanical Engineering Research Institute of the Russian Academy of Sciences, Moscow, 101990, Russia

ns.kovalev@yandex.ru

Abstract. The article is devoted to further researches and developments of the smart prosthesis for transradial amputees. Describes the solutions of thematically related tasks to increase the effectiveness of the amputee's interaction with the external subject-spatial and social environment. One of the tasks is to study the behavior of experimental resistive pressure sensors protective materials in order to optimize sensor contact. Another task is to the provision with the help of conductive materials of the effect of "digital touch" with the aim of creating the amputee through the agency of the screen-compatible prosthesis conditions for active social communication.

1. Introduction

This research work, which is the next stage in the creation of a smart prosthesis, that solution compered in its functions human hand with, was conducted on two directions: protection of resistive pressure sensors acting as receptors and the creation the system for the transradial amputee interaction with projection-capacitive (PC) touch screen devices [1].

The article is conceptual in its nature and is based on the idea of an artificial hand as a mediator providing a wealth of connections between the amputee and the outside world. We emphasized this circumstance earlier, giving the smart prosthesis the "agent" status [3].

2. System approach and complexity - basic principles of smart prosthesis R&D

Recall that the device we are developing is a complex object consisting of two elements - an artificial part of the hand equipped with sensors and a vibration bracelet, connected by wireless transmission of information. At the moment, this complex is supplemented with a conductive structure that allows the amputee to operate with projection-capacitive screens. This design model can be called two-part or binary (figure 1). This is that we meant under the general term 'smart prosthesis' from beginning. Thus, the spatial separation of the smart prosthesis does not negate its integral consideration.

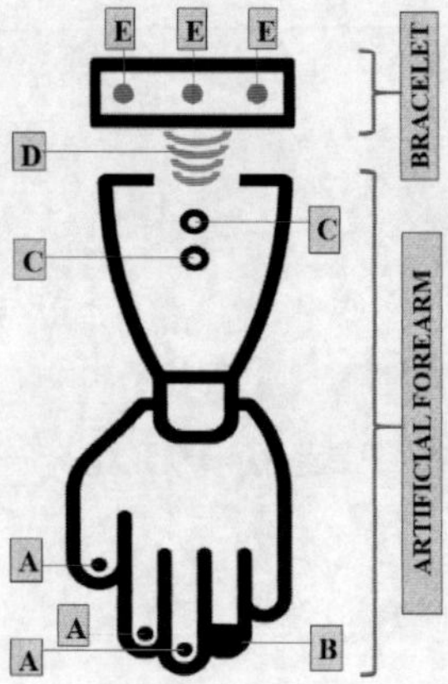

Figure 1. Binary smart prosthesis structure scheme & functions: A – FSRs, B – touch screen compatible finger cot, C – EMG electrodes, D – Bluetooth signal, E – vibrating motors.

The principle of integrity is preserved even when attention is paid to particular components and processes. In this case, the only correct approach is to consider them in the 'artificial organ – human' system. For example, sensing, as a rule, implies the placement of sensors on the palmar part of a myoelectric prosthesis, immediately puts before researchers a number of questions related to the peculiarities of operation. To answer them the biomechatronics theoretical and practical basis is used. This knowledge suggests imitating the most important functions of the human hand using high-tech quasi-organismic elements.

3. The protective aspect of sensorization

Equipping the smart prosthesis working model with resistive pressure sensors (FSRs) necessitates protection by their own outer shell from adverse factors. It is known that the human skin main function is to protect the body, including the receptor system. To a greater extent this applies to the stratum corneum of the epidermis [2]. As applied to the smart prosthesis, the task is to provide the sensors with a local protective layer similar to the epidermis, but at the same time, by its physicochemical properties, it is resistant to mechanical damage, including abrasion. The second task is to preserve the parameters of the sensors declared by the manufacturer or their minimum deviation from the established norms during operation with the applied protective coating.

During the work, a hardware complex was used, including an STM32F429 microcontroller from STMicroelectronics and an FSR 400 sensor from Iterlink Electronics (figure 2).

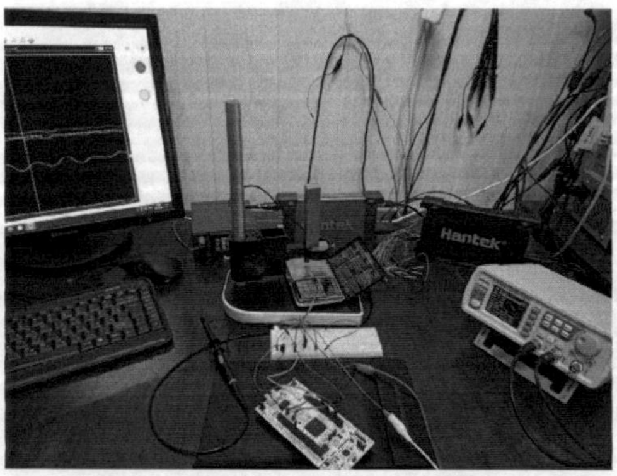

Figure 2. Hardware and software for silicon protection plates study.

The parameters study of the output voltage V_{out} of the silicon protection of resistive pressure sensors with increasing applied forces was carried out using plates of different hardness and thickness provided by Medsil JSC (figure 3).

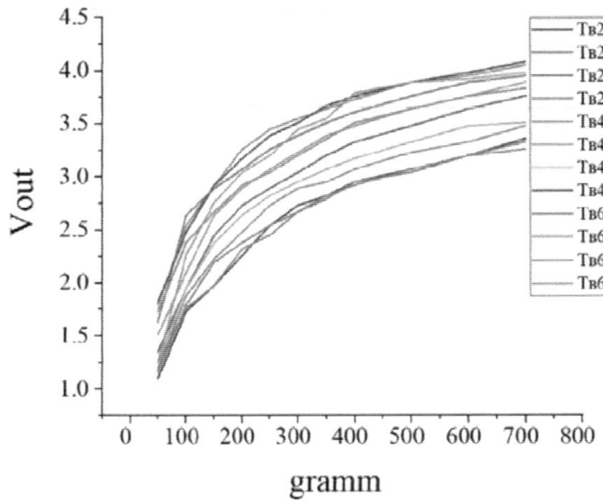

Figure 3. The graph of the output voltage V_{out} of silicone plates at a pressure from 50 to 700 g.

As a result of experiments, have been identified silicone plates that are suitable for stable tactile forces transmission. The threshold parameters of the actuation and the ranges of the distribution of forces are determined in accordance with the metrological sensors properties, as well as the effect of the conductive materials thickness on the transmitted signals.

4. The phenomenon of screen-compatible prosthesis

One of the transradial amputees problems interacting with the surrounding object environment is the impossibility of operating with the screens of communication devices work on projection-capacitive principle. These include mobile phones, tablets and some types of laptops. The reason for this is the dielectric material from which modern hand prostheses are made, in particular the artificial hands fingers.

To date, it is known about several realized attempts to implement the idea of successful interaction of amputees with PC screens. The largest and most famous of them is the creation and promotion of the i-limb quantum hand prosthesis on the world market by the British firm Touch Bionics [3].

Other notable businesses include Point Design, a US-based company that specializes in prosthetics for amputees who have lost fingers. The most noteworthy is the Point Digit II prosthesis, equipped with titanium fingers with conductive pads [4].

The noted shortage of ready-made proposals for screen-compatible prostheses is combined with a shortage of ideas at the professional level. A rare exception is a patent obtained in the USA by Macduff Charles, who developed a finger prosthesis capable of controlling gadgets equipped with PC screens [5].

In order to compensate for the lack of affordable touch screen-compatible prostheses on the market, individual users are trying to make their own removable prostheses for operating with smartphones using a 3-D printer [6].

Attempts to adapt prostheses to work with a PC screen are made on an individual basis in various ways. One of them is the use of conductive silicone buttons Digits invented by biochemist Brian Shy, usually attached to winter gloves [7]. Another way is to use conductive adhesive foam buttons by Gordon Adkins [ibid.].

As for scientific research in the open press dedicated to touchscreen-compatible prostheses, they are rare [8]. While noting this, we express our confidence that such research is needed. This is confirmed by the above facts of an independent solution of the existing problem.

5. The touchscreen-compatible finger cot concept and its implementation

The idea of creating a touchscreen-compatible finger cot arose on the basis of awareness of the modern mobile phones rapidly expanding capabilities, the contact zone of which and the control platform of which are and, probably, will be in the near future, the projection-capacitive screens. This idea organically fit into the overall vision of the smart prosthesis we are developing as a communication system that provides the amputee with the outside world in digitalization conditions. And not only with a subject, but also with a social environment, in which the latest mobile phones have become a necessary condition for comfortable integration of a person into the structure of not only interpersonal, but social relations [9-10].

Based on international experience, it is clear that for successful interaction with such screens, it is necessary to have a conductive element as part of the artificial prosthesis finger. The choice of the finger was due to the placement of resistive pressure sensors on the thumb, index and middle fingers. For ergonomic reasons, the ring finger was preferred. In addition, its proximity to the sensor-equipped middle finger allows to adjust the pressing force on the touchscreen.

At the first stage of creating the finger cot laboratory prototype, we tested three samples of conductive silicones. The minimum electrical resistivity was shown sample of the Russian company RT-Technologies based on the silver nanocomponent. During the design of the prototype, a conductive three-component adhesive by RT-Technologies was also used.

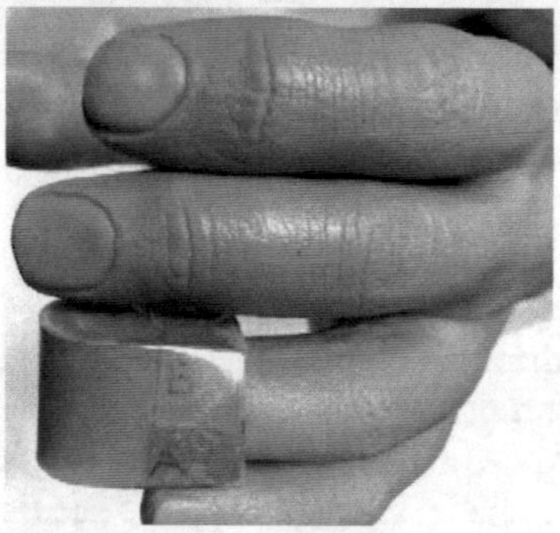

Figure 4. Conductive finger cot experimental prototype.

It is planned for further research to use the finger coat experimental sample with higher electrical conductivity manufactured by RT-Technologies using injection moulding.

6. Conclusions

As a result of the experiments was identified possibilities introducing innovative products in the field of prosthetics, providing transradial amputees with:

- long-term tactile contact with the external object environment in the presence of silicone-protected resistive pressure sensors;
- successful social communication through the use of screen-compatible conductive silicones.

References

[1] Kovalev N S, Solovyev A A and Sergeev A M 2020 The smart prosthesis as the personal agent in the amputee tactile perception process *IOP Conf. Series: Materials Sci. and Eng.* Available at: https://iopscience.iop.org/article/10.1088/1757-899X/747/1/012068/pdf

[2] Kuznetsov S L *et al* 2013 Modern concepts on the structure and functions of the epidermis and derma *J. Skin Vener. Dis.* **2** pp 26-32

[3] Touch Bionics by Össur 2017 *Product Catalog*

[4] Point Digit II. *Installation Guide for the Prosthetist. Point Design* Available at: https://pointdesignsllc.com/media/documentation/PointDigitII-InstructionsforUse.pdf

[5] Macduff Ch C 2014 Bio-mechanical prosthetic full finger Patent US 9,370, 430 B2 Available at: https://patents.google.com/patent/US20140303749A1/en

[6] Terminal socket for keyboard typing & smartphone use 2019 Available at: https://www.gre-nable.fr/en/terminal-socket-for-keyboard-typing-smartphone-use/

[7] Available at: https://mpihouston.com/category/amputee-stories/ Touch screen phone dialing *Amputee Stories*
[8] Day Ph *et al* 2019 Touchscreen usage with upper limb prostheses: initial explorations *Contemporary Ergonomics and Human Factors* (https://publications.ergonomics.org.uk)
[9] Disabled Persons - Disability - Disabled 2018 *Int. Sci. and Practic. Conf.* (Nizhny Novgorod)
[10] Jewitt C *et al* 2020 *Interdisciplinary Insights for Digital Touch Communication* (Springer open) **XVII** p 131

Algebraic harmony in genoms of higher and lower organisms. Non-euclidean biosymmetries

S Petoukhov[1,2], E Petukhova[1] , V Svirin[1,2]

[1] Mechanical Engineering Research Institute of the Russian Academy of Sciences, 4, M. Kharitonyevskiy Lane, Moscow, 101990, Russia
[2] Moscow State Conservatory, 13/6 Bolshaya Nikitskaya Street, Moscow, 125009, Russia

spetoukhov@gmail.com

Abstract. The article is devoted to the study of the relationship between non-Euclidean symmetries in inherited morphogenetic and other bio-structures with algebraic features of information sequences of nucleotides in DNA heredity molecules in the genomes of eukaryotes and prokaryotes. These genomic sequences obey the universal hyperbolic rules of the oligomer cooperative organization, which are associated with the harmonic progression of 1/1, 1/2, 1/3, ..., 1/*n*. This progression has long been known and studied in various branches of mathematics and physics, and has now manifested itself in genetic informatics. The performed analysis of the harmonic progression revealed its connection with the double ratio known in geometry, which is the main invariant of projective geometry. This connection consists in the fact that the magnitude of the double ratio is the same and is equal to 4/3 for any four adjacent members of this progression. This fact of the connection of genetic informatics with the main invariant of projective geometry can be used to explain the implementation of non-Euclidean symmetries in genetically inherited structures of living bodies.

1. Introduction

The phenomena of the symmetric inherited structure of biological bodies have been the subject of research since ancient times. The importance of studying these phenomena is described, for example, in classic books [1, 2]. At the same time, until recently, all attention was paid to the biosymmetries of similarity, which are described using a group of geometric similarity transformations. According to F. Klein's Erlangen program, this group of transformations underlies Euclidean geometry and includes transformations of translations, rotations, specular reflections and scaling. At the same time, the hypothesis by V.I. Vernadsky on the possible important morphological significance of non-Euclidean symmetries in living bodies is widely known.

Clarification of the geometric foundations of morphology entails shifts in various areas of biomechanics and physiology: biomechanics of postures, biomechanics of growth and motor movements, psychophysics, etc. In the variety of organic forms, multiblock structures are of particular interest, the components of which are mutually conjugated on the basis of certain algorithms, repeated on the most diverse lines and branches of biological evolution. The general rule of such varieties, consisting of blocks or motivic units, is that the previous motivic unit is translated into the subsequent one by some fixed Euclidean or non-Euclidean transformation. In other words, so organized segmented varieties have a cyclic group of automorphisms, and their motivic units are arranged along the orbit of the corresponding cyclic group. Earlier, in the works of the Institute of Mechanical Engineering Research Institute of the Russian Academy of Sciences, a wide representation in the inherited configurations of the set of living bodies of non-Euclidean symmetries was revealed, which are based on the groups of projective and Möbius transformations, which generalize the group of similarity transformations and contain it as their subgroup [3, 4].

Genetically inherited morphological structures must be structurally linked to the molecular genetic system for their genetically coding and transmission along the chain of generations. Therefore, it is important to look for traces of symmetries and invariants of non-Euclidean transformations in the genetic informatics of living bodies, including in the information texts of DNA heredity molecules carrying genetic information in the form of long nucleotide sequences. As a result of the work of the laboratory for research of biomechanical systems of the Institute of Mechanical Engineering of the Russian Academy of Sciences in 2020, universal hyperbolic rules for the cooperative organization of genomes of higher and lower organisms were discovered [5]. These rules are associated with harmonic progression (1), which has long been known in various fields of science and has now received the status of an important participant in genetic informatics, capable of influencing the characteristics of inherited biosymmetries:

$$\frac{1}{1}, \frac{1}{2}, \frac{1}{3}, \frac{1}{4}, \dots, \frac{1}{n} \tag{1}$$

In this article, we consider the mathematical connections of this progression with the groups of projective and Möbius transformations, on which non-Euclidean symmetries are based.

2. Computational approaches and models

Our computational approaches and models of biological structures are based on the mathematical properties of harmonic progression (1), which were studied by Pythagoras, Orem, Leibniz, Newton, Euler, Fourier, Dirichlet, Riemann and many other researchers in connection with various problems of physics and mathematics. Nowadays this list is supplemented with the tasks of studying its role in the genetic information fundamentals of living matter and the structuring of inherited biological systems. Its historically known name - "harmonic progression" - sequence (1) was received due to its connection with a number of harmonics in music (or a number of lengths of standing waves when vibrating a stretched string). This article presents the results of an analysis of the conjugation of this progression with the main invariant of projective geometry, called the double ratio (as well as the wurf and anharmonic ratio). In the one-dimensional case, the group of projective transformations and the group of Möbius transformations coincide, and the double relation is simultaneously an invariant of both projective and Möbius transformations. In our research on non-Euclidean biosymmetries, we actively use the provisions of the famous Erlangen program of F. Klein, which defined geometry as the science of invariants of the transformation group. In accordance with this program, each group of transformations has its own geometry, which has its own individual set of invariants. In particular, the groups of projective and Möbius transformations define projective and Möbius geometries (the latter is also called the geometry of reciprocal radii or conformal geometry).

The cross-ratio (or double ratio or wurf) of the four numbers a, b, c, d is determined by the expression (2):

$$(ab, cd) = \frac{c-a}{c-b}\frac{d-b}{d-a} \tag{2}$$

If the four numbers a, b, c, d correspond to the coordinates of four points on the number line, then expression (2) consists of the lengths of the corresponding segments between the indicated pairs of points. Under projective transformations of such a straight line, the lengths of the segments and the simple relations between them change, but the value of the cross-ratio remains unchanged.

3. Results and discussions

One of the properties of the harmonic progression (1) is that the value of the cross-ratio of any of its four neighboring numbers is equal to 4/3. Any four adjacent numbers in a series of natural numbers 1, 2, 3,..., n also have the same value of the cross-ratio. This means that the harmonic progression (1) and the natural series of numbers are projectively equivalent to each other (figure 1).

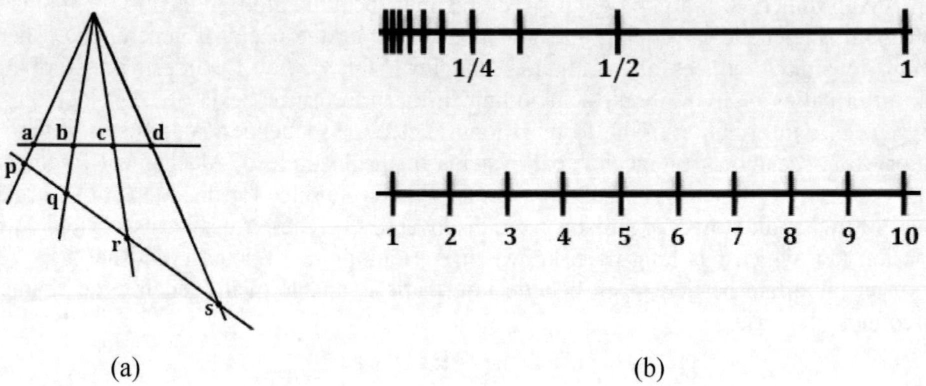

Figure 1. Examples of projective transformation. (a) A central projective transformation changes the lengths of the line segments and their simple ratios, but retains the magnitude of the cross-ratio. (b) The location of the numbers of the harmonic progression (1) and a series of natural numbers on the number line.

The connection of the harmonic progression (1) with the cross-ratio leads to the following conclusions:

1) various fragments of a harmonic progression can be considered as projectively symmetric to each other;
2) in the genomes of eukaryotes and prokaryotes, various hyperbolic sequences of oligomeric sums corresponding to the classes of A-, T-, C- and G-oligomers [5] are projectively similar to each other with the high accuracy with which hyperbolic rules are fulfilled in them;
3) all investigated genomes of eukaryotes and prokaryotes with respect to their sequences of oligomeric sums are projectively similar to each other (with the high accuracy, with which hyperbolic rules are fulfilled in them).

Similar hyperbolic rules associated with harmonic progression (1) are fulfilled not only in genomic DNA sequences of nucleotides, but also in the studied long proteins (such as the Titin protein containing 33423 amino acids and presented at https://www.ncbi.nlm.nih.gov/protein/ ACN81321.1) in relation to oligomer sums in their 20 classes of amino acids. The point is that in the case of protein sequences of 20 types of amino acids, 20 equivalence classes can be analyzed, each of which is determined by the corresponding type of amino acid and combines all n-oligomers (fragments of length n) of a given protein that begin with a given amino acid. For example, the amino acid Ala defines a class of Ala-oligomers, which includes all n-oligomers starting with this amino acid: the subset of Ala-doublets contains all 20 doublets that start with Ala (AlaAla, AlaArg, AlaAsn, ..., AlaCys); the subset of Ala-triplets contains all 400 triplets that start with Ala (AlaAlaAla, AlaAlaArg,…., AlaCysCys), and so on. The quantitative compositions of these subsets of n-oligomers of each of these amino acid classes for different n = 1, 2, 3, ... (here n is much less than the protein length) are mutually related with high accuracy and obey hyperbolic rules associated with harmonic progression (1). Accordingly, taking into account the mentioned connection of this progression with the cross-ratio, we can talk about the projective-geometric (as well as the Möbius-geometric) similarity of the named quantitative characteristics of the structure of long proteins such as the Titin protein [5].

Taking into account that the numerical characteristics of genetic sequences are associated with the main invariant of projective geometry, then it can be expected that projective-geometric and Möbius-geometric symmetries will appear in genetically inherited physiological structures. This allows us to consider in a new light the existence of the non-Euclidean symmetries previously established in the works of the IMASH RAS in a wide variety of morphogenetic structures and growth processes in living bodies [3, 4].

For example, the cross-ratio was applied to the analysis of growth changes in the longitudinal proportions of three-membered kinematic blocks in the human body: three-phalangeal fingers, three-membered limbs (shoulder-forearm-hand and thigh-calf-foot) and the entire three-membered body (in anthropology, the body is subdivided into the upper, trunk and lower segments). The value of the cross-ratio of the lengths of the three members in all these trebled blocks, at least during the entire postnatal period, is grouped around the reference value of 1.3. This is all the more interesting because the growth of the human body is essentially non-linear (figure 2): for example, during a person's life from birth, the upper part of the body grows 2.4 times, the trunk - 2.8 times, and the lower part - 3.8 times. This invariance of the cross-ratio value turns out to be largely true not only for normally developed people, but also for dwarfs and giants. This is also true for a wide range of highly organized animals.

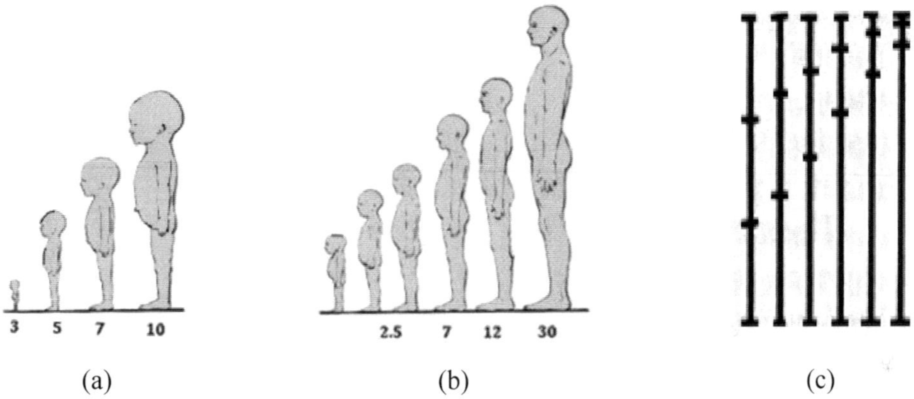

Figure 2. Age-related changes in the proportions of the human body (from [6]). (a), (b) - antenatal and postnatal stages: (a) - in the lunar months; (b) - in years (first newborn on the left); (c) - examples of projectively equivalent three-term segments, the cross-ratio values of which are equal to 1.3.

We note additionally that projective-geometric transformations and the cross-ratio as their main invariant are mathematical tools, on which the study of visual perception and the creation of methods of computer vision, pattern recognition, and artificial intelligence are largely based. A huge number of publications are devoted to this topic (see, for example, [7, 8]). The study of the relationship between the genetic system and genetically inherited biological structures, which are connected with non-Euclidean symmetries and harmonic progression, continues, including for the development of genetic biomechanics and algebraic biology. These issues, in particular, are considered at the International Interdisciplinary Seminar "Algebraic Biology and Systems Theory" [9].

The creators of quantum mechanics P. Jordan and (later) E. Schrödinger pointed out the key difference between living bodies and inanimate ones: inanimate objects are controlled by the average random motion of their millions of particles and the motion of an individual particle is not essential for the whole; on the contrary, in a living organism the chosen - genetic - molecules have a dictatorial (!) influence on the entire organism (see the history of quantum biology [10]). In accordance with this statement, in our research we focus on the properties of DNA information sequences and study their influence on inherited biological structures.

4. Conclusions

Studies of nucleotide DNA sequences in the genomes of higher and lower organisms reveal the existence in them of a system of quantitative hyperbolic rules of a universal nature, which are related to the harmonic progression (1).

The mathematical properties of this historically known progression, used in various fields of science and now found in the quantitative characteristics of molecular genetic informatics, indicate that the algebraic structures of genetic informatics can be a basis of non-Euclidean symmetries in the inherited morphology and biomechanics of movements [11]. On this path, new classes of models in biomechanics, psychophysics of sensory perception and algebraic biology, as well as new approaches in biotechnology and biotechnical systems, including artificial intelligence systems, are being developed.

Additional materials were obtained to the well-known statement based on the achievements of molecular biology: life is a partnership between genes and mathematics [12].

References

[1] Weyl H 2016 *Symmetry* (Princeton, N.J: Princeton Univ. Press)
[2] Shubnikov A V and Koptsik V A 2004 *Symmetry in Science and Art* (Moscow: RCHD)
[3] Petoukhov S V 1987 *Higher symmetries in the biomechanics of shaping* (Moscow: Institute of crystallography)
[4] Petoukhov S V 1988 *Geometries of living nature and self-organization algorithms* (Moscow: Znanie)
[5] Petoukhov S V 2020 Hyperbolic rules of the oligomer cooperative organization of eukaryotic and prokaryotic genomes *Preprints* 2020050471
[6] Patten B M 1959 *Human embryology* (Moscow: Medgiz)
[7] Gonzalez R and Woods R 2007 *Digital Image Processing* (Upper Saddle River, N.J: Prentice Hall)
[8] Forsyth D A and Jean Ponce 2011 *Computer Vision: A modern Approach* (Boston: Pearson)
[9] Available at: https://www.youtube.com/channel/UC8JLsuRzzPsRiHwrwEjMCtw
[10] McFadden J and Al-Khalili J 2018 *Proc. R. Soc. A* **474(2220)** 20180674
[11] Bernstein N A 1966 *Essays on the physiology of movements and the physiology of activity* (Moscow: Medicine)
[12] Stewart I 1999 *Life's Other Secret: The New Mathematics of the Living World* (New York: Wiley)

IOP Conf. Series: Materials Science and Engineering 1129 (2021) 012047 doi:10.1088/1757-899X/1129/1/012047

Overview of possibilities of genometric information systems

I V Stepanyan and M Y Lednev

Mechanical Engineering Research Institute of the Russian Academy of Sciences, 4,M. Kharitonyevskiy Lane, Moscow, 101990, Russia

neurocomp.pro@gmail.com

Abstract. The actuality of the research is due to the development of methods of matrix genetics and the genometric approach that emerged from it. The article presents the results of combining these developments in a single software package in the form of a genometric information system. The developed information system is implemented in C ++ 14 using the Qt library. The software package implements the parameterization of the following variables: scale parameter, offset relative to the beginning of the sequence, overlap step. The following systems of sub-alphabets are implemented: nucleotide, dinucleotide. Visualization types are implemented: structural, frequency, integral. Visualization spaces are implemented: one-, two- and three-dimensional. In the implementations of the algorithms, methods of parallel computations were applied. The introduction of a new parameter - the overlap step - is presented, which allows obtaining clearer graphic displays. An overview of modern bio-information systems and the main directions of their development is given. Genometric bio-informatics tools play a fundamental role in understanding the modulation mechanisms of biological signals and in proposing new hypotheses.

1. Introduction. Bioinformation systems and the concept of a genometric information system

The size of biological databases is growing rapidly, that requires the development of solutions for their accelerated analysis. Bioinformation systems are the tools for analyzing biological data, including the extraction of a new knowledge. Bioinformatics tools play a fundamental role in understanding the modulation mechanisms of biological signals and in proposing of a new hypotheses.

Today bioinformatics resources are scattered across several databases and often do not have a single portal for annotating and analyzing data. The National Center for Biotechnology Information (NCBI) created the dbSNP database [1], the data of which is integrated with other information sources - GenBank, PubMed, LocusLink, and others [2]. Paper [3] discusses several publicly available bioinformatics tools and data resources that are widely used for integrated multivariate analysis and imaging for cancer immunotherapy purposes. Work [4] describes the GENEASE system with access to more than 50 different databases to facilitate the search for genes and their analysis in real time. At the same time, the totality of bioinformatics resources is rapidly expanding, which poses a challenge for scientists to select appropriate tools. To address this challenge, the European Biological Information Infrastructure has provided a portal https: /bio.tools [5], through which scientific communities and developers can describe and share their own bioinformatics resources.

Highly specialized bioinformation systems are developing intensively. Review [6] is devoted to new automated instruments and tools related to mass spectrometry. In [7], a review of about a thousand bioinformatics tools developed for the study of micro-RNAs was made. The review [8] collected about one hundred bioinformatics tools in publications devoted to circular RNAs. In [9], it is shown that bioinformatics tools facilitate the acquisition of biological data for visualization of phosphorylation networks for efficient tracking of biological problems in a clear and concise manner. In [10], bioinformatics tools for CRISPR-Cas9 research are considered to facilitate editing of target genes in eukaryotic genomes. Clustering and classification of biological data on gene expression offers a powerful approach to cancer detection [11].

In doctoral dissertation [12], a multithreaded GPU alignment solution using CUDA based on the NVIDIA TESLA C2070 GPU card with 448 cores is proposed. Work [13] describes a Ribbon alignment visualization tool that uses an intuitive visualization technique to compare genomes.

Bioinformation frameworks are also intensively developing. BioJava [14] uses the Java language to implement bioinformatics applications by implementing independent modules. BioShell [15] is an open source package for processing biological data using a set of command line utilities for the Linux environment (functionality of this package assumes knowledge of C ++ or Python).

However, it should be noted that despite the emergence of new algorithms for studying large biological data, the disadvantage of all these systems is the impossibility of a simple, understandable and convenient visual identification of genetic objects. Thus, there is a need to develop powerful tools for efficient and effective solution of specific genomics issues based on imaging techniques.

Based on the results described in [16], S.V. Petoukhov convincingly showed that the genetic code is an algebraic code, the features of which are associated with hypercomplex double (or hyperbolic) numbers and double stochastic matrices. Genometry is a group of so-called DNA-algorithms for multiscale information display with the using of a system of orthogonal Walsh functions encoding the main chemical characteristics of DNA, with the possibility of visualization in various parametric spaces. These algorithms were developed in [17-20] based on the matrix genetics of S.V. Petoukhov [16, 21-22]. Genometry contributes to gaining new knowledge about the properties of nucleotide sequences through visualization algorithms that display the cluster structure of genetic information.

In this regard, at this work, the concept of a genometric information system is introduced. This is an information system that serves to collect, analyze, store and process information using DNA algorithms. The relevance of the study is due to the development of methods of matrix genetics and the genometric approach that emerged from it. These developments must be combined into a single software package - a genometric information system for practical use.

The aim of the study is to develop an genometric information system. The task of the study is to review the main capabilities of the genometric information systems. The research objective is the accumulation of genometry methods in a single multi-module bioinformation application.

2. Materials and methods: Walsh genetic functions and Hadamard matrices

The biological meaning of the genometric approach lies in the basic DNA-algorithm, which consists of three main stages - scaling, parameterization, and visualization. Each of the stages has different ways of implementation, which affects the final result of genometric processing. Due to the variety of options at each step of the basic algorithm, we are talking about a group of DNA algorithms. Let's consider these stages.

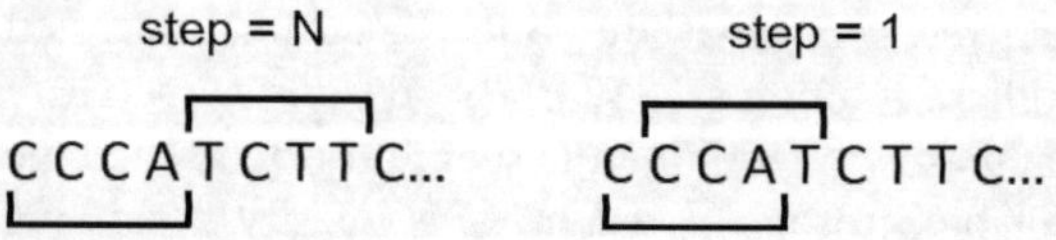

Figure 1. Step of overlapping transformation window.

Scaling stage. A sequence of characters from the set {A, G, C, T / U} is divided into fragments of equal length N. These fragments are called N-measures or N-plets. The development of this stage is the introduction of a new parameter - the overlap step. This parameter provides overlapping of N-plets (figure 1). It turned out that the introduction of this parameter allows to obtain a clearer genometric visualizations, which is associated with an increase in the number of points, the coordinates of which reflect N-plets that displays the nucleotide composition.

Parameterization stage. There is a known system of genetic sub-alphabets that implement the system of Walsh's orthogonal genetic functions (figure 2): G = C "3 hydrogen bonds" / A = T "2 hydrogen bonds"; C = T "pyrimidines" / A = G "purines"; A = C "amino" / G = T "keto". At the stage of parameterization, the nucleotide sequence is represented in the form of three binary sequences (channels) in accordance with the system of these genetic sub-alphabets.

Visualization stage. The resulting binary record of fragments is their representation in the form of three sequences of uniquely identifying values (relative or absolute frequencies, the number of certain symbols in each N-measure, decimal representations and other parameters of N-ples) with visualization in one or another coordinate system.

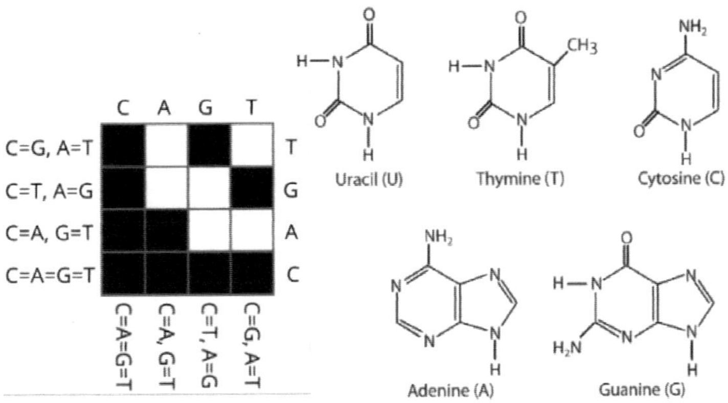

Figure 2. Purine bases: adenine, guanine, pyrimidine bases (from left to right): uracil, thymine, cytosine. Left - a variant of the Hadamard matrix, displaying the encoding of nucleotide sub-alphabets.

There is a variant of the DNA algorithm for dinucleotide imaging based on the alphabet system of 16 doublets. In [21] a system of sub-alphabets are given, which were obtained by analyzing the mosaic matrix of doublets [C A; T G] $^{(2)}$ as variants of the decomposition of this matrix. These sub-alphabets are shown in table 1.

Table 1. Sub-alphabets for systems of 2-block U-complex numbers, which S V Petoukhov obtained in the analysis of the Walsh representation of the mosaic genomatrix [C A; T G]$^{(2)}$.

1 sub-alphabet UC1	3 sub-alphabet UC3
CC = CG = TC = TG = 00e	CC = GC = CA = GA = 00s
TT = TA = CA = CT = 01e	AC = TC = TA = AA = 01s
GG = GC = AG = AC = 10e	GG = CG = GT = CT = 10s
AA = AT = GT = GA = 11e	TG = AG = AT = TT = 11s
2 sub-alphabet UC2	4 sub-alphabet UC4
CC = CG = AT = AA = 00q	CC = GG = AG = TC = 00p
CA = CT = AC = AG = 01q	AA = TT = CT = GA = 01p
GG = GC = TA = TT = 10q	AT = TA = CA = GT = 10p
GT = GA = TC = TG = 11q	CG = GC = AC = TG = 11p

3. Results and discussion. Overview of the main capabilities of genometric information systems

The developed information system is implemented in C ++ 14 using the Qt library. In the implementations of the algorithms, methods of parallel computations were applied. The main feature of the developed genometric information system is the multi-scale visualization of the physicochemical parameters of genetic nucleotide sequences in various parametric spaces, taking into account various systems of genetic sub-alphabets.

Compared to earlier versions of the software implementation in the Python language, the computational speed has significantly increased, new possibilities have appeared for constructing projections of the DNA nucleotide composition (including triangular ones, see figure 4). A convenient graphical user interface has appeared. Switching projection types is implemented in the interface of the software package.

Two additional transformations were implemented - frequency and integral. As noted, when building visualizations, it is possible not only to convert N-ples from binary to decimal representation (structural transformation). With the frequency type of visualization, each coordinate of every point is the frequency of occurrence of the N-plet in the sequence for the corresponding sub-alphabet. In the integral type of visualization, each point corresponds to coordinates, the values of which are calculated by summing the ones (or zeros) of the corresponding N-plet of the corresponding sub-alphabet.

The software package implements the parameterization of the following variables: scale parameter N. The parameter N can be set from the range 1-1023; offset relative to the beginning of the sequence; step of overlapping the transformation window. The following systems of sub-alphabets have been implemented: nucleotide (figures 3-5); dinucleotide, based on alphabets of 16 doublets (figure 6). The following types of visualization are implemented: structural (figures 3-6), integral (figures 8-9), frequency (figures 7, 10). The following spaces are implemented: one-dimensional (figure 7), two-dimensional (figures 3-9), three-dimensional (figure 10).

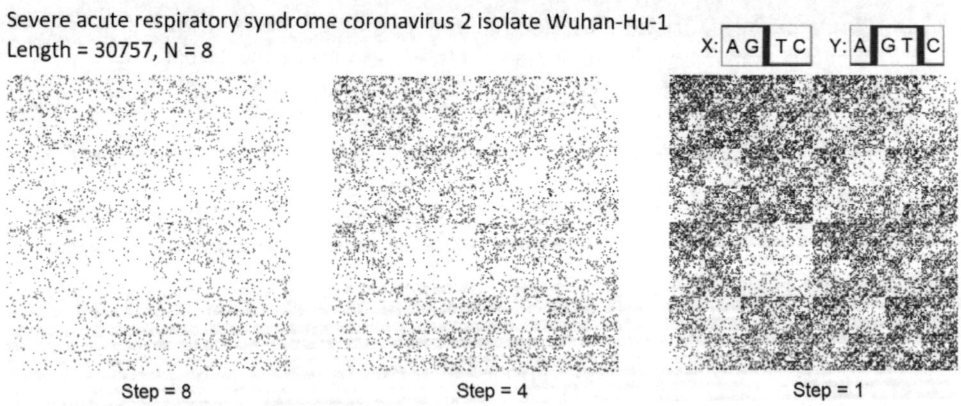

Figure 3. Two-dimensional structural genometrics of the nucleotide composition of the genome of coronavirus at different parameters of the overlap step. The abscissa and ordinate axes correspond to the decimal representations of the binary encoding of each N-plet.

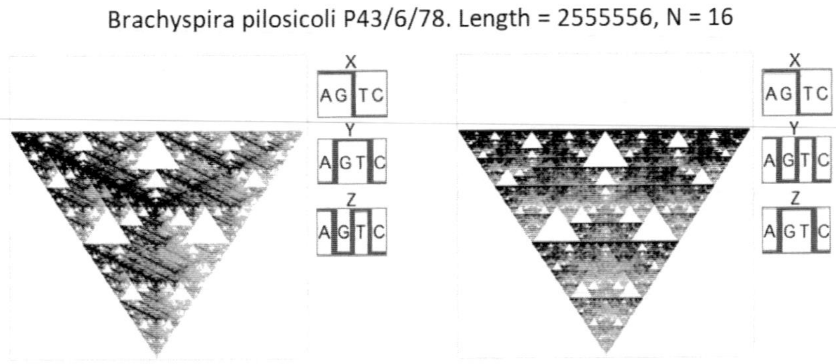

Figure 4. Triangular projections of three-dimensional representation according to the system of nucleotide sub-alphabets of the genome composition of the anaerobic bacterium.

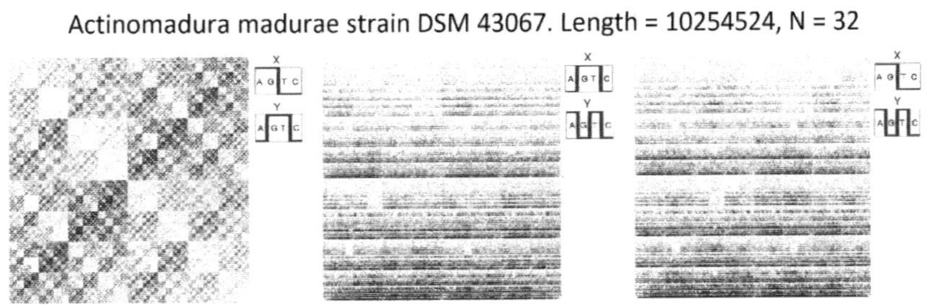

Figure 5. Two-dimensional structural genometrics of 3 projections of the nucleotide composition of the actinobacterium genome. The abscissa and ordinate axes correspond to the decimal representations of the binary encoding of each N-plet.

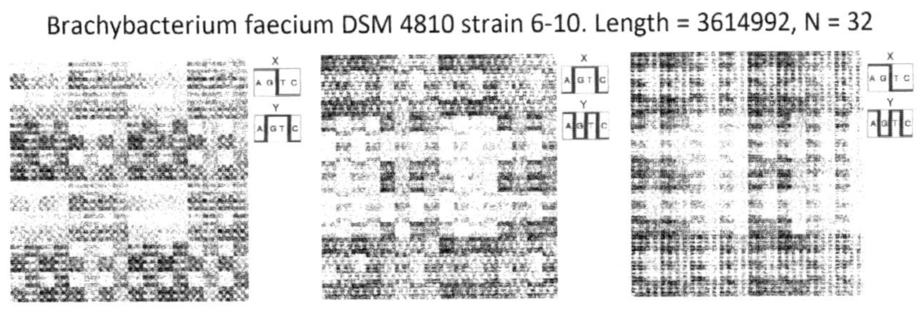

Figure 6. Two-dimensional structural genometrics of 3 projections of the nucleotide composition of the actinobacterium genome. The abscissa and ordinate axes correspond to the decimal representations of the binary encoding of each N-plet.

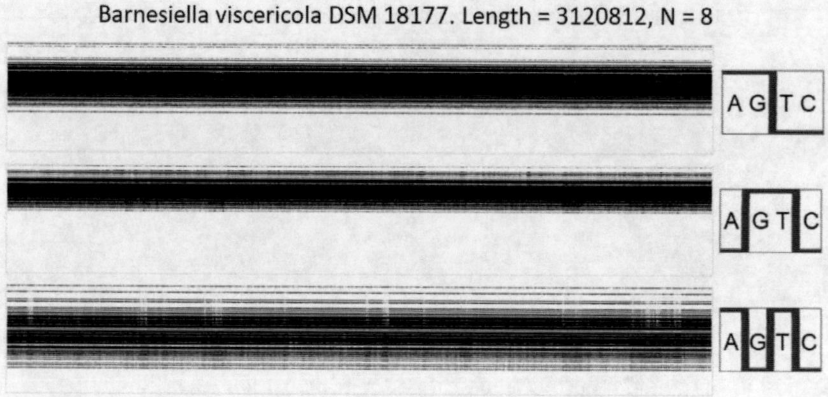

Figure 7. One-dimensional frequency genometrics of the nucleotide composition of the genome of the anaerobic bacterium in three-channel representation.

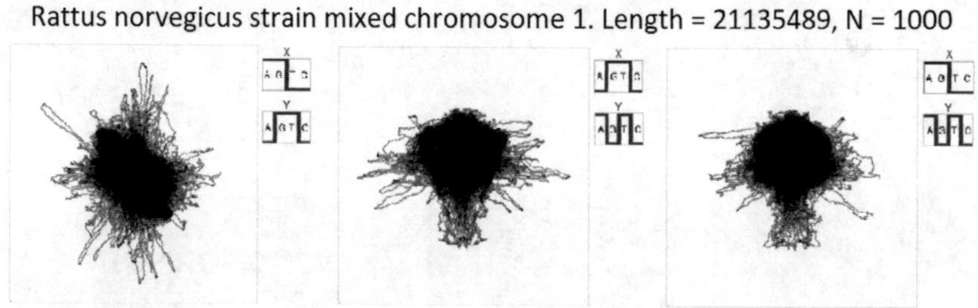

Figure 8. Two-dimensional integral genometrics of the nucleotide composition of the genome of the first chromosome of gray mouse. The coding step is 1. The abscissa and ordinate axes correspond to the number of ones in the binary coding of each N-plet.

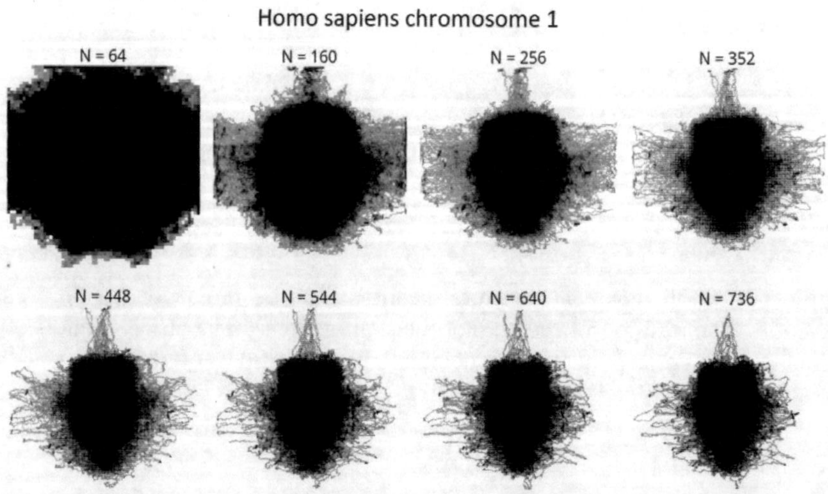

Figure 9. Two-dimensional integral genometrics of the genome of the first human chromosome. The coding step is 1. The abscissa and ordinate axes correspond to the number of ones in the binary coding of each N-plet.

In figure 10 the effect of natural frequency clustering in the three-dimensional space of nucleotide sub-alphabets is shown. Similar clustering effects are inherent in the frequency distributions of N-mers in the DNA of living organisms. The corresponding tools are provided in the developed genometric information system.

A submodule for simultaneous display on a whole range of changes of the parameter N (with a certain step) at once, which facilitates viewing several scales in one window has been developed. An example of such of multiscale visualization of genetic structures is shown at figure 11.

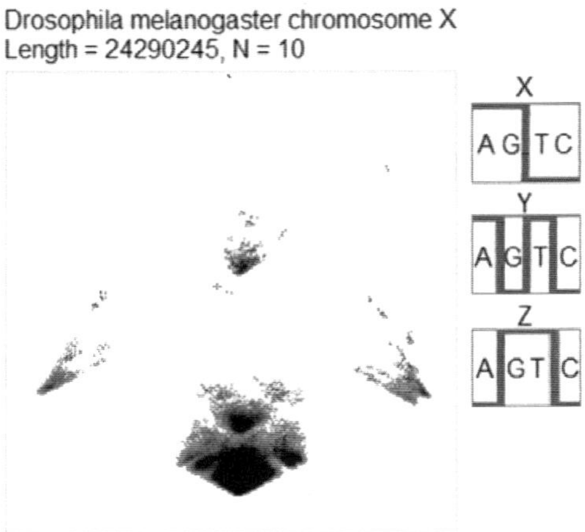

Figure 10. Three-dimensional frequency genometrics of fly genome chromosome X.

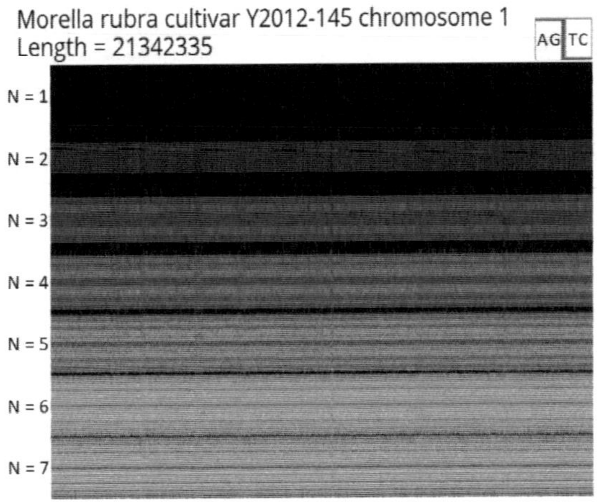

Figure 11. Multiscale one-dimensional structure visualization of the red bayberry.

4. Conclusions

- The developed genometric information system is a complex of computational and logical modules and means of multiscale visualization: structural, frequency and integral in various parametric spaces of various dimensions.
- The work shows some options for visualization. Fractality is a characteristic property of visualizations. This indicates the fractional dimension of genometric objects that characterize the fractal properties of the objects under study - DNA and RNA of various organisms and viruses.
- Implemented in the genometric information system, the toolkit is also applicable for symbolic sequences of various nature, after their transformation to tetra-code. This will allow studying the properties of digital signals with the using of genometric information systems.
- The transition from Python to C ++ allowed us to accelerate the construction of mappings by ~ 10 times and achieve visualization capabilities with large N (up to 1023).
- The concept of genometric information systems contributes to the development of bioinformatics due to the instrumental means of monitoring genetic information, the ability to scale and select the type of DNA algorithms.

References

[1] Sherry S T, Ward M and Sirotkin K 1999 *Genome Res.* **9** pp 677-9
[2] Sherry S T, Ward M, Kholodov M, Baker J, Phan L, Smigielski E M and Sirotkin K 2001 dbSNP: the NCBI database of genetic variation *Nucleic Acids Research* **29(1)** pp 308-11
[3] Palmisano A, Krushkal J, Li M C, Fang J, Sonkin D, Wright G and McShane L 2020 Bioinformatics tools and resources for cancer immunotherapy study *Biomarkers for immunotherapy of cancer* (New York: Humana) pp 649-78

IOP Conf. Series: Materials Science and Engineering 1129 (2021) 012047 doi:10.1088/1757-899X/1129/1/012047

[4] Ghandikota S, Hershey G K K and Mersha T B 2018 GENEASE: real time bioinformatics tool for multi-omics and disease ontology exploration, analysis and visualization *Bioinformatics* **34(18)** 3160-8
[5] Ison J, Ménager H, Brancotte B, Jaaniso E, Salumets A, Raček T and Hancock J M 2020 Community curation of bioinformatics software and data resources *Briefings in bioinformatics* **21(5)** 1697-705
[6] Tsai P L and Chen S F 2017 A brief review of bioinformatics tools for glycosylation analysis by mass spectrometry *Mass Spectrometry* **6(2)** 64
[7] Chen L, Heikkinen L, Wang C, Yang Y, Sun H and Wong G 2019 Trends in the development of miRNA bioinformatics tools *Briefings in Bioinformatics* **20(5)** 1836-52
[8] Chen L, Wang C, Sun H, Wang J, Liang Y, Wang Y and Wong G 2020 The bioinformatics toolbox for circRNA discovery and analysis *Briefings in Bioinformatics*
[9] Shi X X, Wu F X, Mei L C, Wang Y L, Hao G F and Yang G F 2020 Bioinformatics toolbox for exploring protein phosphorylation network *Briefings in Bioinformatics*
[10] Alkhnbashi O S, Meier T, Mitrofanov A, Backofen R and Voß B 2020 CRISPR-Cas bioinformatics *Methods* **172** pp 3-11
[11] Mabu A M, Prasad R and Yadav R 2020 Mining gene expression data using data mining techniques: A critical review *J. of Information and Optimization Sci.* **41(3)** pp 723-42
[12] Said M A M 2017 Parallel acceleration of bioinformatics algorithms *Doctoral dissertation*
[13] Nattestad M, Aboukhalil R, Chin C S and Schatz M C 2020 Ribbon: Intuitive visualization for complex genomic variation *Bioinformatics*
[14] Bhongale M P and Vyas R 2020 Role of BioJava in the Department of bioinformatics tools *Advances in Bioengineering Springer Singapore* pp 29-49
[15] Macnar J M, Szulc N A, Kryś J D, Badaczewska-Dawid A E and Gront D 2020 BioShell 3.0: Library for processing structural biology *Data Biomolecules* **10(3)** 461
[16] Petoukhov S V 2020 The genetic code, algebraic codes and double numbers
[17] Stepanyan I V and Petoukhov S V 2017 The matrix method of representation, analysis and classification of long genetic sequences *Information* **8(1)** 12
[18] Stepanyan I V and Khussein A M 2019 Scaling and visualization of nucleotide sequences *EPJ Web of Conf. IV Int. Conf. Modeling of Nonlinear Processes and Systems* **224** (Russia: Moscow)
[19] Stepanyan I V 2020 Scale-parametric visualization of nucleic acids *Sci. visualization* **12(3)** pp 61-78
[20] Stepanyan I V 2020 Biomathematical system of methods for describing nucleic acids *Computer research and modeling* 12(2) pp 417-34
[21] Petoukhov S V 2017 Genetic coding and united-hypercomplex systems in the models of algebraic biology *Biosystems* **158** pp 31-46
[22] Petoukhov S V, Petukhova E S and Svirin V I 2019 Symmetries of DNA alphabets and quantum informational formalisms *Symmetry: Culture and Sci.* **30(2)** pp 161-79

IOP Conf. Series: Materials Science and Engineering 1129 (2021) 012048 doi:10.1088/1757-899X/1129/1/012048

Cognitive ergonomics of DNA-algorithms

I V Stepanyan

Mechanical Engineering Research Institute of the Russian Academy of Sciences, 4,M. Kharitonyevskiy Lane, Moscow, 101990, Russia

neurocomp.pro@gmail.com

Abstract. To improve the ergonomic qualities of graphical interfaces and optimize the visual perception of large data presented in text form, a method of information processing based on DNA algorithms is proposed. These algorithms were developed by the author in the study of the molecular genetic and structural properties of long nucleotide sequences. The result of processing information presented in text form is an object of discrete geometry, two-dimensional projections of which can serve for visual identification of the internal structure of information, as well as for interpretation as characteristic symbols. At the same time, the considered transformations have the property of visualizing the characteristic cluster structure of information. It is shown that genetic nucleotide sequences, as well as texts in natural language, lend themselves to visual classification, which contributes to an increase in the productivity of mental work when analyzing big data. It is shown that DNA algorithms can reduce the load on the visual analyzer during the initial examination of data presented in text form. This makes it possible to apply this method when operators interact with various information streams and text data of arbitrary nature, as well as to improve the ergonomic qualities of specialized graphical interfaces.

1. Introduction

The quality of information perception depends on the way it is presented. Any information stream can be represented as a sequence of characters of a finite alphabet. A person's ability to perceive information through a visual analyser is closely related to the ability to visualize data. Despite progress in the field of machine learning, the problem of analyzing and visualizing big data has not yet been completely solved. In this regard, in this work, the task is to develop such a computer system that allows to visualize large amounts of information that are not available for quick and primary analysis in their original form due to their significant volume.

This article presents the author's understanding of Gaussian numbers [1] from the standpoint of genetic coding. Gaussian numbers (complex numbers whose real and imaginary parts are integers) have long attracted the attention of researchers. The author initially set the task to represent the DNA molecule with exactly two-dimensional (complex) numbers. This turned out to be possible due to the system of orthogonal Walsh functions encoding the physico-chemical properties of nucleotides when displaying the molecular genetic sequence.

Earlier, to optimize mental work, we developed an algorithm for automatically generated cognitive graphics based on the neurosemantic approach [2-3] for transforming a linear information flow into a hierarchical graph using the mechanism of auto-structuring [4]. This contributes to the measurement, assessment and regulation of information, the achievement of the adequacy of mental stress and prevention of neuro-emotional overstrain of operators interacting with large amounts of information.

In 1990, John Archibald Wheeler proposed the "it from bit" doctrine, according to which information is a fundamental concept in physics [5]. The Hawking paradox about the disappearance of information in black holes is known [6]. At the same time, information is one of the occupational health risk factors. In [7] was mentioned that from 64% to 90% of computer users experience computer syndrome as a reaction of the body to work with a computer (in particular, it is pain in the back, neck and arms). At the same time, new methods of professional ergonomics are being developed for human interaction with various information systems [8]. In [9], data on clinical, pathogenetic, prognostic and therapeutic aspects of computer syndrome were studied. In [10-12], cognitive factors that must be taken into account when developing decision support systems based on large amounts of data are identified, in particular, cognitive limitations of a person, manifested in the context of data analysis, are considered. Thus, the principles of cognitive ergonomics can help to reduce health risks and optimize a mental work when operators interact with information flows with large amounts of data.

From the point of view of cognitive ergonomics, the currently developing methods of automatically generated computer graphics for the analysis of big data are of interest. In [13], a system for automatic text visualization was developed, which receives text in a natural language as input and creates an interactive three-dimensional scene on its basis. In [14], a method for obtaining information from a text using text network analysis is presented. The automatically generated structure, in combination with other indicators, can be used to determine the level of cognitive diversity of the text.

Technologies for the analysis of genetic information are also being developed. OrganellarGenomeDRAW [15] is a tool for drawing graphic maps of plastid and mitochondrial genomes with access to a local copy of the NCBI RefSeq organelle genome database, which allows visualizing large sets of organelle genomes.

2. Calculation model: genometric approach to big data analysis

Genometry is a group of DNA algorithms for multiscale information display developed by the author through the use of a system of orthogonal Walsh functions with the ability to visualize in various parametric spaces. The basic algorithm has three main steps. 1) Scaling. A sequence of characters from the set {A, G, C, T} or {A, G, C, U}, encoding nitrogenous bases, is divided into fragments of equal length N, where N is a free parameter of the algorithm - the length of "semantic units". The resulting fragments of equal length will be called N-measures or N-plets. 2) Parametrization. Taking into account the system of genetic sub-alphabets: G = C "3 hydrogen bonds" / A = T "2 hydrogen bonds"; C = T "pyrimidines" / A = G "purines"; A = C "amino" / G = T "keto" sequence of nitrogenous bases can be represented as three binary sequences, consisting of zeros and ones. The choice of encoding method (what is considered to be zero or one) affects the symmetry transformations of the final rendering. 3) Visualization. The resulting binary record of fragments is their representation in the form of three sequences of decimal or other uniquely identifying values (code frequencies, the number of certain symbols in each N-measure, and other characteristic parameters). So, converting binary N-plets to decimal numbers allows to display them in the selected coordinate system. Other numerical interpretations of N-plets for each of the sub-alphabets are possible, considered in particular at works [16-17] based on the matrix genetics of prof. Sergey Petoukhov [18-19]. Therefore, we are talking about a family of DNA-algorithms. The obtained values set the coordinates of points in visualization parametric spaces.

The study of long genetic nucleotide sequences using a genometric approach can be attributed to the problem of visual perception of big data. A conceptual diagram with options for presenting genetic information to the operator's visual analyser is shown in figure 1.

It should be noted that any information flow can be represented as a sequence of characters of some finite alphabet. An arbitrary signal can be recoded and represented in a quaternary code using the genetic alphabet of nucleotides by analogy with how information is stored in a computer's memory in binary form. The translation of the information signal in the form of a sequence of symbols of an arbitrary finite alphabet into a quaternary code allows the use of DNA visualization algorithms. At the same time, the transformations retain the properties of displaying a characteristic cluster structure of information, if the signal is not chaotic (see figures 2, 3).

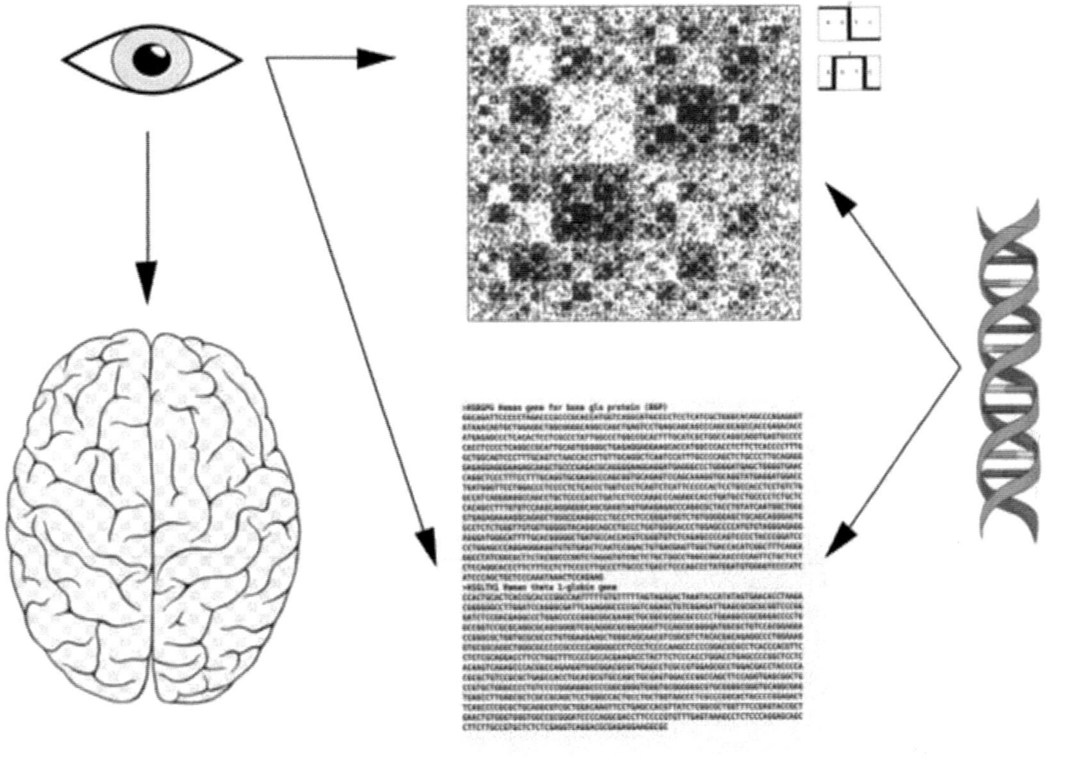

Figure 1. Comparison of the variant of visual perception of the nucleotide sequence in the form of a sequence of letter designations (DNA fragment on the bottom) and visualization of the whole chromosome in a two-dimensional space of binary-orthogonal functions of physicochemical parameters (all DNA on the top). The flow of information to the operator's visual analyzer is shown schematically. The left hemisphere is responsible for the logic and analysis of textual information, the right hemisphere is responsible for the genometric image. The DNA algorithm allows you to additionally use the right hemisphere to increase mental productivity.

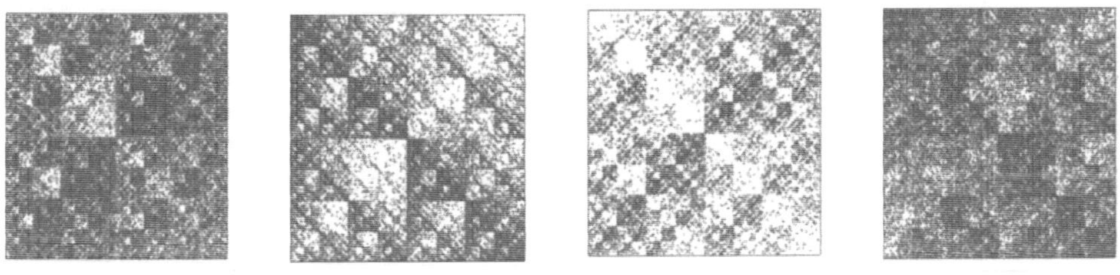

(a) (b) (c) (d)

Figure 2. Genometric infographics. Examples of visualization of genetic information: genome of the bacterium Ralstonia eutropha (a), genome of the bacterium Candidatus Arthromitus (b), genome of the bacterium Actinomadura madurae (c), genome of Emiliania huxleyi (d).

The result of genometric processing of arbitrary information is a visualization, which can be interpreted as a sign or symbol. Various visualization options make it possible to evaluate the symbolic composition of the information signal and track its changes in different sections with the possibility of scaling at different values of N. Also, using scaling, you can evaluate the structure of a particular section of the signal.

Thus, DNA algorithms make it possible to automatically generate cognitive computer graphics for multiscale visualization of the internal structure of signals of various nature. This can be useful for visualizing and identifying areas with different information content within a particular information signal, as well as for comparative analysis of various signals.

3. Results and discussion. Perception of information flows and big data in a text form

At figures 1-4 the results of the transformations that reveal the characteristic structure of the original information is shown. As studies have shown [16], a randomly generated symbolic sequence gives a genometric pattern, all the points of which are scattered chaotically. In this regard, the DNA algorithm can be used to visually evaluate frequency distributions in pseudo-random signals. This can be used, in particular, to evaluate the performance of noise generation algorithms. An example of visualization of the frequency distribution of the signal is shown in figure 3. It can be seen that, according to two sub-alphabets, the frequencies are periodic, which indicates the periodic nature of the original signal.

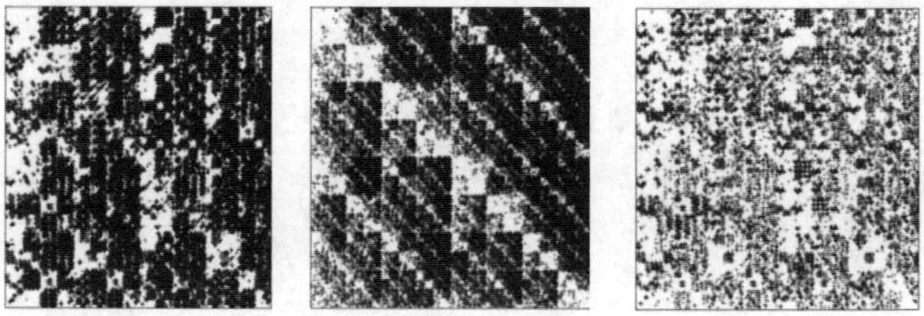

(a) (b) (c)

Figure 3. Genometric infographics. Examples of tetra [A, G, C, T] representations of various texts in Russian: the Bible phonetic composition, different projections (a), (b), "12 chairs" by Ilf and Petrov in representation of Unicode (c).

The ability to find the internal ordering of the information signal in the form of an individual fractal structure of genometric visualization is an important feature of DNA-algorithms. In this regard, the use of DNA-algorithms for the primary study of big data presented in text form can be used as a tool for the cognitive computer graphics. This makes it possible to apply this method to expand the capabilities of operators' interaction with large data of arbitrary nature, as well as to improve the ergonomic qualities of specialized graphical user interfaces (GUIs).

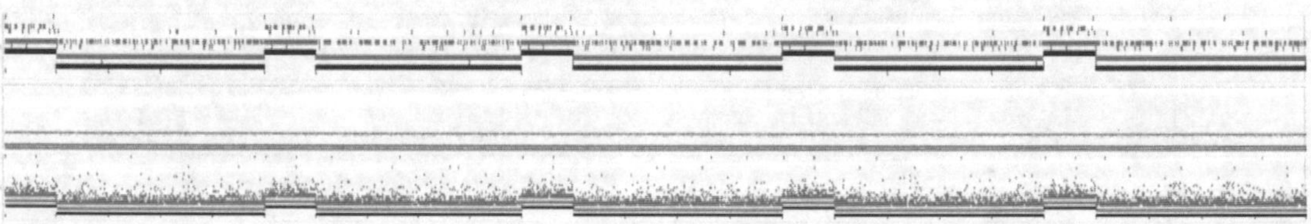

Figure 4. Frequency representation of the signal, decomposed into three genetic sub-alphabets. The abscissa axis is the serial number of the 12-measure of the original sequence, the ordinate is the frequency of the corresponding 12-measure. The first, second and third sub-alphabets are shown in three separate diagrams (top to bottom).

One of the promising options for using the results of research in the field of ergonomics of perception based on DNA-algorithms is the possibility of help of non-specialists to compare the genetic and phenotypic data of living organisms, as well as to search for previously unknown patterns in other subject areas. The methodology for researching visual information with the involvement of a large number of users has been shown to be effective and is known as "crowd surfing". It should also be noted the new possibility of labeling food and seeds (by analogy with a barcode) in addition to simple genometric annotation of genomes.

Due to the fact that the features in the nucleotide composition of DNA/RNA of various organisms are clearly visible, it can be concluded that an ergonomic visualization of molecular genetic data based on the proposed approach is possible. For the analysis of textual information, DNA-algorithms are also applicable, but it seems that, unlike genetic texts, texts in natural languages do not differ from each other so much as different DNAs. At the same time, large volumes of textual data of various subjects in natural or programming languages, as well as binary data stored in the memory of computers, can have genometric representations available for visual analysis in various parametric spaces. Visualizations of sound files, biomechanical, biomedical signals, etc. are of practical interest.

4. Genometric Information System

Genometric information system is an information system used to collect, analyse, store and process information using DNA-algorithms. One of the important properties of genometric information systems is the possibility of using technologies of collective intelligence and knowledge engineering. Mind is viewed as a more general category than intelligence. Mind is inherent only in living organisms. Three directions can be distinguished based on a single technical and organizational solution:

- scientific analysis of big data incl. biological, literary and musical;
- application of DNA visualization technologies in the educational process;
- the use of the results and methods in the field of scientific art to change the way of perception of natural and creative objects. Since the method can be positioned within the framework of the direction of scientific art, it becomes possible to attract humanitarian organizations to conduct virtual exhibitions.

Thus, DNA-algorithms allow the creation of new ergonomic instruments for scientific research. Opportunities are also emerging to popularize scientific fields such as molecular genetics and Big Data through the ability to generate specific digital media content.

5. Conclusion

The considered methodology makes it possible to operate with genometric mappings at the level of semantics and semiotics of Big Data. This is possible because the result of processing any information signals using-DNA algorithms are objects of discrete geometry, the projections of which can carry the function of characteristic identification symbols. This is useful in particular for enhancing the ergonomics of graphical user interfaces. At the same time, the possibility of internal interpretation of genometric representations is remains, since DNA-algorithms allow displaying the internal structure of genetic coding.

It is shown that genetic nucleotide sequences (as sequences of symbols of the genetic alphabet) and texts in natural language (as sequences of symbols of the natural language alphabet) lend themselves to visual perception, which contributes to an increase in mental productivity of operators that are interacting with Big Data.

The question of choosing a method for coding information is actual, since this can affect the result of processing by DNA-algorithms. For example, natural language texts can be analysed at the phonetic and morphological levels. In this regard, for the further systematization of genometric mappings, a coding method is important, taking into account the peculiarities of the internal structure of the information signals themselves. It is of particular interest to take into account the cyclic fluctuations of various biological processes, such as heart rate cycles, circadian rhythms, etc., analysis of the timing structure of musical works, speech intensity, etc. This is possible using the presented approach.

References

[1] Gauss C F 1831 Anzeige von Theoria residuorum biquadraticorum, commentatio secunda *Göttingische gelehrte Anzeigen* **23** pp 169-78
[2] Stepanyan I V and Denisov E I 2015 A method for reducing information loads based on semantic clustering *Occupational Medicine and Industrial Ecology* **9** pp 136-7
[3] Stepanyan I V, Denisov E I, Eremin A L, Bodyakin V I and Saveliev A V 2014 Algorithms for the optimization of intellectual work by methods of information visualization using cognitive semantic graphics *Neurocomputers: development, application* **7** pp 53-9
[4] Bodyakin V I 2012 The concept of building self-learning information management systems based on the neurosemantic paradigm *Managing the Development of Large-Scale Systems (MLSD'2012): Proc. of the VI Int. conf.* **2** pp 289-98
[5] Plastino A 2004 A conceptual framework for the Wheeler program *Physica A: Statistical Mechanics and its Applications* **340(1-3)** pp 85-91
[6] Alonso-Serrano A, Dąbrowski M and Gohar H 2018 Generalized uncertainty principle impact onto the black holes information flux and the sparsity of Hawking radiation *Phys. Rev.* D **7(4)** 044029
[7] Teo C, Giffard P, Johnston V and Treleaven J 2019 Computer vision symptoms in people with and without neck pain *Applied ergonomics* **80** pp 50-6
[8] Radwin R G, Lee S, Li K, Lieblich M and Daniel Park B K 2016 Discussion panel on computer vision and occupational ergonomics *Proc. of the Human Factors and Ergonomics Society Annual Meeting* **60(1)** pp 958-60
[9] Munshi S, Varghese A and Dhar-Munshi S 2017 Computer vision syndrome — a common cause of unexplained visual symptoms in the modern era *Int. J. of Clinical Practice* **71(7)** e12962
[10] Walker G and Strathie A 2016 Big data and ergonomics methods: a new paradigm for tackling strategic transport safety risks *Applied ergonomics* **53** pp 298-311
[11] Araujo C L D O, Santos T J N and Galvão H M 2018 Decision-making by data, an application of data science with cognitive ergonomics *Int. J. of Advances in Agriculture Sci.* **7** pp 14-9
[12] Kalakoski V, Henelius A, Oikarinen, E Ukkonen, A and Puolamäki K 2019 Cognitive ergonomics for data analysis. Experimental study of cognitive limitations in a data-based judgement task *Behaviour & Information Technology* **38(10)** pp 1038-47
[13] Gehrmann S, Strobelt H and Rush A M 2019 GLTR: Statistical detection and visualization of generated text *Preprint arXiv:*1906.04043
[14] Paranyushkin D 2019 InfraNodus: Generating insight using text network analysis *The World Wide Web Conf.* 3584-9
[15] Greiner S, Lehwark P and Bock R 2019 OrganellarGenomeDRAW (OGDRAW) version 1.3.1: expanded toolkit for the graphical visualization of organellar genomes *Nucleic Acids Research* **47(1)** pp 59-64
[16] Stepanyan I V 2020 A multiscale model of nucleic acid imaging *Sci. Visualization* **12(3)** pp 61–78
[17] Stepanyan I V and Khussein A M 2019 Scaling and visualization of nucleotide sequences *EPJ Web of Conf.* **224** 03007
[18] Fimmel E, Gumbel M, Karpuzoglu A and Petoukhov S 2019 On comparing composition principles of long DNA sequences with those of random ones *Biosystems* **180** pp 101-8
[19] Petoukhov S V, Petukhova E S and Svirin V I 2019 Symmetries of DNA alphabets and quantum informational formalisms *Symmetry: Culture and Sci.* **30(2)** pp 161-79

IOP Conf. Series: Materials Science and Engineering 1129 (2021) 012049 doi:10.1088/1757-899X/1129/1/012049

Development of an algorithm for determining the degree of fatigue of a human operator under the influence of vibration

K V Tikunova[1], A N Brysin [1,2], S A Mikaeva [1] and V S Soloviev [2]

[1] MIREA – Russian Technological University, 20, Stromynka St., Moscow, 107996, Russia
[2]Mechanical Engineering Research Institute of the Russian Academy of Sciences, 4, M. Kharitonyevskiy Lane, Moscow, 101990, Russiatikunova@mirea.ru

Abstract. The article discusses the issue of selecting methods for studying human muscle fatigue in the operator-joystick system. The article discusses ways to measure potentials from human muscles. The applicability of the chosen method for evaluating the work of the research object is considered. This article provides an overview of the methods for diagnosing muscle fatigue parameters in the operator-joystick system.

1. Introduction

There are several methods for diagnosing the state of human muscles [1-5]; it is of interest to find a research algorithm that is suitable for diagnosing operator parameters online.

In humans, as in all vertebrates, skeletal muscle fibers have four most important properties [6]:

- excitability - the ability to respond to a stimulus by changes in ion permeability and membrane potential;
- conductivity - the ability to conduct an action potential along the entire fiber;
- contractility - the ability to contract or change the tension when excited;
- elasticity - the ability to develop tension when stretched.

Under natural conditions, muscle excitation and contraction are caused by nerve impulses coming to muscle fibers from nerve centers. Under experimental conditions, electrical stimulation is used to induce muscle contraction.

Direct irritation of the muscle itself is called direct irritation; irritation of the motor nerve, leading to a contraction of the muscle innervated by this nerve (excitation of neuromotor units), is an indirect irritation. Due to the fact that the excitability of muscle tissue is lower than that of the nervous one, the application of electrodes of the irritating current directly to the muscle does not yet provide direct irritation: the current, spreading through the muscle tissue, acts primarily on the endings of the motor nerves located in it and excites them, which leads to a contraction muscles.

In modern medicine, there are many different methods for determining pathologies in the work of the human musculoskeletal system. In work [1], magnetic resonance imaging (MRI) was used as a research method. MRI can show the presence of muscle symptoms, helps to determine the pattern of distribution of muscle involvement, indicating preserved and affected muscles and the extent of their changes, and also resolve suspicions of myopathy due to a specific clinical picture.

The study [2] used muscle sonomyography at rest and under load. The results showed that the ultrasound method can diagnose a sharp increase in blood flow.

In [3], the Doppler myography method was used to further determine the spectral characteristics of the results obtained. It was found that skeletal muscles perform low-frequency vibration movements during contraction. The states of tension and rest of muscle tissues differ markedly from each other both in the frequency characteristics of the registered movements, which is expressed, in particular, in an increase in the spectral component with increasing load.

In work [4] it was revealed that the cutaneous EMG derivation allows in a number of cases a good assessment of the synchronization process in the work of peripheral motor neurons. The author has repeatedly observed a gradual increase in the amplitude of rare rhythmic activity with an increase in the force of contraction. In this case, the oscillation frequency remains practically unchanged.

When analyzing the amplitude characteristics of muscle responses during the repetition of the speed-strength test in [5], a certain tendency was found. It was expressed in a gradual increase in the voltage of the evoked potentials up to the maximum figures by the 5th attempt. At the same time, the degree of voltage increase in different muscle groups of the right and left extremities had different values. It should be borne in mind that the magnitude of the maximum muscle response correlates with the number of functioning motor units in the muscle groups under study. Three types of relationship between the dynamics of myographic parameters and changes in sports results were revealed.

Skeletal muscle is made up of motor units (MUs), the smallest muscle elements that can be activated by willpower. MU consists of the anterior horny cell or motor neuron, an axon, and all muscle fibers excited by this axon. The number of muscle fibers per motor fiber is called the coefficient of innervation. The innervation coefficient for various muscles can range from ten to hundreds of units [6]. Muscle contraction is the sum of the stimulation and contraction of several MUs. The mechanism of muscle contraction is described by the theory of sliding threads, according to which the shortening of each muscle fiber during contraction occurs as a result of the movement of thin threads in the intervals between thick ones without changing their length [7].

Electromyography (EMG) methods are the most autonomous methods for obtaining human biopotentials. There is both invasive and non-invasive EMG. Invasive (needle-like) requires special experimental conditions. Skin EMG is used to assess the state of the body as a whole, while stimulation EMG can show changes online.

It is proposed to use two methods for obtaining EMG. First, a stimulation myogram is taken at rest, after which a "normal" curve is recorded. Then the muscle is examined in a state of overload. Both curves will be called "reference" and they are sent to the control unit. After calibration, a cutaneous EMG measurement is performed. The received data is compared with the reference values and when approaching critical values, a warning about the presence of overloads is displayed.

Thus, a method for studying muscles in the operator-joystick system was determined.

Controlling movements is unthinkable without coordinating the activity of a large number of muscles. Therefore, for the implementation of the movement, a motor program must be formed. The motor or central program is considered as a prepared set of basic motor commands, as well as a set of ready-made corrective subroutines that ensure the implementation of the movement, taking into account the current afferent signals and information coming from other parts of the central nervous system.

The mathematical model of the EMG signal can be simplified by the following equation [8]:

$$u(n) = \sum_{q=0}^{N-1} h(q)e(n-q) + w(n) \tag{1}$$

u(n) – the biopotentials of the EMG signal, h(q) – is the action potential of one MU, e(n) – the stimulus, w(n) – is the additive white Gaussian noise, N – is the number of MU involved in the contraction process.

The action potential extends from the neuromuscular joints in both directions within the muscle fiber through the tubular system [8].

Stimulation of each MU by a neuronal signal causes their contraction, as a result of which an electromyographic signal (EMG) is generated, which is the sum of the action potentials of all cells involved in the process [9-11].

Let's define the muscles for which it is necessary to conduct EMG. To control the joystick, a person uses more than 30 muscles, it will be impossible to examine the myogram for each. It is proposed to use EMG of the biceps of the shoulder as a muscle responsible for controlling the subject's arm.

2. Materials and methods

The problem of discrete control of power linear drives in [12] is considered by the example of solving a one-dimensional control problem for a vibration platform. The proposed circuit solution is shown in figure 1. The task of controlling the presented circuit using a spatial joystick is a very urgent technical problem. We will use the proposed structural diagram and implement on its basis a control system for assessing the capabilities of a human operator of a mobile machine [13, etc.]. As rods 1 and 2, we use linear actuators of variable length. We will ensure their hinge fastening at points 3 and 4, as well as at point 0. We assume that the base is stationary, and at point 0 a laser designator is installed, the movement of which depends on the movement of the Z 3 type joystick. Control is carried out by a human operator. The task is to find the control laws that minimize the "parasitic" acceleration of point 0 when performing manipulation operations using two linear angle sensors and one digital one.

A human operator as a system with delay allows realizing original control in space, since it relies not only on tactile sensors, but also on a more perfect system of spatial stabilization inherent in it by nature.

As input parameters for the control circuit, we use the signals from the "pseudo Z3 type" joystick. A Z3 type joystick is a manipulator capable of transmitting signals along coordinates, X, Y, Z. As a joystick, we will use the JoyStick V1.0 module [14, etc.].

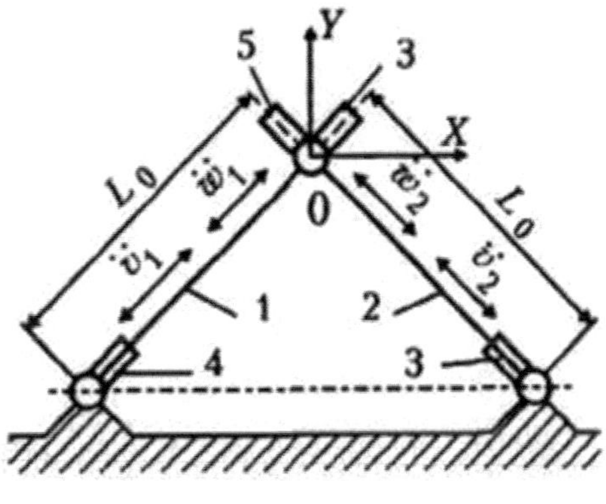

Figure 1. Prototype circuit design.

X and Y movements are monitored by analog potentiometric sensors, Z coordinates are controlled by a digital input. The Z input is digital and provides pseudo-linear control along the third coordinate, and therefore the joystick will be considered "pseudo three-coordinate". The 2D system changes its parameters along the two precision tracking coordinates. No Z movement is required. This idealization eliminates the use of a central analogue processor (DAC) in the signal processing circuit along the Z coordinates. Control by the third Z coordinate is necessary when implementing the mechanism on two coupled systems with 4 linear drives. The proposed technical solution is necessary for debugging the algorithm of interaction of the "microcontroller - joystick - power drives" system.

3. Results and discussion

To control the power drive, we will use a microcontroller system coupled with a motor control system. The microcontroller control system allows realizing control functions, it is equipped with 32 kb of flash memory, 2 kb of which is reserved for the so-called bootloader (storing the program and related static resources) [12-15].

Linear drives 3 and 6 of small power based on a DC motor with a plastic gear 5 and a linear transmission with a rubber stop 4 are used as a power drive.

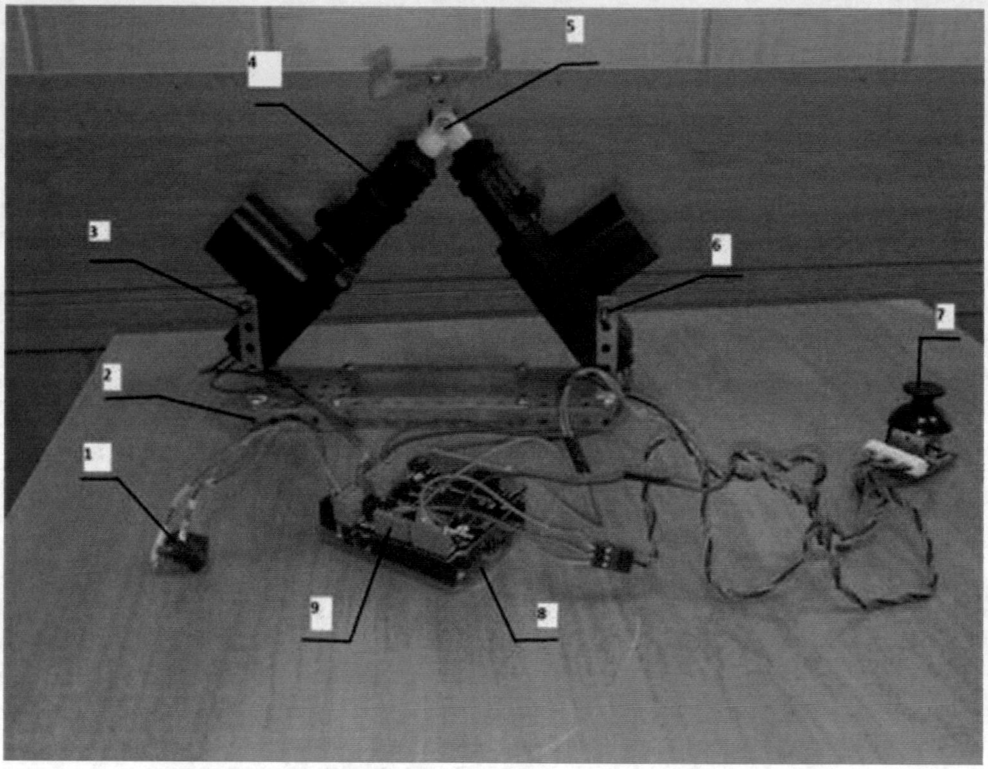

Figure 2. Physical model of a control system for two drives.

The complexity of controlling a system consisting of two drives lies in the need to simultaneously supply two control signals and the need to take into account the complex interaction when moving point 0. Due to the design features of the circuit, it is possible to provide high rigidity with small dimensions, high speed of movement of point 0 in the desired direction in comparison with analogues. To assess the operator's activities, a physical model of the control system was assembled.

The operator, visually controlling the object, sets the direction of movement, using two analog and one digital sensors (potentiometric), structurally implemented in a single control unit (joystick) 7. Information goes to the inputs of the microprocessor system 9 and based on the developed algorithm (figure 3) signals are generated controls coming to the motor control board 8. Power drives 4 are hinged on the base 2 at nodes 3, 6 (in figure 2 they correspond to points 3 and 4) between themselves and the movable platform 5 (point 0 in figure 1) A laser is installed on the platform 5 target designator powered from the power pins of the microprocessor board 9. To power the circuit, voltage is supplied from two sources. The first 6 V power supply, rated for 500 mA current, is connected to the power connector of the microprocessor board 9. The second power supply through adapter 1 with 9 V voltage, rated for at least 1A current is connected to the motor control board 8. Power supplies, switching circuits and a target designator were not shown in figure 2 to make them easier to read.

During the tests, the physical model of the system showed its operability, the control algorithm was debugged, the design possibilities for varying the base between the fasteners 3, 4 and the angles relative to the base were determined. Technological changes were made in the design of the drives, which made it possible to increase their travel from 20 mm to 25. Due to the use of more rigid protective corrugations, the function of automatic exit from the critical upper point is provided.

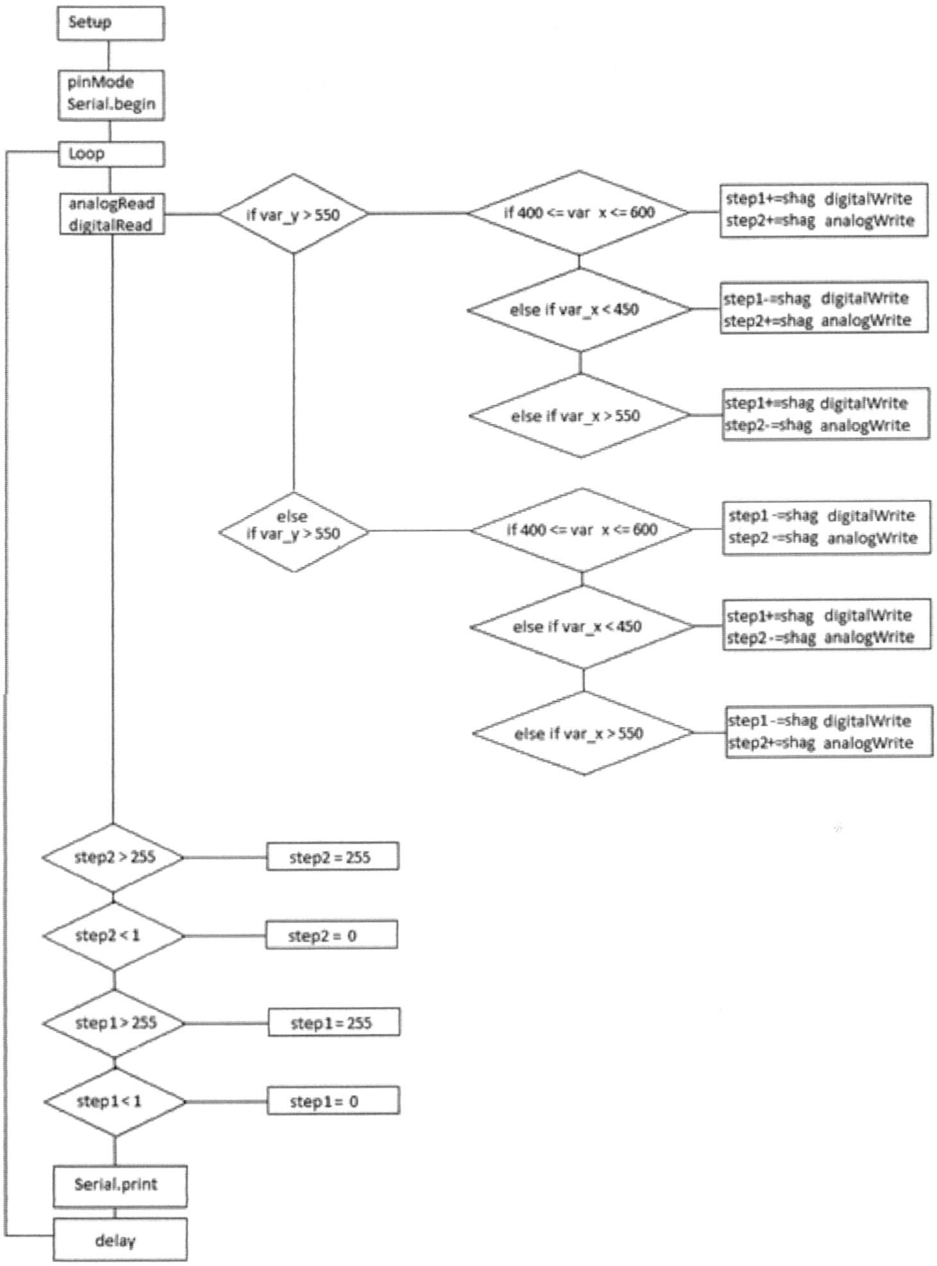

Figure 3. Control algorithm.

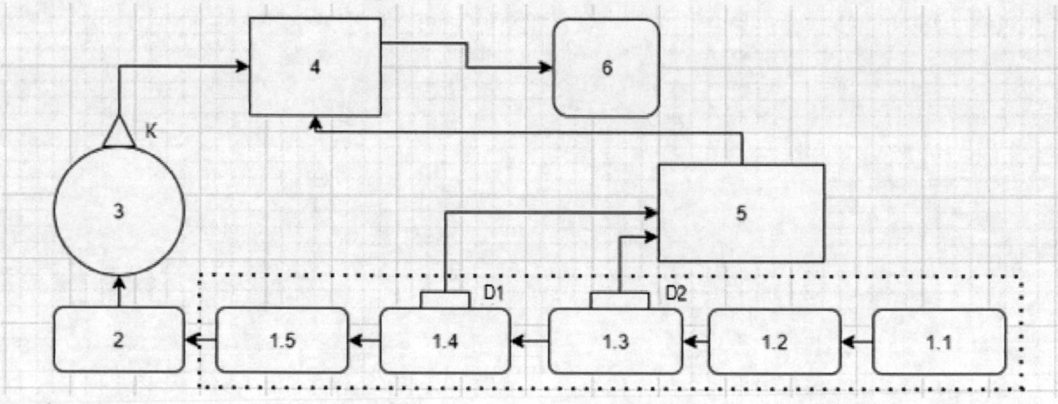

Figure 4. Research algorithm. ^{1} – operator. $^{1.1}$ – seat. $^{1.2}$ – body. $^{1.3}$ – biceps. $^{1.4}$ – hand. $^{1.5}$ – palm ^{2} – joystick. ^{3} – target. ^{4} – control system. ^{5} – data processing unit. ^{6} – PC. $^{D1, D2}$ – EMG sensors. ^{K} – hit controller.

The human operator is on the seat, electrodes are applied to the arm (biceps and forearm) to take an electromyogram. The results obtained during the experiment are sent to the data processing unit with the reference values of the EMG curves. Then the data is transferred to the control system. The palm controlling the joystick is aimed at the target with the controller, which also transmits the results to the control system. All information is transferred to a PC for further analysis. The importance of control and monitoring of dynamic processes is indicated in the work of V.M. Mikhailov [15].

References

[1] Carlier R 2014 MRI of muscles MRI of the whole body in the diagnosis and follow-up of patients with neuromuscular diseases *Neuromuscular DISEASES* **2** 16-27

[2] Delyagin V 2015 Muscle ultrasound examination in health and neuromuscular pathology *SonoAce Ultrasound* **27** 68-73

[3] Kulibaba A, Girnyk S, Tolstoluzhsky D and Barannik E 2008 Doppler myography: Local registration of muscle activity under static loading *Biophysical Bulletin* **1** 20

[4] Lukyanov M 1965 Electromyographic study in discogenic cervical myelopathy *Pathology of the spine and spinal cord* **1** 259-77

[5] Laskov V and Kuleznev V 2008 Objectivation of the results of the speed-force test with the help of stimulating electromyography *Innovative forms and methods in physical culture and sports: Materials of the International scientific-practical conference* **2** 79-81

[6] Orlov R and Nozdrachev A 2010 *Normal physiology:textbook* (Moscow: GEOTAR- Media) p 832

[7] Ostroukh E, Osipov A, Davydova N and Mezhennaja M 2008 *Computer modelling of bioelectrical activity of human musculoskeletal system* (Minsk: Science and technology) p 152

[8] Bonilla F, Lukyanov E, Litvin A and Deplov D 2015 Mathematical modeling of the dynamics of the upper limit *Modern problems of science and education* **1** 27-34

[9] Shayduk A and Ostanin S 2010 Modelling of electromyographic signals by the means of LabVIEW *News of Altai State University. – Barnaul Ed. of Alt. State. un.-ty* **65** 196–201

[10] Kimura J 2001 *Electrodiagnosis in Diseases of Nerve and Muscle: Principles and Practice* (New York: Oxford University Press) p 24-34

[11] Reaz MBI, Hussain MS and Mohd-Yasin F 2006 Techniques of EMG signal analysis: detection, processing, classification and applications *Biol. Proced. Online* **8** 11-35

[12] Eliseev S 1972 Some questions of increasing the efficiency of vibration isolation with the help of motion conversion devices *tr / IPI.* **75** 85-94

[13] Dimentberg F 1982 The theory of spatial hinge mechanisms *Main edition of physical and mathematical literature* **1** 336
[14] Brysin A and Sinev A 2008 New approaches to solving vibration isolation problems *Problems of mechanical engineering and automation* **3** 74-83
[15] Mikhailov V 2018 Monitoring efficiency as a necessary condition for making correct decisions in the sphere of technosphere safety *Russian technological j.* **8** 23-32

Recurrent sequences and the harmonic progression in genetic biomechanics

V Verevkin[1] **and S Petoukhov**[1,2]

[1] Mechanical Engineering Research Institute of the Russian Academy of Sciences, 4,M. Kharitonyevskiy Lane, Moscow, 101990, Russia
[2] Moscow State Conservatory, 13/6 Bolshaya Nikitskaya Street, Moscow, 125009, Russia

spetoukhov@gmail.com

Abstract. The article is devoted to the study of new approaches to the development of mathematical models in genetic biomechanics, which studies the structural relationships of the genetic coding system with genetically inherited biological forms. More specifically, we are talking about models based on the recurrent harmonic progression known in science, the universal connection of which with the information sequences of DNA heredity molecules in the genomes of higher and lower organisms was revealed in the laboratory of biomechanical systems research at the IMASH RAS. This progression is associated with fundamental hyperbolic rules for organizing DNA nucleotide sequences of genomes and is represented on a two-dimensional Cartesian plane by a hyperbolic sequence associated with natural logarithms. The latter have long been used in science to adequately model many inherited biological phenomena. The article describes previously unknown connections of natural logarithms with the structure of the molecular genetic system. The problems of modeling biological phenomena in connection with the revealed structural properties of the molecular genetic system are discussed.

1. Introduction

Investigations of symmetries in inherited biological structures are an important component of biomechanics, biophysics, biochemistry, and other sciences of living matter. Many monographs, international forums, including the Nobel Symposium on Symmetries, etc. are devoted to them. The continuation of these studies is an urgent task, which is associated with the disclosure of many secrets of living nature and the creation of new biotechnologies.

In particular, these studies have long drawn attention to the recurrent structure of many inherited biological configurations, which consist of recurrently repeating blocks related, for example, by similarity symmetries; these configurations are abundant in flowers, shells of mollusks, metameric bodies of animal organisms, etc. [1]. The inheritance of configurations relies on the work of the genetic coding system, which allows the reproduction of such forms in a series of generations. Recently, in the genomes of higher and lower organisms, an important rule for the recurrent organization of DNA nucleotide sequences was discovered; these sequences serve for the transfering genetic information [2, 3]. This hyperbolic rule for the relationship of oligomeric sums is associated with a long-known harmonic progression (1):

$$\frac{1}{1}, \frac{1}{2}, \frac{1}{3}, \frac{1}{4}, \dots \frac{1}{n} \quad (1)$$

Below we consider the use of a number of properties of this progression to build new models in genetic biomechanics and algebraic biology.

2. Computational approaches and models

Our computational approaches and models of biological structures are based on the mathematical properties of harmonic progression (1). Its historically renowned name - "harmonic progression" - derives from its association with a series of harmonics in music (or with a series of standing wavelengths in a vibrating string). It is also indirectly associated with a set of trigonometric functions in the Fourier series. Pythagoras, Orem, Leibniz, Newton, Euler, Fourier, Dirichlet, Riemann and many others were engaged in research of mathematical properties, generalizations and applications of the harmonic progression in various problems. The unexpected discovery of this recurrent harmonic progression in genomes, which are the genetic information basis of living organisms, entails many new scientific issues. The aim and objectives of this study are to study how the mathematical properties of this recurrent genetic progression (1) are able to influence the inherited biomechanical structures bearing the stamp of the structural features of genetic coding. This study leads to the emergence of new model approaches in genetic biomechanics, developing the theme of structural relationships between genetic coding and inherited biomechanical phenomena.

Organisms are cooperative aggregates of a huge number of various molecules that have the ability to reproduce themself in their descendants. Gr. Mendel in experiments on crossing organisms showed that the inheritance of traits to descendants obeys certain algebraic rules. Thus, he actually became the founder of algebraic biology, which uses algebraic methods to model inherited biological phenomena. At present, in this regard, an international interdisciplinary seminar "Algebraic biology and systems theory" ([4]) is functioning in Moscow. According to Mendel's law of independent inheritance of traits, information from the level of genetic DNA molecules dictates the macrostructure of living bodies through many independent channels, despite strong noises and interference. For example, hair, eye, and skin colors are inherited independently of each other. This determinism is provided by unknown algorithms of multichannel noise-immune coding. Accordingly, each organism is a multichannel noise-immune coding machine. Our studies of the properties of genetic inheritance of biological structures and the possibilities of their algebraic modeling are based on the well-known mathematical formalisms of matrix and tensor analysis, recurrent sequences, the theory of information noise-immune coding, as well as on known academic information about genetic and other biological structures.

3. Results and discussion

Let us present some of the possibilities we have studied of using the mathematical properties of harmonic progression (1), which is associated with the organization of genomes of higher and lower organisms, for modeling inherited biological structures and processes.

Harmonic progression is related to the well-known concept of harmonic mean and is a recurrent sequence based on it: knowing two adjacent members of this progression, you can build the entire given progression. Indeed, the harmonic mean x_{n+1} for two numbers x_n and x_{n+2} is expressed by the relation (2):

$$x_{n+1} = \frac{2}{x_n^{-1} + x_{n+2}^{-1}} \rightarrow x_{n+1} = \frac{2x_n x_{n+2}}{x_n + x_{n+2}} \quad (2)$$

Knowing two neighboring numbers x_n and x_{n+1} in this triplet of numbers, the third number x_{n+2} is determined, respectively, by the relation (3):

$$x_{n+2} = \frac{x_n x_{n+1}}{2x_n - x_{n+1}} \quad (3)$$

Successively applying this recurrence relation (3) to the first terms $x_1 = 1$ and $x_2 = 1/2$ of the harmonic progression (1), we obtain the entire series of its terms. It seems interesting and useful to investigate in the future the harmonic mean ratios in inherited biological structures and processes.

Pairs of numbers n and $1/n$ define points on the plane with coordinates $[n, 1/n]$ in the Cartesian coordinate system belonging to the hyperbola $y = 1/n$, which is closely related to the natural logarithm function $\ln x$. As you know, the natural logarithm can be defined for any positive real number "a" as the area under the hyperbola $y = 1/x$ from 1 to a (figure 1 (a)) [5]. In the polar coordinate system $[\theta, r]$, the natural logarithm $\ln x$ - as a function of the variable x - is represented by a spiral if x plays the role of the angular coordinate θ, and $\ln x$ plays the role of the radius vector r (figure 1 (b)).

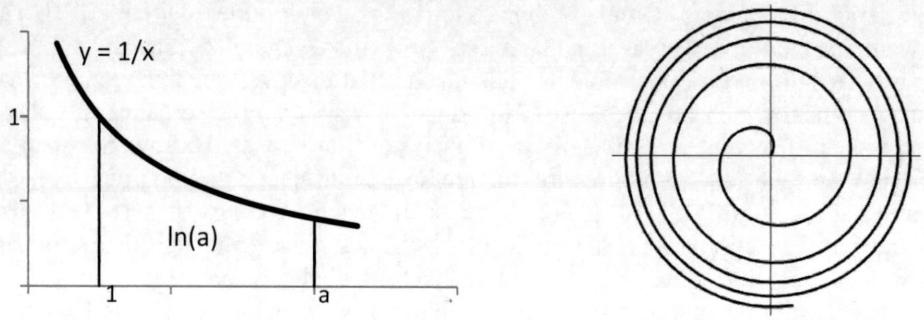

(a) (b)

Figure 1. (a) Illustration of the natural logarithm $\ln x$ as the area under the hyperbola $y = 1/x$. (b) Natural logarithm function in polar coordinates.

This analysis gives evidences that our sensory perception obeys the same structural principles as morphogenesis with its phyllotaxis laws and that these principles can be effectively modelling on the basis of hyperbolic numbers.

The natural logarithm function is widely used in mathematical biology to model inherited biological phenomena. For example, various types of inherited sensory perceptions - sight, hearing, smell, tactile sense, taste - are subject to the basic psychophysical Weber-Fechner law, which is logarithmic. This law states that the intensity of perception is proportional to the logarithm of the intensity of the stimulus. By virtue of this law, for example, the power of sound in engineering is expressed in a logarithmic decibel scale. The logarithmic nature of perception provides the body with a huge expansion of the range of perceived stimuli, which can differ in intensity by many millions or even billions of times: for example, hammer strikes on a steel plate generate noise that is a hundred billion times louder than the quiet rustle of leaves, and the brightness of a volt arc trillions of times the brightness of a faint star, barely visible in the night sky [6]. It can be assumed that organisms use natural logarithms as the basis for their internal counting.

Taking into account the named connection of the natural logarithm function with the hyperbola $y = 1/x$, we note that the recurrent harmonic progression (1) is conjugate, respectively, with a recurrent sequence of natural logarithms (4):

$$\ln 1, \ln 2, \ln 3, \ldots, \ln n \tag{4}$$

Figure 2 shows this recurrent sequence (4) in graphical form in a polar coordinate system, where an integer positive variable n plays the role of the angular coordinate θ, and $\ln n$ plays the role of the radius vector r.

Turning to discrete values of functions in modeling biological phenomena seems to be fundamentally important due to the discrete nature of the organization of both the genetic coding system (discrete DNA alphabets, etc.) and a variety of biomechanical processes (for example, the seemingly continuous movement of a human hand is provided due to the total effect of discrete contractions of a huge variety of muscle units, each of which has its own threshold of excitability and is contracted out of step with the rest). As is known, for all excitable cells of the body, the genetically inherited universal law of "all or nothing" of a discrete nature is valid: the cell does not respond at all to stimuli of a subthreshold value, but reacts to above-threshold stimuli with full amplitude. These facts indicate that biomechanics, if it takes into account these fundamental features of living tissue, should be developed as the mechanics of discrete sequences and processes, taking into account their relationship with discrete genetic coding.

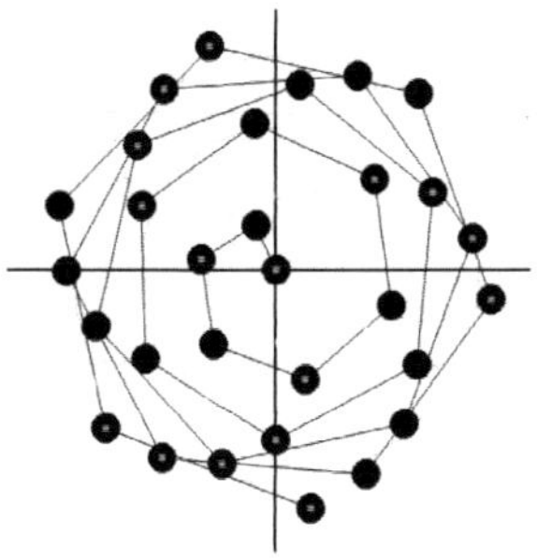

Figure 2. Graphic representation of a sequence of discrete values of the natural logarithm $\ln n$ in a polar coordinate system for $n = 1, 2, \ldots, 30$.

The use of a polar coordinate system for the graphic representation of biological structures as configurations of a spiral type is justified by the fact that spirals have long been considered as "*the curves of life*" [7, 8]: spiral and helical configurations are characteristic of DNA molecules, proteins, bones, tendons, muscles, etc. For example, the heart is a holistic spiral muscle that pushes blood by its action of twisting and untwisting, similar to how a washerwoman squeezes water out of laundry by twisting it.

The natural logarithm function $y = \ln x$, which is important for modeling inherited biological phenomena, is mutually inverse to the exponent function $x = e^y$, which are also used for a long time to model morphogenetic and other bio-phenomena. Accordingly, the recurrent sequence (4) is mutually inverse to the recurrent sequence (5) of exponentials e^n:

$$e^1, e^2, e^3, \ldots, e^n \quad (5)$$

Figure 3 shows a continuous exponential function e^x and a discrete recurrent exponential sequence in a polar coordinate system. These graphs correspond to continuous and discrete logarithmic spirals.

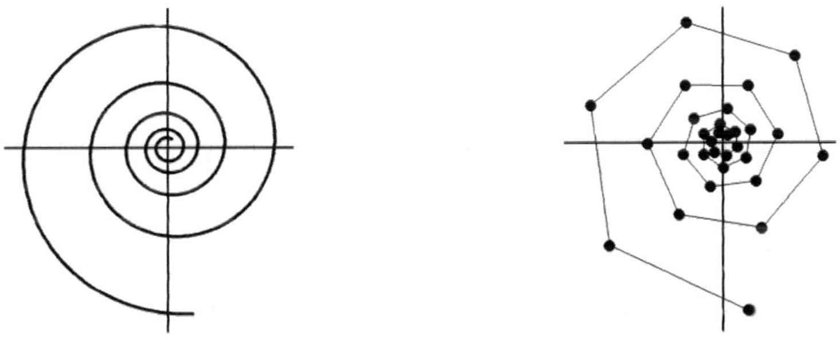

(a) (b)

Figure 3. (a) Graphical representation of a continuous exponential function $e^{0.1x}$ in a polar coordinate system. (b) Discrete sequence $e^{0.1n}$ for $n = 1, 2, \ldots, 30$ in a polar coordinate system.

Logarithmic functions and logarithmic spirals model many genetically inherited biological phenomena, including morphogenetic configurations of phyllotaxis laws [9], which are associated with another recurrent sequence - a series of Fibonacci numbers related to the golden ratio. Figure 4 shows examples of logarithmic spirals in inherited morphological structures.

Let us add that many cases of biomechanical movements - growth and motor - of living bodies can be modeled as a transition from one discrete recurrent logarithmic spiral to another. This corresponds to the statement of the classic of biomechanics N.A. Bernstein: "*The idea that movement is in many respects like an organ seems extremely fruitful*" [10].

(a) (b) (c)

Figure 4. Examples of logarithmic spirals in morphogenetic configurations: (a) sunflower head, (b) two clamshells, (c) aloe.

4. Some concluding remarks

The work of the laboratory for the research of biomechanical systems of the IMASH RAS revealed fundamental hyperbolic rules for the organization of DNA information texts in the genomes of higher and lower organisms (eukaryotes and prokaryotes), associated with harmonic progression (1), which is well-known in science in connection with various mathematical and physical problems. The discovery of the important role of the harmonic progression in genetic informatics leads to the development of new mathematical models in genetic biomechanics, which studies the relationship of inherited biomechanical structures with genetic DNA texts, since DNA molecules play a dictatorial role in living organisms, determining their genetically inherited physiological structures [11]. Additionally, we note that harmonic progression plays a prominent role in aesthetics. This progression (1) is important not only for music with its systems of musical harmony and musical overtones. At least since the time of the Pythagorean doctrine of the aesthetics of proportions, there has been an idea that the composer R. Schumann expressed as follows: "*The aesthetics of one art is at the same time the aesthetics of another art; only the material is different*" [12]. In light of this, architecture has long been interpreted as frozen music and music as dynamic architecture. In connection with our research, assumptions arise about the genetic basis of a number of aesthetic parallels in various arts.

The above features of harmonic progression (1) do not exhaust its rich mathematical features, which can serve to create new approaches to the construction and substantiation of models in genetic biomechanics, as well as contribute to the development of algebraic and quantum biology [13].

References

[1] Shubnikov A V and Koptsik V A 2004 *Symmetry in Science and Art* (Moscow: RCHD)
[2] Petoukhov S V 2020 Hyperbolic rules of the oligomer cooperative organization of eukaryotic and prokaryotic genomes *Preprints* 2020050471
[3] Petoukhov S V 2020 *Symmetry: Culture and Science* **31(2)** pp 222-3
[4] Conway J H and Guy R K 1995 *The Book of Number* (New-York: Copernicus)
[5] Available at: https://www.youtube.com/channel/UC8JLsuRzzPsRiHwrwEjMCtw
[6] Vilenkin N Ya 1985 *Functions in nature and technology* (Moscow: Prosveschenie)
[7] Cook T A 1914 *The Curves of Life* (London: Constable and Co)
[8] Petoukhov S V, Svirin V I and Khazina L V 2015 Bionics of spiral structures *J. of Machinery Manufacture and Reliability* **44(3)** pp 249–53
[9] Jean R V 1994 *Phyllotaxis: A Systematic Study in Plant Morphogenesis* (Cambridge: Cambridge University Press)
[10] Bernstein N A 1966 *Essays on the physiology of movements and the physiology of activity* (Moscow: Medicine)
[11] McFadden J and Al-Khalili J 2018 *Proc. R. Soc. A* **474(2220)** 20180674

[12] Schumann R 1969 *On Music and Musicians* ed Konrad Wolff (New York: Norton)
[13] Petoukhov S V and He M 2010 *Symmetrical Analysis Techniques for Genetic Systems and Bioinformatics: Advanced Patterns and Applications* (USA: IGI Global)

Features of the dynamics of walking machines and robots with cycle type of movers and controlled dissipation in the foot joint

V V Arykantsev, Ya V Kalinin and V V Chernyshev

Volgograd State Technical University, 28 Lenin Avenue, Volgograd, 400005, Russia

arvstu@mail.ru

Abstract. The expediency of introducing an additional damping device (analog of a shock absorber) into the cyclic walking mechanism of a mobile robot, which provides a dissipative connection of a ski-shaped foot with a support link, is discussed. The influence of the damper on the dynamics of the feet changing and the dynamics of the meeting of the foot with an obstacle is considered. It is shown that dissipation in the hinge of the foot leads to a decrease in the intensity of shock processes when changing feet and when meeting an obstacle, provides a decrease in power losses for vertical vibrations of the body, and also expands the propulsion unit's capabilities for self-adaptation to an unorganized surface.

1. Introduction

The use of walking movers based on cyclic walking mechanisms in walking machines and robots makes it possible not to worry about maintaining gait and stability and eliminates the need for a controlled adaptation system. Multi-legged walking robots with cyclic movers have a minimum number of controlled degrees of freedom and are structurally much simpler than analogs with adaptive control [1-3]. High cross-country ability and excellent traction and coupling properties on soils with low bearing capacity, environmental friendliness, simplicity of design, reliability of operation and relatively low cost make it possible to create and operate in extreme conditions walking machines and mobile robotic systems with movers based on cyclic walking mechanisms already at present time.

In Volgograd State Technical University have been developed a number of prototypes of walking machines and robots with cyclic-type movers designed to operate on water-saturated and underwater soils [3, 4]. Their tests have confirmed the superiority of walking machines in terms of traction properties, ground and profile cross-country ability in comparison with traditional vehicles [5, 6]. At the same time, tests revealed a number of problems associated with handling in a disorganized surface. The strict trajectory of the reference points of the cyclic walking mechanisms does not allow, in full measure, to realize the potential of the mover in adaptation to the supporting surface and profile passability. For their more complete implementation, it is necessary to adjust the programmed leg movements - changing the length or height of the step for bumpless obstacle overcoming [4], combining the position of the pivot points when turning to reduce the moment of resistance to turning [7–9], etc. For example, during underwater testing of a walking device MAK-1 (figure 1), due to poor visibility under water (less than 1–3 m), the operator in straight forward modes of movement even at relatively low speeds (about 1.5–2 m·s^{-1}) did not have time to take adequate decisions in the case of the appearance of any obstacle in the direction of travel (the decision time was from 0.5 to 2 s). As a result, when encountering obstacles, there were cases of destruction of the feet and elements of the walking mover (figure 2).

Figure 1. Underwater walking device MAK-1 — meeting with an obstacle.

Figure 2. Feet deformations after meeting with an obstacle.

In the work, with the aim to reduce vibroimpact processes when changing feet and meeting a foot with an obstacle, the possibility of introducing an additional damping device (analogue of a shock absorber) into the cyclic walking mechanism, providing a dissipative connection of the ski-shaped foot with the support link, is discussed.

2. Materials and methods of investigation

The analysis is carried out on the example of the Umnov-Chebyshev cyclic walking mechanism, which is differs by its simplicity and reliability. The mechanism was proposed by N.V. Umnov (A.A. Blagonravov Institute of Mechanical Engineering, Russian Academy of Sciences) and was obtained from the well-known walking mechanism of a foot-walking machine by P.L. Chebyshev [4] by turning it over and giving the support link an L-shaped. In the MAK-1 underwater robot, it was improved (figure 3 (a)) - due to the introduction of an additional degree of freedom, controlled in the gear change mode, into the mechanism, the possibility of adjusting the trajectory of the reference point was provided [4]. Also, a system of passive adaptation of the foot to the supporting surface was implemented, which, due to the kinematics of the mechanism and friction in the hinge of the foot, lifts the toe of the foot in the transfer phase [4, 10]. Moreover, the toe of the foot is lifted both with direct and reverse movement of the machine (figure 3 (b), 3 (c)). To eliminate the influence of random factors on the programmed movement of the hinged foot and reduce the intensity of impact processes when changing feet and when meeting an obstacle, it is proposed to introduce an additional damping device 6 into the walking mechanism (figure 3 (a)), which provides dissipative connection of the foot 4 with the supporting link 2 of the walking mechanism.

The walking mechanism during its kinematic analysis was considered as a flat multi-link mechanism, the individual links of which perform a plane-parallel movement. During compilation of the differential equations of the mechanism links motion, their angular velocities were expressed through the velocities of the points on which external connections are imposed. Expressions for velocities were obtained by consistent, link by link, consideration of the motion of rigid bodies. For points of the mechanism, on which external constraints are imposed, expressions for velocities were transformed into constraint equations. Solving them together with respect to the angular velocities of the links, a system of differential equations of motion was obtained, the integration of which, at given initial values of the angles of rotation of the links, solves the kinematic problem of the movement of the walking mechanism in the body-related frame of reference. The relative accelerations of the featured points of the mechanism, which are functions of the angular velocities and accelerations, were obtained by differentiating their velocity vector with respect to time. The angular acceleration of the links was determined by differentiating the angular velocities. As a result, the relative accelerations of the nodal points of the walking mechanism were determined, including its reference point *C*, which coincides with the center of the foot. Similarly, the kinematic characteristics of the walking mechanism operating in antiphase were determined. When determining the speed and acceleration of the machine body in absolute motion, they were assigned the values of the relative speeds and accelerations of the reference point of the mechanism in the reference phase taken with the opposite sign. It was supposed that the walking mechanism is located in the support, the reference point of which has a smaller coordinate z_C.

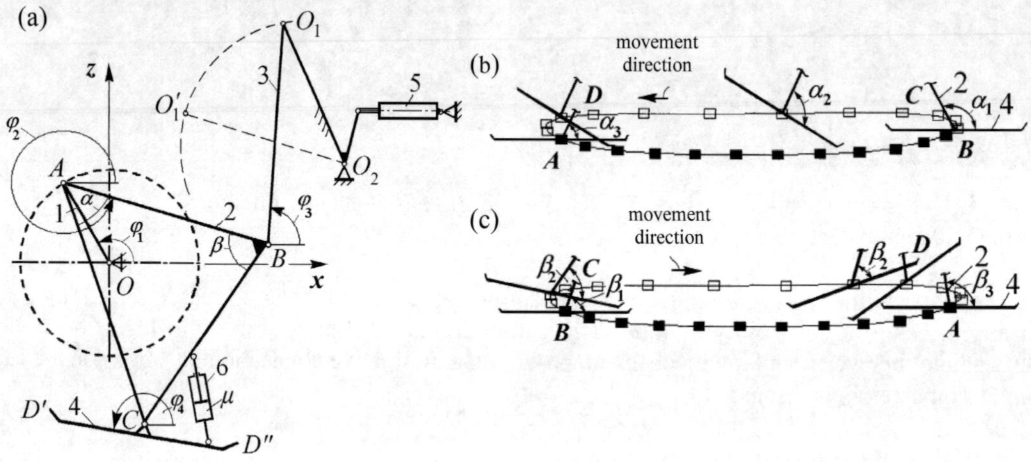

Figure 3. Sketch of the walking mechanism (a) and the trajectory of its support point with special states of a foot in straight (b) and reverse (c) movement: 1 — main crank; 2 — support link (foot); 3 — rocker arm; 4 — ski-shaped foot; 5 — linear electric drive of the rocker arm suspension point shift mechanism; 6 — additional damper; O_1O_2 — controlled turning link (crank of the mechanism for shifting the suspension point of the rocker arm).

3. Results and discussion

Moment of viscous friction at the hinge of the foot $M_R = f(\mu, \Delta\omega_{24}^{(r)})$ determined by the coefficient of viscous friction μ in the damper and the relative angular velocity of the foot in relation to the support $\Delta\omega_{24}^{(r)}$. Program changing of $\Delta\omega_{24}^{(r)}$ of the basic walking mechanism per cycle, obtained from the results of kinematic analysis, is shown in figure 4.

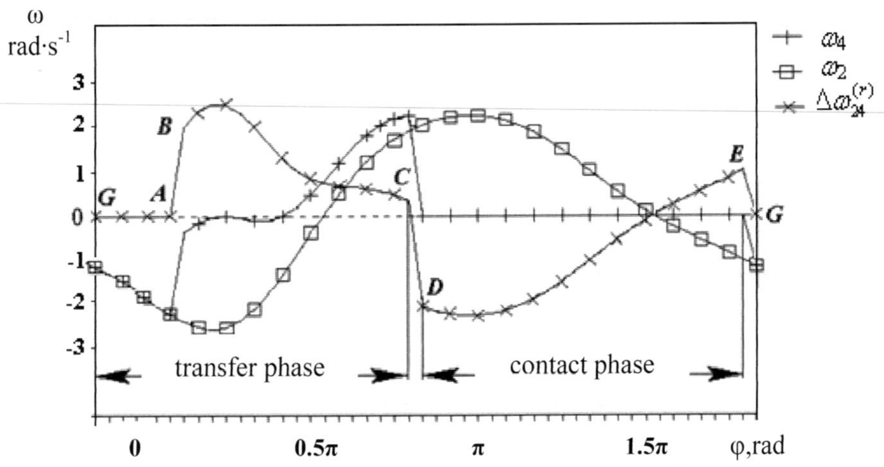

Figure 4. Programmed changing of the angular velocities of the foot ω_4, support link ω_2 and relative angular speed $\Delta\omega_{24}^{(r)}$ at cycle.

In figure 4 section *GABC* corresponds to the phase of foot transfer, section *CD* - to the stage of placing the foot on the ground, *DE* - to the contact phase with the ground, *EG* - to the exit of the foot from the contact phase. The *GABC* transfer phase, can be subdivided into the *GA* section, where the foot moves together with the support link of the walking mechanism, and the *ABC* section, in which the foot in the transfer begins to interact (with the heel) with the ground and moves in a skid.

If, when changing the feet, the coefficient μ of the damping device exceeds the critical value μ_{cr}, at which the normal reaction of the soil N_1 of the foot, which has not yet been established on the ground, begins to exceed N_2 - the reaction of the foot in the reference phase, then in the *CD* section and at the final stage of the transfer phase, at feet coming into contact with the ground, an additional phase of the cycle occurs - the phase of support on the heel. Changing the feet operating in antiphase in this case, see figure 5, where X_C, Z_C - reactions in the hinge of the foot; $P_{\varphi j}$, - adhesion force; $v_{D'}$ and $v_{D''}$ the speed of slipping of the toe and heel of the *j*-th foot ($j = 1, 2$), will take place in 4 stages.

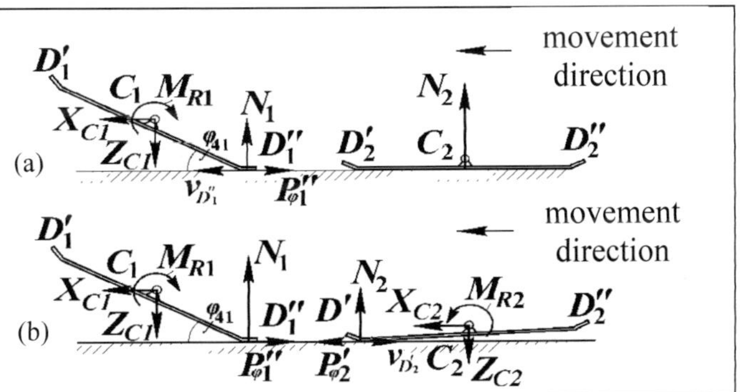

Figure 5. Feet changing when $\mu > \mu_{cr}$: (a) 1-st foot in the end of transfer, 2-nd foot in contact phase ($N_1 < N_2$); (b) 1-st foot in contact phase, 2-nd foot at start of transfer ($N_1 > N_2$).

At the first stage, the 1st foot is in transfer ($N_1 = 0$), and the 2nd foot is in contact with the ground by the entire supporting surface and takes all the load on the mover.

IOP Conf. Series: Materials Science and Engineering 1129 (2021) 012051 doi:10.1088/1757-899X/1129/1/012051

At the second stage (figure 5 (a)), after the heel of the foot (point D'') touched the ground and skidded due to the relative angular velocity $\Delta\omega_{24}^{(r)}$ appears a moment of support M_{R1} and 1-st the foot begins to partially perceive the weight of the machine, while 2-nd the foot remains on the ground and carries the bigger part of the machine's weight ($N_1 < N_2$). Normal reaction of the 1-st foot at this stage, considering that $P''_{\varphi 1} = k_\varphi N_1$, is:

$$N_1 = \frac{J_{C4}\ddot{\varphi}_{41} + M_{R1}}{l_4(\cos\varphi_{41} + k_\varphi \sin\varphi_{41})} \tag{1}$$

where J_{C4} — central moment of inertia of the foot; $l_4 = l/2$; l — ski-shaped foot length; φ_{41} — tilt angle of the 1-st foot; k_φ — adhesion coefficient.

At the third stage (figure 5 (b)), which is an additional phase of the cycle, the support reaction of the 1-st foot becomes larger than the second ($N_1 > N_2$). The first foot carries the bigger part of the machine's weight. The second support comes off the ground, taking part of the load at the beginning of the stage, while contacting the ground with the toe of the foot. At this stage, due to the support moment M_{R1}, a shockless installation of the foot with the entire supporting surface on the ground is ensured. The smooth movement of the foot in this area allows you to neglect the dynamism of the process, then, when determining the support reactions, you can use the following equations of statics:

$$N_1 + N_2 = G_j,$$
$$N_2 = \frac{M_{R2}(\mu, \Delta\omega_{24})}{l_4(\cos\varphi_{42} + k_\varphi \sin\varphi_{42})} \tag{2}$$

where G_j — weight of the j-th mover; φ_{42} — tilt angle of 2-nd foot.

Supporting moment of the second foot $M_{R2}(\mu, \Delta\omega_{24})$ at the third stage, it is determined by the relative movement of the support link in relation to the foot. An additional phase of the cycle, where the machine rests on the ground only on the heels of the feet of in-phase walking mechanisms, delays in time when the feet change, the body "sagging" due to the imperfection of the trajectory of the reference point of the walking mechanisms. This gives the paired foot time to bypass the worst point of the trajectory from the point of view of vertical body vibrations and allows to reduce their amplitude and energy consumption for their maintenance. Moreover, the work of the damper in this mode is carried out at the expense of the energy, irretrievably lost in the basic mechanism of walking.

At the final stage of changing the feet, the 1-st foot is lowered by the entire supporting surface to the ground and perceives the entire side weight of the machine, and the 2-nd has come off the ground and is in the transfer stage. After that, the cycle is repeated.

When driving on an unorganized surface, the additional damper will obviously have a cushioning effect when it encounters an obstacle. In this case, the behavior of the foot will differ from the "programmed" one. The damper begins to work already at the stage of foot transfer - from the moment the toe of the articulated foot contacts the obstacle and the relative angular velocity $\Delta\omega_{24}^{(r)}$ appears. Then, in the second stage, the heel of the foot is connected to the braking - it will slide along the ground in the direction of the car. This ensures the shockless interaction of the foot with the obstacle. At this stage, the braking intensity can be increased by increasing the viscous friction coefficient μ in the damper.

4. Conclusion

The introduction of an additional damping device (analogue of a shock absorber) into the cyclic walking mechanism, which provides a dissipative connection of the ski-shaped foot with the support link, expands the mover's capabilities for self-adaptation to an unorganized surface and leads to a decrease in the intensity of vibroimpact processes when changing feet and when meeting an obstacle. The decrease occurs due to a smoother change in the support reactions when stepping over. There is also a decrease in energy consumption for movement due to a decrease in the amplitude of vertical vibrations of the body. Moreover, the work of the damper is carried out due to the energy that is irretrievably lost in the basic mechanism of walking.

Acknowledgments
The work is supported by RFBR, projects № 19-08-01180, № 19-48-340007.

References

[1] Briskin E S, Maloletov A V and Sharonov N G 2019 Motion planning of a walking robot with rotary movers 2019 *Proceedings of the 22nd Int. Conf. on Climbing and Walking Robots and Support Technologies for Mobile Machines (CLAWAR 2019)* 87-90
[2] Zhoga V V, Terehov S E and Skakunov V N 2019 Algoritm adaptivnogo upravleniya privodom shagayuschego robota na osnove nechetkoy logiki *Izvestiya VolgGTU. Roboty, mehatronika i robototehnicheskie sistemy* **3 (226)** 12-7
[3] Arykantsev V V, Pryanichnikov V E and Chernyshev V V 2018 Informatsionnye vozmozhnosti stopy podvodnogo shagayuschego apparata pri vstreche s prepyatstviem *Informatsionno-izmeritel'nye i upravlyayuschie sistemy* **16 10** 90-6
[4] Chernyshev V V and Arykantsev V V 2017 Tsiklovoy mehanizm shaganiya s transformiruemoy traektoriey opornoy tochki *Teoriya mehanizmov i mashin* **15 2 (34)** 71-9
[5] Sharonov N G, Kirillov A P and Maloletov A V 2020 Issledovanie pokazateley profil'noy prohodimosti shagayuschey mashiny s tsiklovymi dvizhitelyami *Izvestiya VolgGTU. Roboty, mehatronika i robototehnicheskie sistemy* **9 (244)** 89-93
[6] Chernyshev V V and Arykantsev V V 2017 Ispytaniya podvodnogo shagayuschego apparata v usloviyah vodnyh ob"ektov Volgo-Ahtubinskoy poymy *Sistemy kontrolya okruzhayuschey sredy* **8 (28)** 24-9
[7] Briskin E S, Kalinin Ya V and Maloletov A V 2019 On energetically effective modes of walking robots movement *ROMANSY 22 – Robot Design, Dynamics and Control : Proceedings of the 22nd CISM IFToMM Symposium* 425-32
[8] Briskin E S, Gulevskiy V V, Penshin I S and Sharonov N G 2020 About features of management of mobile robots with stepping movers of a rope type in a water environment *IOP Conf. Series: Materials Science and Engineering* **747** 9
[9] Chernyshev V V , Arykantsev V V and Vershinina I P 2019 Minimization of energy costs for movement resistance of ground for walking device by the control of support points motion *Proceedings of the 5th Int. Conf. on Industrial Engineering (ICIE 2019)* **II (2)** 839-48
[10] Chernyshev V V, Arykantsev V V and Kalinin Ya V 2017 Passive foot control in cyclic walking mechanism *Proceedings of the int. conf. Industrial Engineering, Applications and Manufacturing (ICIEAM 2017)* 5 DOI: 10.1109/ICIEAM.2017.8076189

The research of a movement dynamics two-legged robot with spring-loaded weight

D V Bordyugov [1], E S Briskin [1,2] and N G Sharonov [1,2]

[1] Department of Theoretical Mechanics, Volgograd State Technical University, 28 Lenin Avenue, Volgograd, 400005, Russia
[2] Center for Technology Components of Robotics and Mechatronics, Innopolis University, 1 Universitetskaya Str., Innopolis, 420500, Russia

denklopuk@gmail.com

Abstract. The movement of a two-legged robot which moves due to oscillations of a load with a given mass fixed on the robot body with the help of the springs, is considered in this research. By changing parameters, such as mass of the load, stiffness of the springs, size of the robot's body and the initial speed of movement, the various trajectories of movement are executed. In this article mathematical model and calculation results have been presented.

Introduction

Mobile robots have vital differences associated not only with the use of a concrete construction of mover and a schematic structure, but also with the correlation of numbers of propulsion devices and degrees of freedom of a mobile robot as a mechanical system. [1, 2, 3].

For robotic transport platforms, a structural diagram and movers are widely used to ensure static stability. In this case, the number of propulsion devices is equal to or exceeds the number of degrees of freedom of the robot (for example, «Vosminog», «Ortonog») [4, 5].

When moving in aggressive environments, the construction of the robot may undergo changes, because the use of standard types of mover is difficult. The movable mass serves as a mover, being inside the robot and changing its position relative to the body. Such robots are applicable, for example, for movement in pipes [6].

The research of a two-legged robot movement which moves due to a movable mass fixed on springs is considered. The movement is carried out as follows: one of the supports is in continuous interaction with the surface, while the other is in the transfer phase, allowing the rod to rotate around the fixed support, then they change, and the robot changes the direction of rotation, thereby moving. This principle is similar to the movement of the "Celtic stone": if you put it on a horizontal plane and twist it around a vertical axis, then after a certain period of time it will stop rotating, vibrations will occur around other axes, after which the stone will begin to rotate around the vertical axis, but in the opposite direction [7].

Computational model

The movement of a two-legged robot is considered.

Design schedule (figure 1) presents a mechanical system consisting of two weightless supports *A* and *B* (1), connected by a weightless rod 2, moving under the influence of vibrations of a load 3 of mass m, fixed on springs 4 by stiffness $c/2$. The x-axis in the design diagram is aligned with the weightless bar AB, with the origin at point C. At the initial moment of time, the speed is set V_0.

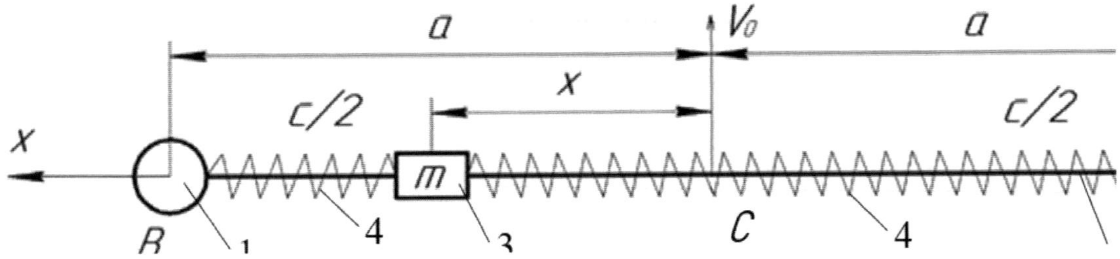

Figure 1. Design schedule: 1 – support, 2 – rod, 3 – load, 4 – spring.

It is assumed that when the two-legged robot under consideration moves, one of the supports at point A or B does not move relative to the surface Oyz, allowing the rod AB to rotate around this support. The condition for immobility of one of the supports is the position M of the load 3 relative to the middle of the rod AB (figure 2 (a), (b)). Point C is in the middle of the road AB and is the origin of the Cx coordinate system. When changing supports, the calculation schemes of the robot's movement change (figure 2 (a), (b)):

$$x \le 0 \qquad \begin{cases} y_A = \text{const}, \\ z_A = \text{const}; \end{cases}$$

$$x > 0 \qquad \begin{cases} y_B = \text{const}, \\ z_B = \text{const}. \end{cases}$$

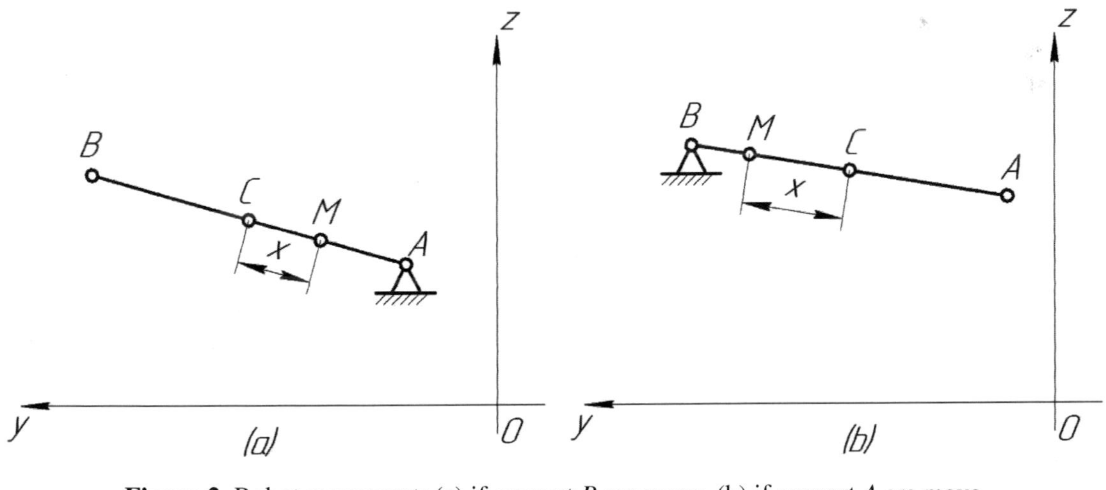

Figure 2. Robot movement: (a) if support B are move, (b) if support A are move.

The effect of frictional forces between the surface and the supports is not taken into account. Lagrange equations of the second kind are used as a solution method:

$$\begin{cases} \dfrac{d}{dt}\left(\dfrac{\partial T}{\partial \dot{x}}\right)-\dfrac{\partial T}{\partial x}=-\dfrac{\partial P}{\partial x}, \\ \dfrac{d}{dt}\left(\dfrac{\partial T}{\partial \dot{\varphi}}\right)-\dfrac{\partial T}{\partial \varphi}=-\dfrac{\partial P}{\partial \varphi}. \end{cases} \quad (1)$$

Two cases are considered: the first – support *A* is motionless, *B* – in the transfer phase (figure 2 (a)), the second - support *B* is motionless, *A* - in the transfer phase (figure 2 (b)). In the first case, expressions for kinetic and potential energies are:

$$\begin{cases} T=\dfrac{1}{2}m\dot{x}^2+\dfrac{1}{2}m\dot{\varphi}^2(a+x)^2, \\ P=\dfrac{1}{2}cx^2; \end{cases} \quad (2)$$

where x – the movement of the load relative to the center O, φ is the angle of transfer, and a is half the length of the robot. Then from (1, 2) the equations follow:

$$\begin{cases} m\ddot{x}-m\dot{\varphi}^2(a+x)=-cx, \\ \ddot{\varphi}m(a+x)^2+2m\dot{\varphi}\dot{x}(a+x)=0. \end{cases} \quad (3)$$

In the second case:

$$\begin{cases} T=\dfrac{1}{2}m\dot{x}^2+\dfrac{1}{2}m\dot{\varphi}^2(a-x)^2, \\ P=\dfrac{1}{2}cx^2; \end{cases} \quad (4)$$

And instead of (4), the following equations take place:

$$\begin{cases} m\ddot{x}+m\dot{\varphi}^2(a-x)=-cx, \\ \ddot{\varphi}m(a-x)^2-2m\dot{\varphi}\dot{x}(a-x)=0. \end{cases} \quad (5)$$

The coordinates of the point *C* of the support *AB* of the robot (figure 3) are determined by the dependencies:

$$\begin{cases} y_c=y_A+a\cos\varphi=y_B-a\cos\varphi, \\ z_c=z_A+a\sin\varphi=z_B-a\sin\varphi. \end{cases} \quad (6)$$

When calculating the movement, the following parameters of the mechanical system are set: $a, m, c, \dot{\varphi}_0$; initial conditions: $x_0, \varphi_0, \dot{x}_0$, as well as the coordinates of the support points *A* and *B* at the initial moment of time.

Then, at the first stage of calculating the robot's movement, system (6) takes the form:

$$\begin{cases} y_c=y_{A_0}+a\cos\varphi, \\ z_c=z_{A_0}+a\sin\varphi; \end{cases} \quad (7)$$

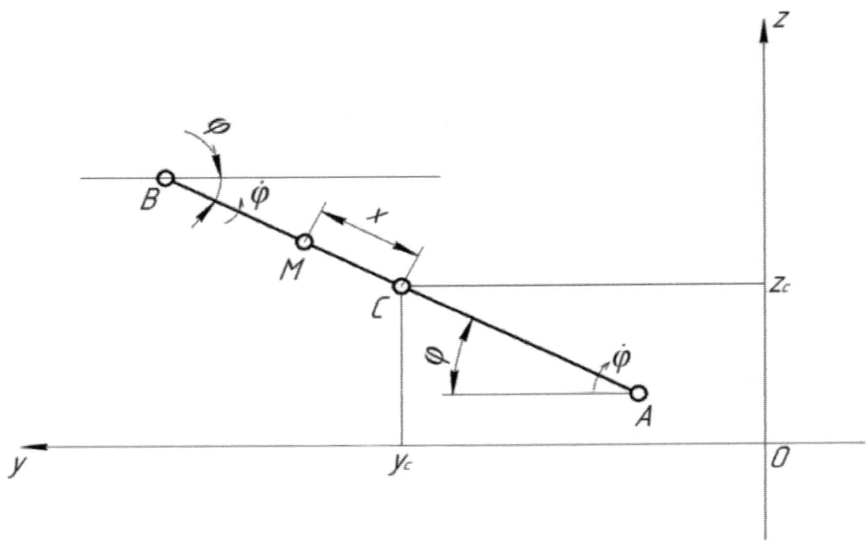

Figure 3. Determining the coordinates of the robot center and supports.

At $x = 0$, at the end of the first stage, the values of the angular velocity $\dot{\varphi}$ **and angle** $\varphi_к$, as well as the coordinates of point B, are determined:

$$\begin{cases} y_{B_1} = y_{A_0} + 2a\cos\varphi_к, \\ z_{B_1} = z_{A_0} + 2a\sin\varphi_к. \end{cases} \quad (8)$$

At the second stage:

$$\begin{cases} y_c = y_{B_1} - a\cos\varphi, \\ z_c = z_{B_1} - a\sin\varphi, \end{cases} \quad (9)$$

then:

$$\begin{cases} y_{A_2} = y_{B_1} - 2a\cos\varphi_к, \\ z_{A_2} = z_{B_1} - 2a\sin\varphi_к. \end{cases} \quad (10)$$

For subsequent stages, calculations are carried out in a similar way.

Results of solving the model problem

Initial conditions: $x_0 = -0.3$ m; $\varphi_0 = 0$ rad; $\dot{x}_0 = 0$ $m \cdot s^{-1}$; $\dot{\varphi}_0 = 5$ $rad \cdot s^{-1}$. The parameters of the considered mechanical system: $m = 5$ kg, $c = 10$ $N \cdot m^{-1}$, $a = 0.5$ m.

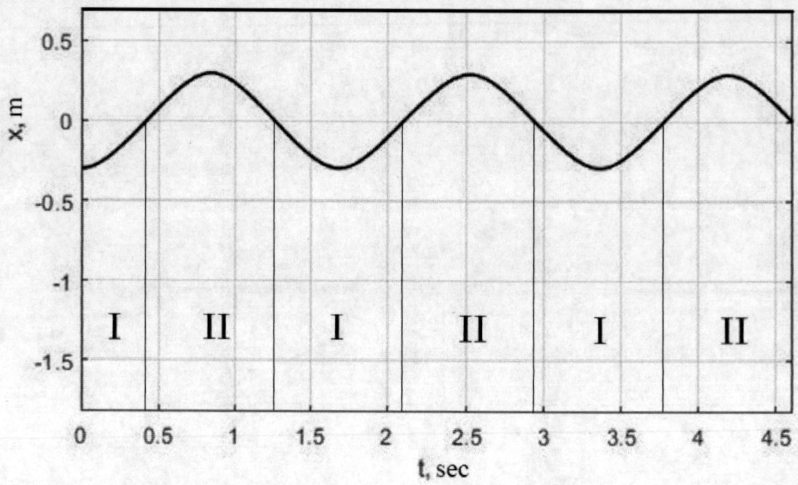

Figure 4. Dependence of the movement of the load *M* relative to the middle of the rod *C* on time: I – rotation around point *A* (figure 2 (a)); II – rotation around point *B* (figure 2 (b)).

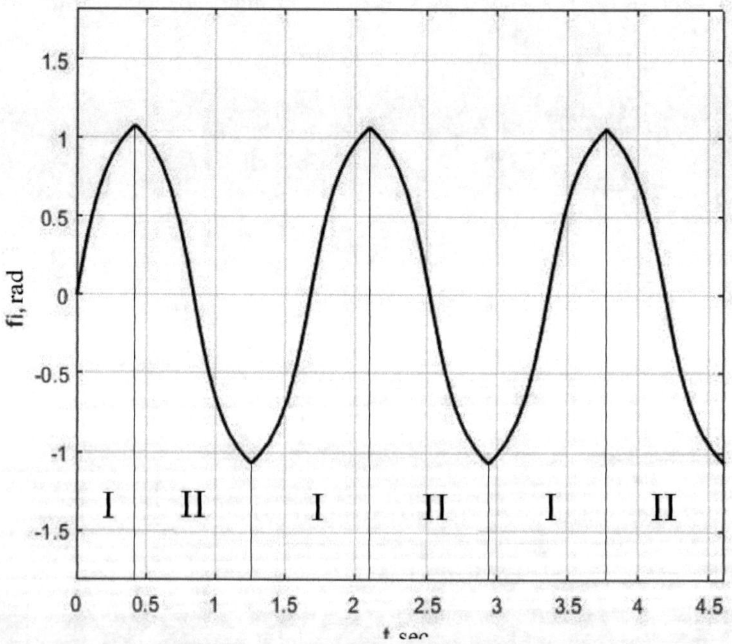

Figure 5. Dependence of the change in the angle of rotation of the robot on time: I – rotation around point *A* (figure 2 (a)); II – rotation around point *B* (figure 2 (b)).

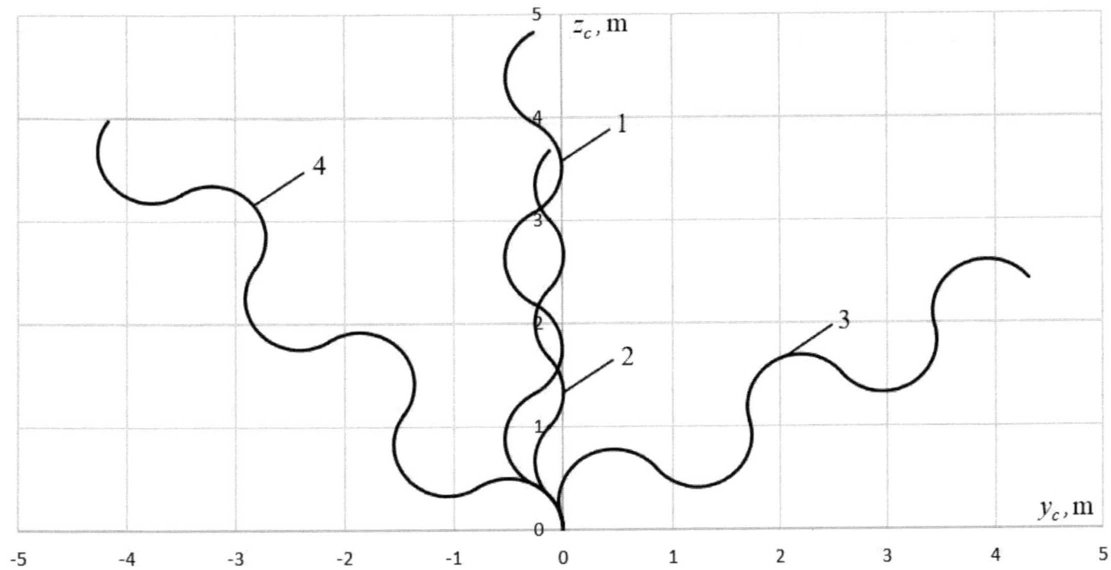

Figure 6. The movement trajectory of the robot center (point C) if: 1) $x_0 = -0.3$ m; $\varphi_0 = 0$ rad; $\dot{x}_0 = 0$ $m \cdot s^{-1}$; $\dot{\varphi}_0 = 5$ $rad \cdot s^{-1}$, m = 5 kg, c = 10 $N \cdot m^{-1}$, a = 0.5 m. 2) $x_0 = -0.3$ m; $\varphi_0 = 0$ $rad.$; $\dot{x}_0 = 0$ $m \cdot s^{-1}$; $\dot{\varphi}_0 = 5$ $rad \cdot s^{-1}$, m = 5 kg, c = 100 $N \cdot m^{-1}$, a = 0.5 m. 3) $x_0 = -0.3$ m; $\varphi_0 = 0$ rad; $\dot{x}_0 = 2$ $m \cdot s^{-1}$; $\dot{\varphi}_0 = 5$ $rad \cdot s^{-1}$, m = 5 kg, c = 10 $N \cdot m^{-1}$, a = 0.5 m. 4) $x_0 = -0.4$ m; $\varphi_0 = 0$ rad; $\dot{x}_0 = -0.5$ $m \cdot s^{-1}$; $\dot{\varphi}_0 = 5$ $rad \cdot s^{-1}$, m = 25 kg, c = 10 $N \cdot m^{-1}$, a = 0.5 m.

Conclusion

Due to the movement of a load with mass m and giving an initial velocity V_0 to the center of the body, occurs the robot's motion, the pattern of which depends on the specified parameters and initial conditions. The movement is undulating. The trajectory and direction depends on the initial conditions.

Acknowledgments

This work was supported by RSF № 18-71-10069.

References

[1] Klimina L A and Formal'skij A M 2020 *Three-link mechanism as a model of a person on a swing* (News of the Russian Academy of Sciences. Theory and control systems) **5** pp 89-105

[2] Fomenko S S and Bordyugov D V 2019 *Automation of the methods for overcoming obstacles by a mobile robot with a section propulsion device* (News of the Volgograd State Technical University) **9** pp 72-5

[3] Chernousko F L 2010 Optimal motion of a two-body system in a resistive medium *J. of Optimization Theory and Applications* vol 147 **2** pp 278-97

[4] Briskin E S, Vershinina I P, Maloletov A V and Sharonov N G 2014 On the control of the motion of a walking machine with twin orthogonal rotary movers *J. of Computer and Systems Sciences Int.* vol 53 **3** pp 464-71

[5] Briskin E S, Kalinin Y V, Maloletov A V and Sharonov N G 2018 *Mathematical modelling of mobile robot motion with propulsion device of discrete interacting with the support surface* (IFAC-PapersOnLine) vol 51 **2** pp 236-41

[6] Chernousko F L 2019 Dynamics and control of motion for systems containing internal moving masses (Dynamics and Control of Advanced Structures and Machines) *Contributions from*

the 3rd International Workshop pp 169-77
[7] Zhuravljov V F and Klimov D M 2008 Global movemen of a Celtic stone *News of the Russian Academy of Sciences. Mechanics of the solid body* **3** pp 8-16

IOP Conf. Series: Materials Science and Engineering 1129 (2021) 012053 doi:10.1088/1757-899X/1129/1/012053

About the geometrical form of the underwater robot platform with non-parallel system the anchor-rope propulsion

E S Briskin[1,2], N G Sharonov[1,2], Y V Kalinin[1,2] and I A Samozheev[1]

[1] Volgograd State Technical University, 28 Lenin Avenue, Volgograd, 400005, Russia
[2] Center for Technology Components of Robotics and Mechatronics, Innopolis University, 1 Universitetskaya Str., Innopolis, 420500, Russia

dtm@vstu.ru

Abstract. The underwater mobile transport robot, which is a platform assembled from the polygonal pontoon modules, is considered. The system are used anchor-rope movers, which ensures high stability and energy efficiency of the entire system due to interaction with the bottom surface. The aim of the study is to select the geometric shape of the module according to such criteria as specific heat losses in the movers, the length of the step and the tension of the ropes. The formulated criteria characterize the energy consumption and the level of dynamic loads in the considered dynamic system, and the achievement of their optimal values contributes to the increase of autonomy. It takes into account not only the exchange rate resistance to movement, but also the influence of the current, which, depending on the direction, can both promote the movement of the platform and counteract it.

1. Introduction

One of the types of underwater transport systems under development is a platform with positive buoyancy - pontoon, the principle of movement of which is based on the use of anchor-cable type engines. [1-3] the Principle of operation is shown in figure 1.

The propulsion unit is an anchor with two cables connected to the drums rotating according to certain laws. In beginning anchor, it occupies a position on one vertical with the drum. When the drum cable is strained and in the absence of slippage of the anchor on the bottom of the mobile robot moves. Then, the anchor moves to the new position by the joint coordinated control of the drums. Its lifting is promoted by the fact that one of the cables is close to the vertical position. Then the cycle repeats. The work of such an engine is similar to the work of a walking [4, 5, 6], and the difference is that the robot does not rely on it, but with its help clings to the support surface.

This type of engines is easy to operate, provides high profile permeability with different bottom topography and" stability" regardless of the distribution of the transported goods on the surface of the pontoon, as well as lower energy costs compared to other types of engines. Another advantage of such pontoon robots may be the ability to build a robot of any configuration using each of the pontoons as an elementary module.

Possible forms of such modules at the top view are shown in figure 2. To ensure the greatest "stability" anchor-cable movers are located on the edges of the modules. The limitation of module shapes is explained by the requirement of tight packing of modules when assembling a robot pontoon of any configuration.

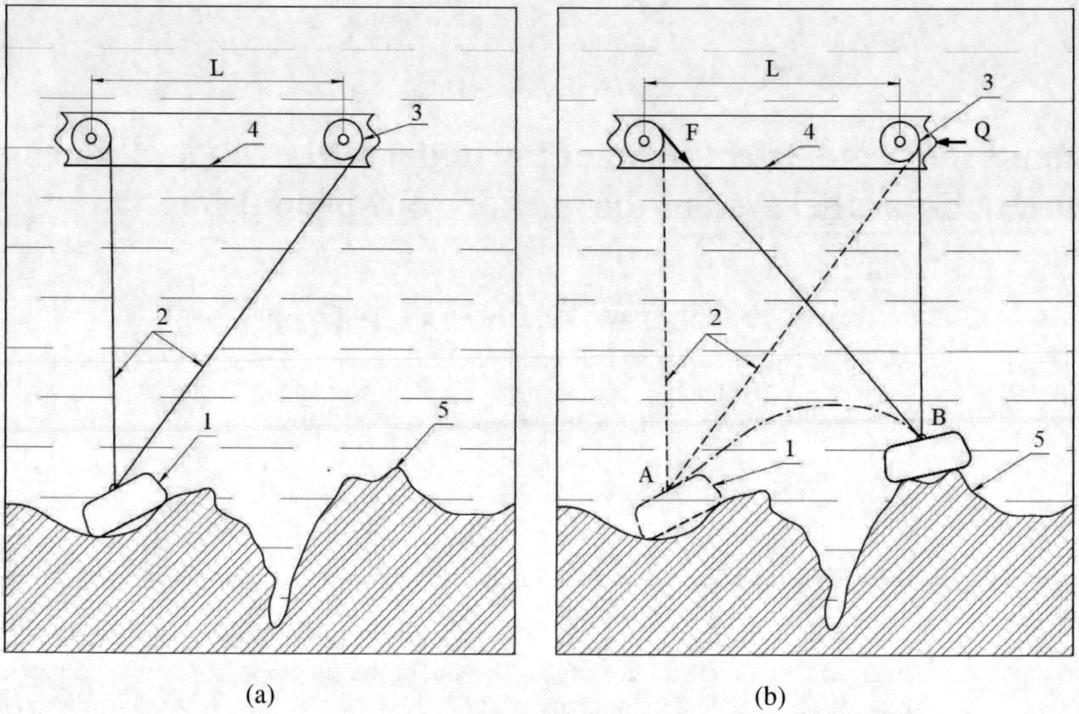

Figure 1. The principle of operation of the anchor-cable motor of the mobile robot-pontoon. a-start position, b-end position (1-anchor, 2-cables, 3-controlled drums, 4-robot pontoon body, 5-bottom profile, AB- the trajectory of movement of the anchor, F-driving force, Q is the resistance of an environment).

Indeed, a robust dense packing is achieved if, in a proper convex polygon, the angle α between the sides satisfies the equation:

$$\alpha = \pi(1 - 2/n) \tag{1.1}$$

Where n is the number of sides of the polygon.

In this case, k is the maximum number of pontoon robots with a common point, determined by the equation:

$$k = 2n/(n - 2) \tag{1.2}$$

Therefore, the number of possible variants of geometric shapes of pontoon robots is limited to three cases: n=3-right triangle; n=4-square; n=6-hexagon.

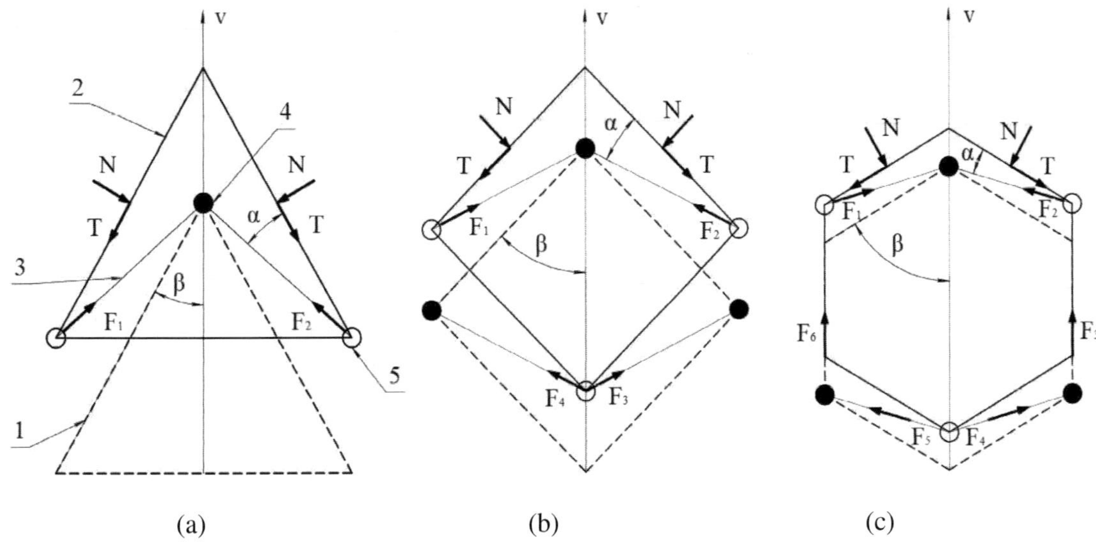

Figure 2. Possible forms of a mobile robot pontoon. a-triangle, b-square, c-hexagon (1-initial position of the body, 2-final position of the body, 3-cable, 4-anchor, 5-controlled drum, F-control force, Q-resistance force of the medium).

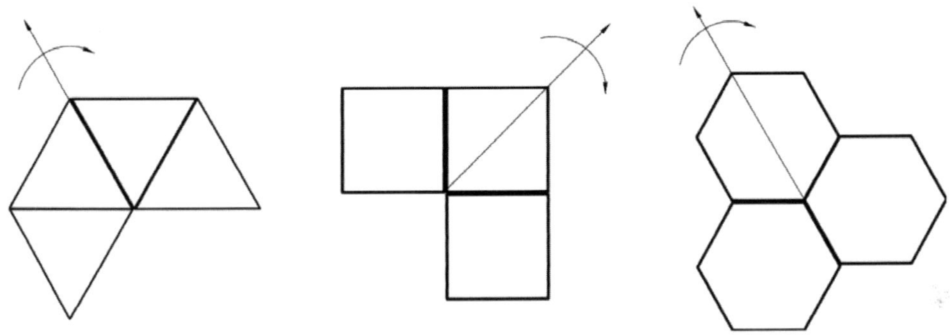

Figure 3. Possible configurations of pontoon robots.

The disadvantages of such pontoon robots include the difficulty of controlling the exchange rate movement because the efforts developed in the anchor-cable movers in the direction, as a rule; do not coincide with the required direction of the robot-pontoon program movement.

Another drawback is that the movement of the robot pontoon in one step (one permutation of the anchors) is limited to the ability of the cables to work only in tension. With their help, you can only be attracted to the anchors, but not repelled. In mathematical calculations, this circumstance is a limitation on the amount of forces, the numerical values of which can only be positive.

2. Problem statement

We consider three forms of pontoon robot modules with the same area *S*, which at the same height of the platform *h* corresponds to the same load capacity. Then the lengths of the sides l for the modules of triangular, square and hexagonal shapes are respectively equal:

$$l_{tr} = (4S)^{0.5} / 3^{0.25} \qquad l_{sq} = S^{0.5} \qquad l_{hex} = (4S)^{0.5} / 108^{0.25} \tag{2.1}$$

All robots make a translational motion with the same speed v. The propellers are installed on all edges of the robot-pontoon. Then, at low speeds v, the normal N and tangent T forces of resistance to motion acting on each face of the robot - pontoon, with a sufficient degree of accuracy are determined by expressions [2].

$$N = \rho h l v^2 \sin\beta \quad T = hl\mu v \cos\beta \tag{2.2}$$

Where ρ is the density of the liquid; μ is the coefficient of viscous friction; β is the angle of attack, v is the velocity of the body.

The problem of determining the optimal shape of the robot pontoon by criterion I, consisting of additive indicators, including the maximum possible length of the L_{max} step and the level of heat loss W. the condition of commensurability - the equality of the areas S of modules of all types is taken into account.

3. Method of solution

The solution is reduced to the geometric problem of establishing dependence for each type of robot-pontoon:

$$\beta = \beta(L) \tag{3.1}$$

And solutions of this equation for the determination of L_{max} at $\beta=\pi / 2$. Preparation of equations of translational uniform motion of pontoon robots. Based on nonlinear equations definition of laws of change of forces in anchor-cable movers depending on the passed way. Assessment of the level of heat loss per unit path based on the obtained laws of change of forces in the anchor-cable thrusters and characteristics of the driving engines. Obtaining dimensionless complex criterion I and determination of Pareto-optimal boundaries of the considered indicators.

$$I = k_1 L_{\max} + k_2 L_{\max} \tag{3.2}$$

3.1. The definition of the movement of the robot-pontoon in one step (one installation of anchors).

The maximum displacement of the pontoon corresponds to the angle α, determined from the equation:

$$\alpha + \beta = \pi/2$$

Herewith:

$$L_{\max} = l\cos\beta \tag{3.3}$$

3.2. The equations of the translational uniform motion and estimating the efforts of the ropes propulsion

The type of equations of motion depends on the number of engines performing work on the movement of the robot pontoon.

For a triangular robot, the required number of engines is two. Then the forces developed by the drive motors $F1$, $F2$ and resistance forces must be equal. The equation of motion is:

$$0 = -2N\sin\beta - 2T\cos\beta + (F_1 + F_2)\cos(\alpha+\beta) \tag{3.4}$$

Where:

$$F_1 = F_2 = (\rho h l v^2 \sin^2\beta + hl\mu v\cos^2\beta)/\cos(\alpha+\beta) \tag{3.5}$$

Where:

$$\beta = \pi/6; \alpha = arctg(vt\sin\beta)/(l - vt\cos\beta) \tag{3.6}$$

t - time of movement, changing in the interval:

$$0 < t < l\cos\beta / v \tag{3.7}$$

For a square robot, the possible number of propellers is four. The flat movement of the robot pontoon is provided by four driving forces *F1*, *F2*, *F3*, *F4* (figure 2) satisfying the system of equations:

$$(F_1 + F_2 + F_3 + F_4)\cos(\alpha+\beta) - 2T\cos\beta - 2N\sin\beta = 0$$

$$-(F_1 + F_3)\sin(\alpha+\beta) + (F_2 + F_4)\sin(\alpha+\beta) - (N_1 - N_2)\sin\beta + (T_1 - T_2)\cos\beta = 0 \tag{3.8}$$

$$(l/2)(F_2 - F_1)\cos\alpha + (l/2^{-0.5})(F_4 - F_3)\cos(\pi/4 - \alpha) = 0$$

Where β=π / 4; *T1*, *T2*, *N1*, *N2* – are determined from the equations (2.2); α is determined from the equation (3.6).

Execution of the last equation of the system (3.8) ensures the absence of rotational motion of the robot-pontoon relative to the center of mass. However, a feature of this system of equations is the presence of four control forces F_j in the three necessary conditions of the studied motion. Therefore, to uniquely determine the control forces it is necessary to involve an additional equation. It can be arbitrary, and its execution is provided by the control system [3]. It is most simple to provide a linear connection between F_j. [7]

$$\sum_{i=1}^{4}\gamma_i F_i = \sigma_i \tag{3.9}$$

Where γ_j, σ_j is an arbitrary constant. It is also possible to solve the optimization problem, for example, to require a minimum of heat losses. If all drives are identical, this reduces to an additional equation [5, 8].

$$\int_0^{\tau}\sum_{i=1}^{4}F_i^2\,dt \Rightarrow \min \tag{3.10}$$

Solved together with the equations of the system (3.8).

For a robot-pontoon in the form of a hexagon, the difference from the previous case will be that β=60°, and the number of control forces is six.

3.3. An example of the numerical solution.

To solve the model problem, the control equations are of the following form: [9]

$$\gamma_1 F_1 = F_2 \tag{3.11}$$

square pontoon S=10; the height of the pontoon h=1; the density of the liquid ρ=1000; speed pontoon v=2; the coefficient of viscous friction μ=0.0894; control coefficients: γ_1= 1.7 ; γ_2=20; γ_3=4

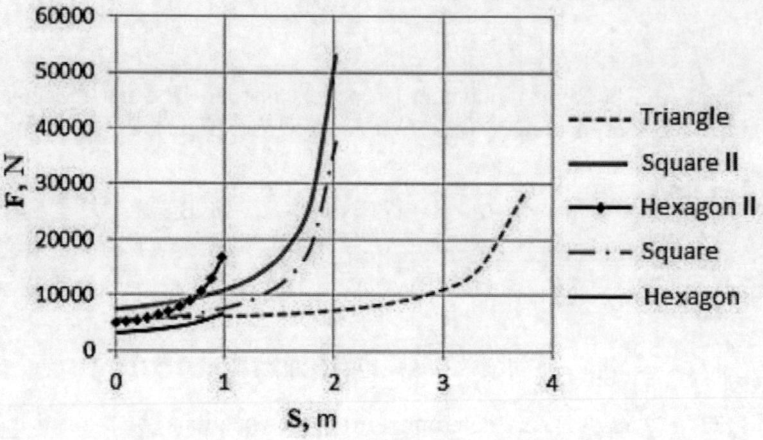

Figure 4. Dependence of the maximum cable tension force on the path.

I: γ_1= 1.7 ; γ_2=20; γ_3=4

II: γ_1= 1.2 ; γ_2=25; γ_3=8

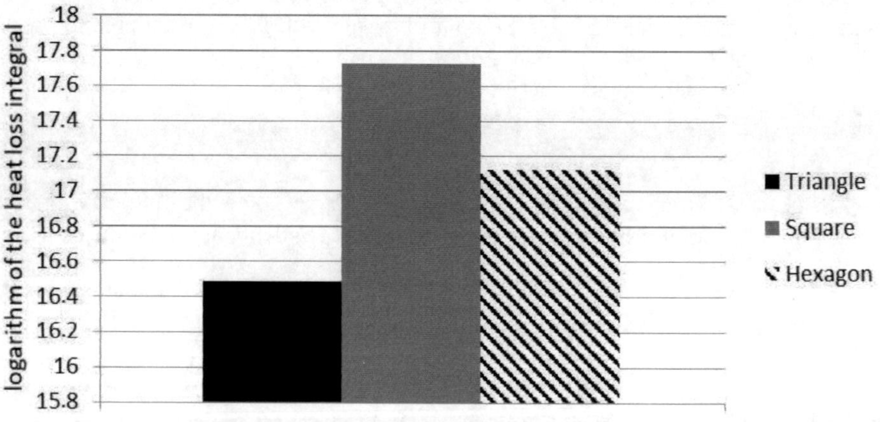

Figure 5. Specific heat losses for all forms of modules, expressed in a logarithmic scale.

4. Result

Based on the results of the calculation, the triangular shape has a maximum step length and minimum heat losses in the engines, however, such a module has no parallel directional propellers and the tension forces are relatively high. The square module has average stride length, Maximum heat loss and maximum tensile forces. The hexagonal shape has a minimum tension force, average heat loss and a minimum step length; all the propellers are directed in parallel, which opens up the possibility of movement of tacks in case of need. At this stage of development, the hexagonal shape of the robot-pontoon is the most suitable.

Acknowledgements

This work was supported by RSF № 18-71-10069.

References

[1] Briskin E S, Kalinin Y V, Maloletov A V, Serov V A and Ustinov S A 2017 On controlling the adaptation of orthogonal walking movers to the supporting surface *J. of Comp. and Syst. Scien. Intl.* **3** T56

IOP Conf. Series: Materials Science and Engineering 1129 (2021) 012053 doi:10.1088/1757-899X/1129/1/012053

[2] Briskin E S and Leonard A V 2016 Energy profile and the open-loop control of the translational motion of the walking machine cyclone *J. of Comp. and Syst. Scien. Intl.* **6** T55
[3] Zhoga V V 1998 *Computation of walking robots movement energy expenditure* (Belgium, IEEE International Conference on Robotics and Automation) p 163-8
[4] Sychev V V and Bashkin V A 2003 *Lectures on theoretical hydrodynamics studies. manual for University students in the direction of Butt. mathematics and physics* vol 2 (Moscow: Phys.-tech. state University)
[5] Vasiliev A V 1989 *Controllability of the ships* (Leningrad: Shipbuilding) p 62
[6] Golubev Yu F and Pogorelov D Yu 1998 *Computer simulation of walking robots* vol 4 (Fundamental and applied mathematics) p 525-34
[7] Briskin E S, Sharonov N G, Efimov M I, Gulevsky V V and Penshin I S 2020 Some problems of controlling the cable propulsion devices of mobile robots *23rd Int. Conf. CLAWAR Robots in Human Life* p 321-8
[8] Kalinin Y V and Miroshkina M V 2018 Dynamics walking robot with cyclic walking movers with minimum energy costs of moving *29-th Int. Conf. MICYSS* p 283-5
[9] Chernyshev V V, Arykantsev V V, Kalinin Y V, Gavrilov A E and Sharonov N G 2015 *Development of the walking mover for underwater walking vehicle* (Vienna: Int. Conf. DAAAM) p 1143-8

IOP Conf. Series: Materials Science and Engineering 1129 (2021) 012054 doi:10.1088/1757-899X/1129/1/012054

Development and research of adaptive control algorithm for road train lateral motion

D N Demyanov and R V Khamatyanov

Kazan Federal University, 18 Kremlyovskaya street, Kazan, 420008, Russia

demyanovdn@mail.ru

Abstract. In the presented work, the problem of control of a road train when it performs various maneuvers while moving along the highway is considered. An adaptive algorithm for controlling the lateral displacement of road train elements is proposed, based on the use of a PD-controller, the values of the coefficients of which depend on the current value of the longitudinal velocity. Computer simulation has been carried out, which has shown the correctness of the proposed algorithm.

1. Introduction

One of the main trends in the modern stage of development of automotive technology is the widespread introduction of various kinds of electronic driver assistance systems [1]. Such systems are intended, first of all, to increase efficiency when performing typical maneuvers while driving: ensuring directional stability, maintaining a constant speed or distance to the vehicle in front, etc. However, there are also promising developments that provide a much higher degree of automation. These include systems that provide autonomous and semi-autonomous movement along the route (including movement in a column) with automatic route selection and its rebuilding when external conditions change [2]. According to a number of experts, in the next 10-15 years there will be a significant increase in the share of vehicles that can perform transportation without a driver [3]. Moreover, from a commercial point of view, trucks and road trains are of considerable interest – their full automation will significantly reduce operating costs and increase efficiency. Thus, the development of algorithms for automatic control of vehicle traffic can be considered extremely relevant.

When building control systems for automotive equipment, as a rule, a hierarchical principle is applied, in which 3 levels are distinguished [4]. At the lower (local) level, the issue of managing individual nodes and units is being resolved; at the middle (tactical) level, the coordinated operation of subsystems and the implementation of typical maneuvers by the car as a whole are considered; at a high (strategic) level, the tasks of choosing a route and determining modes of movement are solved in interaction with third-party information systems.

In this paper one of the tasks of the tactical level of road train control is considered – the implementation of typical maneuvers when moving along the highway. It is required to develop an algorithm that would ensure a high quality of control processes in all permissible modes of motion. It should be taken into account that the dynamics of the control object changes significantly depending on its speed.

2. Calculation model used

To describe the movement of the elements of a road train, consisting of a truck and a semitrailer connected by a fifth wheel coupling, we will use the mathematical model proposed in the work [5]. It is assumed that the road train makes a flat motion, the deformations of the supporting structures of the truck and the semitrailer can be neglected, the steering angles of the steered wheels and the slip angles of each of the wheels for all axles are pairwise equal to each other (thus, each axle can be replaced by one conventional wheel, which is affected by longitudinal and transverse components of the reaction force). The corresponding calculation scheme of the model is shown in figure 1.

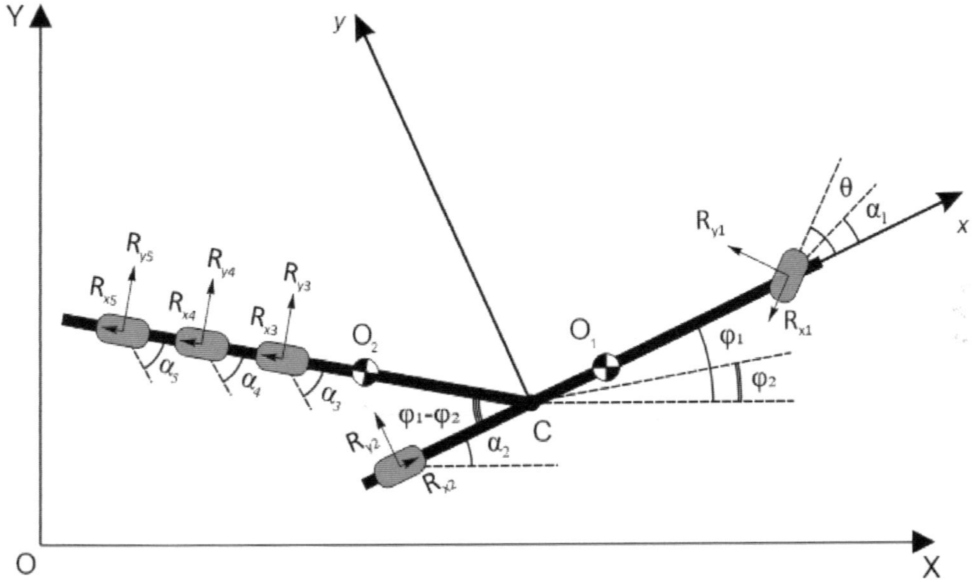

Figure 1. Calculation road train scheme.

Taking into account the accepted assumptions, the movement of the truck and semitrailer can be described by the following system of equations:

$$\begin{bmatrix} \sigma(m_1+m_2) & 0 & 0 & -m_2d_2\sin(\varphi_1-\varphi_2) & 0 & 0 & 0 & 0 \\ 0 & m_1+m_2 & m_1d_1 & -m_2d_2\cos(\varphi_1-\varphi_2) & 0 & 0 & 0 & 0 \\ 0 & m_1d_1 & J_1+m_1d_1^2 & 0 & 0 & 0 & 0 & 0 \\ 0 & -m_2d_2 & 0 & J_2+m_2d_2^2 & 0 & 0 & 0 & 0 \\ 0 & 0 & 0 & 0 & 1 & 0 & 0 & 0 \\ 0 & 0 & 0 & 0 & 0 & 1 & 0 & 0 \\ 0 & 0 & 0 & 0 & 0 & 0 & 1 & 0 \\ 0 & 0 & 0 & 0 & 0 & 0 & 0 & 1 \end{bmatrix} \times \begin{bmatrix} \dot{v}_x \\ \dot{v}_y \\ \dot{\omega}_{z1} \\ \dot{\omega}_{z2} \\ \dot{X} \\ \dot{Y} \\ \dot{\varphi}_1 \\ \dot{\varphi}_2 \end{bmatrix} = \begin{bmatrix} F_1 \\ F_2 \\ F_3 \\ F_4 \\ F_5 \\ F_6 \\ F_7 \\ F_8 \end{bmatrix}. \quad (1)$$

$$\begin{cases} F_1 = (m_1 + m_2) v_y \omega_{z1} + m_1 d_1 \omega_{z1}^2 - m_2 d_2 \omega_{z2}^2 \cos(\varphi_1 - \varphi_2) - F_{\hat{a}} - R_{x1} \cos\theta + R_{x2} - \\ \quad - R_{y1} \sin\theta - (R_{x3} + R_{x4} + R_{x5}) \cos(\varphi_1 - \varphi_2) + (R_{y3} + R_{y4} + R_{y5}) \sin(\varphi_1 - \varphi_2), \\ F_2 = -\sigma (m_1 + m_2) v_x \omega_{z1} + m_2 d_2 \omega_{z2}^2 \sin(\varphi_1 - \varphi_2) - R_{x1} \sin\theta + R_{y1} \cos\theta + R_{y2} + \\ \quad + (R_{x3} + R_{x4} + R_{x5}) \sin(\varphi_1 - \varphi_2) + (R_{y3} + R_{y4} + R_{y5}) \cos(\varphi_1 - \varphi_2), \\ F_3 = (\sigma - 1)(m_1 + m_2) v_x v_y - m_1 d_1 v_x \omega_{z1} - R_{x1} l_1 \sin\theta + R_{y1} l_1 \cos\theta - R_{y2} l_2, \\ F_4 = m_2 d_2 v_x \omega_{z1} - R_{y3} l_3 - R_{y4} l_4 - R_{y5} l_5, \\ F_5 = v_x \cos\varphi_1 - v_y \sin\varphi_1, \\ F_6 = v_x \sin\varphi_1 + v_y \cos\varphi_1, \\ F_7 = \omega_{z1}, \\ F_8 = \omega_{z2}. \end{cases} \quad (2)$$

Here v_x and v_y – are values of the longitudinal and transverse components of the velocity of the fifth-wheel coupling in the moving coordinate system associated with the vehicle body; ω_{z1} and ω_{z2} – are angular speeds of rotation of the body of the truck and semi-trailer; X and Y – are coordinates of the fifth-wheel coupling in immovable coordinate system associated with the road; φ_1 and φ_2 – are angles of rotation of the body of the truck and semitrailer; m_1 and m_2 – are truck and semitrailer weights; J_1 and J_2 – are moments of inertia of the truck and semitrailer relative to the axes passing through their centers of inertia perpendicular to the road plane; σ – is rotational inertia coefficient; d_1 and d_2 – are distances from the fifth-wheel coupling to the centers of inertia of the truck and semitrailer; R_{xi} and R_{yi} – are longitudinal and transverse components of the reaction force acting on the i-th axle; F_a – is aerodynamic resistance force; θ – is steering angle of the front wheels of the truck; l_1 and l_2 – are distances from the fifth-wheel coupling to the front and rear axles of the truck; l_3, l_4, l_5 – are distances from the fifth-wheel coupling to the corresponding axles of the semitrailer.

When solving the system of equations (1) – (2) it is assumed that the value of the longitudinal component of the reaction force is directly proportional to the rolling resistance coefficient of the i-th axle, and the transverse component of the reaction force is directly proportional to the side-slip angle of the i-th axle.

It is shown [5], that the presented model is relatively simple, but at the same time it allows to describe with a sufficiently high accuracy the behavior of the road train elements when performing typical maneuvers while moving along the highway.

To ensure the desired dynamics of the road train, it is proposed to use a control system, the block diagram of which is shown in figure 2. It includes a longitudinal velocity controller and a lateral displacement controller. The longitudinal velocity controller determines the required torque depending on the difference between the required and the actual value of the longitudinal velocity component. The lateral displacement controller determines the required steering angle of the front wheels of the truck depending on the difference between the required and the actual value of the lateral displacement of the fifth-wheel coupling. Since the dynamics of a road train at different speeds can differ significantly, it is proposed to make the lateral displacement controller adaptive: its coefficients will change depending on the actual value of the longitudinal velocity, providing a higher quality of control processes.

The described control system, which includes a model of road train dynamics (1) – (2) and models of controllers, was implemented in the Simulink environment of the Matlab computer mathematics system. The numerical values of the coefficients used in the model of the dynamics of the road train were chosen close to the characteristics of a real KAMAZ truck and semitrailer KRONE SDP 27.

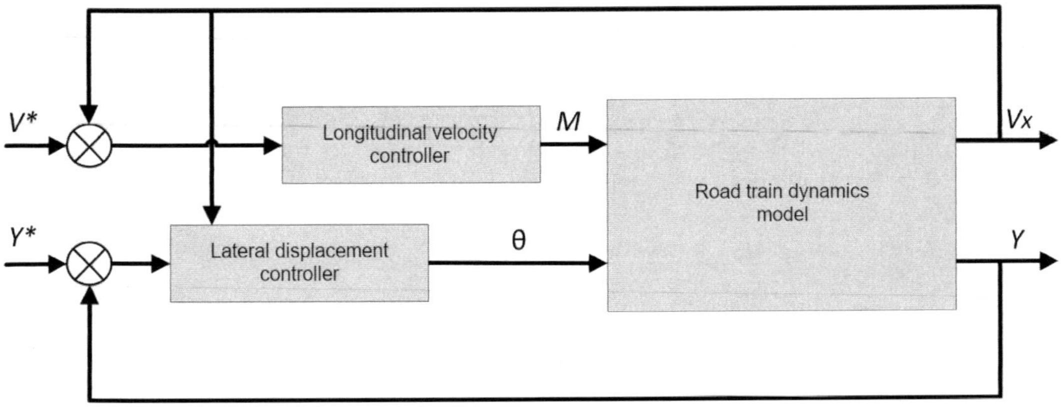

Figure 2. Block diagram of the road train dynamics control system.

3. Determination of parameters of the adaptive controller

To control the value of the longitudinal velocity and lateral displacement, in this work it is proposed to use a PID-controller. Controllers of this type are widely used in practice due to their rather high efficiency combined with ease of practical implementation [2].

To determine the optimal parameters of the controller we will use the standard Check Step Response Characteristics block of the Simulink environment [6]. At the same time, we will take into account that the maximum and minimum values of the output signal of the controller are limited both from above and from below – this is due to the physical nature of the signals M and θ.

At the initial stage, we will adjust the parameters of the longitudinal velocity controller using the previously mentioned tools. In the course of a series of computational experiments, it was determined that the optimal settings of the PID-controller are practically independent of the current motion speed. In this case, the values of the PID-controller coefficients averaged over the operating range were: $K_P = 2.006$; $K_I = 1.014$; $K_D = 1.113$. Thus, the use of an adaptive controller to control the longitudinal velocity of a road train is not advisable.

The computer simulation showed that for the selected values of the coefficients the quality indicators of the transient process remain sufficiently high in the speed range from 5 m·s^{-1} to 30 m·s^{-1}, which corresponds to the real modes of motion of a road train along the highway. As an example, figure 3 shows the test results corresponding to the acceleration of a road train from 5 m·s^{-1} to 10 m·s^{-1} and from 15 m·s^{-1} to 20 m·s^{-1}.

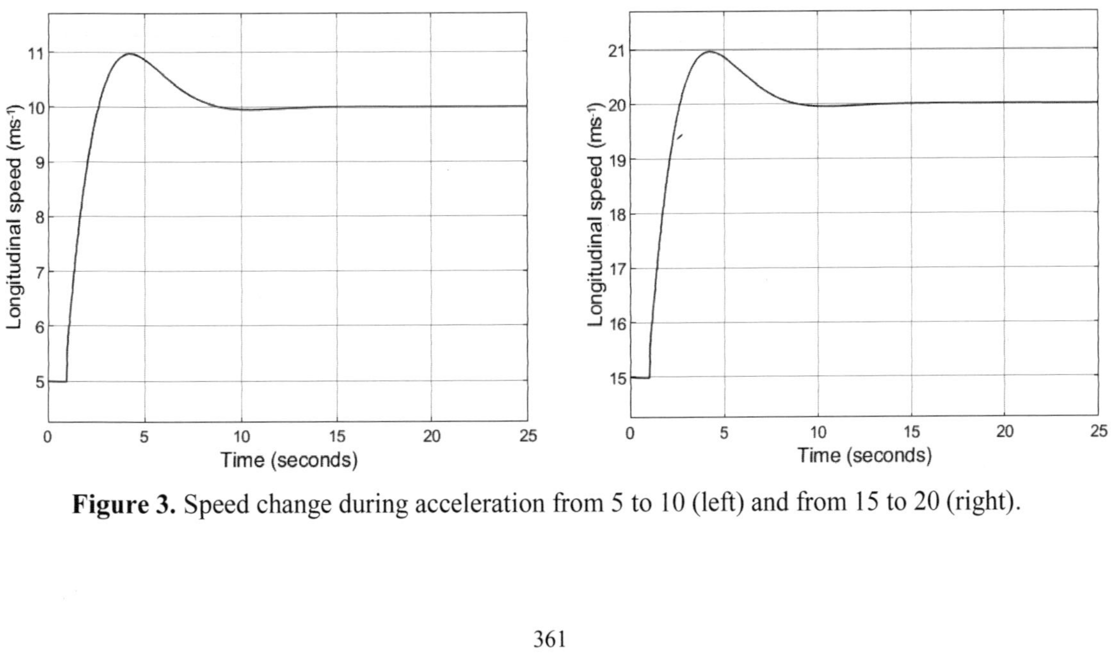

Figure 3. Speed change during acceleration from 5 to 10 (left) and from 15 to 20 (right).

As we can see, the selected controller settings allow obtaining results similar in efficiency.

When developing the lateral displacement controller, we will take into account the physics of processes occurring in the control object. Since the value of the lateral velocity component is integrated when determining the value of the lateral displacement, it is not advisable to use the integral component of the controller. To confirm this assumption, a series of optimization experiments was carried out using the capabilities of the Simulink environment for tuning P-, PD-, PI-, PID-controllers at various values of the longitudinal component of the speed. The results obtained have confirmed that the use of a PD-controller allows providing the best quality indicators of control processes over the entire operating speed range.

Table 1 shows the values of the PD-controller coefficients obtained using the Check Step Response Characteristics tool, corresponding to various values of the longitudinal velocity of motion. It can be seen that the parameters of the controller differ greatly at different speeds; therefore, averaging the settings does not allow ensuring high quality of the control processes.

Table 1. Values of the controller coefficients.

Longitudinal velocity, m·s^{-1}	Coefficient K_P	Coefficient K_D
1	36.528	25.845
5	12.904	8.384
10	3.278	2.568
15	0.524	0.485
20	0.296	0.402
25	0.223	0.364

As an example, figure 4 shows the results of tests simulating the motion of a road train to an adjacent traffic lane at longitudinal velocities of 2 m·s^{-1} and 20 m·s^{-1}. In this case, the parameters of the controller were determined by averaging the optimal values tuned for the reference speeds of the specified range.

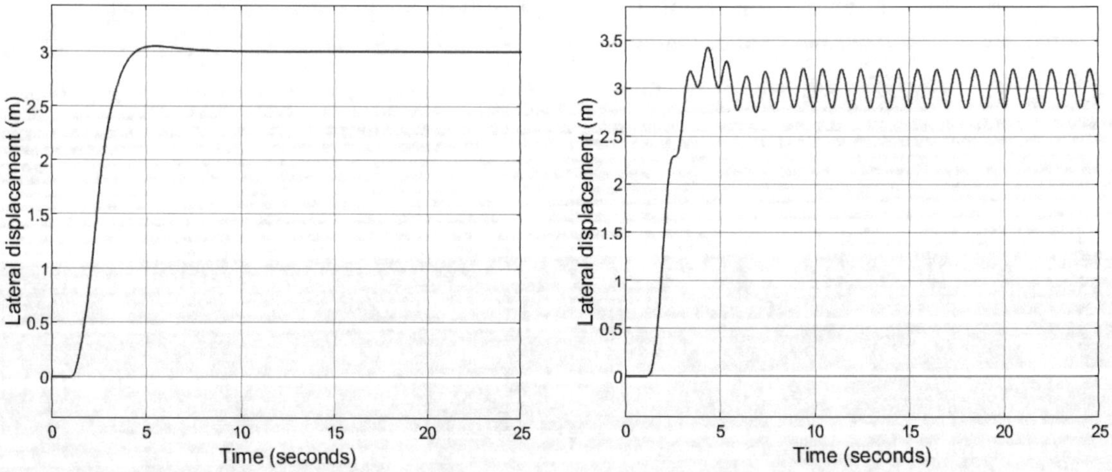

Figure 4. Change in lateral displacement when driving at a speed of 2 m·s^{-1} (left) and 20 m·s^{-1} (right).

As we can see, with the selected values of the coefficients at high speeds oscillations occur that are unacceptable in driving conditions on a real highway. Later a series of computational experiments was carried out for other ranges of speeds and values of the controller coefficients. The analysis of its results showed that the high quality of lateral displacement control processes over the entire operating range of longitudinal velocities cannot be ensured using a PD-controller with constant coefficients.

Thus, to ensure the required quality indicators, it is necessary to use an adaptive controller, the coefficients of which change depending on the current value of the longitudinal velocity of the road train. To construct such a regulator, we use the previously obtained optimal values of the coefficients, an example of which is given in table 1. To calculate the values of KP and KD, corresponding to the intermediate values of the velocity, we will use the known interpolation methods.

To select the best type of interpolation, a series of computational experiments was carried out, during which the quality indicators of control processes were compared for various intermediate values of the velocity with linear, quadratic and spline interpolation. The resulting values of the coefficients and quality indicators of the transient process turned out to be quite close to each other. This can be explained by the relatively small velocity step that was used to prepare the initial data.

After analyzing the results obtained, it was decided to use the linear interpolation method for calculating the values of the coefficients in the construction of the adaptive controller, since, with a comparable quality of control processes, it is easier to implement and requires less resources of the onboard information and control system.

4. Results and discussion

To assess the quality of control using the developed adaptive controller, a series of tests was carried out that simulated typical road train maneuvers (lane change, overtaking, etc.) at different speeds. As an example, figure 5 and figure 6 show the graphs of the lateral displacement change during the lane change.

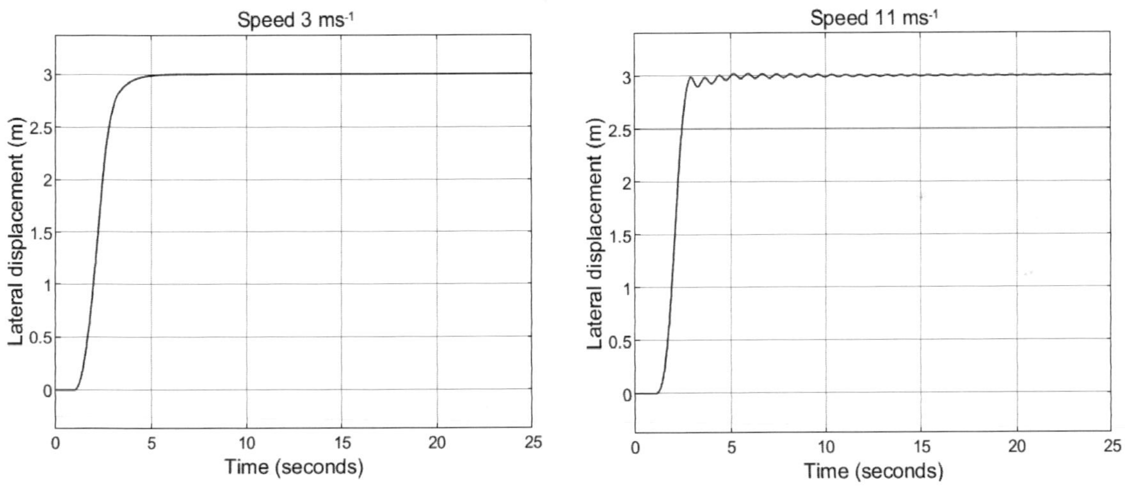

Figure 5. Change in lateral displacement when using adaptive controller at low speeds.

The analysis of the results obtained shows that the time of the transient process in all cases is acceptable, oscillation and overshoot are insignificant. Thus, the resulting controller makes it possible to ensure the set quality of the transient process at all speeds of the operating range.

It should be noted that in this work, a significant simplification was made regarding the nature of the road train movement: each transmission value corresponded to its own speed range, and these ranges did not overlap. Obviously, for a real car, this assumption is not fulfilled. Therefore, within the framework of further research, it is planned to improve the developed algorithm by taking into account the selected transmission when calculating the controller coefficients.

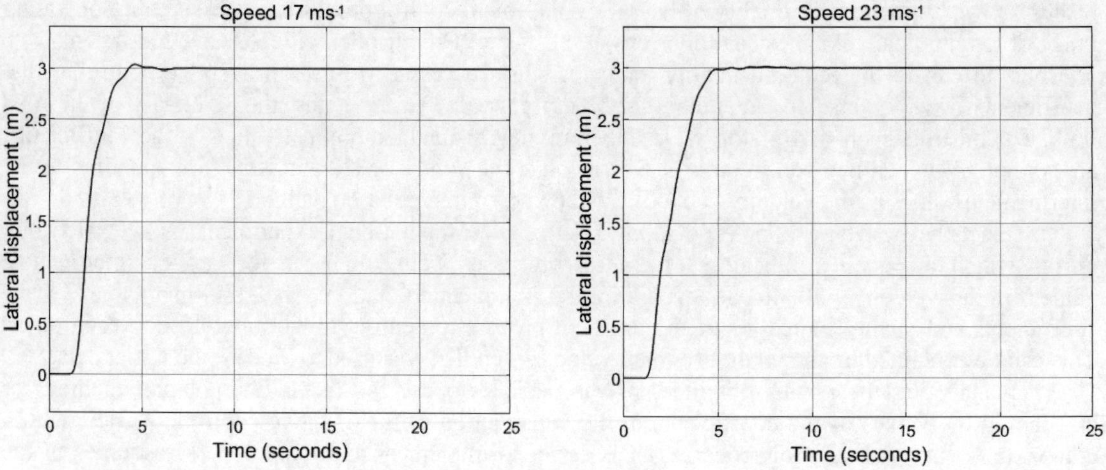

Figure 6. Change in lateral displacement when using adaptive controller at medium speeds.

5. Conclusions

In the presented work, a block diagram and algorithm for the operation of the system for controlling the motion of a road train are proposed. At the same time, it is proposed to use a classical PID-controller to control the longitudinal motion, and an adaptive PD-controller to control the lateral displacement, the coefficients of which change depending on the value of the longitudinal velocity. Computer simulation has been carried out, which has shown the high quality of the lateral motion control processes of the road train at the entire operating range of longitudinal velocities. In the future, it is planned to improve the proposed algorithm for calculating the controller coefficients, supplementing it with a dependence on the current value of the transmission gear ratio (this will allow to correctly work out situations of movement with the same speed in different transmissions).

References

[1] Kukkala V K, Tunnell J, Pasricha S and Bradley T 2018 Advanced driver-assistance systems: a path toward autonomous vehicles *IEEE Consumer Electronics Magazine* **7(5)** pp 18–25

[2] Gul F, Alhady S S N and Rahiman W 2020 A review of control algorithm for autonomous guided vehicle *Indonesian J. of Electrical Engineering and Computer Sci.* **20(1)** pp 552–62

[3] Schoitsch E 2016 Autonomous vehicles and automated driving: Status, perspectives and societal impact *24th Interdisciplinary Information Management Talks: Information Technology, Society and Economy Strategic Cross-Influences IDIMT* (Podebrady) pp 405–23

[4] Asanov A Z 2017 Modern architecture board information and control systems of heavy vehicles *Russian Technological J.* **5(3)** pp 106–13

[5] Volkov V G, Demyanov D N and Karabtsev V S 2018 Development and research of the mathematical model of planar motion of a vehicle with a semitrailer *Mathematical Models and Computer Simulations* **10(1)** pp 99–110

[6] Klee H and Allen R 2018 *Simulation of Dynamic Systems with MATLAB and Simulink* (New York: CRC Press) p 832

Synthesis of algorithm for stabilization of spatial position of manipulator gripper of parallel structure

V V Dyashkin-Titov[1] and V V Zhoga[2,3]

[1] Volgograd State Agricultural University, 26, University avenue, Volgograd, 400002, Russia
[2] Volgograd State Technical University, 28, Lenin avenue, Volgograd, 400005, Russia
[3] Innopolis University, 1, Universitetskaya Street, 420500, Russia

c_43.52.00@mail.ru

Abstract. Method of synthesis of kinematic algorithm of stabilization of spatial position of manipulator gripper of parallel-serial structure relative to given position is described in work. Coordinates and speeds and accelerations of generalized coordinates of manipulator are taken as control parameters. Analytical ratios of the control signals of the drive electric motors are obtained, which ensure the movement of the working body of the manipulator in the vicinity of the designated point. Numerical examples of the implementation of the obtained control laws are given.

1. Introduction

Manipulators of parallel-sequential structure find application in technological processes, during which high speeds and accelerations of the working element are required. Tripod manipulators are used in the assembly of chips, sorting and packaging of products. The operation time is up to 0.5-2 seconds [1-5]. Invention describes kinematic scheme and structure of manipulator consisting of two series-connected manipulators - manipulator-tripod, parallel structure and manipulator of serial structure. The sequential manipulator plays the role of a gripper. The task is divided into two stages. At the first stage, the configuration of the manipulator in the specified final position is determined. At the second stage laws of variation of generalized coordinates are formed when attachment point of gripping device moves from initial position to specified neighborhood of final state in preset time.

2. Object of research

Manipulator in the form of triangular pyramid with active actuating links of variable length l_1, l_2, l_3 is considered. One end of these links is connected by means of two-movable hinges located on fixed base *ABC*. Other ends of manipulator links are connected at point *M* by means of special spherical hinge assembly [6]. Photo and kinematic diagram of the manipulator under consideration are given in figure 1, 2 [7]. As actuators, links with electric DC drive are used. Actuating links have analogue displacement sensors.

Controlled gripping device consisting of three links connected in series with each other by means of cylindrical hinges is mounted on spherical hinge unit. Generalized coordinates of manipulator are lengths of links $l_i(t)$, angles of relative turns of links of gripping device $\alpha(t)$, $\psi(t)$, $\beta(t)$. The manipulator has six degrees of freedom.

IOP Conf. Series: Materials Science and Engineering 1129 (2021) 012055 doi:10.1088/1757-899X/1129/1/012055

The spatial position of the manipulator grip (figure 1) is determined relative to the *Oxyz* absolute coordinate system. Movable coordinate systems $O_jx_jy_jz_j$, (j=1-3) [8, 9] are connected to housing of each actuating link of gripping device.

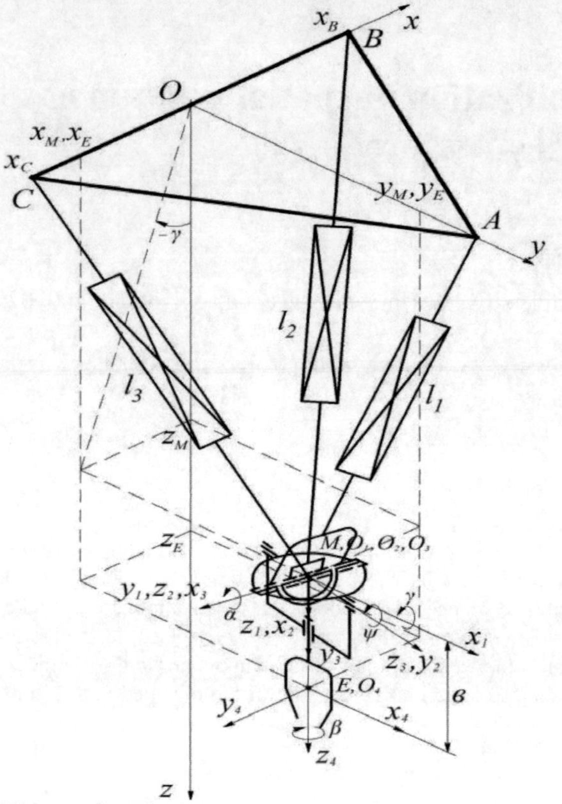

Figure 1. Kinematic diagram of manipulator - tripod on fixed base.

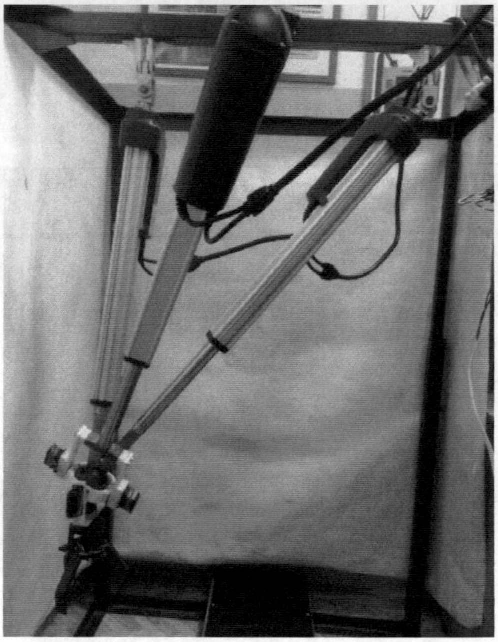

Figure 2. General view of the manipulator.

3. Algorithm for synthesizing the laws of moving the grip to a given position

We believe that when implementing the control algorithm, requirements for control accuracy are imposed only on the end point, without imposing restrictions on the trajectory of movement of the characteristic capture point, which is not predetermined. As generalized coordinates of the manipulator we take $q_1=l_1(t)$, $q_2=l_2(t)$, $q_3=l_3(t)$, which describe the tripod configuration. The coordinates $q_4=\alpha(t)$, $q_5=\psi(t)$ describe the state of the three-link gripper. Angle $q_6=\beta(t)$ does not affect the orientation of the working element, and the law of its change is determined by the nature of the technological operation.

For each actuating link, control loops are known, closed by the generalized coordinates $q_i(t)$, i=1-6. In this case, the solution of the problem is reduced to the definition of laws for the formation of control actions $q_i(t)$ for actuators that ensure the movement of the working element to a given position.

At the initial time t=0, the configuration of the manipulator is known and is characterized by the lengths of the actuating links $q_i(0)=l_{i0}$, i=1-3 and the angles $\alpha(0)=0$, $\psi(0)=0$, $\beta(0)=0$. Since the manipulator-tripod has great rigidity, Cartesian coordinates of the characteristic point of the working member are determined through the lengths of the actuating links with great accuracy [10].

$$x_M = \frac{l_3^2 - l_2^2}{4OB}, \quad y_M = \frac{L}{2OA}, \quad z_M = \left(l_1^2 - \frac{\left(l_3^2 - l_2^2\right)^2}{16OB^2} - \frac{L^4}{4OA^2} \right)^{0.5}, \tag{1}$$

$$L^2 = \left(-l_1^2 + 0.5l_2^2 + 0.5l_3^2 - OB^2 - OA^2\right).$$

At specified coordinates of characteristic point $M(x_M, y_M, z_M)$ and guiding cosines of gripping device α_{pq} in assigned point, length of actuating links l_i^0 and angles α, ψ, γ are located from solution of positioning problem [10]. The method of control system synthesis for laws of change of lengths $l_i(t)$ of actuating links ensuring movement of grip in vicinity of point with specified coordinates in specified time T is described below.

$$\left| l_i^0 - l_i(T) \right| \le \varepsilon ,$$

where ε is a small positive number characterizing the specified accuracy.

The angle β does not change, and the laws of forming the stabilization of the angles $\psi(t)$ and $\alpha(t)$ are formed according to the same method.

We accept that on the trajectory of motion the deviations from the final state of the actuating links $\Delta l_i(t) = l_i^0 - l_i(t)$ change in accordance with the solution of the differential equation

$$\Delta \ddot{l}_i(t) + a_{1i}\Delta \dot{l}_i(t) + a_{2i}\Delta l_i(t) = 0 \tag{2}$$

where a_{1i}, a_{2i} are constant positive numbers.

Laws of generation of driving effects $l_i(t)$ of actuators defined by equations (2) correspond to differential equations

$$\ddot{l}_i(t) + a_{1i}\dot{l}_i(t) + a_{2i}l_i(t) = a_{2i}l_i^0, \tag{3}$$

at initial conditions describing the manipulator state at t_0=0

$$l_i(0) = l_{i0}, \; \dot{l}_i(0) = \dot{l}_{i0}. \tag{4}$$

The general solution of equation (3) is:

$$l_i(t) = l_i^0 + C_{1i}e^{p_{1i}t} + C_{2i}e^{p_{2i}t}, \tag{5}$$

where the roots p_{1i}, p_{2i} are the roots of the respective characteristic homogeneous equation.

Permanent integration is found from (5) taking into account (4)

$$C_{1i} = \frac{\Delta l_i(0)p_{2i} + \dot{l}_{i0}}{\left(p_{1i} - p_{2i}\right)}, \quad C_{2i} = -\frac{\Delta l_i(0)p_{1i} + \dot{l}_{i0}}{\left(p_{1i} - p_{2i}\right)}, \tag{6}$$

where $\Delta l_i(0) = l_i^0 - l_i(0)$.

The coefficients of equations (3) are determined through the roots of the characteristic equation p_{1i}, p_{2i}

$$a_{1i} = -\left(p_{1i} + p_{2i}\right), \; a_{2i} = p_{1i}p_{2i}.$$

To ensure the requirement $l_i(t) \to l_i^0$ at t=T, the roots of the characteristic equation are negative different. Values of roots of characteristic equation should be selected from permissible parameters of manipulator mobility [11] (maximum permissible speeds of grip movement in working zone) and required time of getting to target point with required error.

The law of change of lengths of executive links (5) is program. To construct the control law with feedback, it is necessary to calculate the operating speedups on the current values of the generalized coordinates and speeds. In formulas (6) take the current values of the generalized coordinates for initial values of the generalized coordinates.

$$\ddot{l}_i^*(t)=\frac{\left[l_i^0-l_i(t)\right]p_{2i}+\dot{l}_i(t)}{\left(p_{1i}-p_{2i}\right)}p_{1i}^2e^{p_{1i}(T-t+\Delta T)}-\frac{\left[l_i^0-l_i(t)\right]p_{1i}+\dot{l}_i(t)}{\left(p_{1i}-p_{2i}\right)}p_{2i}^2e^{p_{2i}(T-t+\Delta T)}. \tag{7}$$

Feedback in control circuits of drive motors is performed by variables $l_i(t),\dot{l}_i(t)$. The values of these variables are used to evaluate expressions included in expressions (7). Structural diagram of control loop corresponding to (7) is given in figure 3.

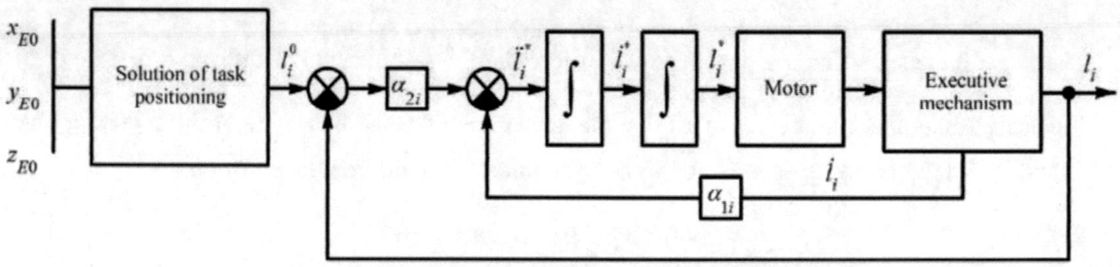

Figure 3. Structural diagram of the control system.

Figure 4, 5 shows the results of calculations when the gripper moves from the position determined by the coordinates of the point E x_{E0}=88.3 mm, y_{E0}=-6.9 mm, z_{E0}=1073 mm to the point with coordinates x_E=250 mm, y_E=-6.9 mm, z_E=1300 mm, p_{1i}=-3, p_{2i}=-2:

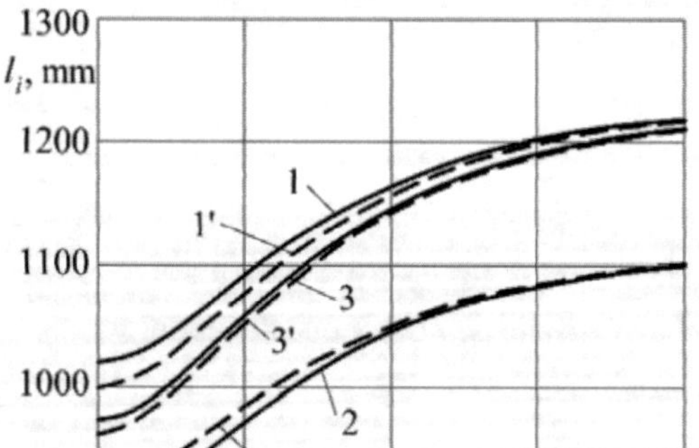

Figure 4. Laws of change of length of actuating links (Solid line - zero deviations, dashed line - with initial deviations).

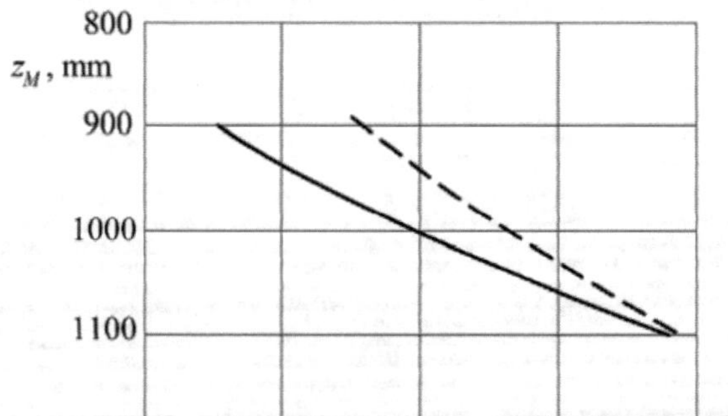

Figure 5. Trajectories of capture point M.

4. Conclusion

The article proposes an algorithm for terminal control of manipulator gripping. The difference of the proposed algorithm is to correct the program path so that it passes at each time point through the assigned point. This method of constructing a program path allows you to automatically adjust the speed when moving to a given point. The need to correct the program path can be caused by external disturbances, the problem of weak terminal control is solved. Analysis of the closed system showed that over time the mobile object falls into the final region δ_y of the target point, however, the magnitude of this region depends on the dynamic properties of the control system.

Acknowledgments
The reported study was funded by the grant of the President RF number MK-210.2020.8, and RFBR number 19-38-90067 and the Volgograd region according to the research project number 19-48-340013.

References

[1] Glazunov V A, Koliskor A Sh and Krainev A F 1991 *Spatial mechanisms of parallel structure* (Moscow: Science)
[2] Bushuev V V and Hol'shev I G 2001 The mechanisms of parallel structure in mechanical engineering *STIN* **1** pp 3–8
[3] Afonin V L, Podzorov P V and Slepcov V V 2006 *Processing equipment on the basis of mechanisms of parallel kinematics* (Moscow: Mashinostroenie)
[4] Liu N and Wu J 2014 Kinematics and application of a hybrid industrial robot Delta-RST *Sens. Transducers* **169**(4) pp 186–92
[5] Tanev T K 2000 Kinematics of a hybrid (parallel-serial) robot manipulator *Mech. Mach. Theory* **35**(9) pp 1183–96
[6] Gerasun V M, Pyndak V I, Nesmiyanov I A, Dyashkin-Titov V V and Pavlovsky V E 2012 Manipulators for mobile robots. The concept and design principles *(Preprints IPM of M V Keldysha)*
[7] Zhoga V V, Dyashkin-Titov V V, Dyashkin A V and Vorob'eva N S 2017 *RU Patent* 2616493
[8] Korendesev A I, Salamandra B L and Tyves L I 2006 *Theoretical basis of robotics. In 2 books* (Moscow: Nauka)
[9] Kolovskii M Z and Sloushch A V 1998 *Foundations of Industrial Robot Dynamics* (Moscow: Nauka)
[10] Zhoga V V, Dyashkin-Titov V V, Nesmiyanov I A and Vorob'eva N S 2016 Manipulator of parallel- serial structure with a controlled gripper positioning task *Mechatronics, automation, contro.* **8(17)** pp 525–30
[11] Hapkina I K 2016 The control algorithm for manipulation robots, built on the equations of dynamics *Izvestija TulGU. Technical science* **2** pp 296-304

IOP Conf. Series: Materials Science and Engineering 1129 (2021) 012056 doi:10.1088/1757-899X/1129/1/012056

Evaluation of the accuracy of the recognition algorithm of an autonomous robotic device to control weeds

A I Dyshakov and M A Mirzaev

Federal State Budgetary Scientific Institution "Federal Scientific Agricultural Engineering Center VIM", 1st Institutsky proezd 5, Moscow, Russia

a.i.dyshekov@gmail.com

Abstract. The main problem in the agro-industrial sector is the use of harmful herbicides, resulting in some of the weeds becoming more resistant to common herbicides and therefore using more and more powerful and expensive options. To overcome this problem, t is necessary to reduce the amount of herbicide use. The proposed study aims to develop a system that can detect and map weeds and perform mechanically applied spot spraying. The weed detection techniques described in this study include several aspects: imaging, a new algorithm for detecting plants through contour and colour segmentation. The method developed can detect weeds at early stages of growth. Further development of the methods is aimed at solving problems related to complex background without soil treatment, faster image acquisition speed and shorter processing time for spot injection in real time. The results of the experiment show that the proposed hybrid method can be used as a tool for weed control.

1. Introduction

Weeds adversely affect the yield of beet crops by taking away their nutrients. At present, chemical weeding with herbicides is widely used as a traditional method of weed vegetation control. However, herbicides are not considered as an effective method of weed control because of their negative environmental impact. Moreover, consumers are increasingly choosing natural, organic foods that have not been exposed to toxic chemicals. There are also non-chemical weeding methods, ranging from electrical to mechanical or physical control methods.

Many companies introduced a robot sprayer for weed control. Such selective spraying methods are still depend on the use of herbicides but can reduce their use [1].

The purpose of this study was to create and test in the laboratory a new prototype for the recognition of weeds and to examine its feasibility for successful application in agriculture. For this purpose, we have developed a universal platform for mechanical weeding of weed vegetation with point application of herbicides.

2. Materials and methods

To carry out the study, a system with one camera was used, which was previously installed at a certain angle, to calculate the coordinates of objects (figure 1). The limitation of this installation is that the robot is suitable for use on a flat surface, as due to irregularities, the focal length and angle of the camera will greatly. The stand itself consists of a platform with guides and a carriage installed on it which is driven by a worm gear by a step motor which provides movement along all three planes. On the Z-axis there is a camera Logitech HD Webcam C930e, resolution 1920x1080, speed of capture 120 psi. Images are imported into the weed recognition module through the bus and processed in System Visual Studio. The camera was calibrated in such a way that pixel coordinates of the weeds could be converted into real ones. The camera was fixed and the carriage changed its position to obtain 10 calibration photos which were used to get the camera's own parameters and distortion coefficients. Besides, after the calibration of the system, the parameters of the starting point set for the coordinate conversion. High image processing speed and system simplicity are also important aspects of this robotic weeding system in real time [2].

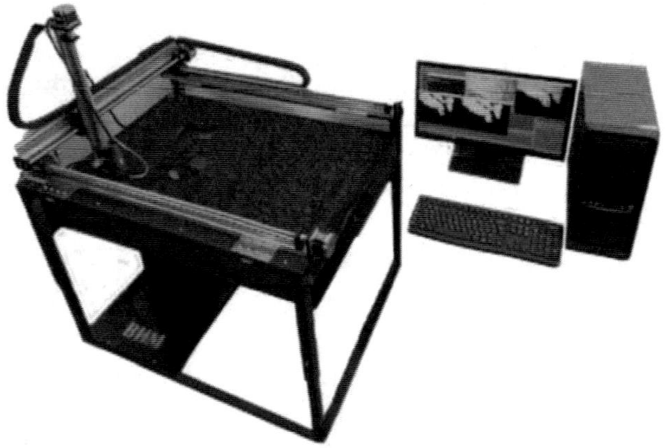

Figure 1. Laboratory test bench for weed vegetation control.

3. Image processing and weed recognition

The processing of images was carried out in two stages. The first step was to detect the culture (beet) and weeds which was aimed at obtaining 4 images, where the crop and weeds were defined as red and green (1) and the background was gray (soil) (2). Then, based on the difference in colors, an algorithm was used which compared the hue, saturation and value (HSV) of each pixel with a predetermined color (figure 2) [3].

At the beginning we get a picture of the whole landscape (figure 2 (a)). We set tolerances (± 0.1) for all values that in future work can be used for differentiation of plants and background in all other obtained images. This is followed by colour segmentation to identify green plants. Figure 2 (b) shows the result of the extraction method, in which large green leaves (considered beets) are clearly visible, while other plants (weeds) are less distinct and are separated from grey shades of soil and black tray. Since there was a lot of visual noise in the soil (stones, lumps of soil, etc.), the image was filtered and a black mask was applied. In addition, due to the influence of light, it was possible to see darkened plants which would affect the calculation of plant centroids. Therefore, the image was filled with contour completion to fill the voids. The next step was to segregate the culture so that weeds could be identified. Three segmentation methods were proposed, tested and compared. The first two segmentation algorithms suggested obvious differences in area and color between weeds and beet. The first of these algorithms was based directly on the difference in area. The area was calculated using Python function and compared with the contour of the already developed base to define it as belonging to a culture (beet).

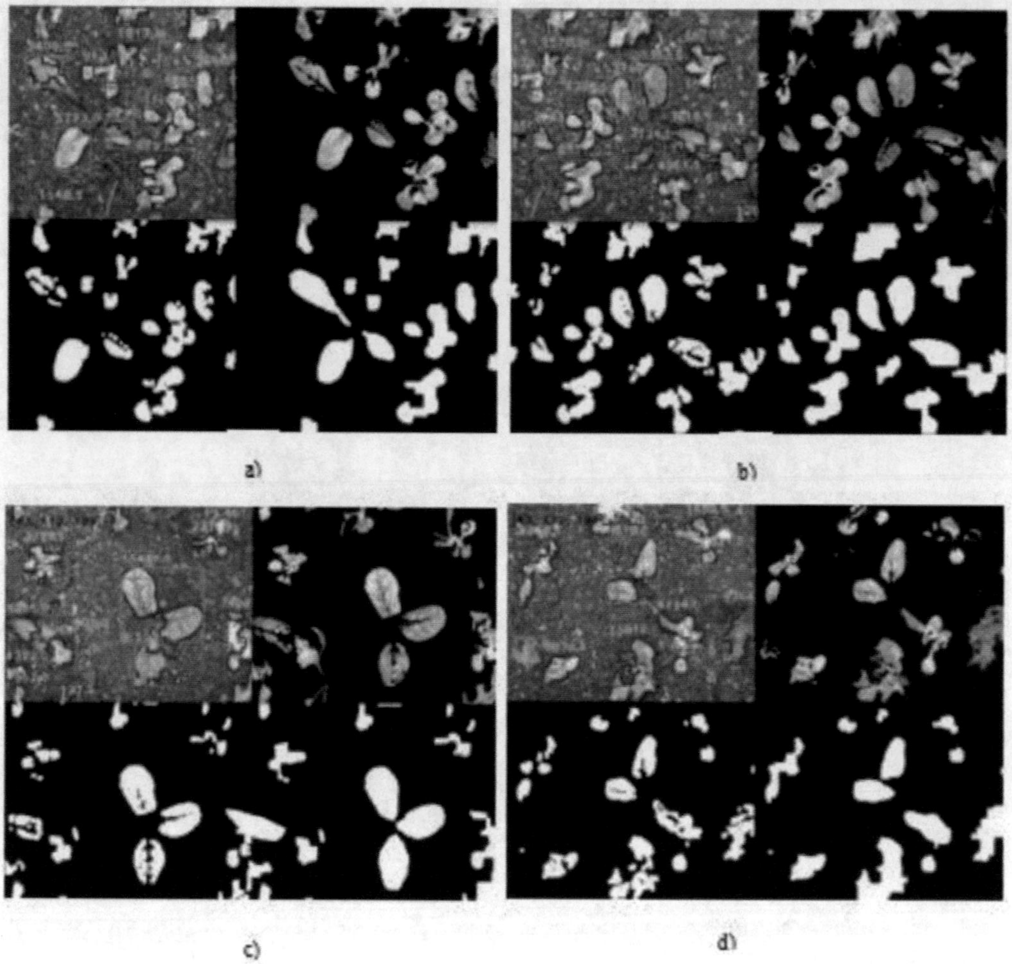

Figure 2. Image processing protocol: (a) original image; (b) use of colour segmentation to determine green plants in the background; (c) obtaining a binary image; (d) identification of a cultivated plant.

The problem of weed and culture intersection is solved by methods of comparison of the size of foxes and crop contours and unwanted plants [4]. Therefore, the second method based on erosion and dilatation operations was implemented. Using this method, images were eroded until all weeds were detected, as shown in figure 2 (d). Then smoothing took place as shown in figure 2 (c) with increasing erosion coefficient. Thereafter, an operation was performed to superimpose the obtained contours on the original image to determine the culture, as shown in figure 2 (a). The following is the result of segmentation - highlighting in red and building weed centroids. The processing time of the two algorithms for figure 2 was 0.0968 s (in area) and 0.0346 s (based on erosion and dilatation), respectively, on our equipment as described above [5]. Table 2 below shows the results of image processing the number of errors where the system made an incorrect determination of weed culture. The analysis was performed on 50 photographs by visual processing and comparison. Plants were divided into two groups - weeds and useful cultures [6, 7].

However, due to changes in light intensity in the environment, fixed thresholds were not sustainable enough. To correct this deficiency, a color reproduction algorithm was adopted [8, 9].

Table 1. Experimental values.

No. photo	A number of the recognizable beet	In total beets	A number of the recognizable weeds	In total weeds	Quantity of errors of. %	Number of errors during beet removal	Number of weed spreading errors
1	3	4	13	10	14.29	25.00	30.00
2	2	3	14	12	6.67	33.33	16.67
3	2	3	11	9	8.33	33.33	22.22
4	4	4	11	12	6.25	0.00	8.33
5	2	3	10	12	6.67	33.33	16.67
6	5	5	13	14	5.26	0.00	7.14
7	3	3	11	11	0.00	0.00	0.00
8	2	3	13	13	6.25	33.33	0.00
9	2	3	14	12	6.67	33.33	16.67
10	3	3	11	11	0.00	0.00	0.00
11	3	3	12	10	15.38	0.00	20.00
12	4	3	11	13	6.25	33.33	15.38
13	3	2	13	13	6.67	50.00	0.00
14	2	2	14	12	14.29	0.00	16.67
15	2	2	11	11	0.00	0.00	0.00
16	4	3	12	12	6.67	33.33	0.00
17	3	3	12	11	7.14	0.00	9.09
18	4	4	13	12	6.25	0.00	8.33
19	4	3	14	14	5.88	33.33	0.00
20	3	3	12	12	0.00	0.00	0.00
21	2	2	14	14	0.00	0.00	0.00
22	3	2	13	12	0.00	50.00	8.33
23	2	3	12	13	0.00	33.33	7.69
24	1	1	11	12	7.69	0.00	8.33
25	3	2	14	15	0.00	50.00	6.67
26	2	2	14	12	14.29	0.00	16.67
27	4	3	12	14	5.88	33.33	14.29
28	2	2	13	13	0.00	0.00	0.00
29	1	1	14	14	0.00	0.00	0.00
30	1	1	14	12	15.38	0.00	16.67
31	3	3	10	11	7.14	0.00	9.09
32	3	3	12	13	6.25	0.00	7.69
33	4	4	11	13	11.76	0.00	15.38
34	4	4	10	11	6.67	0.00	9.09
35	1	2	12	10	8.33	50.00	20.00
36	1	3	13	10	7.69	66.67	30.00
37	2	2	10	10	0.00	0.00	0.00
38	2	3	10	9	0.00	33.33	11.11
39	2	2	12	11	7.69	0.00	9.09

40	4	3	11	11	7.14	33.33	0.00
41	1	2	13	12	0.00	50.00	8.33
42	1	1	14	13	7.14	0.00	7.69
43	3	3	10	11	7.14	0.00	9.09
44	4	3	10	12	6.67	33.33	16.67
45	2	2	14	15	5.88	0.00	6.67
46	1	1	13	13	0.00	0.00	0.00
47	2	2	13	14	6.25	0.00	7.14
48	3	2	11	11	7.69	50.00	0.00
49	2	2	14	15	5.88	0.00	6.67
50	4	4	11	13	11.76	0.00	15.38
				Total	**5.87**	**16.50**	**9.10**

Using the results obtained, it is possible to compare the obtained result with the actual one. The final result of segmentation can be seen in figure 2 (a) in which the beet centrids were marked. The processing time was 1 s. The methods of combining shape and colour were simple and fast [2, 10].

4. Conclusions

In this study, a machine vision-based algorithm was implemented and tested to recognize weed vegetation by combining contour and color detection of the beet crop with further marking of centroids, followed by removal. Tests have shown that the accuracy of the recognition algorithm autonomous robotic device to control weeds using the proposed method is up to 83.5.The analysis of the dependence of the recognized crop in the image on the real amount of weeds was also carried out. Various areas of treated soil were taken to assess the operability of the algorithm. Other parameters such as soil to lens height, the same recording mode, permanent lighting and others were also constant, only the cultivated areas were changed.

References

[1] Mao W, Hu X and Zhang X 2008 Weed detection based on the optimized segmentation line of crop and weed *Computer And Computing Technologies In Agriculture* vol II The International Federation for Information Processing ed D Li (Boston, MA: Springer US) pp 959–67

[2] Dyshekov A I, Smirnov I G, Mirzaev M A and Shereuzhev M A 2020 Principles of functioning of the autonomous device for weed control for precision agriculture *IOP Conf. Ser.: Mater. Sci. Eng.* **747** 012108

[3] Andreasen C, Rudemo M and Sevestre S 1997 Assessment of weed density at an early stage by use of image processing *Weed Research* **37** 5–18

[4] Weis M and Sökefeld M 2010 Detection and Identification of Weeds *Precision Crop Protection - the Challenge and Use of Heterogeneity* ed E-C Oerke, R Gerhards, G Menz and R A Sikora (Dordrecht: Springer Netherlands) pp 119–34

[5] Woebbecke D M, Meyer G E, Bargen K V and Mortensen D A 1993 Plant species identification, size, and enumeration using machine vision techniques on near-binary images *Optics in Agriculture and Forestry* **1836** 208–19

[6] Chi Y –T, Chien C –F and Lin T –T 2003 Leaf shape modeling and analysis using geometric descriptors derived from Bezier curves *Transactions of the ASAE* **46** 175

[7] S El-Faki M, Zhang N and E Peterson D 2000 Weed detection using color machine vision *Transactions of the ASAE* **43** 1969–78

[8] Muangkasem A, Thainimit S, Keinprasit R and Isshiki T 2010 Weed Detection over Between-Row of Sugarcane Fields Using Machine Vision with Shadow Robustness Technique for

Variable Rate Herbicide Applicator *Energy Research Journal* **1** 141–5
[9] Rumpf T, Römer C, Weis M, Sökefeld M, Gerhards R and Plümer L 2012 Sequential support vector machine classification for small-grain weed species discrimination with special regard to Cirsium arvense and Galium aparine *Computers and Electronics in Agriculture* **80** 89–96
[10] Thorp K R and Tian L F 2004 A Review on Remote Sensing of Weeds in Agriculture *Precision Agriculture* **5** 477–508

Development of a control system for a mobile robot based on Markov processes

A A Fedorov, S Y Orekhov, P I Veysman and N Y Vasilchuk[4]

Kaluga Branch of the Bauman Moscow State Technical University, Bazhenova street 4, Kaluga, 248004, Russia

sf090@yandex.ru

Abstract. The purpose of this work is to develop and study a control system for mobile robots based on Markov processes. The paper investigates the adequacy of the control system algorithms based on Markov processes as applied to a mobile tracked robot.

1. Introduction

At this stage of development, there is an acceleration of scientific and technological progress, this is due to the fact that there is an improvement in the old and the introduction of new methods of planning and managing enterprises or individual technological processes. The widespread use of automated control systems can increase the pace of development, since they pay special attention to the issue of decision making [1].

One of the parts of the general theory of decision making that studies stochastic [2] and dynamical systems [3] that are subject to control is called the theory of Markov processes [4]. Markov processes, or rather Markov chains, are used in modeling the process of maintaining the operability of an automated control system [5].

2. Controlling robotic systems using Markov processes

If the following parameters: the set of states $i=\overline{1,m}$, the vector of probabilities of the initial states $\overline{p_0}=\|p_i\|_m$, the set of decisions $k=\overline{1,K}$, the matrix of conditional probabilities of the transition of the process for the iteration $R^k=\|r_{ij}^k\|_{mm}^K$, the matrix of conditional payments for the iteration are known, then the Markov process is taken as given [6].

Among the set of strategies S in the course of solving the Markov problem, the optimal strategy f^* will be obtained. The column vector $f^s=\left[k_1^s\, k_1^s \ldots k_m^s\right]^T$ is an arbitrary strategy with the index $s=\overline{1,S}$. The selection of the optimal strategy is performed criterionally. The main criterion is the average maximum income for an iteration or accumulation of the maximum value of payments. Payment functions are considered initially specified when developing a system based on Markov processes. At the current moment, the process goes through the iteration n and is in the state i, and in this strategy, k_i^s is accepted as the decision. Matrices P^k and R^k are replaced by working matrices P^s and R^s.

3. Solution of the direct problem of the Markov decision-making process

During the process of solving the direct problem of the Markov process, a comparison is made based on the size of the average payment for the iteration, then a working algorithm is built, in which V^s is taken as the average payment for the iteration, P^s and R^s are working matrices.

The algorithm used applies the matrix form of the $f^s = \left[k_1^s\, k_1^s \ldots k_m^s\right]^T$ strategy to construct and configure matrices. The construction of the general matrix occurs layer by layer or line by line. The configuration parameters of the first layer (rows) of the P^s matrix are formed from the first layer of the $P^{k_1^s}$ matrix, for the second layer a similar principle will be applied, applied to the $P^{k_2^s}$ matrix. The principle of formation of subsequent layers is similar to the principle presented above. Also, all layers of the R^s matrix are formed by analogy with the P^s matrix.

To form the matrix of transition probabilities P^s and payments for the iteration R^s, it is required to form the matrices P^k and R^k in advance. The average payment for an iteration in the i-th iteration:

$$r_i^s = \sum_{j=1}^{m} p_{ij}^s r_{ij}^s \tag{1}$$

The unconditional average payment is calculated using the vector of state probabilities. The unconditional average payment for this case will look like this: $\overline{f}^N = \left[p_1^N\, p_1^N \ldots p_m^N\right]^T$. The N index in the formula presented earlier will denote the step number in the established strategy. The stationary strategy for calculating the average payment V^S per iteration applies the following formula:

$$V^S = \sum_{j=1}^{m} p_i^N r_i^s = \sum_{i=1}^{m} p_i^N \sum_{j=1}^{m} p_{ij}^s p_{ij}^s. \tag{2}$$

Provided that the payment is profitable, the criterion for choosing the optimal strategy will look like this:

$$s^* = \arg\max_{s\in\{\overline{1,S}\}} V^S. \tag{3}$$

The vector of the limiting probabilities of the states of the Markov process $\overline{p}^N$ tends to correspond to the matrix equation $(p^S)^T \overline{p}^N = \overline{p}^N$. Also, the probabilities of the states must satisfy the normalization condition, which is the following: $\sum_{i=1}^{m} p_i^N = 1$. The current value of the coordinates of the vector is obtained by finding the roots of a system consisting of two equations.

4. Solution of the inverse problem of the Markov decision-making process

The decision-making procedure in the Markov process at the input must have data on the state of the chain i_n and the decision k_n, which satisfy $i_n, k_n \in \{1;2\}$. The result of the operation of the algorithms incorporated in the decision-making procedure in the Markov process will be the value of the payment v. Within the framework of the problem under consideration, a useful payment will be the length of the path traveled by the robot. In this case, only the useful path is taken into account. To achieve the result, it is required to obtain an intermediate solution element, which is an implementation that includes the total payments and process states. The output of the inverse problem is a matrix of probabilities and payments. This task is solved in three stages:

Stage 1. For each implementation, a strategy $f^s \in \{f^1, f^2, f^3, f^4\}$ is identified. The complete set of strategies: $f^1 = [1\ 1]^T$; $f^1 = [1\ 2]^T$; $f^3 = [2\ 1]^T$; $f^4 = [2\ 2]^T$. According to the frequency of implementation, the closest one is evaluated, which is put in accordance with this implementation.

Stage 2. For each implementation, the entire set of transition probability matrices is estimated: P^1, P^2, P^3, P^4. For each implementation, the frequencies of the transition from one state to another are calculated taking into account the decision. The frequencies are estimates of the conditional transition probabilities in one step, the probabilities are placed in the corresponding places of the conditional transition matrices.

Stage 3. Construction of payment estimates - using the parameters of the conditional transition matrices and payments in each observation, calculate the estimates of the payment vector r_i^s using the least squares method, the recurrent form of which, connecting the previous $(q-e)$ estimates of observations with the current $((q+1)-th)$, has the following form:

$$\begin{aligned} \hat{r}_{q+1} &= \hat{r}_q + Q_q P_q \left[P_q^T Q_q P_q + 1 \right]^{-1} \left[v_{q+1} - P_q^T \hat{r}_q \right], \\ Q_{q+1} &= Q_q - Q_q P_q \left[P_q^T Q_q P_q + 1 \right]^{-1} P_q^T Q_q \end{aligned} \quad (4)$$

where $Q_q = (P_q^T P_q)^{-1}$; v_{q+1} is the payment in the $(q+1)$ observation;

P_q - transition matrices obtained at stage 2 in the q-th observation.

Markov chain can be divided into several stages, each of which is called the process step. In turn, the sequential execution of the steps of the process leads to the implementation of the process, the new implementation recursively refines the estimate of the payment vector. This process can be described as the formalization of positive experience based on the decision-making in the Markov process. Also, the Markov chain approximates current preferences.

Correction of strategies taking into account vectors of payments taken from current observations of actions, as well as rejection of payments, goals, preferences and elimination of uncertainty arising from the lack of information about matrices and, is due to the recurrent estimation algorithm.

5. Markov process with discrete time

The model of a Markov process is presented in the form of a graph, the vertices of which are interconnected by connections. The vertices of such a graph are states, and links are transitions between them. Figure 1 shows the graph of the model under study.

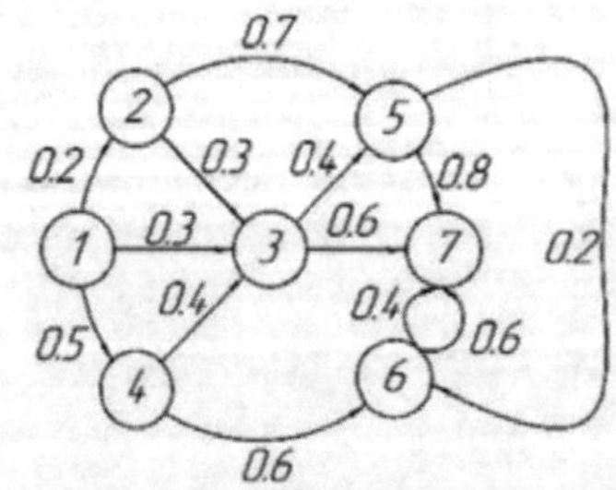

Figure 1. Markov transition graph.

Each of the transitions is characterized by its own transition probability P_{ij}. The probability shows how often the system changes from state i to state j. The probability of crossing is the probability of passing a section for its full cost, but also with a certain probability a penalty may fall on this section, which will halve the payment [7].

Figure 2 shows a tree of possible system transitions - possible trajectories of movement.

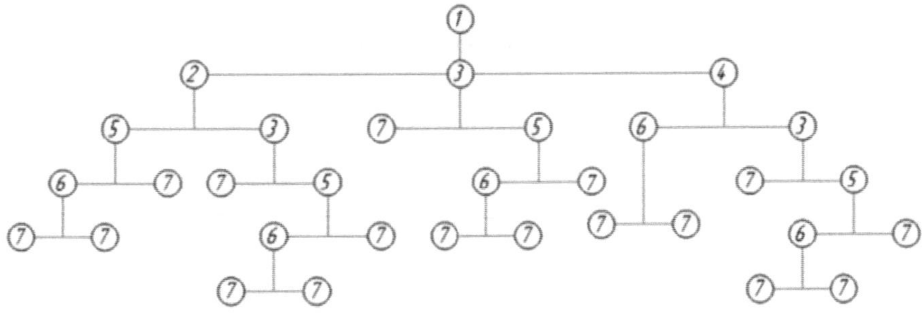

Figure 2. Tree of possible system transitions.

The matrix of transition probabilities for this graph is shown in table 1. The sum of transition probabilities is always equal to one, since the system must go into some state.

Table 1. Probability matrix.

	In S_1	In S_2	In S_3	In S_4	In S_5	In S_6	Sum of probabilities
of $\boldsymbol{S_1}$	0.2	0.3	0.5	X	X	X	0.2+0.3+0.5=1
of $\boldsymbol{S_2}$	X	0.3	X	0.7	X	X	0.3+0.7=1
of $\boldsymbol{S_3}$	X	X	X	0.4	X	0.6	0.6+0.4=1
of $\boldsymbol{S_4}$	X	0.4	X	X	0.6	X	0.4+0.6=1
of $\boldsymbol{S_5}$	X	X	X	X	0.2	0.8	0.2+0.8=1
of $\boldsymbol{S_6}$	X	X	X	X	X	0.4/0.6	0.4+0.6=1

Table 2 containing the values of all payments and penalties for each transition.

Table 2. Table of payments and penalties.

	12	13	14	25	23	46	43	35	37	56	57	67(1)	67(2)
Payment	20	30	50	40	30	60	50	20	70	10	40	30	10
Fine	10	15	25	20	15	30	25	10	35	5	20	15	5

In total, there are 17 possible trajectories for this graph:

1)1 – 2 – 5 – 6 – 7(1);
2)1 – 2 – 5 – 6 – 7(2);
3)1 – 2 – 5 – 7;
4)1 – 2 – 3 – 7;
5)1 – 2 – 3 – 5 – 6 – 7(1);
6)1 – 2 – 3 – 5 – 6 – 7(2);
7)1 – 2 – 3 – 5 – 7;
8)1 – 3 – 7;
9)1 – 3 – 5 – 6 – 7(1);
10)1 – 3 – 5 – 6 – 7(2);
11)1 – 3 – 5 – 7;
12)1 – 4 – 6 – 7(1);
13)1 – 4 – 6 – 7(2);
14)1 – 4 – 3 – 7;
15)1 – 4 – 3 – 5 – 7;
16)1 – 4 – 3 – 5 – 6 – 7(1);
17)1 – 4 – 3 – 5 – 6 – 7(2).

After carrying out 10000 experiments of the control system based on Markov processes, the following graphs were obtained (figure 3, 4). The first graph shows the distribution of travel along a particular trajectory for all experiments, and the second graph shows the average amount of payments earned for all passes along the trajectories.

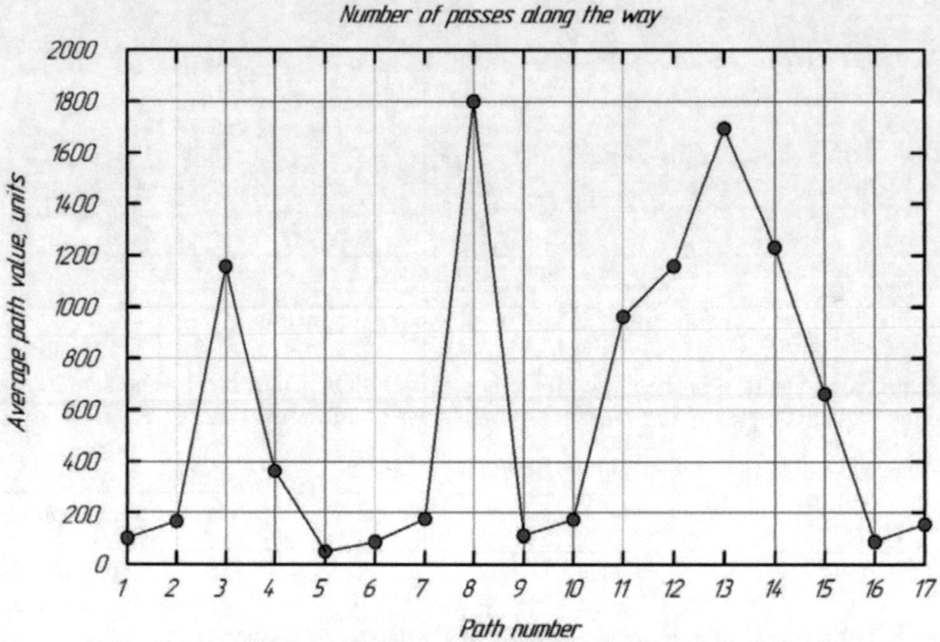

Figure 3. Number of passes along trajectories.

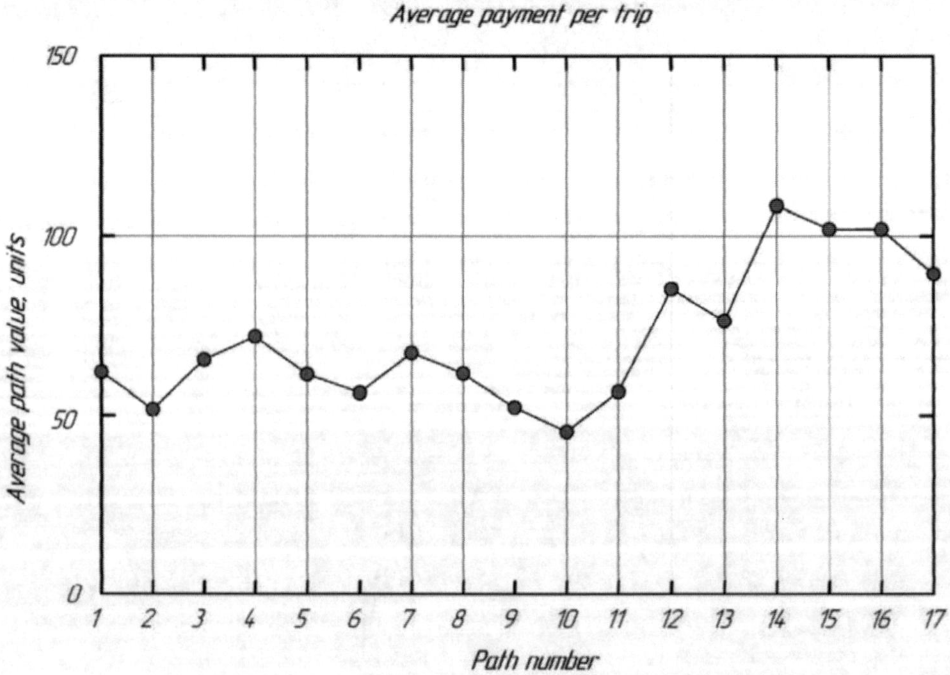

Figure 4. Average payment for the passage of each trajectory.

Having carried out a series of experiments and obtained the values of the average payments for all trajectories, it is possible to compose a Markov chain reflecting the trajectory with the greatest value - number 14 (figure 5).

Figure 5. Markov chain formed by the Markov graph.

During the modeling process, it is also possible to obtain a time diagram of transitions in the Markov graph. The diagram shows how the process of transition between states occurs over time (figure 6).

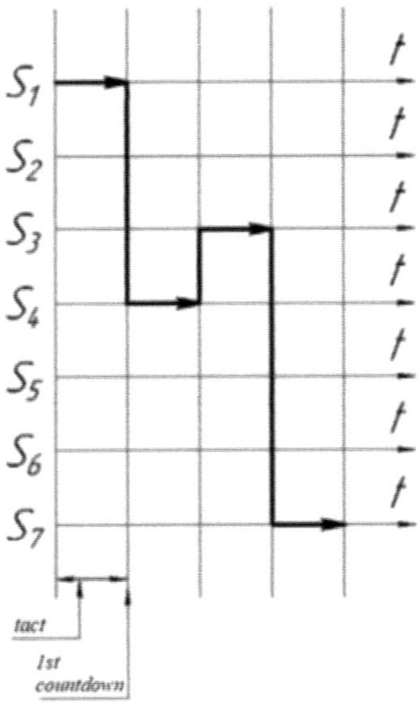

Figure 6. Timing diagram of transitions in the Markov graph.

The transition from the initial state to the final state occurred in four clock cycles.

Figure 7 shows a view of the proposed test site, as well as the optimal trajectory of movement in figure 8.

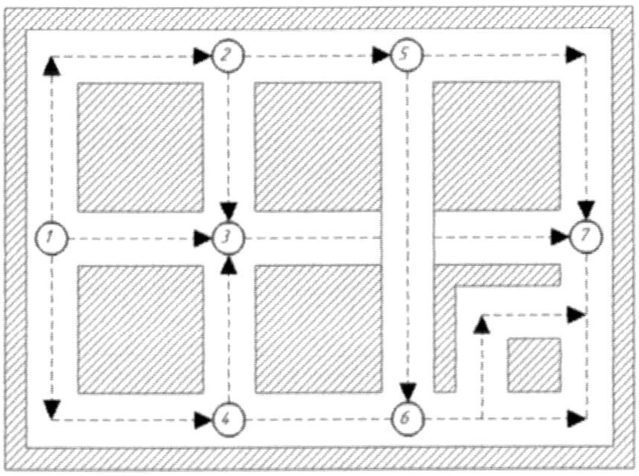

Figure 7. Proving ground.

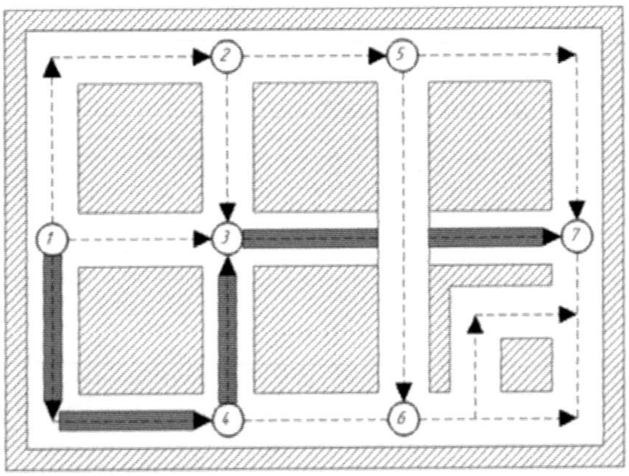

Figure 8. Optimal trajectory of movement.

IOP Conf. Series: Materials Science and Engineering 1129 (2021) 012057 doi:10.1088/1757-899X/1129/1/012057

To demonstrate the operation of the control system, the design of a tracked platform was chosen, since this type of mobile robot is more convenient in terms of motion stabilization, since the movement is limited by the plane [8]. Such a design is convenient for placing orientation sensors and other required components (figure 9).

Figure 9. Construction of a tracked robot.

6. Navigation system

The mobile robot's navigation system allows it to determine its position in the environment and its orientation. This information is used to determine the direction of movement of the mobile robot.

A laser rangefinder and an optical slit infrared sensor in the amount of two pieces are responsible for the navigation system.

During the experiment on the positioning accuracy over the mobile robot, the obtained data were presented in figure 10 and 11, which include a comparative characteristic of the real and theoretical trajectory of movement, as well as the positioning error.

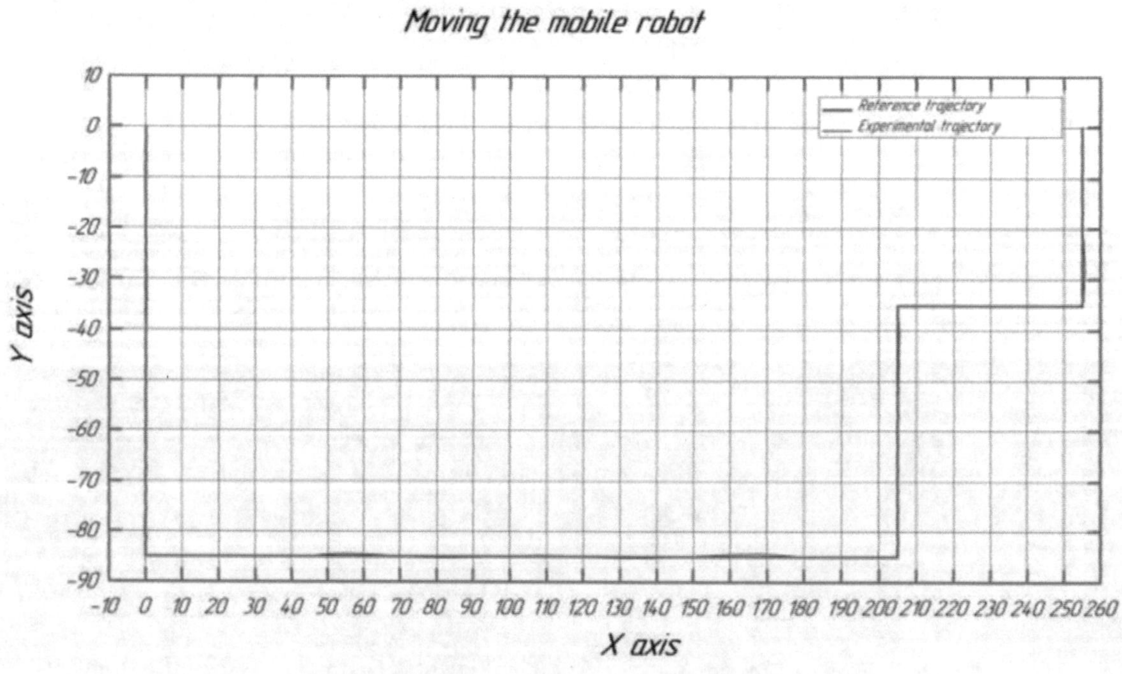

Figure 10. Real and theoretical trajectory.

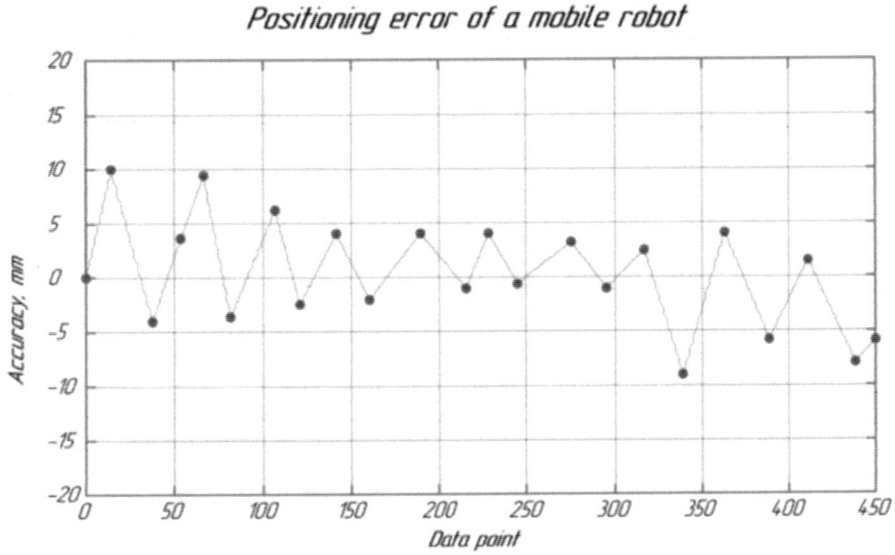

Figure 11. Positioning error.

7. Conclusions

The purpose of this research is to study intelligent control methods for mobile robotics. This study examines a control system for a mobile robot based on Markov processes.

For control systems for mobile robots based on Markov processes: an experimental setup based on a tracked robot has been created; algorithms for the operation of the control system based on Markov processes have been developed. Data were obtained confirming that this type of control system is adequate and the errors do not exceed the permissible limits. The next step of the study is to create a system for joint control of a group of robots based on Markov processes.

References

[1] Ibrahimov A A 1984 Markov decision-making processes in the development of an algorithmic control system for technological objects *PhD thesis* (Tashkent: NPO Cybernetics) p 196

[2] Vilisov V Y 2014 Analysis of the robot learning efficiency under target nonstationarity *Collection of scientific articles based on the materials of the XI Int. Sci. and Tech. Conf. Kursk: Vibration technology, mechatronics and controlled machines Vibration-2014* vol **2** pp 282-7

[3] Vilisov V Y 2014 Analysis of robot dynamics learning under nonstationarity criteria *Moscow Univercity of Technology: Information Technology Bulletin* vol **2** pp 34-9

[4] Dynkin E B 1959 *Foundations of the theory of Markov processes* (Moscow: Science) p 227

[5] Emelyanov A A, Shilnikova O V and Emelyanova N Z 2015 Modeling the process of maintaining the operability of a developing ACS *Applied Informatics* vol **10** 5(59) pp 93-108

[6] Main H and Osaki S 1977 *Markov decision-making processes* (Moscow: Science) p 176

[7] Markov A A 1951 *Selected works: Theory of numbers. Probability theory* (Moscow: USSR Academy of Sciences) p 717

[8] Zhdanov A A 2008 *Autonomous artificial intelligence* (Moscow: Binom) p 359

IOP Conf. Series: Materials Science and Engineering 1129 (2021) 012058 doi:10.1088/1757-899X/1129/1/012058

The results of the impact theory application in modeling the gear shifting processes in automatic transmissions

G K Korendyasev, K B Salamandra and L I Tyves

Mechanical Engineering Research Institute of the Russian Academy of Sciences (IMASH RAN), Maliy Kharitonievsky pereulok, 4, Moscow 101000, Russia

ksalamandra@yandex.ru

Abstract. The process of gear shifting in vehicles automatic transmissions is considered in the paper as the shock interaction of transmission elements. Two mathematical models are proposed and considered for simulation the gear shifting processes. Computer simulation of the models' movement during acceleration of a vehicle, as well as modes of low- and high-frequency gear shift cycling (self-oscillations of the output shaft of the gearbox) is carried out. Methods for decreasing the amplitude of the gearbox output shaft oscillations caused by the impact nature of gear shifting are proposed.

1. Introduction

Gear shifts in automatic transmissions are performed without interrupting the power flow. This process involves two clutches. The off-going clutch of the previous gear and on-going clutch of the next gear are working simultaneously at gear shifting [1]. The gear shift time duration is 0.2 - 0.5 seconds in modern gearboxes [2, 3]. Obviously, the fast gear shifting process leads to an abrupt change in the velocities of the transmission elements even with a small difference in the speeds of the engine and the output shaft of the gearbox. The values of inertial torques as a result of the gear shift significantly exceed the torques of the engine and the torques of the resistance to motion forces, which in turn can worsen the comfort of the vehicle passengers, reduce the service life and reliability of the transmission.

Works [1-6] in particular are devoted to dynamic studies of gear shifting processes, while in most of them this process is divided into stages and considered sequentially. The short time of gear shifting makes it possible to represent the gear shifting as instantaneous process and apply the theory of impact for its analysis. Considering gear shifting as an interaction with an impact is especially relevant for dual clutch transmissions, which have recently become quite widespread. There is no torque converter between the engine and the gearbox, which acts as a damper of torsional vibrations of the gearbox input shaft at gear shifting.

2. Model to calculate

The model of the vehicle transmission showed in figure 1 consists an engine, a two-speed gearbox with gear ratios i_1, i_2, two friction clutches $c1$, $c2$, engaging corresponding gears, an output shaft of a gearbox with elasticity c and damping b, differential and semi-axles, connected to the drive wheels and the vehicle body.

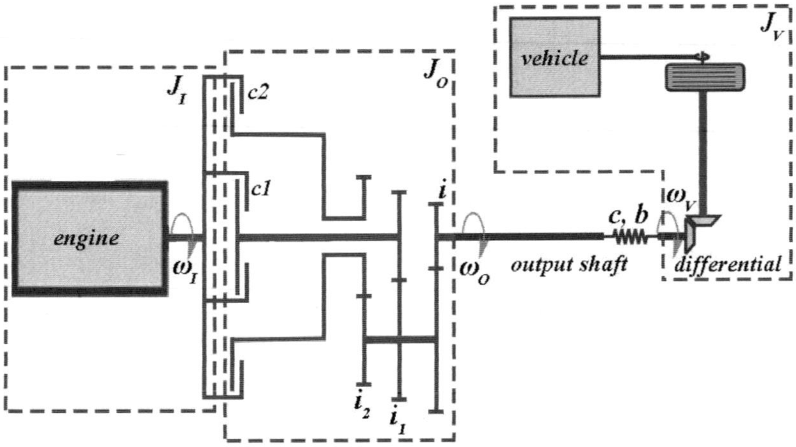

Figure 1. Vehicle transmission with 2-speed dual-clutch gearbox.

The moment of inertia of the moving parts of the engine reduced to the body of the clutch being switched on will be denoted by J_I; J_O – the moment of inertia of the gearbox moving parts to the link with the maximum flexibility reduced to the output shaft; element J_V is the moment of inertia of the transmission links after the link with the maximum flexibility, reduced to the output shaft of the gearbox, including the differential, half-shafts, drive wheels and the vehicle body. Thus, the resulting dynamic model contains three inertial elements: J_I and J_O, which are rigidly connected by means of clutches $c1$ or $c2$, and the J_V element, connected to the J_O element by elastic connection.

The inertial elements J_I and J_O are rigidly connected to each other by one of the clutches ($c1$ or $c2$). J_I and J_O between gear shifts form a single inertial element with the moment of inertia reduced to the output shaft of the gearbox $\left(J_I i_j^2 + J_O\right)$, where i_j is the gear ratio of the engaged gear. The inertial element $\left(J_I i_j^2 + J_O\right)$ angular velocity after each of the two considered gear shifts is determined using the theorem on conservation of the angular momentum at impact [7].

The movement of the model in figure 1 between gear shifts when one of the clutches c1 or c2 is switched on is a torsional vibration of inertial elements $\left(J_I i_j^2 + J_O\right)$ and J_V relative to the center of mass. The model center of mass rotates under the action of the engine torque $M_{\mathrm{I}} = const \neq 0$ and moments of resistance forces $M_{\mathrm{R}} = const \neq 0$. Thus, the model in figure 1 between gear shifts has 2 degrees of freedom and is described by a system of two linear differential equations:

$$\begin{aligned} \left(J_I i_j^2 + J_O\right)\dot{\omega}^{(j)} + b\left(\omega^{(j)} - \omega_V^{(j)}\right) + c\left(\varphi^{(j)} - \varphi_V^{(j)}\right) = iM_I \\ J_V \dot{\omega}_V^{(j)} + b\left(\omega_V^{(j)} - \omega^{(j)}\right) + c\left(\varphi_V^{(j)} - \varphi^{(j)}\right) = -M_R \end{aligned} \qquad (1)$$

In system (1): φ_O are the angular coordinates of the inertial element $\left(J_I i_j^2 i^2 + J_O\right)$; φ_V - the angular coordinate of the output shaft after the link with maximum flexibility; superscript (j) indicates the gear engaged; i is a gear ratio equal to i_1 or i_2.

Free vibration frequencies of the model in the absence of damping: $k_1 = 0$; $k_2^{(j)} = \{c(J_I i_j^2 i^2 +$ *JO+JVT·JIij2i2+JOJV−10.5*. Frequency of damped oscillations: *k2∗(j)=k2(j)2−h(j)20.5*, where $h^{(j)} = b\left(J_I i_j^2 i^2 + J_O + J_V\right) \cdot \left[2\left(J_I i_j^2 i^2 + J_O\right) J_V\right]^{-1}$.

The relative motion of the elements $\left(J_I i_j^2 i^2 + J_O\right)$ and J_V are their synchronous torsional vibrations in antiphase, and the magnitudes of the oscillation amplitudes are inversely proportional to the moments of inertia [8]. The general solution of the obtained equations (1) has the form:

$$\begin{aligned}\varphi_O^{(j)} &= A^{(j)} \cdot \exp(-h^{(j)}t) \cdot \sin(k_2^{*(j)}t + \beta^{(j)}) + C_1^{(j)} + C_2^{(j)}t + a^{(j)}t^2/2 \\ \varphi_V^{(j)} &= \mu^{(j)}A^{(j)} \cdot \exp(-h^{(j)}t) \cdot \sin(k_2^{*(j)}t + \beta^{(j)}) + C_1^{(j)} + C_2^{(j)}t + a^{(j)}t^2/2\end{aligned} \tag{2}$$

where $\mu^{(j)} = A_V^{(j)}/A^{(j)} = -(J_I i_j^2 i^2 + J_O) \cdot J_V^{-1}, a^{(j)} = (iM_{\text{I}} - M_{\text{R}}) \cdot (J_I i_j^2 i^2 + J_O + J_V)^{-1}$.

The constants A, β, C_1 and C_2 are determined through the initial angular coordinates and velocities of the inertial elements of the model. Parameters C_1, C_2 and a - characterize the motion of the center of mass of the model in (3), parameters A and β - characteristics of the relative motion.

After differentiating equations (2) and substituting $\sin\alpha^{(j)} = h^{(j)} \cdot \left(h^{(j)2} + k_2^{*(j)2}\right)^{-0.5}$ and $\cos\alpha^{(j)} = k_2^{*(j)} \cdot \left(h^{(j)2} + k_2^{*(j)2}\right)^{-0.5}$ we obtain the general form of the equations for the velocities of inertial elements in the j-th gear between gear shifts:

$$\omega^{(j)} = A^{(j)} \cdot \exp(-h^{(j)}t) \cdot \left(h^{(j)2} + k_2^{*(j)2}\right)^{0.5} \cos(k_2^{*(j)}t + \beta^{(j)} + \alpha^{(j)}) + C_2^{(j)} + a^{(j)}t$$

$$\omega_V^{(j)} = \mu^{(j)}A^{(j)} \cdot \exp(-h^{(j)}t) \cdot \left(h^{(j)2} + k_2^{*(j)2}\right)^{0.5} \cos(k_2^{*(j)}t + \beta^{(j)} + \alpha^{(j)}) + C_2^{(j)} + a^{(j)}t$$

$C_2^{(j)}$ is the initial velocity of the center of mass of the model at the j-th speed.

Consideration of the gear shifting process as an impact interaction with zero-speed recovery factor allowed [9] to present the considered in figure 1 transmission model in a simplified form presented in figure 2.

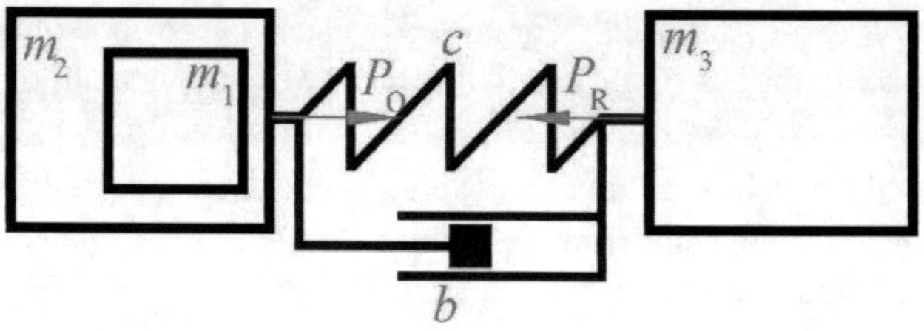

Figure 2. Simplified impact model of gear shifting with elastic element and damper.

The simplified impact model in figure 2 reflects the analogy of the physical process of gear shifting. In the model in figure 2 elements of rotation of the diagram in figure 1 are replaced by elements of translational linear movement. The next correspondences take place: $m_1 \sim J_I$, $m_2 \sim J_O$, $m_3 \sim J_V$. The P_O force corresponds to the torque on the transmission output shaft; force P_R corresponds to the torque of rotation resistance forces. Elements with stiffness c and damping coefficient b connect masses m_2 (frame in figure 2) and m_3. The process of gear shifting in the model is reduced to "soft" separation of the mass m_1 from one border of the frame m_2 and the subsequent impact connection with its other border with a zero-speed recovery factor. Thus, there is a "violent" transfer of mass m_1 from one border of the frame m_2 to the other, which reflects the effect of quick connection the inertial transmission elements reduced to the input and output shafts of the clutch of the switched gear. The m_1 and m_2 represent a single mass $(m_1 + m_2)$ in the intervals between gear shifts.

The movement of the model in figure 2 in the intervals between impacts ("gear shifts") is described by a system of two linear differential equations [9]:

$$\begin{aligned}(m_1 + m_2)\ddot{x} + b(\dot{x} - \dot{x}_3) + c(x - x_3) &= P_\text{O} \\ m_3\ddot{x}_3 + b(\dot{x}_3 - \dot{x}) + c(x_3 - x) &= -P_\text{R}\end{aligned} \tag{3}$$

where x, $\dot{x}$, $\ddot{x}$ - coordinate, velocity and acceleration of the mass $(m_1 + m_2)$; x_3, $\dot{x}_3$, $\ddot{x}_3$ - coordinate, velocity and acceleration of the mass m_3.

The speed $\dot{x}$ of the mass $(m_1 + m_2)$ calculation for the model in figure 2 is carried out through the momentum conservation law.

The mathematical model (3) was considered in textbooks on vibration theory [8, 10], monographs [11, 12], as part of a vibro-impact system [13], as well as in the study of mechanical systems with strong threshold nonlinearities [14]. Further more complex model is studied in [15], which consists of a chain of masses connected by elastic elements and each mass containing inside itself a serial connection of another mass and a damper.

The model with translational linear movement in figure 2 is put in correspondence with the model with rotation of inertial elements showed in figure 1. Such correspondence greatly simplifies the understanding of the impact process of gear shifting. The frequency of free vibrations of the model in figure 2 at the absence of damping in this case

$$k_1 = 0;\ k_2 = \{c(m_1 + m_2 + m_3) \cdot [(m_1 + m_2)m_3]^{-1}\}^{0.5}.$$

Damped oscillation frequency:

$$k_2^* = (k_2^2 - h^2)^{0.5}, \text{ where } h = b(m_1 + m_2 + m_3) \cdot [2(m_1 + m_2)m_3]^{-1}.$$

Thus, the vibration frequency of the model elements in figure 2 does not depend on the gear ratio, which is important when analyzing gear shifting. The frequency k_2^* for the model in figure 1 is different for each gear because the moments of inertia of the driving links (engine, etc.) reduced to the output shaft of the gearbox depend on the gear ratio of the engaged gear. The assumption of equality of vibration frequencies can be justified in the model in figure 2 by the fact that with a large number of gears in the gearbox the gear ratios of adjacent gears are close in magnitude [16].

Similarly, the solution to the system of equations (3) is:

$$\begin{aligned} x &= A \cdot \exp(-ht) \cdot \sin(k_2^* t + \beta) + C_1 + C_2 t + at^2/2 \\ x_3 &= \mu A \cdot \exp(-ht) \cdot \sin(k_2^* t + \beta) + C_1 + C_2 t + at^2/2 \end{aligned} \quad (4)$$

where $\mu = A_3/A = -(m_1 + m_2^*) \cdot m_3^{-1}$, $a = (P_O - P_R) \cdot (m_1 + m_2^* + m_3)^{-1}$.

After differentiating equations (4) and substituting $\sin\alpha = h \cdot \left(h^2 + k_2^{*2}\right)^{-0.5}$ and $\cos\alpha = k_2^* \cdot \left(h^2 + k_2^{*2}\right)^{-0.5}$ we obtain the equations for masses motion in the j-th gear between gear shifts:

$$\dot{x}^{(j)} = A^{(j)} \cdot \exp(-ht) \cdot \left(h^2 + k_2^{*2}\right)^{0.5} \cos\left(k_2^* t + \beta^{(j)} + \alpha\right) + C_2^{(j)} + a^{(j)} t$$

$$\dot{x}_3^{(j)} = \mu A^{(j)} \cdot \exp(-ht) \cdot \left(h^2 + k_2^{*2}\right)^{0.5} \cos\left(k_2^* t + \beta^{(j)} + \alpha\right) + C_2^{(j)} + a^{(j)} t$$

3. Computer simulation

A computer simulation of the elements' movement of the model in figure 1 during acceleration of a vehicle equipped with a 6-speed dual-clutch gearbox are carried out in the section.

The gear ratios of the gearbox considered as an example and the corresponding velocities of the output shaft of the gearbox $\omega^{(j)}$ at the assumed maximum speed of rotation of the engine shaft $\omega_I = 6000\ \text{rev} \cdot \text{min}^{-1} = 628{,}3\ \text{rad} \cdot \text{s}^{-1}$ (gear shifting velocity) are given in table 1.

Table 1. Gear ratios of a dual-clutch 6-speed gearbox.

Speed	i	Step	$\omega^{(j)}$ at $\omega_I = 628.3\ \text{rad} \cdot \text{s}^{-1}$
1.	4.12		152.5
2.	2.48	1.66	253.2
3.	1.64	1.51	382.2
4.	1.17	1.40	535.2
5.	0.91	1.29	690.5
6.	0.75	1.22	842.3

The rest of the model's parameters correspond to a modern middle-class passenger car. The dependence of the output shaft angular velocity over time is shown in figure 3.

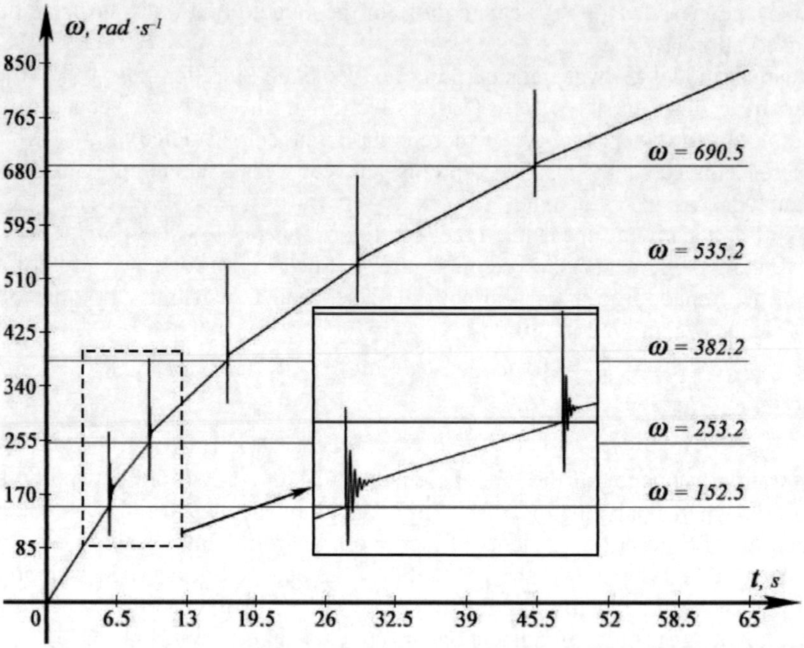

Figure 3. Graph of changes in the output shaft angular velocity of the gearbox.

The results in figure 3 show that in the case of an instantaneous gear shift significant impact loads arise in the transmission. The angular velocity increases for a short time by 133.3 $\mathrm{rad \cdot s^{-1}}$ at gear shifting from second to third speed. And it increases by 121.6 $\mathrm{rad \cdot s^{-1}}$ at gear shifting from fifth to sixth speed. The amplitude of the velocity change depends on the gear ratio step and decreases with decreasing step accordingly.

It should be noted that an extreme shift process is considered in figure 3, which makes it possible to estimate the maximum velocity changes in the case of an instant gear shift. Unacceptably high torques act for a short time on the transmission links with such sharp jumps in velocity. It can lead to premature wear of gears, their breakdown and failure of transmission elements. It is important to optimize the shift control in transmissions therefore.

An increase in the number of speeds in gearboxes makes it possible to reduce the range of engine velocities (which increases the fuel economy [17]), while the gear shifting velocities between neighbor speeds become closer [2-5]. A sharp change in the gearbox output shaft velocity caused by the shifting the next gear leads to the fact that its velocity at the moment exceeds the velocity of gear shifting to the next speed as shown in the graph in figure 3. The proximity of the gear shifting velocities to neighbor speeds therefore can cause the excitation of a self-oscillating process and, as a consequence, high-frequency gears shifting from the engaged gear to the disengaged and vice versa.

A computer simulation of the high-frequency looping process at close values of the gear shifting velocities to neighbor gears was carried out by using a simplified impact model in figure 2. The computer simulation results showed in figure 4 [9].

The gear shifting velocities from 1st speed to the 2nd $\dot{x}^{(12)}$ and back $\dot{x}^{(21)}$ (see in figure 4) are not the same according to the gear shifting map [5]. It is assumed that the model in figure 2 has a relatively small damping coefficient in this case and the velocities $\dot{x}^{(12)}$ and $\dot{x}^{(21)}$ are quite close in magnitude. The model moving in 1st speed at $t = 0$ and reaches the 2nd speed gear shifting velocity $\dot{x}^{(12)}$. The impact caused by gear shifting leads to the excitation of the gearbox output shaft oscillations and the amplitude of the change in the output shaft velocity exceeds the difference between the gear shift velocities of neighbour speeds. The gearbox output shaft velocity decreases to the gear shift velocity $\dot{x}^{(21)}$ from the 1st speed to the 2nd as a result. The gear shifting occurs again, then the self-oscillating process is repeated and for the selected parameters of the model is damped as seen in figure 4.

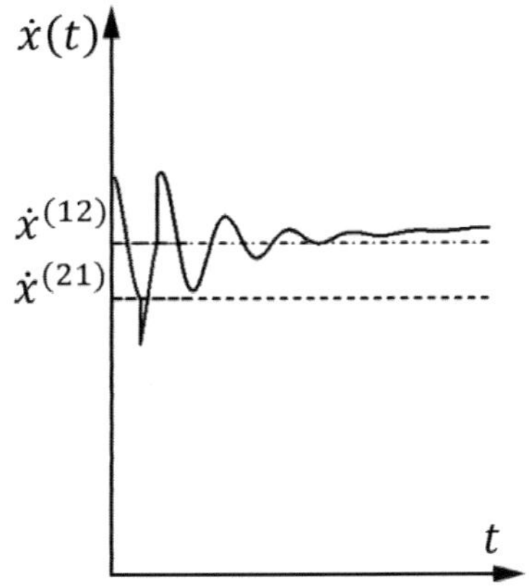

Figure 4. Computer simulation results of high-frequency gear shifts looping.

The solid lines 1-2, 2-3, 3-4, etc. on the shift maps of an automatic transmission showed in figure 5 are gear shift lines from low to high speed, and dashed lines 2-1, 3-2, etc. are lines of gear shifting from upper to lower speed.

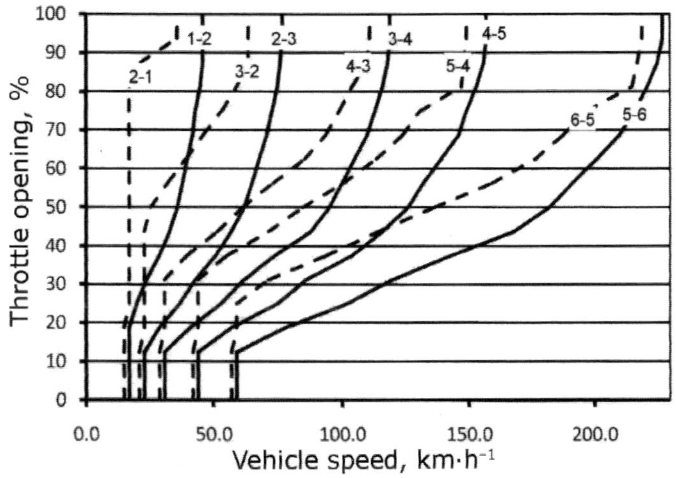

Figure 5. Gear shifting map for 6-speed automatic transmission.

The gear shifting velocities to neighbour speeds at low vehicle velocity are quite close as seen from figure 5. Therefore, the problem of low-frequency looping (also a self-oscillating process) may arise. Self-oscillations are excited here due to periodic changes in the external conditions of the vehicle, caused by factors such as the topography of the road or the driver's behaviour. For example, at a frequent change of ups and downs: an upshift will occur when driving downhill, and a downshift will occur when driving uphill. Another example is on winding roads: an upshift will occur before a turn and downshift after a turn. Simulation of the low-frequency looping process performed for a simplified model in figure 2 is shown in figure 6.

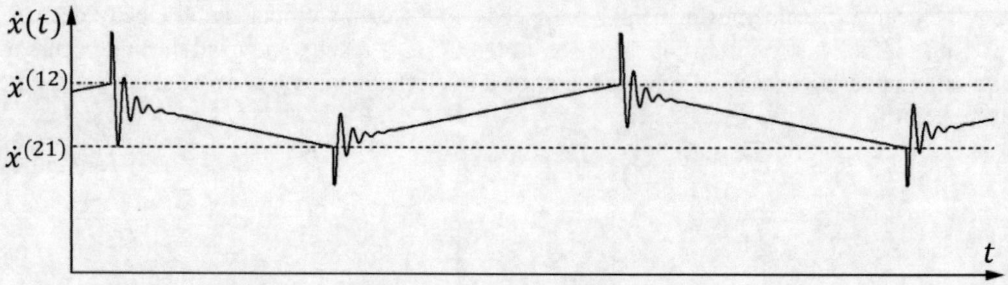

Figure 6. Computer simulation results of low-frequency gear shift looping.

It is assumed that a vehicle in the computer simulation of the low-frequency looping process in figure 6 moves along a hilly road where ups and downs alternate with equal frequency. The vehicle velocity is maintained constant $\approx$ 50 $\mathrm{km \cdot h^{-1}}$ due to an automatic speed control system (cruise control). Computer simulation shows in that self-oscillations and low-frequency looping occur, that is when vehicle ride down the hill an upshift occur, and when uphill - a downshift one.

4. Results and discussion

The article discusses the process of gear shifting in automatic transmissions as the impact interaction of transmission elements. Quite simple models of vehicle transmission have been developed and mathematically described. The process of gear shifting in various modes of vehicle movement is considered on the models.

It was shown as a result of computer simulation that velocity changes of the gearbox output shaft during instantaneous gear shifting are very significant and can exceed the value of the gear shift velocity to the next speed.

Computer simulation of self-oscillating modes of low-frequency and high-frequency looping of gear shifts are carried out. The first is associated with the movement of the vehicle at low speeds with close to gear shifting velocity and changes in external driving conditions, such as a periodic change of ups and downs, winding road topography or driver actions. High-frequency looping is associated with the impact nature of gear shifting, low damping and with a large amplitude of velocity change of the gearbox output shaft, which exceeds the gear shifting velocity of the next speed.

5. Conclusion

The short duration of gear shifts in automatic transmissions allows to consider this process as an impact. The vehicle transmission in this case at a short time interval is a closed self-oscillating mechanical system, for which the theorem on the change in the angular momentum upon impact is applicable.

The assumption made in the development of mathematical models that there is no slippage between the discs of the engaging clutch greatly simplifies the modeling, but at the same time allows one to estimate the ultimate loads arising in the transmission.

The main way to prevent high impact torques acting on the transmission elements is to decrease the engine shaft velocity before upshift and, conversely, to increase the engine shaft velocity before downshift [5]. Such an algorithm is provided in automatic transmissions for cars. Among the additional recommendations aimed to reduce the vibration amplitude of the gearbox output shaft caused by gear shifting, the following can be distinguished:

- introduction of a delay for reading the gearbox output shaft speed sensor during gear shifting;
- increasing the rigidity of the gearbox output shaft and increasing damping;
- changes in the ratios of the gearbox moments of inertia and transmission links;
- increasing the difference between the velocities of neighbor gear shifts.

The mathematical models described here can be used in the development of control systems for gear shifting in automatic transmissions and for calibration of the engine operating modes and transmission under various conditions of vehicle movement.

Acknowledgments
The research was supported by Russian Science Foundation (Project No. 19-19-00065).

References

[1] Sharipov V M, Dmitriev M I, Zenin A S *et al* 2015 On the issue of slipping of friction clutches when shifting gears without interrupting the power flow in gearboxes of cars and tractors *Tractors and agricultural machinery* **6** pp 5-9
[2] Naunheimer H, Bertsche B, Ryborz J and Novak W 2011 *Automotive Transmissions* (Berlin: Springer) p 740
[3] Fischer R, Küçükay F, Jürgens G, Najork R and Pollak B 2015 *The Automotive Transmission Book* (Cham: Springer) p 372
[4] Genta G and Morello L 2009 *The Automotive Chassis. Mechanical Engineering Series* (Dordrecht: Springer) p 627
[5] Bai Sh, Maguire J and Peng H 2013 *Dynamic analysis and control system design of automatic transmission* (Warrendale: SAE International)
[6] Pfeiffer F 2008 *Mechanical System Dynamics. Corrected Second Printing* (Berlin Heidelberg: Springer-Verlag) p 575
[7] Korendyasev G K and Salamandra K B 2021 Gear shift analysis in dual-clutch transmissions using impact theory advances in mechanical engineering *MMESE 2020. Lecture Notes in Mechanical Engineering* pp 92-101
[8] Panovko Ya G 1991 *Introduction to the Theory of Mechanical Vibrations: Textbook 3rd Ed.* (Moscow: Nauka) p 256
[9] Salamandra K and Tyves L 2018 Modeling of cyclic gearshifts in automatic transmissions for vehicles *Engineering for rural development* pp 2071-8
[10] Timoshenko S 1955 *Vibration problems in engineering* (Toronto: D. van Nostrand Company Inc)
[11] Vulfson I I 1976 *Dynamic calculations of cycle mechanisms* (Leningrad: Masinostroenie)
[12] Veyc V L, Kolovsky M Z and Kochura A E 1984 *Dynamics of controlled machine units* (Moscow: Nauka)
[13] Babitsky V I 1978 *Theory of vibroimpact systems* (Moscow: Nauka) p 372
[14] Babitsky V I and Krupenin V L 2001 *Vibration of strongly nonlinear discontinuous systems. Foundations of engineering mechanics* (Berlin: Springer) p 404
[15] Kolesov D A *et al* 2020 *IOP Conf. Ser.: Mater. Sci. Eng.* 747 012053
[16] Salamandra K and Tyves L 2016 Application of the Angular Momentum Conservation Law at the Dynamic Analysis of Gearshifts *FISITA 2016 World Automotive Cong.–Proc.* F2016-THBG-002
[17] Salamandra K 2019 Perspective planetary-layshaft transmissions with three power flows *Advances in Mechanical Engineering. Lecture Notes in Mechanical Engineering* pp 153-67

IOP Conf. Series: Materials Science and Engineering 1129 (2021) 012059 doi:10.1088/1757-899X/1129/1/012059

Automated robotic assembly of complex workpieces from tiped components forms

P Kozlov and A Klimchik

Center for Technologies in Robotics and Mechatronics Components, Innopolis University, Innopolis, Russia

p.kozlov@innopolis.ru

Abstract. This paper deals with the automated robotic assembly of complex casting molds from components of regular shape. It addresses the issues of model decomposition and assembly sequence taking into account particularities of the gluing process and shape features. The decomposition parameters were determined by several approximation experiments. The obtained results validated via simulation and experimental study dealing with the assembly of particular workpieces with UR10e robot.

1. Introduction

Lean manufacturing progressively replace conventional materials by new ones that provide essential advantages from the point of view of mechanical properties of the final products and material re-use. One of such process is producing casting molds from model plastic instead of commonly used wood. Producing casting molds from the wood requires permanent presence of skilled an employee in the working zone while production process and that production can be hardly automated. Process automating is a common trend in industry given a possibility to decrease product manufacturing cost [1], which is very important for heavy metallurgy and engineering industry. This paper will be devoted to automation of casting mold production from re-usable model plastic.

Using robots in the casting mold production gives us a possibility to decrease production time and, consequently, product cost. Firstly, it is possible to use reusable material like special model plastic. Plastic will be a reasonable choice because manufacturers can melt shavings from milling, old casting molds and other plastic garbage into new forms. Secondly, an automated robotic complex is capable of building such molds independently without any employee present during the production process; the only requirement is providing enough material to the system. Also, the automated robotic complex may work faster because its performance can be much higher than a human's one due to robot sustainability, hence their higher production accuracy. Thus, using reusable material and automated complex will optimize manufacturing, decrease production costs and increase production accuracy.

This papers deals with casting mold approximation with typical components and further assembly of complex workpieces [2], which will be subject to machining [3] at the final stage to get desired form. It propose an approach that relies of the automated control program design for industrial manipulator based on the specified technological task and sequence of motions.

2. Materials and research methods

There is one of the ways to create the automated robotic complex to solve the problem. The whole solution exploits regular shape components from the model plastic [4] with further milling proccssing [5].

This problem can be divided into several tasks:

- Scanning the shape of the complex workpiece;
- Determination components positions at the workpiece according to the shape and size of regular components. Obtained components should cover all workpiece;
- Determining assembly sequence according to the prepared model;
- Assembling workpieces by determined sequence. For the assembly, it is reasonable to use a manipulator and also we should determine the initial position of each regular component.
- Determining faces which should be covered with glue;
- Milling assembled model in order to obtain the final shape of the form.

There has been used industrial manipulator by Universal Robots [6]. Universal Robots UR10e is used as industrial manipulator. This is collaborative robot, hence it is user-friendly. Therefore humans might interact with robot with decreased risk. This manipulators can be simply accommodated with external control. They also move convenient to control from PC than KUKA or Fanuc and easy to integrate with ROS. After approximation is over robots stats manipulation typical components sequentially. Manipulation starts on their initial positions and end on the target places.

The kinect sensor was also used in work. It was desired to use because it combines camera and depth sensor. These sensors require to determine initial positions of the typical components. Camera is used as multi colour filter to remove external objects and also remove some noise. Depth sensor is used to determine typical components positions and the heights layer of them.

The robotic work-cell for assembly complex workpieces consists of industial manupulator (UR10e manupulator), camera and depth sensor (Kinect) and laptop to control this system. Control system will select an appropriate algorithm taking into account current and final states of assembly.

To cover required tasks the system requires the following steps:

- Obtain target 3d model of complex workpiece (example is given in figure 1a). Target 3d model should be presented as stl file. There are no limitations to the form of the initial object, but there might be some limitations based on typical component shape. The shape of the typical component during described implementation will be a box.
- Approximate it with a set of typical components combined in a 3D binary array;
- Determine assembly sequence and specify the planes where the glue is required;
- Dynamically get positions of typical components in the stock by Kinect;
- Program manipulator according to the determined typical components position and assembly sequence. Figure 1 (b) presents final assembly for the example of workpiece given in figure 1 (a)

(a) Initial 3D model of the workpiece. (b) Assemblied workpice from typical components (without glue).

Figure 1. Example model conversion.

2.1. 3D object approximation

The 3D object approximation consists of several steps. First step requests stl file with an object shape (figure 2 (a)). The object shape is a mesh with consists of several triangles. For every triangle in the mesh there is a normal vector. This vector is used to determine internal and external surfaces.

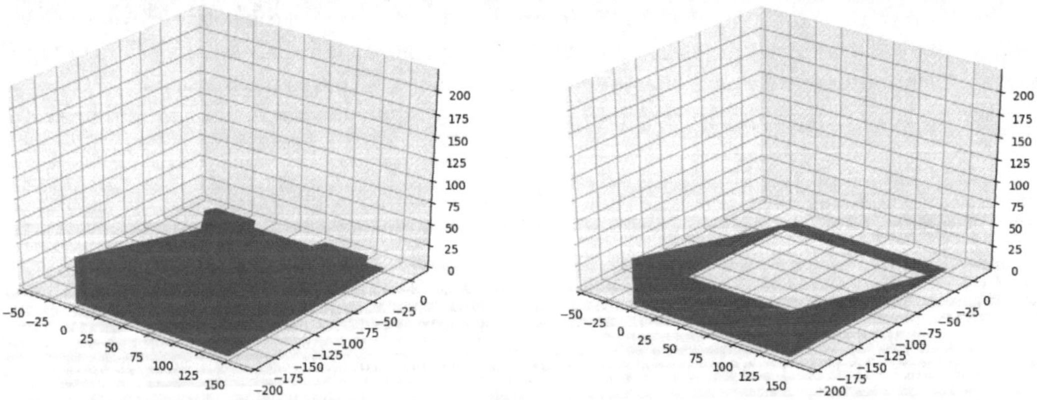

(a) Input 3D object model. (b) Picked triangles for given height.

Figure 2. Input mesh object.

Now we are able to approximate given model. Model commence slicing from bottom. It happens because higher layers should know base to determine glue. There are several layers (Z) of the approximated model:

$$Z = |\frac{H}{h}| \tag{1}$$

Where H — height of the 3D object, h – height of the typical component. For every layers there is several (z) slices. Several slices requested to reduce error of approximation. It allows us to detect some protrusions in one layer. Hence, the 3D object approximation will be have better quality.

There is an algorithm to make each slice. Firstly, we are going to determine the triangles (figure 2 (b)), which are intersects with horizontal plane at given height.

Next step is get intersection of the all picked triangles with horizontal plane. It may done by simple intersection two planes. After that, slice becomes a 2d body. This body consists of several lines (figure 3 (a)).

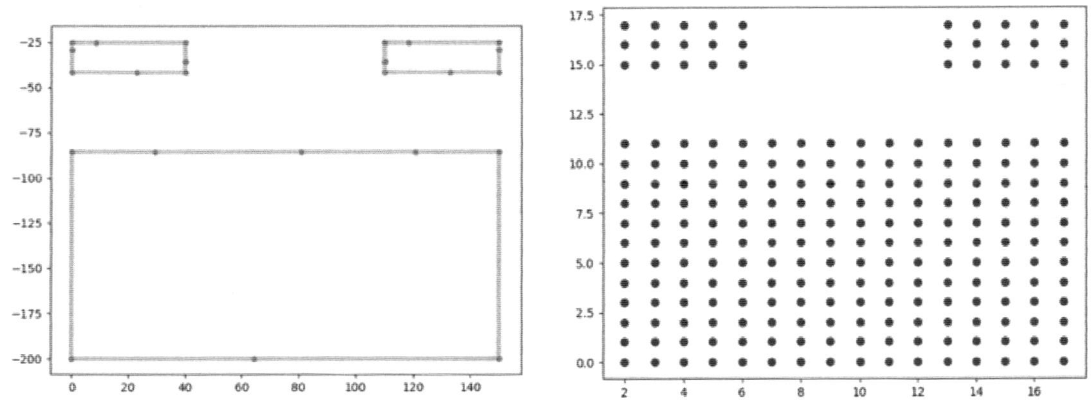

(a) Intersection of triangles and plane. (b) Binary matrix for one layer.

Figure 3. One layer convertions.

Based on the shape which is box and dimensions of the typical component there is a possibility to convert intersection. Intersection is going to be converted in to binary matrix, where point means that this is down left point of the square. The square is a projection of the typical component to the horizontal plane. The slice for given height is done.

Another operation is to combine z slices for one layer to get binary layer (figure 3 (b)). There is an "or" operation by elements during combining z slices.

After all layers are approximated there is a circumstance to combine them in to one 3D array (figure 4) which represents approximated 3D object. The 3D object approximation is finished.

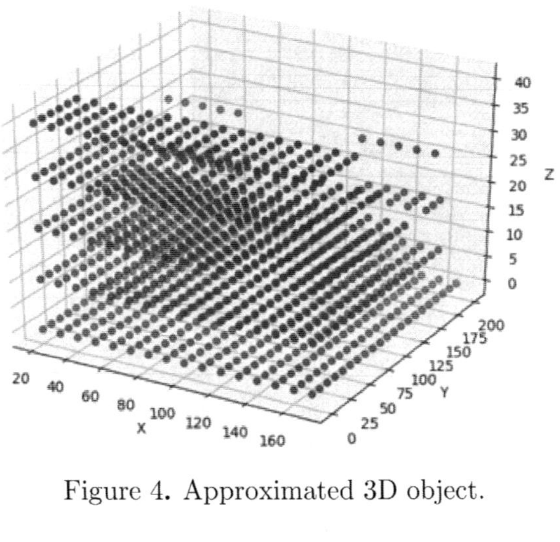

Figure 4. Approximated 3D object.

Sequence generation
Sequence generation is necessary for get information about typical components order. Based on the order it might be easily determined sided of the components which should covered with glue.

In order to simplify realization we begin from bottom level, top left corner in the level. This approach provides us the simplest algorithm. The figure 5 demonstrates graphically described above algorithm.

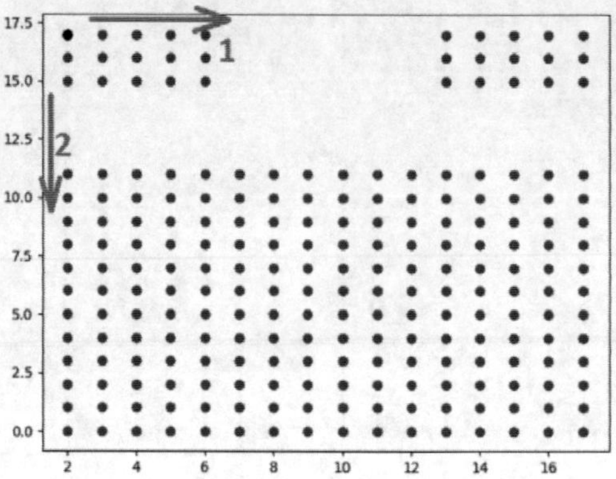

Figure 5. Building sequence direction.

For the second and higher layer there is one more requirement: typical components should be installed firstly on the cubes which already installed. In other words, there is no possibility to install components in air, without any connections.

Table 1. Part of the building sequence.

x	y	z	back	front	left	right	top	bottom
20	200	0	False	False	False	False	False	True
30	200	0	False	False	True	False	False	True
40	200	0	False	False	True	False	False	True
50	200	0	False	False	True	False	False	True
60	200	0	False	False	True	False	False	True

According to the table 1, the coordinates (x, y, z) demonstrates position of the center of the typical component in centimetres. The "back", "front", "left", "right", "top", "bottom" demonstrates the faces of the component to get knowledge about glue.

3. Results and discussion
After loading model and setting all requested parameters, the approximation is stated. After this, robotic system starts implementing required model. This was done by using desired building sequence. Moreover, it demonstrates some shifting of the component. This shifting indicates that this is place to apply force. This force is used to achieve better adhesion which has been made with glue.

In this study there will not considered "scanning the shape of the complex workpiece" and "milling assembled model in order to obtain the final shape of the form" items for general algorithm.

In order to implement and proof and algorithm typical components is presented as a cube. Therefore approximation algorithm is also has been simplified in term of rotation of the components is not necessary. It also do not affect to the anything.

Another simplifications were made for typical components detection. There is a requirement that Kinect should be installed horizontally and cube projection on the image should be parallel to the borders of image. These simplifications will be fixed during modifying initial positions determinations. It might be done as a cloud of points and adding the highest plane determination.

4. Conclusion

Over the work on this paper, an automated robotic assembly of complex workpieces from tiped components forms was developed. This robotic system included universal robot UR10e and Kinect. The main limitation is connected with typical component detection. It can be solved by using cloud of points and calculating planes dynamically. Another limitation is connected with gripper type. The possible improvement is modify gripped to neglect described problem.

References

[1] Ji W and Wang L 2019 Industrial robotic machining: a review *Int J Adv Manuf Technol* **103** 1239-55
[2] Kozlov P and Klimchik 2019 A Automated robotic assembly of complex workpieces from regular components *3rd School on Dynamics of Complex Networks and their Application in Intellectual Robotics* 92-3
[3] Klimchik A, Ambiehl A, Garnier S, Furet B and Pashkevich A 2016 Experimental study of robotic-based machining *IFAC-PapersOnLine* **49** 174-9
[4] Greer C, Nycz A, Noakes M, Richardson B, Post B, Kurfess T and Love L 2019 *Introduction to the design rules for Metal Big Area Additive Manufacturing* **27**
[5] Tankova T and Silva L 2020 Robotics and Additive Manufacturing in the Construction Industry *Current Robotics Reports* **1**
[6] Universal Robots Available at: https://www.universal-robots.com/

IOP Conf. Series: Materials Science and Engineering 1129 (2021) 012060 doi:10.1088/1757-899X/1129/1/012060

Automation of robot programming in milling

P Kozlov, S Mikhel and A Klimchik

Center for Technologies in Robotics and Mechatronics Components, Innopolis University, Innopolis, Russia

p.kozlov@innopolis.ru

Abstract. The paper demonstrates a method of automatic robot programming on the example of milling. The following problems are considered: format of the robot commands, analyze of an obtained trajectory, translation into the robot-specific language. The results confirm the ability of practical application for the proposed technique.

1. Introduction

Automatic programming can be used to simplify the application of the industrial manipulators for common operations [1]. The result of such programming is typically a script file. This script differs for every manufacturer of industrial manipulators. Sometimes it differs for manipulators even in one company, for example, KUKA IIWA requires a different programming method than other KUKA models. At the same time, the number of operations that an industrial manipulator has to do is limited to a small set of commands [2]. Therefore the algorithm of processing can be described in a robot-independent form and then translated to the specific platform.

Some developers introduce their languages to describe the robot operations [3, 4]. This makes sense when complex or collaborative tasks have to be done. But for most of the monotonic industrial operations, it is enough to define the sequence of commands that can be translated into the robot source code almost one-by-one. In this paper, we are going to describe such a system for the automatic programming of the milling process.

2. Materials and research methods

Let us look at the process of generating a control program on the example of milling. It contains the following steps.

- Find an area to process. In the case of milling, one can subtract the approximated model and initial model. It provides mesh difference that should be milled.
- Create a milling sequence. Generally, the milling process will use linear, spline, and circular movements.
- Configure translator using the manipulator parameters. Different manipulators have unique geometrical parameters that must be taken into account in addition to the robot language syntax.
- Translate the result into the robot specific language. It produces a program that can be easily uploaded to the robot.

IOP Conf. Series: Materials Science and Engineering 1129 (2021) 012060 doi:10.1088/1757-899X/1129/1/012060

The milling sequence can be saved independently from the generated robotic script. It allows regeneration without creation milling sequence and hence it will reduce production time. Milling sequence stores two arrays: points and operations. First one, contains all point (table 1) which the robot end-effector will attend. Sometimes end-effector may attend the same coordinates several times. Hence storing coordinates as variables will save a bit of memory. Second one, contains all operations (table 2) with information about requested points for current operation. The manipulator moves sequentially from on point to another, it means that the last point of the robot state in the previous operation is the first point of the new motion, and only one (last) position has to be defined. The exception is circular motion where one intermediate point must be set to define the shape of the curve. Parameter "proc" stores information about the milling process. It indicates is there any contact with the material, that can be used for decision making. For example, if the robot is not in contact, it can move faster or turn off the instrument.

Table 1. Robot end-effector points.

Name	X	Y	Z
p0	0	0	0
p1	15	15	50
p2	15	15	45
p3	35	15	45
p4	35	15	50
p5	15	20	50

Table 2. Robot generated commands.

Type	Proc	Points
ptp	false	p0
lin	false	p1
lin	true	p2
lin	true	p3
lin	false	p4
lin	false	p5
lin	true	p3
lin	true	p4

In general, 3 parameters of the end-effector orientation have to be defined. For example, it can be roll-pitch-yaw angles. Most of the time robot moves with a constant orientation of the end-effector, and it is practical to simplify the trajectory representation via skipping the orientation parameters like in table 1.

The list of points is generated in the frame of workpiece because the generator does not have any information about the robot to be used, it deals with the tool motions only. In practice, the workpiece can be located in any part of the robot workspace and has an arbitrary orientation. Therefore it should be transformed into the most convenient frame for the current task. For example, if we are going to use the robot base frame, and the workpiece has displacement P and rotation R with respect to the robot, then each i-th point of the workpiece p_w must be translated into the robot frame p_r via transformation $p_r^i = Rp_w^i + P$.

The found points can be translated into the robot program directly. It is the simplest approach that does not require any information about robot kinematics. However, some points can be out of the robot workspace, or the path can pass through a singular state. Therefore the found trajectory should be verified previously. Such verification requires solving the inverse kinematics problem. We used the Jacobian-based technique [5]. The method is selected for the following reasons:

- it is universal and can be used for different manipulator structures;
- it allows to find singularity, for example, from the manipulability parameter $\sqrt{det(JJ^T)}$, where J is the manipulator Jacobian matrix, $det(\cdot)$ is the determinant [6].

In general, the Jacobian based approach does not allow to find multiple solutions, but in the case of milling the sequential points are typically close to each other and the manipulator has to move smoothly, then it is enough to have the closest solution only. If the singularity is found, it can be avoided using the control in joint space, in this case, the "lin" and "circ" commands have to be transformed into the equal "ptp" form. If the trajectory cannot be executed, the operator sees a warning message before starting execution on the robot. Such an approach minimizes the possibility to have an error during real workpiece processing. It should be noted that in the process of verification the tool transformation must be taken into account.

The final step is generating the robot script. We used Python for the following reasons:

- powerful linear algebra library for working with Jacobians and kinematics;
- xml library for parsing the Universal Robot Description Format (URDF);
- simplicity of working with data collections and strings.

The translator works as follows. It takes two files as command-line parameters: URDF-file with the robot kinematics and the obtained trajectory. Usage of the URDF file allows us to easily change the format of the generated program and the verification settings. It is used in the constructor of the class "Robot" that reads kinematic parameters of the manipulator and uses it when the methods "FK" (forward kinematics) or "Jacobian" are called. The points and commands of the trajectory are represented in such a way that can be directly translated into the Python lists and dictionaries which allows us to avoid the explicit data parsing. Then the robot model virtually moves from one point to another and check its reachability and the robot manipulability. If there are no problems, the commands can be translated into the robot language format. Typically it requires us to define some obligatory instructions in the beginning and the end of the file, in particular, to wrap the control program into the procedure or function.

Example of the generated program is shown in table 3. It contains a fragment of the generated *ls*-file for a Fanuk manipulator, that corresponds to the sequence of instructions in table 2. Each command contains several parameters additionally to the point index, such as velocity, acceleration, and other motion-specific elements. In our implementation they were defined in translator, but in general they should be generated together with the path points.

3. Results and discussion

This approach provides a possibility to execute one algorithm on several and sometimes different industrial manipulators. It only requests an industrial manipulator itself, generated program, and the translator.

The described conversion is also applicable to the other stages. For example, it is possible to create a program for industrial manipulators for assembling complex workpieces from typical components. The assembling is also included an automated glue application for desired sides. Thus, a laptop is only requested during model approximation and program generation. It produces the opportunity to use the environment more efficiently.

Table 3. Output code example.

Type	Point	Parameters
4:J	P[1]	10% FINE ACC10;
5:L	P[2]	500 cm min^{-1} CNT80 ACC65 ;
6:L	P[3]	500 cm min^{-1} CNT80 ACC65 ;
7:L	P[4]	500 cm min^{-1} CNT80 ACC65 ;
8:L	P[5]	500 cm min^{-1} CNT80 ACC65 ;
9:L	P[6]	500 cm min^{-1} CNT80 ACC65 ;
10:L	P[4]	500 cm min^{-1} CNT80 ACC65 ;
11:L	P[5]	500 cm min^{-1} CNT80 ACC65 ;

Program generation for building complex workpiece uses several moving types. Moving types might be in joint space or linear (Cartesian) space. Linear space motion is using when there is a requirement to move linearly. Also, there might be some environment restrictions for example limited environment or finally placed components. Joint space motion is not linear in Cartesian space but it provides faster manipulation.

During workpiece production linear moving is required for:

- Typical component picking (figure 1 (a));
- Glue application;
- Component final positing with force applying (figure 1 (b)).

The joint space motion is used during moving between component picking and placing operations. It is allowable because there is no requirements on the accuracy or any other environment limitations.

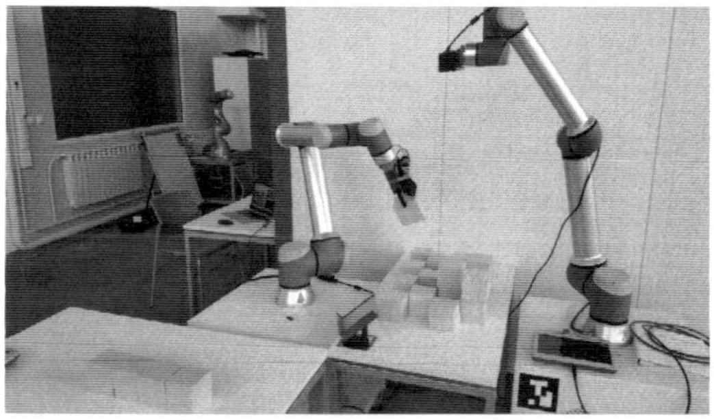

(a) Picking typical component

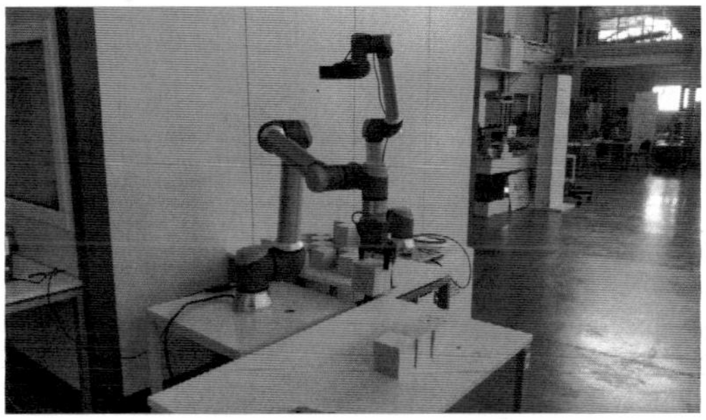

(b) Placing typical component

Figure 1. Building complex workpiece demonstration.

The limitation of the proposed approach is its focus on sequence of predefined motions, therefore it cannot be used in cases, when the robot has to choose different motions depending on the sensory information or other factors. It can be solved by increasing the number of commands and complexity of translator, but in this case introduction of an intermediate language instead of the command set looks more advisable.

4. Conclusion

The paper demonstrates the method of robot programming, based on the automatic trajectory generation and process description in terms of basic motion commands in robot-independent form. This approach was applied for generating script for milling on the Fanuc and KUKA industrial manipulators. The results confirm the possibility of its practical application. The proposed technique can be used in other robot tasks, such as assembling operations.

In future work we are going to improve the trajectory generator to estimate all the required parameters of the process during the path planning and remove this part from the translator. Also we are going to introduce a variable angle for the milling process. It can be achieved because there is some concave parts which cannot be milled by horizontal configuration. It will produce a possibility to increase an accuracy for the final model.

References

[1] Mitsi S *et al* 2005 Off-line programming of an industrial robot for manufacturing *The Int. J. of Advanced Manufacturing Technology*

[2] Nilsson K 1996 Industrial robot programming *Department of Automatic Control, Lund Institute of Technology*

[3] Scripts in SprutCAM Available at: https://kb.sprutcam.com

[4] RCML manual Available at: http://robotct.com

[5] Buss S R 2004 Introduction to inverse kinematics with jacobian transpose, pseudoinverse and damped least squares methods *IEEE J. of Robotics and Automation*

[6] Wang S L and Waldron K J 1987 *A study of the singular configurations of serial manipulators*

Research of the process of creating a vacuum on air-permeable surfaces

M A Lapikov, Y V Kalinin, A M Makarov and Y P Serdobintsev

Volgograd State Technical University, Lenin avenue, 28, Volgograd, 400005, Russia

remax777@yandex.ru

Abstract. A research of the process of vacuuming on air-permeable surfaces is carried out. A mathematical model is presented for calculating the parameters of the gas filtration process through porous surfaces during their vacuumation. This model allows you to calculate the parameters of vacuum systems for gripping devices necessary for reliable capture of breathable materials.

1. Introduction

Vacuuming on air-permeable surfaces is directly related to the filtration of gas through these surfaces. Filtration theory is considered in the section of hydrodynamics devoted to the study of the movement of liquids and gases through porous media [1, 2]. These studies have been conducted for more than 100 years, since filtration phenomena occur both in everyday life and in industry. However, this topic has not been fully studied and requires solving a number of tasks. The study of gas filtration is related to solving problems such as accidents in gas fields, processing inhomogeneous surfaces, as well as the use of vacuum grippers in industry and robotics. Therefore, it is necessary to develop mathematical models for describing and calculating the process of gas movement through air-permeable surfaces at pressure drops

2. Mathematical model

"Porous medium" refers to a solid material with a complex branched system of connected voids. These voids allow gases and liquids to be distributed unevenly. As a result of the inhomogeneous pressure distribution in porous media, the process of non-stationary filtration occurs [2, 3]. Since the pore, structure has a complex system with channels of different sizes and their random distribution over the volume, to study filtration processes, a porous material is considered as a continuous medium with averaged porosity characteristics. The porosity of a medium is defined as the pore volume per unit volume of a solid:

$$a = \frac{V_p}{V}$$

where V_p is the pore volume and V is the total volume of the medium.

To build a mathematical model, we take an air-permeable material as a stationary homogeneous porous medium, and then from the filtration laws, the movement of gases in a porous medium [4] can be described by the following equations:

Continuity equation:

$$\frac{\partial \rho}{\partial t} + \vec{\nabla} \cdot (\rho \vec{v}) = 0$$

The equation of motion of gas:

$$\rho \left(\frac{\partial \vec{v}}{\partial t} + (\vec{v} \cdot \vec{\nabla})\vec{v} \right) = -\vec{\nabla} p + -\vec{F} + \rho \vec{g}$$

The equation of energy of a solid medium:

$$(1-a)\rho_s c_s \frac{\partial T_s}{\partial t} = \alpha(T - T_s) + (1-a)\lambda \vec{\nabla}^2 T_s$$

The energy equation of gas:

$$\rho c_V \left(\frac{\partial T}{\partial t} + (\vec{v} \cdot \vec{\nabla})T \right) = \frac{p}{\rho} \left(\frac{\partial \rho}{\partial t} + (\vec{v} \cdot \vec{\nabla})\rho \right) + \vec{\mathrm{F}} \cdot \vec{v} + \lambda_g \vec{\nabla}^2 T - \alpha(T - T_s)/a$$

The equation of state:

$$p = \rho R T$$

where p – the gas pressure, $\vec{v}$ – the gas velocity, ρ – the gas density, ρ_s – the density of the solid medium, $\vec{g}$ – acceleration of free fall, $\vec{F}$ – the strength of interfacial friction, c_v – the heat capacity of the gas c_s – the heat capacity of the solid medium, T, T_s – temperatures of gas and solid media, λ_g, λ – thermal conductivities of gas and solid media, α – constant of intensity of interphase heat exchange, R– the gas constant, and $\vec{\nabla} = \partial / \partial x_i$ – the gradient operator for the spatial coordinates.

These equations are derived from filtration laws using a number of simplifications that do not affect the result of modeling:

1. The absence of phase transitions.
2. The interfacial friction force [5, 6] is determined by the gas velocity $V^{\rightarrow}$relative to the porous medium, the size or porosity a, the permeability coefficient of the porous medium k [m^2], and the dynamic viscosity of the gas mixture μ:

$$\vec{F} = a \frac{\mu}{k} \vec{v}$$

3. In the solid phase, the work of internal forces, due to the assumption of immobility of the solid medium, is zero.
4. The thermal conductivity and temperature of a gas and a porous solid medium are constant.

To calculate the gas energy, we take into account the heat capacity of the gas at constant pressure c_p, then:

$$\frac{p}{\rho} \frac{\partial \rho}{\partial t} = \frac{\partial p}{\partial t} - R \frac{\partial T}{\partial t}$$

where $R = c_p - c_v$. Hence, the gas energy is calculated from:

$$\rho c_p \left(\frac{\partial T}{\partial t} + (\vec{v} \cdot \vec{\nabla})T \right) = \left(\frac{\partial \mathrm{p}}{\partial t} + (\vec{v} \cdot \vec{\nabla})p \right) + a \frac{\mu}{k} \vec{v}^2 + \lambda_g \vec{\nabla}^2 T - \alpha(T - T_s)/a$$

The ability of a porous medium to pass a liquid is characterized by permeability. Its definition is closely related to the basic law of fluid motion in a porous medium, called Darcy's law. Under certain conditions, this law is valid for both liquids and gases with a steady viscous flow through porous surfaces. Darcy's law is also called the linear filtration law, which means that the average flow rate of a gas or liquid through a channel is proportional to the pressure difference at the inlet and outlet at

Reynolds numbers Re << 1 [4]. Then the stationary gas filtration process can be described by a system of equations:

$$\vec{\nabla} \cdot (\rho \vec{v}) = 0$$

$$\vec{\nabla} p = \rho \vec{g} - a \frac{\mu}{k} \vec{v}$$

$$p = \rho R T$$

3. Results and discussion

A number of devices have been developed and investigated [7 - 10] for automatically opening, holding and closing flexible containers using vacuum grippers (figure 1). In the presented devices, the product is captured due to the difference between atmospheric pressure and vacuum pressure, which is formed between the internal capture cavity and the product plane using a vacuum pump or based on the ejector principle. A vacuum is most often called a region of space filled with gas at a pressure below atmospheric pressure [11].

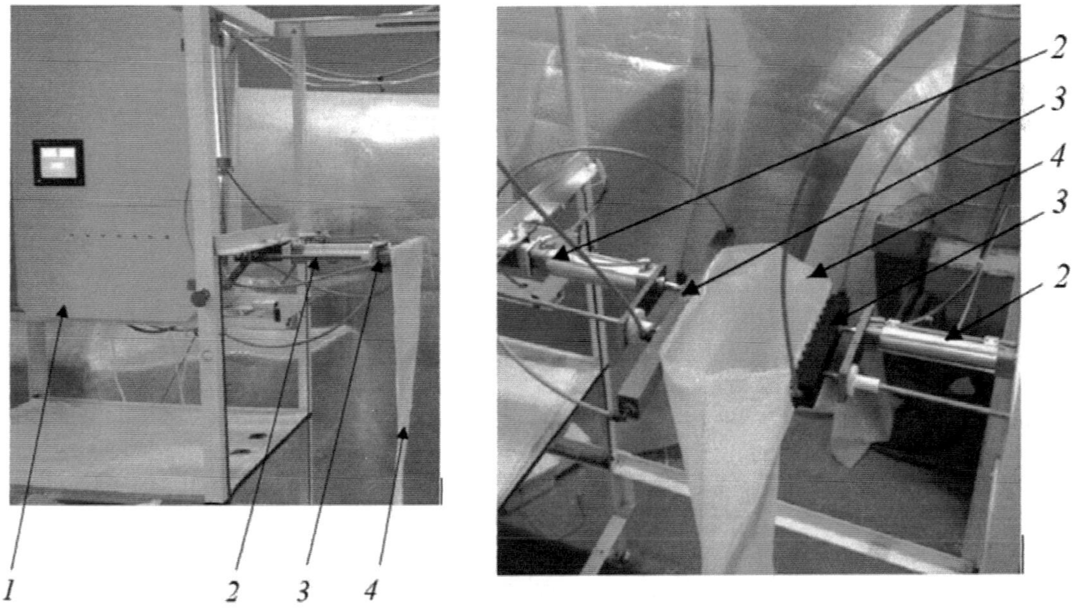

(a) (b)

Figure 1. The prototype of a modular vacuum gripping device produced by the additive technology method: (a) flexible container is captured by all sections, (b) flexible container is captured by the central sections, 1 - control cabinet, 2 – pneumatic cylinder, 3 - vacuum gripping device, 4 - flexible container.

Vacuum gripping devices can be suckers or gripping devices with vacuum chambers (figure 2). One of the ways to manufacture such devices can be additive technologies [8]. They allow you to make gripping devices with vacuum chambers made in the form of truncated cones, oriented with their large base to the working surface. The chambers can be positioned in a staggered manner or in a honeycomb to make full use of the area of capture. A large number of vacuum suckers are used to increase the reliability of retention in the event of failure of some of them due to insufficient contact with the gripping surface.

(a) (b)

Figure 2. Vacuum grippers: (a) – vacuum suction Cup, (b) – gripper with vacuum chambers, manufactured by additive technologies.

For mathematical modeling of the filtration process by the area of the vacuum suction Cup, let's assume that the process of vacuuming on an air - permeable surface is time-established and has a constant speed, pressure, and other parameters of the filtration process. We project the vacuuming process, where P_0 - the vacuum pressure, P_a - the atmospheric pressure, u - the gas filtration rate per unit of the porous medium, and H - the height of the porous surface (figure 3).

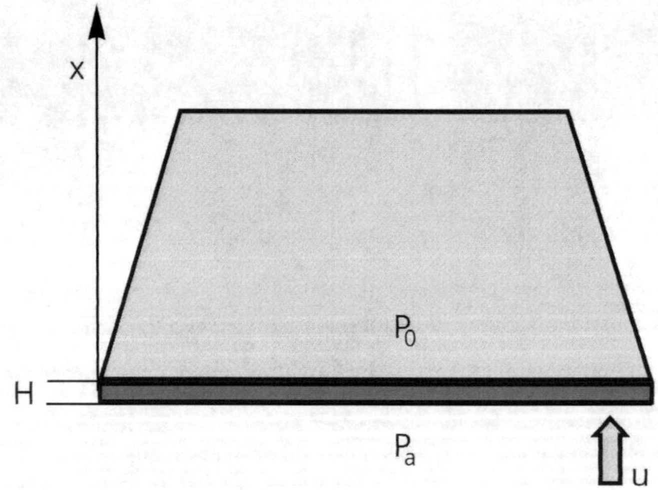

Figure 3. Projection of the filtration process during vacuuming of a porous surface.

Then the mathematical description of the vacuum is reduced to:

$$u = av$$

$$\frac{d(u\rho)}{dx} = 0$$

$$\frac{dp}{dx} = -g\rho - u\frac{\mu}{k(x)}$$

$$p = \rho RT$$

$$q = u\rho = const$$

where q - the gas flow rate – the volume flowing through the cross – section of the flow per unit time.

Pressure at the boundaries of the porous layer:

$$q = \frac{P_a^2 - P_0^2 \exp\left\{x\frac{2g}{RT}\right\}}{2\mu qRT\int_0^H exp\left\{\frac{2g}{RT}\xi\right\}\frac{d\xi}{k(\xi)}}$$

$$\rho = \frac{p}{RT}$$

$$u = \frac{q}{\rho}$$

When a vacuum is formed inside the suction Cup, atmospheric pressure exerts an effect on the surface of the suction Cup pressing it against the plane of the object. The pressure force acts evenly on the suction Cup and on the surface on which it is located on the reverse side, and is proportional to the vacuum level inside the suction Cup. Having vacuum chambers with an area of S and creating rarefied air in their cavities, it is possible to determine the clamping force by the formula [12]:

$$F = S(P_a - p_0)$$

However, this formula is only feasible for sealed connections of the vacuum to the surface, that is, in the absence of air suction along the boundary of the vacuum chamber. If this phenomenon is taken into account, it is necessary to Refine and Supplement the mathematical model, which is the goal of further research.

4. Conclusions

Modeling of gas filtration under vacuum on a breathable surface, namely the gas flow in micro channels of woven material of the container, allow to calculate the flow and speed of gas flowing through the breathable surface, the maximum achievable force of the vacuum pressure of the vacuum chamber at known parameters of the vacuum system, as well as to identify the characteristic flow regimes in micro channels.

Acknowledgments

The work was supported by the RFBR grant "Postgraduate student" № 20-38-90290\20.

References

[1] Leibenzon L S 1947 *Movement of natural liquids and gases in a porous medium* (Moscow: Gostekhizdat) p 244
[2] Barenblatt G I, Zentov V M and Ryzhik V M 1972 *Theory of non – stationary liquid filtration* (Moscow: Nedra) p 288
[3] Nield D A and Bejan A 2006 *Convection in Porous Media 3rd edition* (New York: Springer Science & Buisiness Media) p 640
[4] Leontiev N E 2017 *Fundamentals of filtration theory 2nd edition* (Moscow: MAX Press) p 88
[5] Nigmatulin R I 1978 *Fundamentals of mechanics of heterogeneous media* (Moscow: Nauka) p 336
[6] Nigmatulin R I 1987 *Dynamics of multiphase media* vol 1 (Moscow: Nauka) p 464
[7] Makarov A M, Kristal M G, Kovalev A A and Suleymanov D A 2015 Device for automatic opening, holding and closing of flexible containers (utility model 155000 RF, IPC B65B5/00, B65B7/02) (VSTU)
[8] Makarov A M, Mushkin O V, Drobotov A V, Serdobintsev Y P, Kuhtik M P and Lapikov M A

2019 Device for automatic opening, holding and closing of flexible containers (utility model 186346 RF, IPC B65B7/02, B66C1/02) (VSTU)

[9] Makarov A M, Mushkin O V, Lapikov M A, Kuhtik M P and Serdobintsev Y P 2019 The quality control of the automatic manipulating process of a flexible container when bulk materials are packaged *Machines* vol 7 chapter 4 p 24

[10] Makarov A M, Mushkin O V and Serdobintsev Y P 2020 Research of process of automatic opening of air-penetrating flexible containers for free-flowing products *Springer Nature Switzerland AG* ed A A Radionov and A S Karandaev pp 147–57

[11] O'Hanlon J F 1980 *A User's Guide to Vacuum Technology* (New York: Wiley Interscience)

[12] Hradetsky V G, Knyazkov M M, Fomin L F and Chashukhin V G 2010 *Mechanics of miniature robots* (Moscow: Nauka) ed V G Hradetsky p 265

Effect of recording device parameters on the operation of the sugar beet recognition algorithm in the early stages of growth

M A Mirzaev and A I Dyshakov

Federal State Budgetary Scientific Institution "Federal Scientific Agricultural Engineering Center VIM", 1st Institutsky proezd 5, Moscow, Russia

mirza.pochta@gmail.com

Abstract. This experiment makes it possible to estimate the influence of the distance from the recording device lens to the ground on the recognition algorithm by comparing the area of the recognized part of the plant. The experiments are based on the fact that the main error of the algorithm is to reduce the area of the recognized part of the plant with a significant deterioration of parameters. The study carried out a number of experiments including a gradual change of height from the minimum structurally possible value to the maximum at which there is a significant loss of quality under different lighting conditions ensuring normal operation of the vision algorithm.

1. Introduction

Machine vision is a section of robotics that uses image analysis to solve industrial problems. This technology is widely used in production and in particular in agriculture. Machine vision technologies together with positioning systems are often fundamental in the development of devices for agricultural operations in precision farming. For example, by using unmanned aerial vehicles topographic maps of the terrain are created and the application of image processing technologies to them makes it possible to obtain 3D models of the Earth's surface with the ability to determine any geometric dimensions. The error of geometric measurements does not exceed tens of centimetres. Machine vision systems can also be used to count the number of berries and leaves of grapes. In case if the vine has few leaves, the quality of grapes will be low, if it too much – the berries will not get enough water, so it is extremely important for agricultural producers to know their exact quantity it determines when and how to water the plant [1, 2].

The most important features of the vision system are performance, ease of use, easy installation, complete solution (including software), price, reliability, technical support, versatility of settings, speed, complete set of diagnostics. Image processing and analysis are mainly focused on working with 2D images. These operations include contrast enhancement, edge detection, noise elimination or geometric transformations such as image rotation. These operations assume that image processing/analysis operates independently of the content of the images themselves [3].

This paper deals with the application of machine vision based on image recognition, an area that uses various methods to obtain information from video data, mainly based on a statistical approach. The process of weed vegetation works according to the following principle – identification of cultivated plants, anything that is not the last one to be removed. This approach is used in the selective application of herbicides and mechanical removal of weeds while avoiding damage to cultivated plants [4].

2. Materials and methods

The experiment was conducted on a laboratory stand (figure 1) that consists of:

- the platform with guides;
- carriage;
- step engines;
- video camera;
- working body;
- system of mechanical and chemical removal of weed vegetation;
- container with herbicide;
- pump;
- control package;
- the electronic computer with specialized software [5].

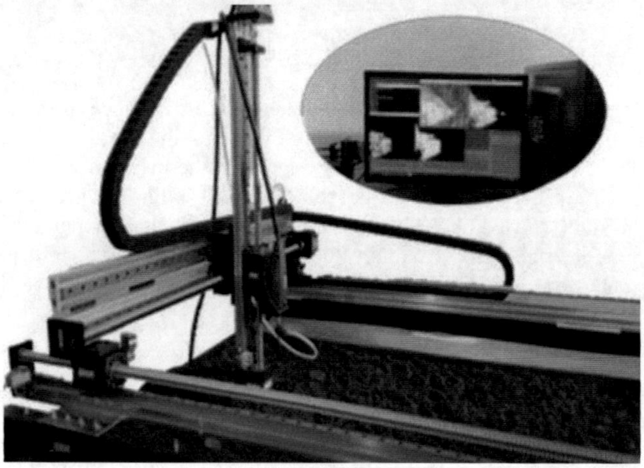

Figure 1. The robotic weed control complex developed during the research work.

It simulates the working conditions of an autonomous robotic complex for point control of weed vegetation. The movement of the platform with the guideways on which the carriage is mounted is installed by stepping motors within a limited volume. The carriage is equipped with a video camera and a working body which contains a system of simultaneous mechanical and chemical removal of weeds vegetation. The working device is connected through a pump to a container with herbicide. Outside the working area there is a control unit and an electronic computer with specialized software for image analysis [6].

Plant elements are characterized by values above the threshold ζ, when subtracting *ExR* from *ExG*.

Here, $ExR = (2r - g - b)$ excess of red, and $ExG = (2g - r - b)$ – excess of green. Where $r = R^*/(R^* + G^* + B^*)$, $g = G^*/(R^* + G^* + B^*)$ and $b = B^*/(R^* + G^* + B^*)$ are chromatic coordinates, and R*, G*, and B* are normalized values of RGB (0-1). The principle is illustrated in figure 2.

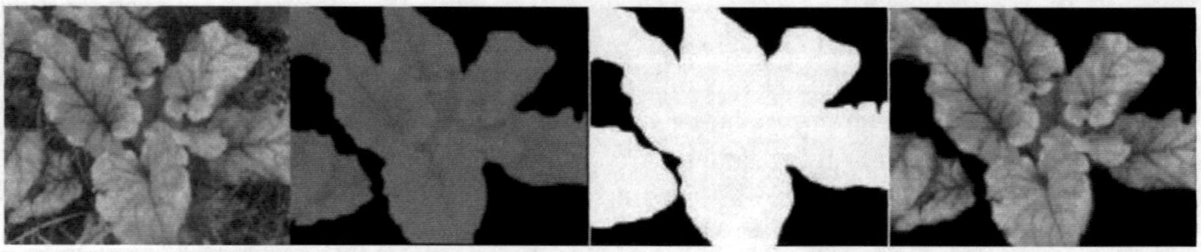

Figure 2. Plant identification using the ExG - ExR method.

As a recording device various options were considered. There are 1D-, 2D-, 3D-cameras. 1D-cameras also called linear or line cameras are a kind of CV-cameras which differ from usual that images in them are formed by scanning the subject. 2D-cameras this is the cameras creating images in two-dimensional space in width and height. 3D cameras are used when it is necessary to analyze the volume of objects, their shape or their position in three-dimensional space. In this work were taken 2D cameras due to the wide availability and ease of use of the algorithm.

Cameras are also divided into analog and digital cameras. The former are for creating video surveillance network, as they are cheap, easy to install and use, resistant to hacking and have no delays in data transmission. If you use cameras of new standards with high resolution pictures such as HDCVI, HD-TVI or AHD, you can get good image quality including moving and remote objects. In a digital camera the signal is not converted back from digital to analog for transmission but is sent digitized form. Before transmission it can be encoded and compressed. This happens in IP cameras or transmitted uncompressed and unencrypted so in HD-SDI cameras. Optimal for this test bench is a digital camera [7].

CMOS was selected by matrix type. Among the advantages of this matrix stand out is high speed, low power consumption, cheaper and simpler production. The disadvantages are low light sensitivity, pixel filling factor and dynamic sensitivity as well as high noise level [8]. Table 1 below shows the remaining characteristics of the recording device selected for laboratory testing.

Table 1. Characteristics of the Recorder.

Matrix type	CMOS
Matrix resolution	3 MPix
Maximum frame rate	30 frame per sec
Maimum permission.	1920x1080 Pix
Used optics	Carl Zeiss Tessar
Lens material	Glass
Number of lenses	1
Digital zoom	4x
Auto Focus	Yes
Auto. white balance adjustment	Yes
Low light photography	Yes
Indication of power on	Yes
Case material	Plastic
Horizontal rotation	90°
PC Communication Interface	USB 2.0
USB cable	Integrated
Cable Length	1.8 m
Color	black/silver
Overall size	94x43x71 mm
Video camera weight	162 g

This choice provides the fulfillment of all necessary requirements for the algorithm's work but it has a number of disadvantages. Testing the operation of the algorithm depends heavily on the quality of the recorder. There are specialized solutions for technical vision technology this is machine vision cameras. The latter shoot images in high resolution and send them uncompressed to your computer. Can have a resolution from VGA to 86 Megapixels for line scan or 4K for single-line scan and shoot up to 200 frames per second. The machine vision camera produces images that are optimal for computer processing, analysis, measurement, diagnostics, recognition and control. In further wor, based on the results of the study is planned to move to this type of recording device but at the stage of testing the developed algorithm is sufficient and convenient solution is the camera specified in the table.

The figure below shows an image of the process of light measurement obtained by the robotic complex recording device while performing the recognition algorithm.

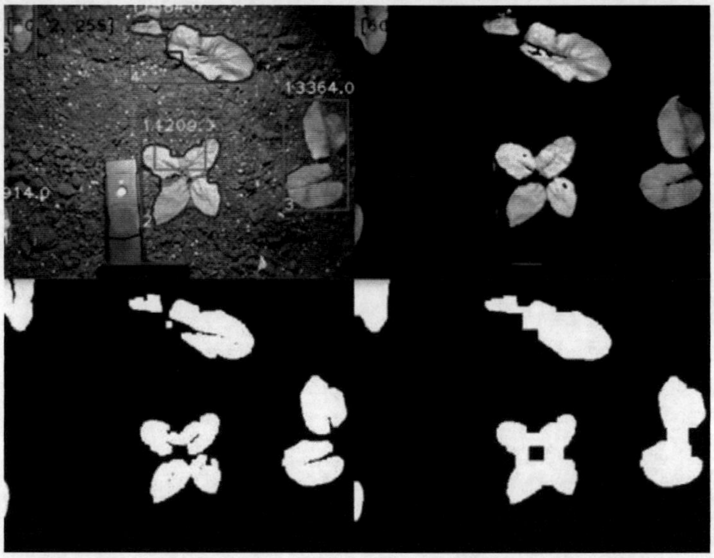

Figure 3. Measurement of illumination of stand of autonomous robotic complex for control of weed vegetation.

3. Results and discussion

To assess the performance of the algorithm at different values of illumination the agricultural crop - sugar beet was chosen.

Figure 4 shows the result of the previous work is a graph of the dependence of the area of the recognized surface on a natural light source which simulates the field conditions of the robotic complex.

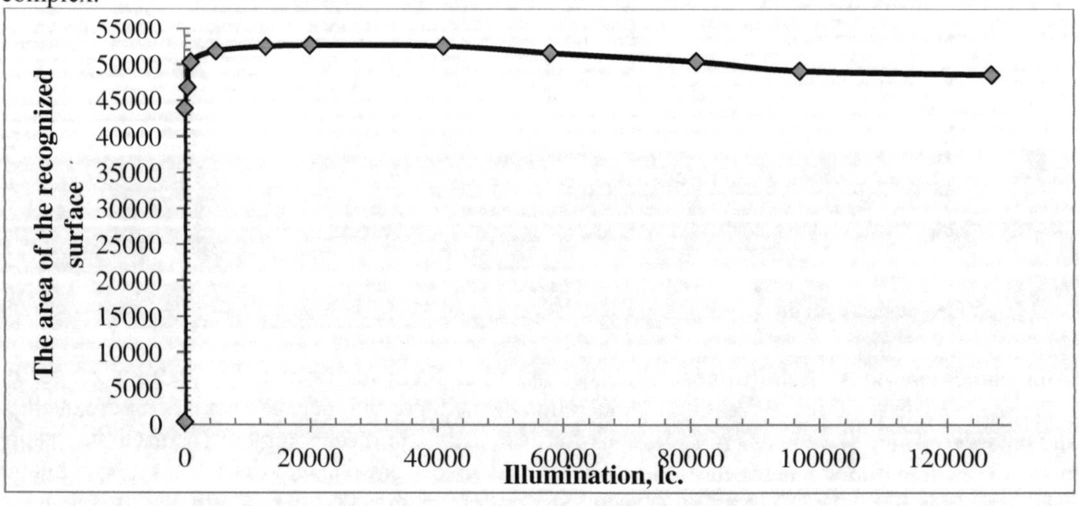

Figure 4. Plot of independence of area of the recognized surface on a natural light source.

At an illumination value of 5000 to 6000 lux the recognition of sugar beet with maximum quality was achieved. This illumination range extends over most of the day. To ensure that the algorithm works in low light conditions it is possible to use artificial light sources similar to natural light in the average range of experiments. The possibility of selection up to spectral distribution in particular imitation of sunlight will allow to use the complex during the whole time of day.

In total this study will examine the effect of the distance from the lens to the soil in the optimal illumination range - from 5000 lux to 60000 lux in 5 experiments gradually increasing the value from minimum to maximum specified.

Below is a number of photographs with successive magnification from 20.5 cm to 40.5 cm. 20.5 cm is the minimum permissible design value of the unit and ensures optimal operation of the complex in the field [9].

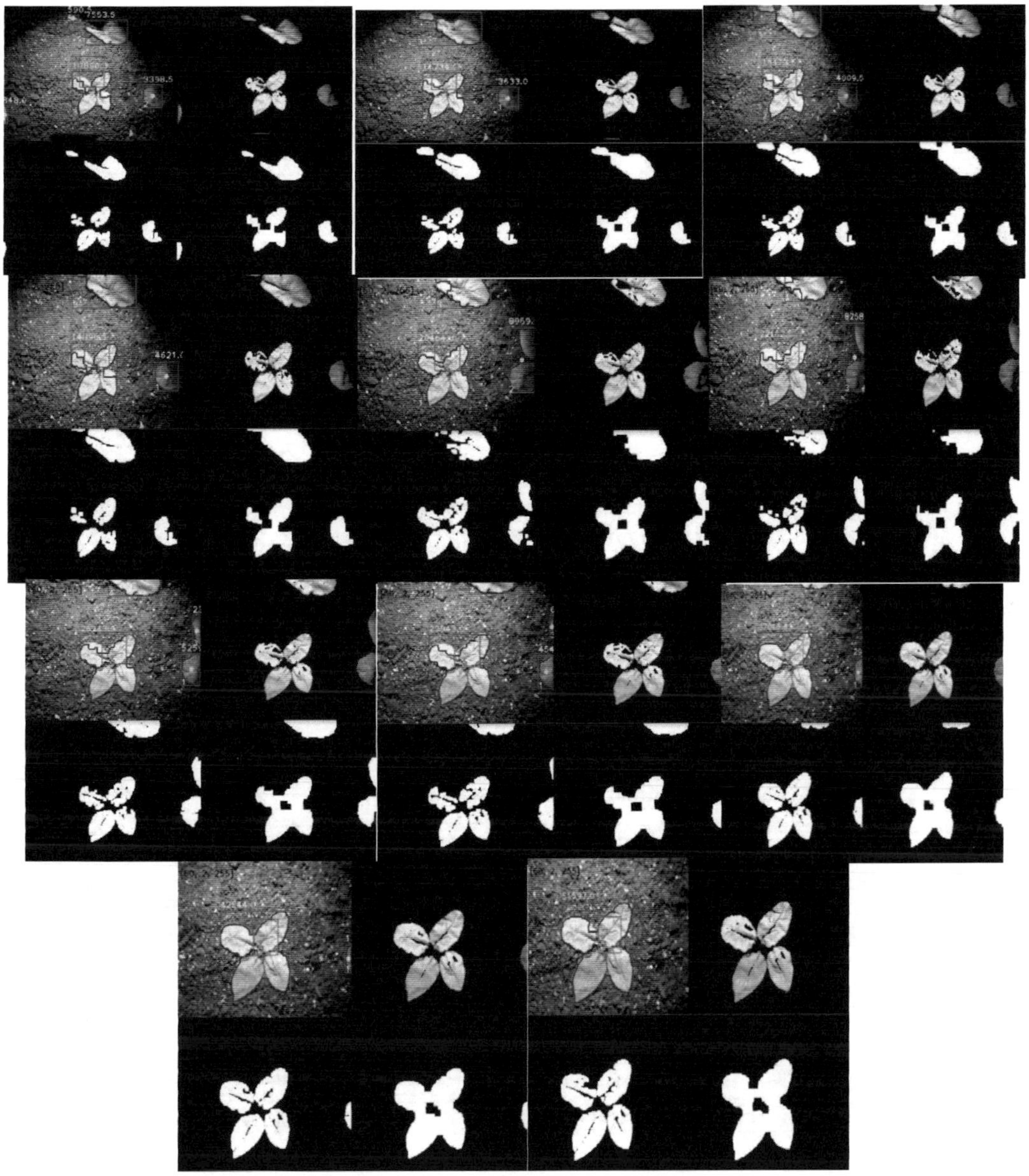

Figure 5. Images of the vision algorithm recognition results from 20.5 cm to 40.5 cm in 2 cm increments for an illumination of 41000 lux.

Figure 5 shows the results of the first experience. The numerical values are shown in table 2 and are shown in figure 6.

Table 2. First experience results.

	h1	h2	h3	h4	h5	h6	h7	h8	h9	h10	h11
	20.5	22.5	24.5	26.5	28.5	30.5	32.5	34.5	36.5	38.5	40.5
S'	51597.0	42544.0	37015.0	29943.0	26234.0	22152.6	20464.0	14896.5	15472.5	14236.0	10890.5

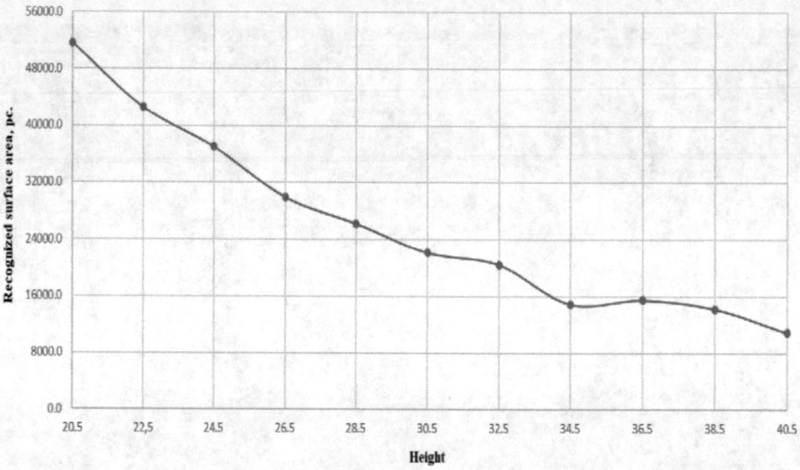

Figure 6. Results of the first experience.

It is obvious from the photographs that the number of pixels characterizing the plant decreases as the height increases. In order to actually evaluate the effect of height on the quality of the algorithm it is necessary to process the results so as to exclude the reduction of the plant in the frame. For this purpose, we will use the formulas:

$$S_1 = S_1'; \ \bar{S}_2 = S_2' \frac{h_2^2}{h_1^2}; \ S_3 = S_2' \frac{h_3^2}{h_1^2}; \ S_4 = S_2' \frac{h_4^2}{h_1^2}; \ \ldots\ldots, \text{ and the like.}$$

After normalizing the values of all experiments, table 3 was obtained.

Table 3. First experience results

	h1	**h2**	**h3**	**h4**	**h5**	**h6**	**h7**	**h8**	**h9**	**h10**	**h11**
	20.5	22.5	24.5	26.5	28.5	30.5	32.5	34.5	36.5	38.5	40.5
`S1	51597.0	51250.2	52869.1	50035.6	50704.5	49036.2	51433.9	42190.5	49049.9	50211.3	42506.0
`S2	52458.2	51195.2	51192.6	50617.0	51823.9	49769.1	48512.8	49483.3	45755.7	41918.8	46916.6
`S3	51344.5	52510.1	52440.3	51081.5	51868.9	50367.9	47344.3	44930.8	42902.4	46737.1	39604.0
`S4	52888.7	52899.5	50878.8	51641.4	50995.5	49374.4	46520.4	47219.8	43044.6	45082.0	40378.5
`S5	52690.3	51372.7	51952.4	51917.3	50811.7	50134.1	49716.5	44992.1	49755.5	44966.7	46975.8

4. Conclusions

This experiment allows to estimate the influence of the distance from the recording from the lens to the soilon the recognition algorithm operation by comparing the area of the recognized part of the plant [10–12]. As you can see the main error of the algorithm is that at a significant deterioration in the parameters of the system partially does not recognize the plant. This fact was the basis for the experiment. Below in figure 7 there is a diagram of the results of five experiments after data processing in particular the normalization of parameters

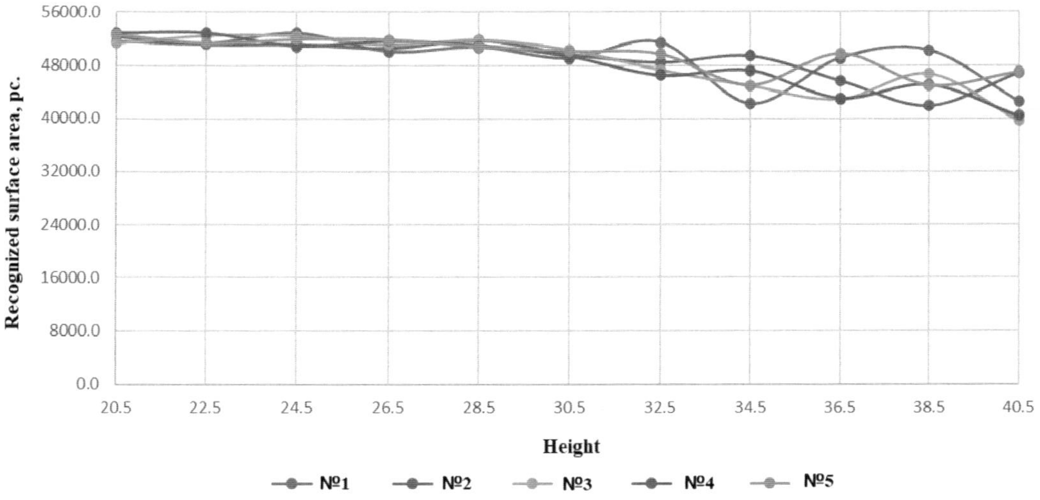

Figure 7. Results of all experiments after normalization of indicators.

As you can see from the charts after 30.5 cm the algorithm failures begin and resulting in the data spread shown in figure 7.

In addition, the total decrease of the recognized plant area can be seen taking into account all experiments.

In total we can conclude that increasing the distance to the soil has a negative impact on the work of the algorithm,especially it manifests itself from 30 cm. At the same time optimal work is provided in a wide range from 20 cm to 30 cm.

References

[1] Jain R, Kasturi R and Schunck B G 1995 *Machine vision* (New York: McGraw-Hill)

[2] Vernon D 1991 Machine vision - Automated visual inspection and robot vision *NASA STI/Recon Technical Report A* **92**

[3] Woebbecke D M, Meyer G E, Bargen K V and Mortensen D A 1995 Shape features for identifying young weeds using image analysis *Transactions of the American Society of Agricultural Engineers* **38** 271–81

[4] Guyer D E, Miles G E, Schreiber M M, Mitchell O R and Vanderbilt V C Machine Vision and Image Processing for Plant Identification *Transactions of the ASAE* **29** 1500–7

[5] Filippov R A, Hort D O and Kyterev A I 2017 Robotic vehicle with machine vision system *Innovazionie tehnologii v nayke I obrazovanii Materialy V mejdunarodnoy naychno-practicheskoy konferenzii* pp 475–8

[6] Dyshekov A I, Smirnov I G, Mirzaev M A and Shereuzhev M A 2020 Principles of functioning of the autonomous device for weed control for precision agriculture *IOP Conf. Ser.: Mater. Sci. Eng.* **747** 012108

[7] Wang L, Wu F C and Hu Z Y 2007 Multi-Camera Calibration with One-Dimensional Object under General Motions *2007 IEEE 11th Int. Conf. on Computer Vision* pp 1-7

[8] Torr P H S and Zissermann A 1997 Performance characterization of fundamental matrix

estimation under image degradation *Machine Vision and Applications* **9** 321–33
[9] Gazieva E E, Vasiliev A V, Syriamkin V I and Yurchenko V I 2010 Lighting systems in vision equipment for monitoring electronic devices and systems *Tezisy docladov 8 naychno-tehnicheskoy konferenzii Electronie i electromehanicheskie sistemy I ystroistva* pp 339–40
[10] Malevan K M and Kalachev A V 2019 Pattern Recognition Systems *Gorizonty obrazovaniya* **21**
[11] Masich I S 2015 Recognition of leaf plant species by machine learning methods *Nauka i obrazovanie: opit problem I perspectivy razvitiya* 294–6
[12] Bondarenko V Y, Barkovskiy A V, Shashko A Y and Demidchik V V 2018 Development of technology for image recognition of higher plants using machine learning methods *Kletochnaya biologoya biotehnologoya rasteniy* 108

Loading and transport unit with a tripod manipulator and hybrid drive

M E Nikolaev, I A Nesmianov, A G Ivanov and V V Dyashkin-Titov

Volgograd State Agricultural University, University avenue, 26, Volgograd, 400002, Russia

mr.maks.nikolaev.1994@mail.ru

Abstract. The analysis of loading and transport operations during the harvesting of vegetables Packed in bags or nets was carried out. The geometric parameters of the pincer grip are justified and analytical and graphical dependences of the required holding force for different bag masses are obtained. The device of a developed and manufactured loading and transport unit with a tripod manipulator is described. The tasks of further research are defined.

1. Introduction

In agro-industrial production, a large volume of piece cargo is represented by bags, boxes, soft bags and nets with packed vegetable products weighing from 20 to 80 kg. For example, harvesting onions, carrots, and sometimes potatoes is most often done by packing them in nets directly on the field, followed by loading onto vehicles and delivery to the place of sale. In the technological process of harvesting, this is one of the most labor-intensive operations, besides, during loading operations on harvesting vegetables, manual labor predominates [1, 2].

Despite the development of means of mechanization in agricultural production, the volume of manual labor in the procurement of fruits and vegetables reaches up to 40% ... 50%. It is possible to eliminate this drawback only by comprehensive mechanization and automation of loading and unloading operations.

An important indicator of the operation of automatic and semi-automatic lifting mechanisms is the positioning accuracy of the lifting device. Usually, according to the conditions of the technological process, the positioning accuracy of the load-gripping body is lower than the positioning accuracy of the actuators. For agricultural loaders, a low positioning accuracy is required, depending on the value of the permissible deviations of the load positions at the initial position and the unloading position (in the initial and final configuration) [3].

The prospect of creating a lifting and transport robotic unit based on a bag loader is determined by the following criteria:

1) implementation of portable and orienting coordinate degrees of mobility;
2) the use of a load-gripping device in the form of a pincer grip;
3) installation of a manipulation device transfer mechanism on a mobile chassis;
4) the relative uniformity of manipulation objects (bags) in their shape and mass;
5) relatively low required accuracy of positioning of loads;
6) the possibility of organizing a strictly regulated technological process.

2. Computational model

Developed handling unit has one portable global coordinate S, the change which happens with the speed of movement of the chassis V_{chh}, the angular coordinate of rotation of the base of the tripod of φ, varying with speed ω and the three generalized coordinates of the tripod l_1, l_2 and l_3 providing the displacement of the point M suspension grip with the speeds V_{MX}, V_{MY} and V_{MZ}, which is a projection of velocity V_M on the coordinate axis (figure 1).

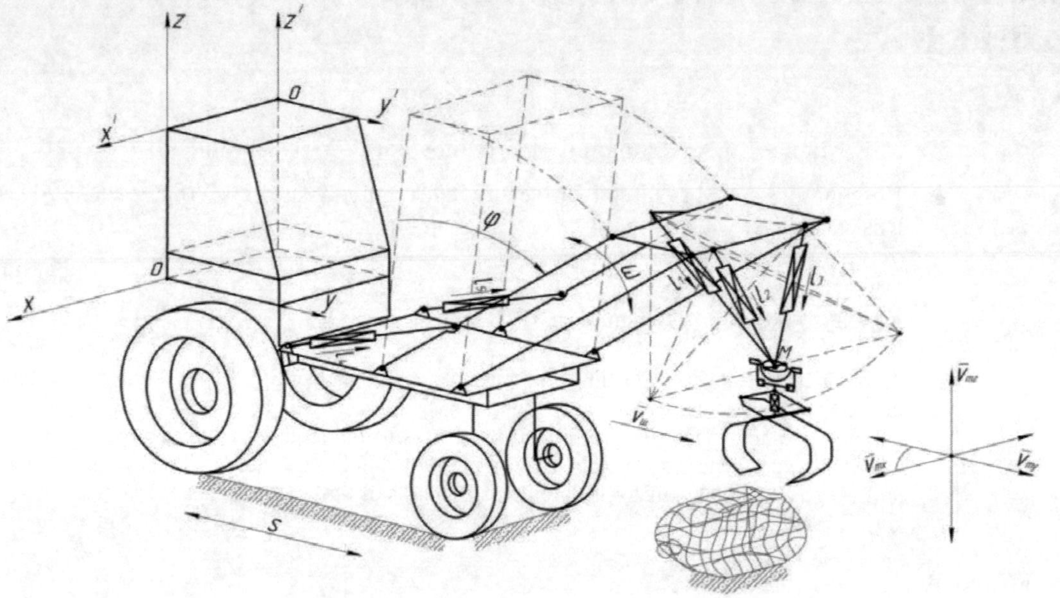

Figure 1. A design scheme for loading a transport unit with a manipulator-tripod.

The productivity of loading and transport operations when selecting nets with onions depends on the speeds of the actuators, the technological speed, the uniformity of the distribution of nets (bags) across the field, and, not least, the orientation of the nets in relation to the self-propelled loader. Based on the analysis of technological processes of loading and transport operations during harvesting of root crops Packed in nets or bags, a design and technological scheme of a self-propelled loading vehicle with a robotic manipulator-tripod for loading bags of vegetables is proposed and developed [4, 5, 6].

The structure and required degrees of mobility, geometric and kinematic parameters that ensure the necessary movement of the cargo are justified for the developed manipulator of the loading vehicle. Analytical dependences are obtained between the design parameters of the manipulator, which make it possible to ensure a constant speed of the loading unit [7, 8, 9].

3. Define and build a service area configuration

Guaranteed retention of the load by the gripper is caused either by a geometric closure, or by a force closure, which ensures the impossibility of translational and rotational movements of the captured bodies (figure 2).

The minimum required holding force is determined from the expression

$$P(\varphi) \geq \frac{2 \cdot \cos\varphi(N \cdot l - 0.25 \cdot Q \cdot a)}{b} \tag{1}$$

The resulting dependence (1) allows you to select the drives to provide the necessary holding force at the design stage of the gripper. Here, however, it should be remembered that the holding weight is vegetables, which, if a significant force is applied to them, can be damaged.

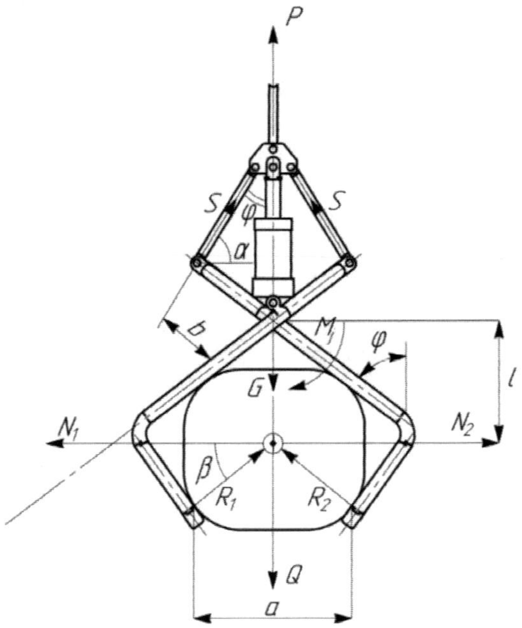

Figure 2. Calculation scheme of tick capture.

Depending on the size of the mesh, their gross weight also varies within different limits, the graphic dependences of the holding force on the bag weight and the angle of the gripper lever position are shown in figure 3.

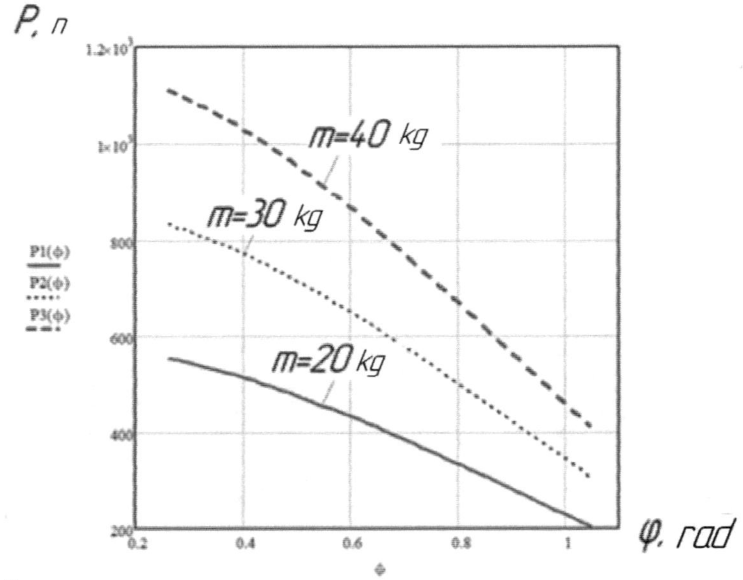

Figure 3. Dependence of the holding force on the mass of the bag and the angle of the grip lever position.

Figure 4 shows the dependence of the holding force of the pincer grip on the angle of the position of the levers φ and their length b. The values of the holding forces in a pincer grip with curved jaws are 5 times less than in a grip with flat sponges. As a result, the capture scheme in figure 2 was chosen, which is preferable from the point of view of less force on the grid with vegetables and, accordingly, less damage to vegetables.

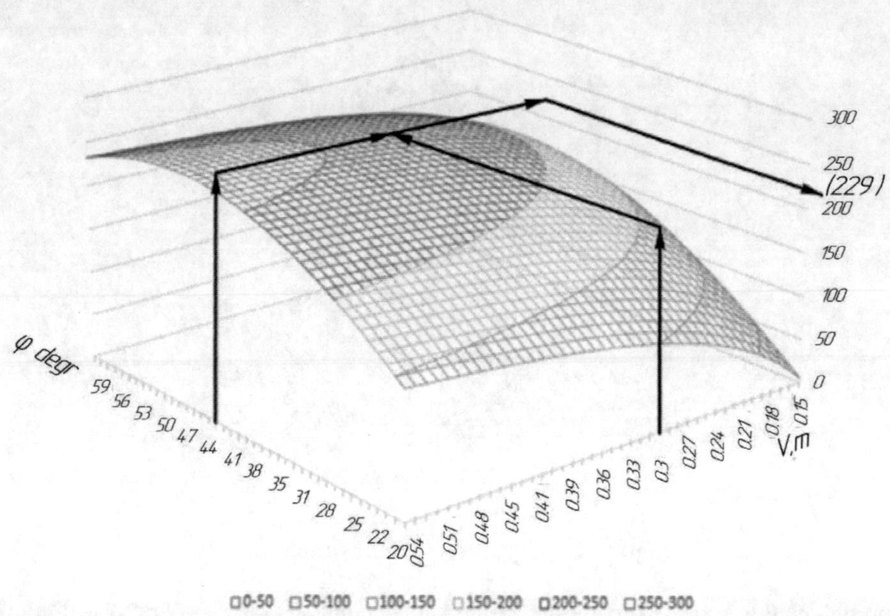

Figure 4. Dependence of the holding force P on the angle of position φ and its length b for gross mass.

Taking into account the obtained dependencies, a pincer grip for bags of vegetables was made (figure 5 (a)), mounted on a manipulator-tripod of a loading and transport unit on a self-propelled chassis (figure 5 (b)).

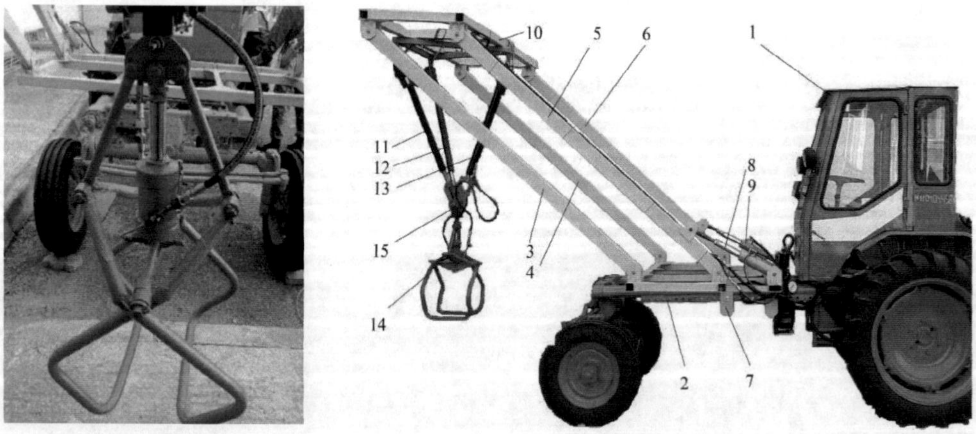

(a) pincer grip (b) Space-parallelogram mechanism

Figure 5. Pincer gripper and loading and transport unit.

The loading and transport unit with a tripod manipulator consists of a self-propelled T-16MG chassis (figure 5 (b)) 1, to the spars 2 of which a composite boom is attached, which is a spatial parallelogram mechanism, which consists of four racks 3, 4, 5, 6, fixed by means of cylindrical hinges to the frame 7, allowing using Executive hydraulic cylinders 8, 9 to implement the rotation angle so that the platform 10 is always kept in a horizontal position. A manipulator-tripod is attached to the platform 10, which has three linear Electromechanical drives 11, 12, 13, the ends of which are attached to the platform 10 with hinges on one side, the rods of the opposite ends of the three linear actuators-actuators are connected to each other and with a two-stage controlled grip 14 by means of a five-moving spherical hinge 15. The drive of the Executive mechanisms is made hybrid, the rotary parallelogram mechanism and the pincer grip are driven by a hydraulic drive, the drive of the Executive links of the manipulator-tripod is electric (supply voltage 24 V).

When performing technological operations for loading bags, the self-propelled chassis 1 moves across the field, the chassis operator controls the tilt of the boom, while positioning the pincer grip to the bag is performed automatically according to the specified program laws [10, 11]. To recognize bags on the field and then determine the capture coordinates, the technical vision system is used (figure 6). Bags are loaded into the body of a self-propelled chassis (pictured in figure 5 (b) the body is removed).

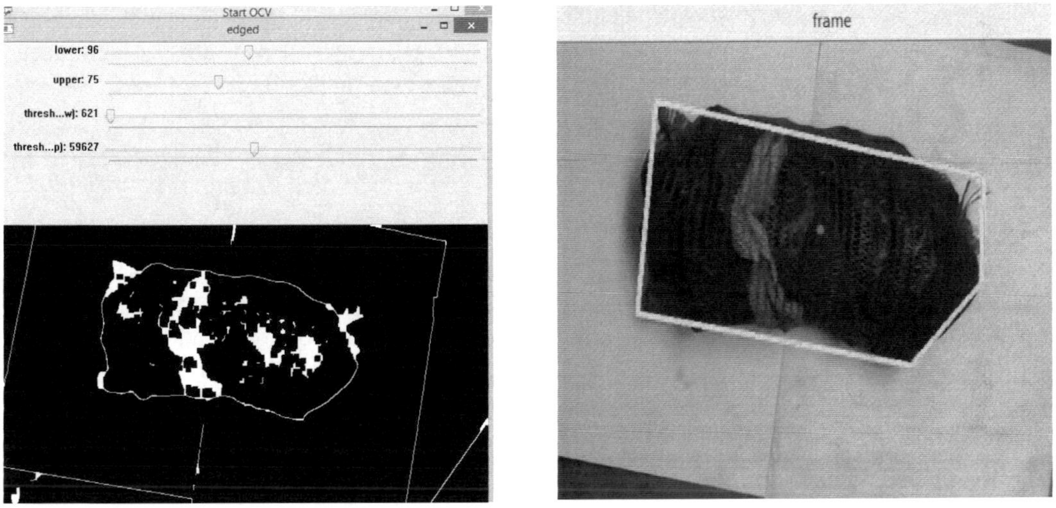

Figure 6. Screenshot of the program of the technical vision system for recognizing bags on the field.

It is planned to evaluate the effectiveness of the developed robotic loading and transport unit for collecting vegetable nets from the field in terms of productivity, positioning accuracy and movement along specified trajectories, and energy consumption per cycle of technological operations for several control modes:

- fully manual operator control mode;
- semi-automatic control mode, loading and transport unit for collecting bags of vegetables from the field.

In the future, it is planned to fully automate the control of Executive drives by replacing manual hydraulic valves with electrically controlled ones.

4. Conclusion

The design of the manipulator and the parameters of the pincer grip of bags are justified. The design of the manipulator was developed and manufactured, on the basis of which an experimental loading and transport unit was assembled. The developed loading and transport unit with a parallel-sequential manipulator structure can be installed on a self-propelled chassis "Agromash" SH-50 or vzt-30SSH. The resulting working area of the robot loader arm provides full coverage of the entire body area along the width of the chassis. To control the manipulator actuators, a control system has been developed that allows implementing the specified program laws of motion. The manipulator is equipped with a technical vision system for recognizing the coordinates of grids with vegetables relative to the coordinate system directly connected to the chassis body.

Acknowledgments
The work was supported by the RFBR grant no. 19-38-90067 "Postgraduate students" and the RF President Grant no.MK-210.2020.8.

References

[1] Gojaev Z A, Grishin A P, Pehalsky I A, Grishin A A and Grishin V A 2016 Development of work on the creation of agricultural robotics *Polythematic Network Electronic Scientific J. of Kuban State Agrarian University* pp 488-502

[2] Godzhaev Z A, Horoshenkov V K, Goncharov E I and Il'chenko E N 2018 Trends in the development of mechanotronics, automation and robotics for the synthesis of new high-performance technologies, machine production *Fundamental and Applied Problems of Engineering and Technology* pp 4-16

[3] Vorobieva N S, Zhoga V V, Dyashkin-Titov V V and Dyashkin A V 2017 Development of a database of models of a parallel-serial structure manipulator *Izvestia SFedU. Engineering Sciences* pp 143-52

[4] Nesmiyanov I A, Nikolaev M E, Vorob'yova N S, Dyashkin A V, Dyashkin-Titov V V and Ivanov A G 2019 *Pat. 2700304 Russian Federation MPK51 V66S 23/44 Robot-manipulyator Byul. 26*

[5] Nesmiyanov I A, Nikolaev M E, Vorob'yova N S, Dyashkin A V, Dyashkin-Titov V V and Ivanov A G 2019 *Pat. 2651781 Russian Federation, MPK51 V66S 23/44 Manipulator-tripod for purposes Byul. 12*

[6] Nesmeyanov I A, Zhoga V V, Vorobyova N S, Dyashkin-Titov V V, Dezhkin A V and Nikolaev M E 2018 *Pat. 183553 Russian Federation, MPK51 V66S 23/44 Crane manipulator*

[7] Nikolaev M E, Nesmiyanov I A and Fomin S D 2019 Substantiation of the structure of the manipulator of a mobile mesh loader with vegetables *Izvestia Nizhnevolzhsk Agro-University Complex: Science and Higher Professional Education* pp 389-97

[8] Nikolaev M E, Nesmianov I A and Zaharov E N 2020 Definition of service area of agricultural loading robot with manipulator of parallel-serial structure *IOP Conf. Series: Materials Sci. and Engineering. Int. Conf. of Young Scientists and Students Topical Problems of Mechanical Engineering* pp 012125

[9] Nesmeyanov I A, Nikolaev M E and Vorobyova N S 2020 Justification of parameters of the loading and transport robot manipulator *Proc. of the Volgograd State Technical University* pp 65-8

[10] Zhoga V V, Dyashkin-Titov V V, Nesmiyanov I A and Vorobyov N S 2016 The task of positioning a manipulator of a parallel-serial structure with a controlled gripper *Mechatronics, Automation, Control* pp 525-30

[11] Dyashkin-Titov V V and Pavlovsky V E 2014 The problem of optimal control of the movement of the grip of the manipulator-tripod *Proc. of Lower Volga Agro-Engineering Complex: Science and Higher Professional Education* pp 241-7

IOP Conf. Series: Materials Science and Engineering 1129 (2021) 012064 doi:10.1088/1757-899X/1129/1/012064

Precise positioning of mechanisms based on dynamics

S Y Orekhov, V M Masyuk, V S Kuznetsov, S P Dolgolenko and K G Kislov

Kaluga Branch of the Bauman Moscow State Technical University, Bazhenova street 4, Kaluga, 248004, Russia

kuznetsov850@ya.ru

Abstract. The purpose of this work is to study the factors that affect the positioning accuracy and find a way to determine the positioning accuracy of a mechatronic system. The paper investigates a model of a hexapod manipulator with six degrees of freedom. A mathematical model of the manipulator has been built. The study of the dependence of the positioning accuracy on the influence of external forces was carried out. It is concluded that manipulators of a parallel structure are applicable for light loads.

For today, mechatronic systems are widely used in such areas as: mechanical engineering, instrumentation, nanotechnology, medicine, photo and video, etc. This list is only expanding every day. In any production, considerable attention is paid to the quality of details. Complex automated systems are used to comply with all norms and obtain sufficient quality. A manipulator occupies the main place in these systems. High demands are placed on the accuracy of these mechatronic systems.

While working on a given program, it is difficult to achieve the required accuracy in the absence of corrective actions. The large mass of the structure, in particular, the working link, does not allow obtaining the required accuracy due to the forces and torques arising in the system.

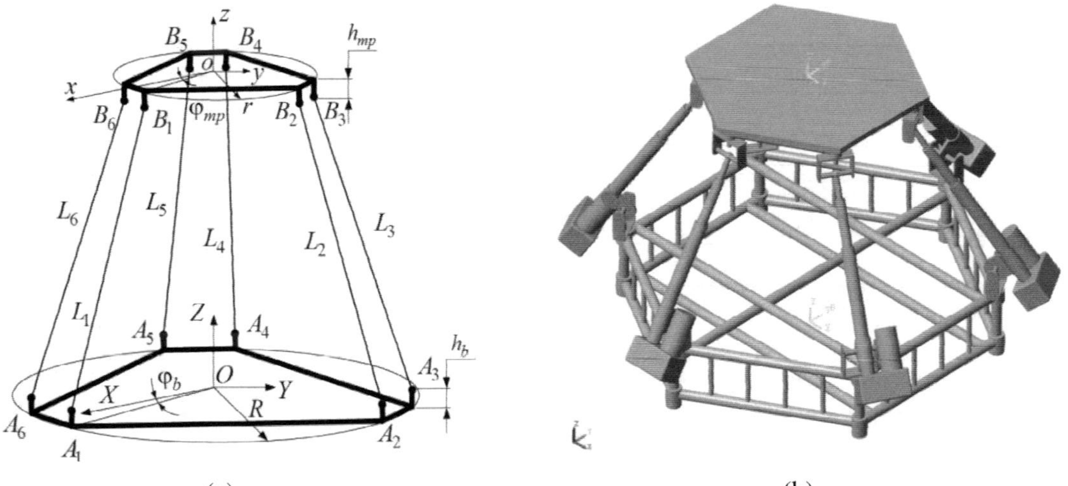

(a) (b)

Figure 1. Block diagram of the manipulator (a) and model in CAD Solidworks (b).

The positioning accuracy study will be based on a six-degree-of-motion controlled hexapod with actuators attached to a lower platform, with motors in parallel (figure 1). The construction of this model was based on the works [1], [2]. According to the Somov-Malyshev formula we made the choice of kinematic pairs and the type of hinge joints. When using the mode with six degrees of mobility, the robot has an RRTS structure (rotational-rotational-translational-spherical), in which two rotational pairs are used at the base, equivalent to a gimbal drive, a prismatic pair, a spherical gearbox. The rotary joints are passive, with only six linear actuators operating. The robot has 9 kinematic pairs of the fifth class (12 turning pairs and 6 rectilinear sliding pairs) and 6 kinematic pairs of the third class (6 spheres). The number of degrees of freedom according to the Somov-Malyshev formula is 6 [1].

This hexapod model describes the dynamic processes in the unloaded state and the method for finding the error in the displacement of the output link under the action of external forces [3].

To study this manipulator, it is necessary to solve the direct dynamic problem. Consider the forces that act on the entire system: the force of gravity acting on the movable platform, and the force of gravity acting on each of the rods, the force applied to the points and directed along the rods [4].

Having compiled the expression of work and substituting the values of the acting forces, we find the values of the generalized forces:

$$Q_j = \sum_{i=1}^{6}\left(F_i^x \frac{\partial x_i}{\partial q_i} + F_i^y \frac{\partial y_i}{\partial q_i} + F_i^z \frac{\partial z_i}{\partial q_i}\right),\ j = 1,2,3,4,5,6. \quad (1)$$

Solving this system of differential equations, which describes the dynamics of a mechanical system we can find the law of variation of the generalized coordinates q_i, according to the given values of the control forces F_i [5].

The rigidity of a parallel manipulator determines its technical characteristics, such as accuracy, reliability, productivity, durability. The work performed by an external force and three forces in the rods are forces multiplied by the corresponding displacements [4]. Let's write the matrix equation.

$$F_P^T \times \Delta P = f^T \times \Delta q\,, \quad (2)$$

where F_P^T – resultant of system of forces;

ΔP – vector of infinitesimal displacement of the output link arising under the action of the applied force;

f^T – vector of forces in the rods;

Δq – vector of infinitesimal deformations arising under the action of forces in the rods.

The vector of infinitesimal displacement of the output link can be represented as the product of the direct Jacobi matrix (J) and the error vector of the rod drives [2]. After transformation, we get the expression:

$$F_P = J^{-T} \cdot f\,. \quad (3)$$

The stiffness of the translational and rotational mechanisms does not depend on each other, since the structure of this manipulator with a parallel structure has a kinematic decoupling of the actuators. According to Hooke's law:

$$\varepsilon = \frac{\Delta q_i}{q_i} = \frac{\sigma}{E} = \frac{f_i}{A \cdot E}, \quad (4)$$

where ε – the relative deformation;

Δq_i – change in the length of the rod under the action of the applied force;

q_i – rod length;
E – elastic modulus in tension and compression;
A – the cross-sectional area of the rod;
f_i – strength in the rod;

$\frac{A \cdot E}{q_i}$ – the stiffness of the rod in tension and compression.

We express from equation (4) the force arising in the rod and substitute it into equation (2):

$$F_P = J^{-T} \cdot K_s \cdot J^{-1} \cdot \Delta P = K_c \cdot \Delta P, \tag{5}$$

where $K_c = J^{-T} \cdot K_s \cdot J^{-1}$ – the spatial matrix of the rigidity of the mechanism in the orthogonal coordinates system of the base.

Let us express from equation (4) the error of displacement of the centre of the platform under the action of an external force:

$$\Delta P = K_c^{-1} \cdot F_p, \tag{6}$$

where K_c^{-1} – the matrix of static stiffness of the parallel robot;

F_p – applied force.

The algorithm for finding the stiffness and displacement error of the output link under the action of an external force is implemented in the Matlab environment and is shown in figure 2. Based on the data obtained, dependencies were built (figure 3).

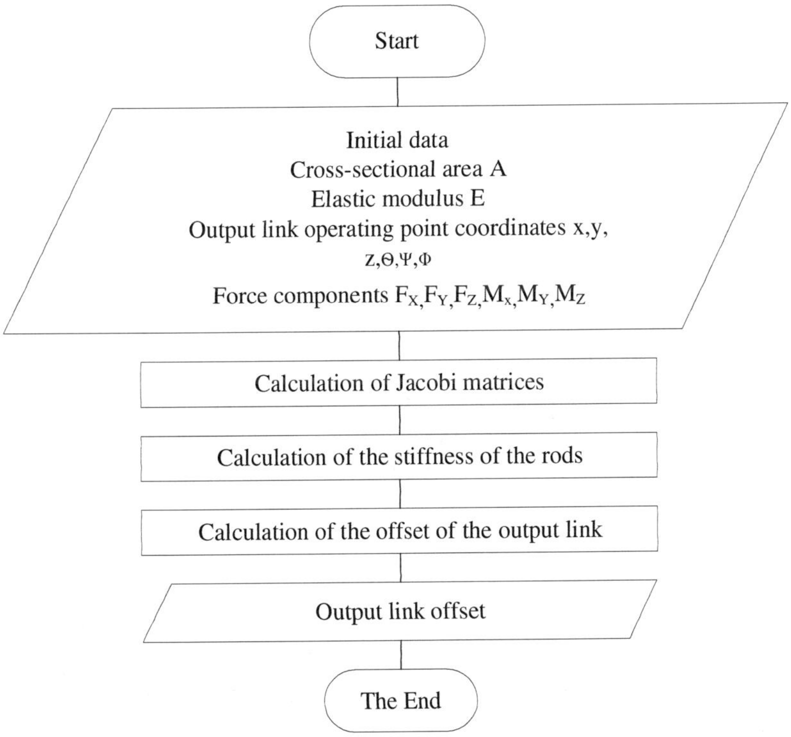

Figure 2. Algorithm for calculating the error.

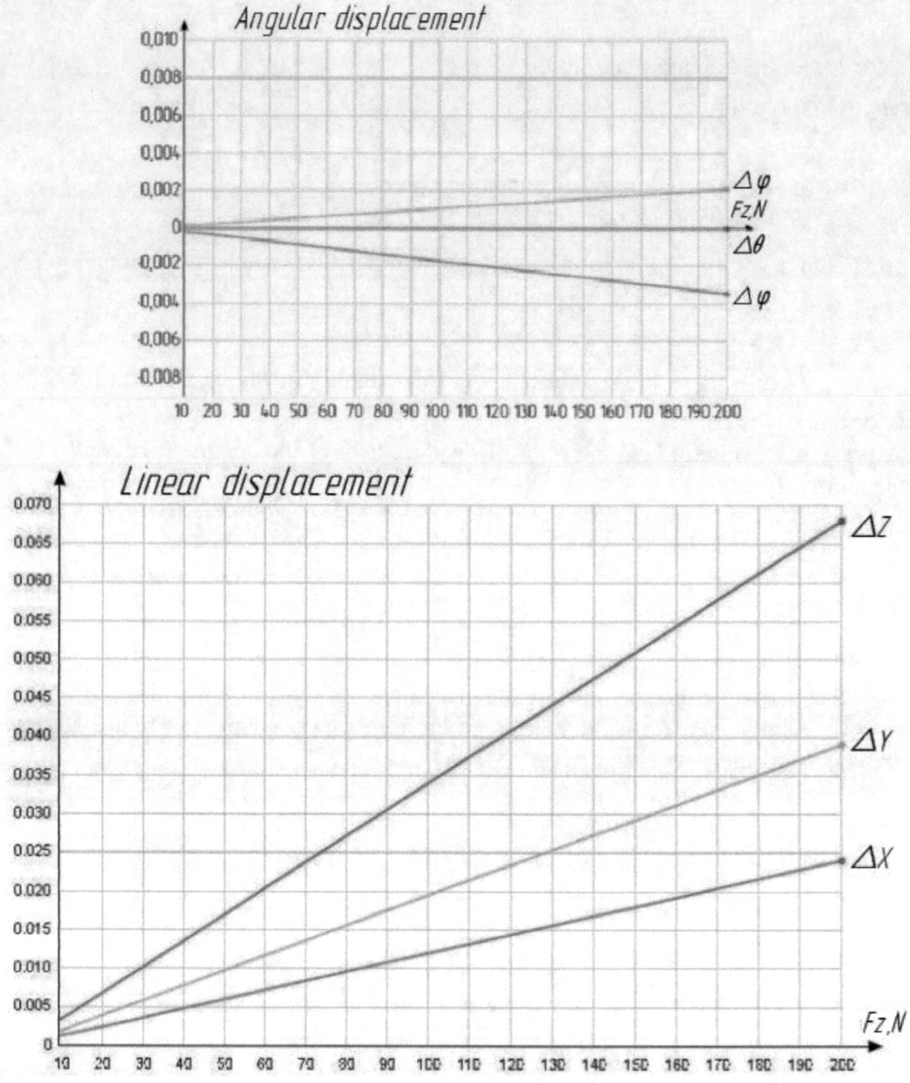

Figure 3. Dependences of the linear and angular displacements of the output link on the component of the applied force.

The experiment will be carried out on a controlled robot-hexapod with actuators fixed on the lower platform. The measurements will be taken from a platform that is acted upon only by gravity. The obtained measurements will be displayed on graphs in the Matlab environment.

As can be seen from the graphs, the offset error of the output link increases linearly with increasing force. It follows from this that manipulators of a parallel structure should be used at low dynamic loads. Also, from the graph of the angular displacement it can be seen that their values are so small that they can be ignored.

The data obtained are of a theoretical nature, the measurements were obtained on the basis of experiment modeling in SolidWorks and Matlab environments. The experiment was carried out according to the well-known methods [6]. At the moment, the choice of technical and hardware means is being made for the actual assessment of these parameters for the mechanism of the parallel structure and verification of the obtained theoretical basis.

As a result of the research, the influence of external forces on the positioning accuracy of mechanisms of a parallel platform-type structure with six degrees of freedom was determined. It was determined that the positioning accuracy of mechanisms of a parallel structure, arising from the error of drives, requires additional study and finding new methods to improve accuracy. The dependences of linear and angular displacements on external forces were experimentally determined.

References

[1] Artemov D V, Masyuk V M, Orekhov S Y and Pchelkina I V 2018 Analysis of a robot of a parallel structure with three drive pairs *XXX Int. Innovative Conf. of Young Scientists and Students* pp 424-7

[2] Bushuev V V and Kholoshev I G 2001 Mechanisms of parallel structure in mechanical engineering *STIN* **1** pp 3-8

[3] Zuev S M 2014 *Stabilization of equilibrium positions of loaded modifications of the Stewart platform* (Saint Petersburg) p 115

[4] Gaponenko E V 2014 *Dynamics of controlled movement of a tripod robot with six degrees of mobility* (Belgorod) p 219

[5] Glazunov V A 1995 *Using the program for calculating plane mechanisms of parallel structure in the educational process* (Ivanovo) p 12

[6] Merlet J P 2006 *Parallel Robots* (Dordrecht, Netherlands: Springer)

Increasing the level of the metrological supply of large-sized windows blades (from 60m) at the production stages

Sinitsa M O

Metrology and Interchangeability Department, Bauman Moscow State Technical University, Moscow, ul. 2-nd Baumanskaya, 5/1, 105005, Russia

Sinmaria@yandex.ru

Abstract. The development of the field of renewable energy sources today is happening at a fairly rapid rate, especially in the field of wind energy. In the last year alone, several large wind farms have been built in Russia. The fast pace of development of wind energy in Russia leads to the fact that in the near future there will be a need to produce our own wind turbines, according to the projects of Russian developers. Also, the development of technologies, the rapid introduction of new production systems, an increase in the share of autonomy of industrial enterprises requires the introduction of new modern approaches in the field of energy, energy efficiency and energy conservation. Against this background, the need for innovative solutions in the field of metrological support and measuring technologies in the energy sector is sharply increasing. The paper presents a complex for carrying out control and measuring procedures, which will improve the quality of metrological support in the production process in Russia. Also, the metrological and economic parameters of the described stand have been calculated....

1. Introduction

The development of the field of renewable energy sources today is happening at a rapid pace. According to data from the website for statistics on world energy in 2019, the share of wind and solar energy in the global energy balance increased (+ 12% for wind and + 24% of solar energy), while the generation of electricity from thermal energy decreased.

Renewable energy currently accounts for 35% of the energy balance in the EU, 27% in China, 21% in India and about 18% in the USA, Russia and Japan. Electricity generation from offshore wind turbines increased by 20%, thanks to the commissioning of 5.5 GW of plants in countries such as Belgium, Germany and the United Kingdom.

The use of renewable energy sources influenced the accelerated growth of capacities in wind and solar energy, which led to a sharp increase in electricity generation by wind and solar power plants in China (an increase of + 10% and 31%, respectively, almost 9% of the total energy balance), USA (+ 9% and 15%, respectively, almost 10% of the total energy balance), the EU, Japan, India, Australia and Latin America (strong growth in Chile, Brazil, Mexico and Argentina) [1].

Wind energy, given its low environmental impact and internal energy source, is by far the most demanded resource. Despite the high initial cost, wind energy leads to an increase in the profitability of the economy.

By Decree of the Government of the Russian Federation No. 335 of March 28, 2019, the state program "Energy efficiency and energy development" was renamed "Energy development", and the implementation period was extended until 2024.

The goal of the program is two-fold:

- maximum assistance to the socio-economic development of the country, which plays a significant role in the formation of budget revenues of the Russian Federation;
- strengthening and maintaining the position of the Russian Federation in the world energy sector.

Achieving this goal requires an accelerated transition to more efficient, flexible and sustainable energy [2]. To achieve the best results, it is necessary to introduce new modern technologies into production.

Already today, the world's leading companies are introducing modern energy management technologies. It should be noted that the modern level of automation has reached such a scale that it is impossible to control technological processes without the introduction of measurement, control and management systems. Today, instrumentation is not just a set of sensors, but measuring complexes combined with modern measurement and control methods, as well as advanced measurement techniques [3].

The development of technologies, the rapid introduction of new production systems, an increase in the share of autonomy of industrial enterprises require the introduction of new modern approaches in the field of energy, energy efficiency and energy saving.

Against this background, the need for innovative solutions in the field of metrological support and measuring technologies in the field of energy is sharply increasing.

Research of the last decades is largely devoted to the issues of increasing the reliability and diagnostics of industrial facilities. The operational reliability of power generating capacities distributed throughout the territory of Russia for thousands of kilometers must be guaranteed by means of monitoring their technical condition [15].

2. The blades

Today, on the eve of innovative technologies in the field of renewable energy, especially in the wind energy of the Russian Federation, a special place is occupied by metrological support of the entire structure of a wind turbine. Since the production of its own wind turbines in Russia began relatively recently, control problems occupy important positions in the life cycle of wind turbines.

Carrying out control and measuring operations of the large-sized products are widely used all over the world today. So, in the work "Coupled system of 3D laser scanning and correlation of digital images for obtaining geometry and monitoring deformations of a railway tunnel" results of research and measurement using 3D scanning of a section of a railway rock tunnel in Brazil are presented [4].

In the authors' study "Monitoring the alignment of the wind turbine shaft in real time using laser measurement", an example of the use of laser measurements to determine and control the fixation of positional changes in wind turbines during operation and alignment of the drive lines of the shafting in wind turbines is given. The measuring system is a set of lasers of the installation; based on the results of the control, the average and periodic amplitudes are determined from the ensemble of the averaged signal [5].

Thus, the problem of measuring control of large objects, including blades, is of great importance, and many modern solutions are based on the use of non-contact measuring systems, including scanning and lasers.

2.1. The features of the blades

For the structural design of wind turbine blades, a compromise must first be found between aerodynamic and structural efficiency. The choice of materials and manufacturing process will also affect how thin (aerodynamically perfect) the blade can be made and the price point. Consequently,

IOP Conf. Series: Materials Science and Engineering 1129 (2021) 012065 doi:10.1088/1757-899X/1129/1/012065

the structural design process plays a critical role in bringing together all design and manufacturing disciplines to create the optimal solution in terms of performance and cost.

When manufacturing a blade, the following criteria must be considered:

1. The blade must be strong at the base. This is necessary to create sufficient lift to accommodate the lower wind speed near the hub. Unfortunately, the thickness required to provide blade rigidity and strength is greater than that required for aerodynamic efficiency, so a trade-off must be found between the weight of the structure and the loss of aerodynamic efficiency.
2. Towards the end of the entire piece, the blade should be tapered. The narrowest part of the blade is needed at the apex where drag is most critical, also for power regulation in strong winds, the thinner part breaks off more easily, therefore useful at the tip.

Thus, the shape of the blade itself, its thickness directly affects the efficiency of the wind turbine. If the blade is made thinner, its aerodynamic performance is improved, but stronger spars will be required, which will increase the cost of the blade. Optimal geometry is achieved iteratively, taking into account turbine design, loads, design and manufacturing costs. As mentioned earlier, different product thicknesses are required to create lift on the blade, which in turn drives the turbine and is distributed along the blade approximately in proportion to the local radius, that is, the tip has more lift than the hub. The lifting force acts on the blade, bending it. This effect is called bending moment. At the base, the bending moment is most significant, but at the tip, the bending moment drops to zero. [6]

Today, statistical tests are carried out on wind turbine blades. This is done to confirm the required load profiles and to check the blade design. Typically subjecting the blades to 150% of their rated load power. Accurate blade failure testing requires high strength, high precision and reliability. Impact Resistant Test Equipment. Checking the profile of the blade section, its shaping, plays an important role in the quality and efficiency of the wind turbine. We need innovative technical solutions based on a scientific and modern technological base. Our country already has accumulated potential, scientific groundwork and staffing [7-14].

3. Control and measuring operations of wind turbine blades using the test board

Today, when the production process is automated, when it is equipped with modern progressive technologies, production is being improved, which means that the role of metrology is increasing. It becomes relevant to carry out measurements in dynamics with a significant influence of external factors. With the introduction of new technologies in the production process, today measurements are more often carried out at all stages of the product life cycle, while the share of measurement procedures often exceeds the share of production.

Therefore, a promising direction in control and measurement operations in metrology is considered to be such measurements using the methods of non-contact measuring procedures based on the use of modern laser and optical measuring systems.

Since most of the generated electricity from the wind turbine depends on the quality of the manufactured blade, the increase in such parameters as reliability, accuracy of mold manufacturing, namely, compliance with the specified aerodynamic parameters, largely depends on the quality of the control performed.

3.1. The stand

Wind farms are actively being built in Russia; on May 1, 2020, the Kamenskaya wind farm with an installed capacity of 100 MW was commissioned. Until the end of the first half of the year, it is planned to launch another wind farm, Gukovskaya with a capacity of 100 MW. Also, by the end of 2023, it is planned to build a wind farm with a capacity of 1.8 GW. And what is remarkable, Russian companies are engaged in the construction of wind farms. But, unfortunately, the wind turbines themselves are not manufactured on the basis of Russian developments, but on the basis of joint ventures with foreign firms. Rusnano and Rosatom are actively localizing production from towers to gondolas, and in May this year, supplies of wind turbine blades to Denmark began.

IOP Conf. Series: Materials Science and Engineering 1129 (2021) 012065 doi:10.1088/1757-899X/1129/1/012065

The rapid pace of development of wind energy in Russia leads to the fact that in the near future there will be a need to produce own wind turbines, according to the projects of Russian developers. This is not only a cost-effective strategy, but it will also allow for the creation of wind farms, depending on our climatic and landscape conditions.

That is why, when producing our own wind turbines, the presented test bench will be the solution to many problems in measuring and controlling blades.

The test bench is a simple structure equipped with an electric drive for automatic movement of the control device. Thus, an example of a design is shown in figure 1. The stand consists of:

1. the base, which is also used for painting the blade;
2. a mechanism equipped with an electric drive for moving the carriage with a scanning device;
3. carriage on which the measuring device is installed;
4. measurement tool - 3-D scanner;
5. computer for receiving and processing the information received.

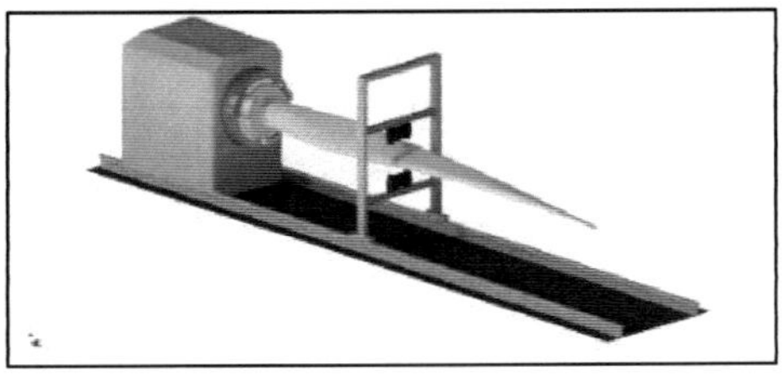

Figure 1. Design of the stand.

The device is moved along the slats. The rail serves to adjust the position of the device attached to the bracket. The measuring instrument is selected taking into account the design features, the shape and dimensions of the measured part, the required measurement accuracy, the metrological characteristics of the instrument, the control performance, etc.

3.2. Means of measurement used in control

The device under development must perform various measurements on the surface of the object in the process of scanning an object. These measurements must then be transferred to a microcontroller to process and prepare the data in the required format. Consequently, the main requirement for a laser meter is the availability of a data interface, through which information about each measurement can be obtained and stored. This interface must be supported by the controller.

The second important requirement for the measuring device is the ability to control the operation through the data interface. This function will allow transmitting data in the form of a command from the controller to the laser device and thereby completely control the process of scanning the object. The microprocessor must at least be able to send a command for one measurement. Therefore, the laser must support the “measure” command transmitted to it via a specific data interface.

For the purposes of our work, the scanner 3D scanner Creaform HandySCAN 307 is suitable.

The blade will be scanned in two sections simultaneously, i.e. to scan the upper and lower parts of the blade, two scanners will be used, which will subsequently carry out the measurement process synchronously.

After scanning and processing digital images, the system receives data from a part of the object's surface in the form of an array ("cloud") of points. For each point, its three-dimensional coordinates are known.

To perform one scan, the time and speed of the carriage is set by the operator sitting at the computer and monitoring the progress of the process.

To automatically combine data, a system of markers is used that are glued to the surface of an object or next to it. During scanning, the coordinates of the center of the markers are calculated automatically.

After the scanning of the entire object is completed, the system processes the received array of points, after which you can start analyzing the results.

3.3. Metrological characteristics

For correct processing of the results of measurements carried out at the stand, it is necessary to take into account the error created by the installation itself. Without taking into account this error, the data obtained should be considered fictitious.

To calculate the total error, we will use formula 1, obtained by the stand, and summarize the errors that affect the measurement result.

$$\Delta = ({\Delta_{instr}}^2 + {\Delta_{meth}}^2 + {\Delta_{rand}}^2 + {\Delta_{sub}}^2 + {\Delta_{stand}}^2 + {\Delta_{fact}}^2 + {\Delta_{design}}^2)^{1/2} \quad (1)$$

The formula shows that the error consists of the following components:

1. Δ_{instr} - instrumental error. In our case, when using a 3D scanner Creaform HandySCAN 307
2. Δ_{meth} - is a methodical error.

Methodological error arises from the deviation Δt_1 and temperature fluctuations Δt_2. It is due to the difference in the coefficients of linear expansion of materials and parts.

$$\Delta_t = d \cdot \theta \cdot 11{,}6 \cdot 10^{-6}, mm \quad (2)$$

Where θ is the temperature regime.

3. Δ_{rand} -Random error - a component of the measurement error that changes randomly during repeated measurements of a given value. Calculated as follows:

$$\Delta_{rand} = tS \quad (3)$$

4. Δ_{sub}- the subjective error in this design due to the use of an automated device with a non-contact measurement method is negligible, so we do not take it into account.
5. Δ_{fact} – the error due to external factors. External error occurs due to non-compliance with normal conditions when measuring in the room GOST 15150-69. In the room where the experiment will take place, it is necessary that the room meets the requirements prescribed in GOST (temperature, humidity, air cleanliness, lighting, sound and vibration isolation, radiation protection, electricity, water, air, heat, refrigerant, etc.).
6. Δ_{stand}- the error of the workpiece. Deviations from the geometric shape and dimensions that occur during the processing of the workpiece must be within the tolerances that determine the maximum permissible values of errors in the size and shape of the part. When machining, ensuring the specified accuracy depends on the choice of technological bases and the installation scheme of workpieces.

The blank error can be written as follows:

$$\Delta_{stand} = (\Delta_{basing}^2 + \Delta_{work\ piece}^2)^{1/2} \quad (4)$$

Δ_{basing} – the basing error. The error of basing occurs as a result of basing the workpiece in the device on technological bases that are not related to measurement bases. When basing on the design base, which is also the technological base, the basing error does not occur.

$\Delta_{work\ piece}$- the accuracy of fixing, the accuracy of the result directly depends on the installation of the part on the base. The fixing error is formed from the surfaces that occur before the application of the clamping force and during clamping.

7. Δ_{design} – the error structure. This error directly affects the measurement results. The error of movement of the carriage. An important component of the test stand design is the carriage, which is used to move the scanning device. This error must be taken into account, as it has a direct impact on the accuracy of the measurement result;

The design error is calculated using the formula (5).

$$\Delta_{design} = ({\Delta_{moving\ thr\ carriage}}^2 + {\Delta_{linear\ movement}}^2)^{1/2} \quad (5)$$

Thus, with more detailed calculations, the error that may occur when using a test bench varies depending on the conditions of external factors from 40-45 microns, which is relatively small in a large-scale production process [16-20].

3.4. The economic benefit

General industrial economic requirements are based on the conditions of competitiveness and economic feasibility in industrial production. The main indicator that determines the economic feasibility of costs for the creation and implementation of the test bench is the annual economic effect.

When calculating technical and economic indicators, the following results were obtained:

- Cost of 3 939 322 rubles.
- The profitability index is equal to 2 (more than 1) at a given discount rate Eн = 10%.
- The payback period of the project is 1.5 years with a productivity of 240 units per year.

As a result of the two main calculations, the calculation of the error and economic indicators, it follows that the proposed solution for monitoring blades over 60 meters using a test bench is not only economically profitable, but also effectively used in the manufacturing process of the blade itself [21-22].

4. Conclusion

As a result of the two main calculations, the calculation of the error and economic indicators, it follows that the proposed solution for monitoring blades over 60 meters using a test bench is not only economically profitable, but also effectively used in the manufacturing process of the blade itself.

The results obtained can be used online at the enterprise. The test bench and the program for processing the results are able to analyze the state of the object in real time, measure and control the required geometric parameters.

Acknowledgments

This work was supported by a grant from the President of the Russian Federation for state support of young Russian scientists, doctors of sciences MD-1209.2020.8. Separate results were supported in the framework of the state task №0705-2020-004 in the field of scientific activity.

References

[1] Statistical Yearbook of world energy 2020 Available at: https://yearbook.enerdata.ru

[2] State program of the Russian Federation Energy Development Available at: https://minenergo.gov.ru

[3] World energy issues monitor 2020 Available at: https://www.worldenergy.org

[4] Farahani B V, Barros F, Sousa P J, Cacciari P P, Tavares P J, Futai M M and Moreira P 2019 A coupled 3D laser scanning and digital image correlation system for geometry acquisition and deformation monitoring of a railway tunnel. *Tunnelling and Underground Space Technology* **91** 102995

[5] Mankowski O and Wang Q 2013 Real-time monitoring of wind turbine generator shaft alignment using laser measurement *Procedia CIRP* **11** 291-5

[6] Wind Turbine Blade Structural Engineering *WE Handbook- 3- Structural Design* 1-16

[7] Boldasov D D, Komshin A S and Syritskiy A B 2019 Method of lathe tool condition monitoring based on the phasechronometric approach *International Russian Automation Conference* (Springer, Cham) pp 753-62

[8] Kiselev M I, Pronyakin V I and Tulekbaeva A K 2018 Technical diagnostics functioning machines and mechanisms *IOP Conference Series: Materials Science and Engineering* vol 312 (IOP Publishing) **1** 012012

[9] Komshin A S and Orlova S R 2016 Control of degradation of construction materials in the course of use for the case of string elements *Measurement Techniques* **59(6)** 589-94

[10] Kolesnikov A G, Cherepanov D S, Chekulaev A V and Mironova M O 2018 Analysis of drive mechanisms for the working stand in periodic cold-rolled pipe mills *Metallurgist* **61(11-12)** 1102-7

[11] Galushkin M G, Yakunin V P, Grigoriantz A G, Dyachkov R G, Funtikov V A and Tretiakov R S 2016 Creation and experimental study of a planar solid-state laser based on a YAG: Yb 3+ crystal with diode edge pumping *Bulletin of the Russian Academy of Sciences: Physics* **80(4)** 440-3

[12] Boldasov D D, Potapov K G and Syritskiy A B 2016 Phase–chronometric diagnostics of metal-cutting lathes *Russian Engineering Research* **36(8)** 668-72

[13] Kolyada Y B, Korolev Y S, Krushnyak N T, Razgulin V G and Yanushkin V N 2010 Prospects and problems of automation, improving response speed, accuracy, and reliability of devices for measuring linear and angular quantities *Measurement Techniques* **53(8)** 858

[14] Kolyada, Y B, Korolev Y S, Krushnyak N T, Razgulin V G and Yanushkin V N 2008 Efficiency gained with the use of scanning in precision photoelectric systems for measurement of displacement *Measurement Techniques* **51(3)** 242-4

[15] Komshin A S, Potapov K G , Syritsky A B and Fomin A E 2020 A system for monitoring hydro and wind power equipment based on intelligent measurement systems and neurodiagnostics *IOP Conference Series: Materials Science and Engineering*

[16] Golubev E A and Isaev L K 2005 Measurements, control, quality *GOST R ISO 5725: Basic provisions. Questions of development and implementation* (Moscow: FSUE STANDARTINFORM) p 136

[17] On fundamentals of labor protection in the Russian Federation *Federal law No. 181-FZ of 17.07.1999*

[18] On technical regulation *Federal law No. 184 of 27.12.2002*

[19] Stepanov S N, Tabenkin A N and Tarasov S B 2012 *Metrological support of production. Valuation parameters and methods of measurement of surface texture* (SPb: Publishing house of Polytechnical University) p 147

[20] On investment activity in the Russian Federation implemented in the form of capital investments *The Federal law from 25.02.1999 N 39-FZ (ed. from 25.12.2018)*

[21] Methodical recommendations on estimation of effectiveness of investment projects and their selection for financing 1994 (Russia Goskomprom) 7-12/47

[22] Patrichny V A Methods for evaluating the economic efficiency of metrological support of production in industries 1985 *Measuring equipment* **8** pp 55-7

IOP Conf. Series: Materials Science and Engineering 1129 (2021) 012066 doi:10.1088/1757-899X/1129/1/012066

Developing of transport airship for delivery oversized cargo to Vostochny Cosmodrome

A A Didkovskij, E D Mukhina and A K Chernov

Bauman Moscow State Technical University, 2-d Bauman Street, 5, Russia, Moscow, 105005

ikruzov@bk.ru

Abstract. The analysis of the world experience in the creation of airships has been carried out and the most actual tasks for such devices in the field of rocket and space technology have been revealed. The appearance of a transport airship with a carrying capacity of 100 tons has been developed and its main technical parameters have been defined.

1. Introduction

It is common in the modern world that spaceports and rocketry manufacturing plants are located at a great distance from one another. Therefore, there is a logistics problem of delivering goods (launch vehicle blocks, spacecraft parts, etc.) from one location to the other. Rocketry goods have a series of inherent distinctive features that only complicate the delivery process. Such features are: oversize and overweight as well as high sensitivity to vibrations, impacts and overloads.

Currently, most widely spread delivery methods are railroad, automobile transportation and water/avia carriage.

Each method has its own strengths and weaknesses and is suitable only for a certain type of goods to be transported. When choosing the appropriate delivery method, one must consider transportation length and time, infrastructure availability (presence of the railroads, highways, rivers, airfields, etc.), mass-inertial properties of the cargo, its responsiveness to impacts, vibrations and overloads.

Due to the development of the lunar program [1] in 2020-2040, heavy and super-heavy rocket blocks would require transportation from the construction plants to the "Vostochny" spaceport. The launch frequency is expected to be low, each cargo would be unique in its characteristics, which leads to economic inefficiency of airplane delivery. Moreover, the sheer size of the cargo, bearing in mind that the construction plants lack seaport availability, usage of water and railroad transportation is deemed to be limited at best [2-4].

Overall, there exists an upcoming problem for the space industry: how to deliver oversized cargo (super-heavy launch vehicle parts, on-ground industry objects, etc.) to the spaceport site. We offer a possible solution to this problem – a transport airship.

The aim of the research: to determine the parameters and to develop the draft construction blueprints of the transport airship suitable for the transportation of the space industry goods from the construction plants to the "Vostochny" spaceport.

The objectives of the research:

- to analyze the existing experience in the transport airship development and to summarize the key construction insights that determine the shape and form the airship (power structure type, aerodynamics, gas to be used as a fuel, engine type, etc.);
- to prepare the calculation of the main airship parameters (the volume of the shells, airship mass, engine power, flight altitude, cruise speed etc.);
- to calculate, using the above-mentioned project parameters, the core characteristics of the main systems (the power structure, steering elements, thermostat system, docking equipment, etc.);
- to develop the main systems design;
- to prepare the updated calculations based on the precise main systems characteristics;
- to determine the economic feasibility of the project as a heavy transport aircraft.

2. Calculation model

The main task of the transportation airship will be the delivery of space rocket parts: first launch vehicle stage (2 units) and third launch vehicle stage (1 unit), with a combined mass of over 90 tons. Secondary task will be to deliver ground infrastructure as well as to perform search and extract missions in the used stage collapse locations. Moreover, the airship would also be used as an emergency rescue vehicle for the evacuation of the returned spacecraft. Radar equipment would be mounted on the airship.

After careful analysis of the existing global experience in the development of such airships [5-12] as well as the assessment of the cargo to be delivered, we have decided upon the main design and technical parameters of the airship. Planned flight altitude is 6-8 km, cruise speed – 100-150 km·h^{-1}, cargo bay measurements – 11.5×12×55.4 m, payload - 100 tones.

It is necessary to state that we have derived the above measurements from the planned mass/volume parameters of the whole airship. At this point, the precise mass/volume parameters of the separate systems are to be calculated.

Atmospheric conditions were claimed from the analysis based on the radio probing in between 0 and 8 km alongside the route Moscow – Samara – "Vostochny" spaceport.

Using the planned velocity, required to reach the cruise speed, we have derived an equation for the airship's engine power:

$$P_{\mathrm{E}} = \frac{4 \cdot (3M_{\mathrm{PL}})}{\eta_{\mathrm{ad}}} \frac{V_{\mathrm{K}}}{t_{\mathrm{V}}} \tag{1}$$

We have also derived the equations for the bearing framework (BF) and functional shell (FS) areas basing on the model shell analysis:

$$S_{\mathrm{BF}}(V) = 1.29 \cdot V^{\frac{2}{3}} \tag{2}$$

$$S_{\mathrm{FS}}(V) = 6.46 \cdot V^{\frac{2}{3}} \tag{3}$$

The fuel mass, as well as the mass of the airship itself, can be calculated using the following equation:

$$M_{\mathrm{F}}(V, v_{\mathrm{K}}, h) = \frac{L}{v_{\mathrm{K}}} \left(C_x \frac{\rho_{\mathrm{air}} v_{\mathrm{K}}^2}{2} V^{\frac{2}{3}} \right) \tag{4}$$

$$M(V, v_{\mathrm{K}}, h) = M_{\mathrm{PL}} + P_{\mathrm{E}}\rho_{\mathrm{E}} + M_{\mathrm{F}}(V, v_{\mathrm{K}}, h) + S_{\mathrm{BF}}\rho_{\mathrm{BF}} + 2S_{\mathrm{FS}}\rho_{\mathrm{FS}} + \rho_{\mathrm{air}} V \tag{5}$$

After that we have solved the question of the optimal flight altitude and the cruise speed. Due to the fact the maximum flight altitude increases with the increase of the hoisting force, maximum altitude should correspond with the maximum speed. The equations of the optimal altitude and the cruise speed are as following:

$$\begin{cases} M\left(V, v_{\mathrm{K}}, h)g = C_y \dfrac{\rho_{\mathrm{air}} v^2}{2} V^{\frac{2}{3}} + \rho_{\mathrm{air}} g V\right) \\ v \cdot C_x \dfrac{\rho_{\mathrm{air}} v^2}{2} V^{\frac{2}{3}} = P_{\mathrm{E}} \end{cases} \tag{6}$$

Aerodynamics scheme is a hybrid one. Thrust is achieved by exploiting the tangential shell overblow. The air leaves the shell in the direction, opposite to the airship's movement. This system allows to reduce the drag and, hence, the fuel usage. Yawing is managed by the difference in the right- and left-side thrust openings, pitching – by the stabilization system.

Wrapping parameters were derived by using the calculations from the Software Complex (SC) SolidWorks Flow Simulation.

First, we have settled the parameters for the shell: quantity of the thrust openings N, width of the openings δ and impeller diameter D_{im}. Flow separation distance is equal to:

$$x_{\mathrm{cr}} = \frac{10^6 v_{\mathrm{of}}}{v_{\mathrm{of}}} \tag{7}$$

The amount of the thrust openings is calculated by the following formula:

$$N = \frac{1}{2}\left[\frac{D_{\mathrm{aao}}}{x_{\mathrm{cr}}}(\pi - \varphi_0) - 1\right] \tag{8}$$

The width of the opening is determined by the major of three factors: the velocity of achieving the cruise speed (thrust), doubled thickness of the alloy and the minimum technologically achievable width:

$$\delta = \max\left[\frac{4Mv_{\mathrm{airship}}}{\pi t_{\mathrm{v}} \rho_{\mathrm{air}} D_{\mathrm{aao}} v^2 \sum_{i=\left(\frac{\varphi_0}{\Delta\varphi}\right)+1}^{N} \sin^2(i\Delta\varphi)}, 50\frac{v_{\mathrm{K}}}{v_{\mathrm{of}}}, \delta_{\mathrm{tech}}\right] \tag{9}$$

The impeller diameter could be determined by the thrust of the entire opening cascade:

$$D_{\mathrm{im}} = 2\left[\frac{D_{\mathrm{aao}}}{v_{\mathrm{im}}} \sum_{i=\left(\frac{\varphi_0}{\Delta\varphi}\right)+1}^{N} v_i \delta_i \sin(i\Delta\varphi)\right]^{\frac{1}{2}} \tag{10}$$

This is used to derive the area corresponding to the impeller. This area, in particular, could be achieved by utilization of several impellers. It is also necessary to make sure that there exists laminar airflow in between the shells, which can be explained by:

$$Re < Re_{\mathrm{cr}} \tag{11}$$

The power scheme consists of two longitudinal gas shells attached on top of the supporting truss. In the middle of the truss, there is the transversal cargo bay. The shells are quasi-rigid. The spiral air-filed pipes support the inner side of the shell. The shell is placed in check by the cords. Therefore, such scheme allows for the minimum mass with a considerable structural rigidity. Moreover, air-filed pipes also serve as an insulator. Shell deflation system, supported by the outer cords, helps to diminish the aerodynamic pressure during docking. When the air fills the pipes, normal forces start to affect the spirals, which aim to enlarge the shell. When the target volume of the shell is achieved, the inner cords, attached to the center axis of the ballonet, prevent further enlargement.

We have taken the random spiral convolution as a calculation basis for the force factors, that appear in the system when the pressure is applied to the said convolution. Thus, we have decided upon the main system parameter – the rigidity. It decreases up to a certain point of the cord movement. Achieving rigidity minimum means that any further decrease of the contact area would not influence the rigidity. In the further project development, it is necessary to bear in mind that the initial cord tension should be chosen greater than the minimum one so that the rigidity of the system would grow with the growth of strain applied to this system.

We have taken into account two situations: when the strain is applied to the center axis of the gas shell and when the strain is distributed to the outer shell surface. We have calculated a preliminary durability of the spiral pipes taking into account a possibility to use two layers of spirals.

In order to assess the insulation properties of the functional shell, we have also taken into account two possible variants of the shell layer: with one or two spiral convolutions.

We have based our airship shell development upon the idea of a "double-glazed window". In the gas shell layer, the air-filled spiral pipe serves as the double-glaze, therefore, when calculating the heat flows inside the spiral pipe, one needs to take into account calculation methods applicable for the heat losses in the barriers with the air-filled cavities.

The calculations were carried out in the MS Excel using the step-by-step approximations. Theoretical calculations were proved worthy in the SolidWorks Simulation software.

To select the optimal size and design of the gas shell layer with the aim of maintaining long-term temperature control, we have calculated the dependence of the heat transfer coefficient from the layer thickness.

3. Results and Discussion

Using the above-mentioned algorithms, we have determined the characteristics and core parameters of the transport airship main systems.

The main carrying gas is the heated air; supporting one is helium. The obvious advantage of using the heated air is its availability and ease of use – the shell could be folded while docked. Helium, on the other hand, could help to increase the payload, making the airship more versatile.

Overall, the main airship parameters are stated in the table 1 below.

Table 1. Results of the calculations.

Parameter	Value
Volume (m^3)	815370
Mass (kg)	394920
Cruise speed ($km \cdot h^{-1}$)	108
Optimal flight altitude (km)	6.28
Length×Width×Height (m)	225.2×120.5×76.6

After calculating the tangential shell overblow, we have derived the following: the flow separation distance and the number of the openings are: $x_{cr} = 0.33$ m, $N = 798$, the width of the opening: $\delta = 1$ mm, the equivalent impeller diameter and the respective area are: $D_{im} = 14.283$ m, $S_{im} = 160$ m^2.

This area is a combined group of 10 impellers, each of 4.3 meters diameter, which in total form the main thrusting engine. The length of the opening k is taken as 20 mm. From one side, the length of the opening allows for a better flow direction, but from the other – increases the friction.

During the assessment of the two-layered spiral pipes durability compared to a single-layered variant, we have derived that final rigidity of the system (compared to initial single-layered calculation) has increased. The radius of the spiral convolution is equal to 0.975 m; the pipe radius is 0.65 m.

After that we have found the optimal cord system when $K_z = 1.2$. The optimal variant is as follows: 20 cords for each ballonet convolution with 71 attachment positions (the distance between the attachments is 3.17 m).

The insulating characteristics of the functional shell were derived from the SolidWorks Simulation software. To select the optimal size and design of the gas shell layer with the aim of maintaining long-term temperature control, we have calculated the dependence of the heat transfer coefficient from the layer thickness $k_l(\delta_l)$. The most suitable scheme appears when using 2 spiral pipes, when the heat transfer coefficient and the layer thickness are $k_l = 0.784$ m and $\delta_l = 0.97$ m.

So, in a crude way, we have derived the main specifications of the transport airship systems. The preliminary airship design is shown in the figure 1.

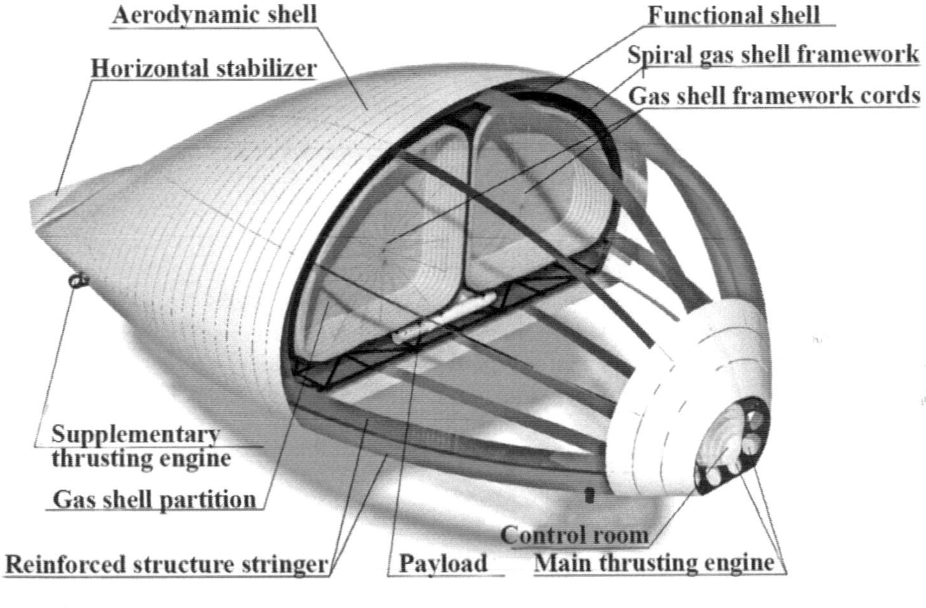

Figure 1. Airship design.

4. Conclusions

Airships have never seen use in the transportation of the space industry cargo. In our project we suggest innovative approach to the power and aerodynamic properties of an airship: quasi-rigid power structure based on the spiral pipe usage, inflated with air under pressure, as well as tangential shell overblow system.

Results of our research:

- current problems in the space industry are analyzed;
- explicit weather conditions alongside the route Moscow – Samara – Vostochny spaceport are reviewed;
- the design of the airship is created, its main aerodynamic characteristics are examined;

- the core parameters of the main systems (the power structure, steering elements, thermostat system, docking equipment, etc.) are derived;
- the structure of the above-mentioned systems is elaborated.

References

[1] 2015 The main principles of space industry state policy of the Russian Federation up to the year of 2030 and further, signed by the President of the Russian Federation on 19.04.2013 *Defence-Industrial complex of Russia* vol 11 pp 502-12

[2] Zolin A and Didkovsky A 2019 Development of the transport airship for cargo delivery to the Vostochny Cosmodrome *Proc. Academic Space Conf. dedicated to the memory of academician S.P. Korolev and other outstanding Russian scientists – Pioneers of space exploration* (Moscow)

[3] Kirilin A 2015 Do new generation airships change a paradigm in transport logistics? *Proc. of the 10th Int. Airship Convention & Exhibition* (Germany: Friedrichshafen) pp 1-14

[4] Kirillin A N 2019 The creation of the new transportation system based on the modern airships suitable for development of the furthest Russian regions *Monthly national aviation magazine «Kryliya Rodini»*

[5] Prentice B E and Knotts R 2014 Cargo airships: international competition *J. of Transportation Technologies* **4(3)** pp 187-95

[6] Kirillin A N 2013 *Airships* (Moscow: MAI-Print)

[7] Grumondz V T, Semenchikov N V and Yakovlevskiy O V 2017 *Airship's aeromechanics* (Moscow: Nauka)

[8] Azouz N, Bestaoui Y and Lemaitre O 2002 Dynamic analysis of airships with small deformations *Proc. of the Third Int. Workshop on Robot Motion and Control* (United States) pp 209-15

[9] Bolkcom C 2006 Potential military use of airships and aerostats *Library of Congress Washington DC Congressional Research Service*

[10] Colozza A 2004 Airships for Planetary Exploration *NASA Technical report*

[11] Dick H and Robinson D 2014 The golden age of the great passenger airships: Graf Zeppelin and Hindenburg *Smithsonian Institution Press*

[12] Kirilin A N, Boldyreva A A, Timushev S F and Tsipenko A V 2016 Thermal calculation of airship hull protection from snow *Global J. of Pure and Applied Math.* pp 602-15

Development of a skid for manual contact welding with the machine movement in the working areas

P V Boslovyak and V V Dvoretsky

Bauman Moscow State Technical University, ul. Baumanskaya 2-ya, 5/1, Moscow Postcode: 105005

boslovyak89@mail.ru

Abstract. There is represented the creation of a skid for manual contact welding. It consists of a metal structure of the skid itself and a movement carriage. Strength and stiffness calculations of the metal structure of the skid and of the carriage were performed. A full-scale sample of the skid assembly result is shown.

1. Introduction

In the process of assembling a panel, which consists of two stainless steel sheets by manual contact welding, there was a need to optimize the movement process of the manual contact welding machine (figure 1) along a given trajectory, without any preparation of a separate template for each panel [1]. It is also necessary to ensure the mobility of this skid and the ability to perform welding works in any suitable place.

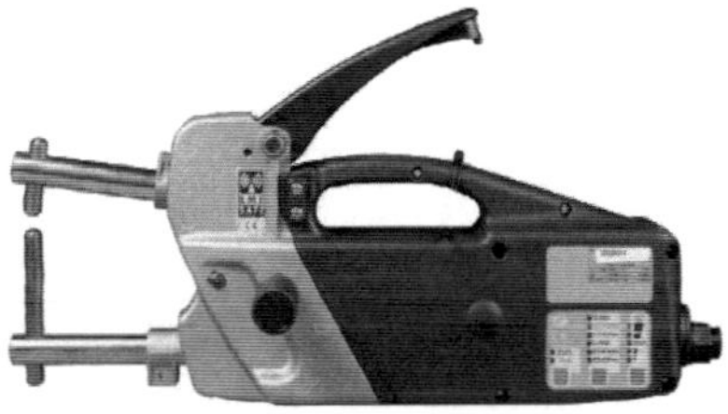

Figure 1. Machine for manual contact welding [1].

To perform the specified work by manual contact welding, the platform must move in the horizontal plane with a certain, adjustable step.

Based on the above, the creation of an experimental skid for a manual contact welding machine is an urgent task. It would give the possibility to identify the specifics of the operation of all skid mechanisms, conduct field studies of its design and in case of necessity - to carry out its upgrading, including replacing the materials employed with other materials with higher physical mechanical characteristics.

2. Design model

The experimental skid for manual contact welding is a metal structure, along the guides of which the carriage moves. During the development process of the design of this skid, it is advisable to take metal

IOP Conf. Series: Materials Science and Engineering 1129 (2021) 012067 doi:10.1088/1757-899X/1129/1/012067

structures of belt conveyors [2-4] and roller conveyors [5-7], which are used in warehouses, as an analogue. In the process of an engineering research, it is also necessary to take into account the development of optimal design [8-13] of existing designs of other machines in relation to the metal structure of this skid.

The experimental skid would be designed for a manual contact welding machine by TECNA [1], the main characteristics of which are represented in table 1.

Table 1. Characteristics of the contact welding machine TECNA.

№	Device characteristics	Value
1	Maximum current	5000 A
2	Power	13000 W
3	Welder plank	4 mm
4	Total welding sides	two
5	Compression drive	manual
6	Refrigerating system	air
7	Dimensions (LWH)	370x100x230 mm
8	Mass	11 kg
9	Electrical stickout	125-500 mm

The creation process of an experimental skid for manual contact welding consists of several stages:

1. Calculation and design of the structure of the skid for contact resistance welding.
2. Creation of a 3D-model with subsequent strength and stiffness calculation in the SolidWorks IT system.
3. Assembling the skid based on the obtained results.

Design and calculation of the structure of the skid for manual contact welding. At the initial stage, a kinematic diagram of the carriage movement mechanism [14] with a central location of the gearbox between the running wheels was preliminarily adopted (figure 2).

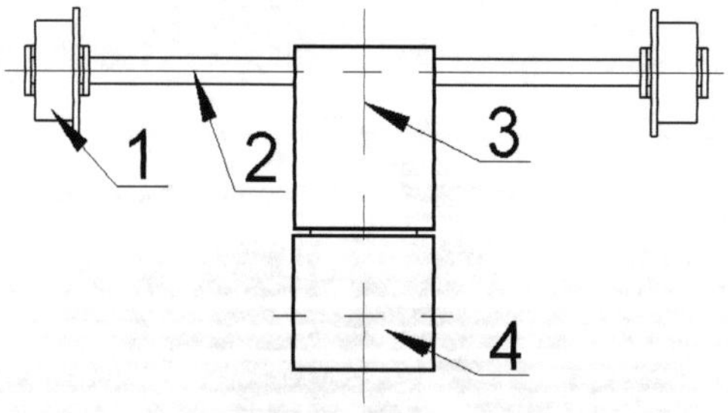

Figure 2. Kinematic diagram of the carriage movement mechanism: 1 - wheel, 2 - drive shaft, 3 - gearbox, 4 - electric motor.

The main characteristics of the approved kinematic scheme are as follows:

- the number of carriage wheels - four;
- let us place the drive mechanism between the drive travel wheels;
- the carriage does not need a braking device;
- the running wheels connected to the drive are driving wheels, the rest wheels are idle.

Further, the values were determined, allowing the selection of mechanisms for the movement of the carriage.

The resistance W to the movement of the carriage during the movement period, reduced to the wheel rim, consists of the friction force between the wheel rim and the surface of the angle W_{fr} = W. It is determined by the formula [14]

$$W = G_c + \frac{2\mu + fd_j}{D} k_{pf}, \tag{1}$$

W_{fr} – frictional drag;
G_c – carriage weight;
μ – coefficient of rolling friction of a wheel on a rail;
f – reduced coefficient of sliding friction in wheel bearings;
d_j – journal diameter;
k_f – flange friction coefficient.
An electric motor was selected, depending on the value of the power N [15].

$$N = \frac{W v_c}{\eta_{pr} \psi_{s.av}}, \tag{2}$$

v_c – carriage speed;
η_{pr} – preliminary value of the efficiency of the mechanism;
$\psi_{s.av}$ – multiplicity of the average starting torque of the engine in relation to the nominal;
The choice of the gearbox was carried out according to the gear ratio [15]

$$i = \frac{n_m}{n_{rw}}, \tag{3}$$

n_m – motor shaft speed;
n_{rw} – rotation frequency of carriage running wheels.
Wheel speed n_{rw} [15]

$$n_{rw} = \frac{v_c}{\pi D_c}, \tag{4}$$

D_c – carriage running wheel diameter.

The 3D-model creation with subsequent strength and stiffness analysis in the SolidWorks software package. The usage of the complex metal structures requires preliminary experimental research and experimental design [17, 18]. Initially, a 3D-model of the skid for manual contact welding was designed in the SolidWorks software package [16] (figure 3), which consists of the skid metal structure and a carriage, moving along it.

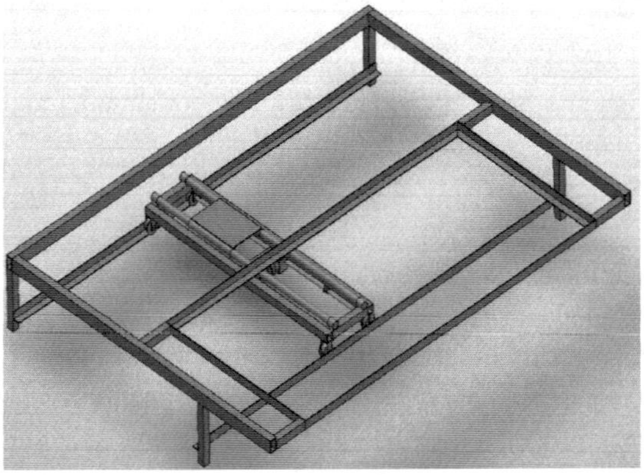

Figure 3. 3D-model of the skid for manual contact welding.

Let us consider and analyze the carriage and metal structure separately.

The constructed initial carriage model (figure 4) was subsequently transformed into a finite element model. The design scheme was formed on the basis this model (figure 5).

The metal structure of the carriage is under the load of 108 N from the contact welding machine acting on the plate. The welding machine is attached to this plate.

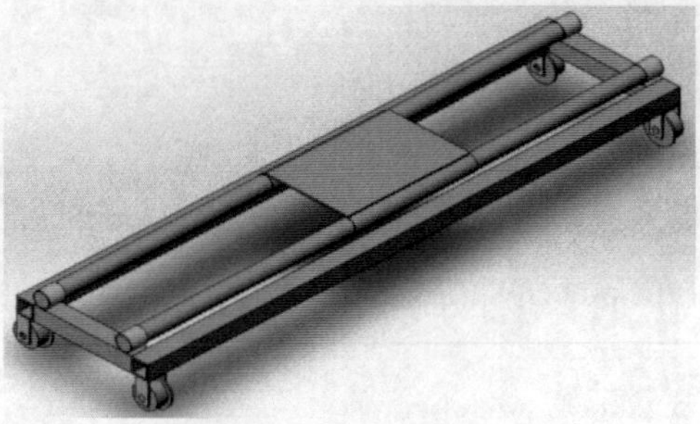

Figure 4. Carriage 3D model.

Figure 5. Design diagram of the carriage.

At the next stage, the stress-strain state of the carriage was calculated. The calculation results are shown in figure 6 [19].

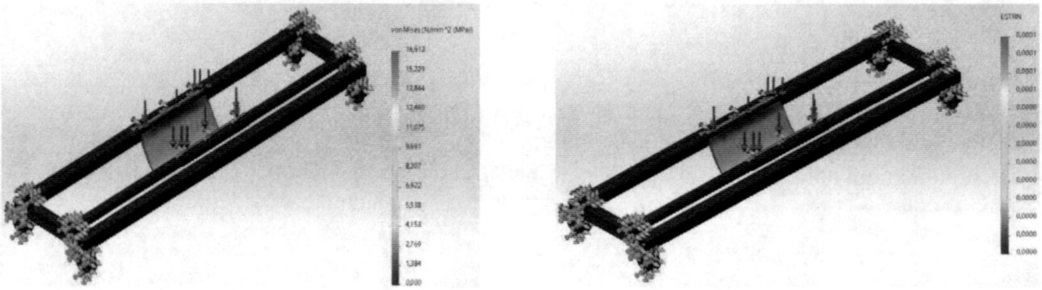

(a) (b)

Figure 6. The result of calculating the carriage: (a) - stress distribution; (b) – strain distribution.

Further, the initial model of the metal structure (figure 7) was transformed into a finite element model and the design scheme of the metal structure was obtained (figure 8).

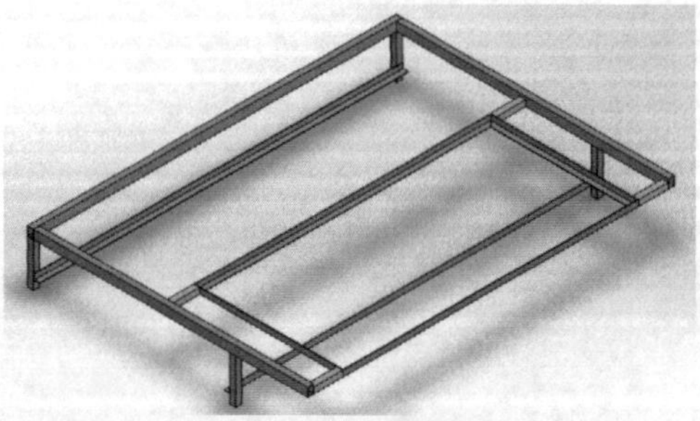

Figure 7. 3D-model of metal structure.

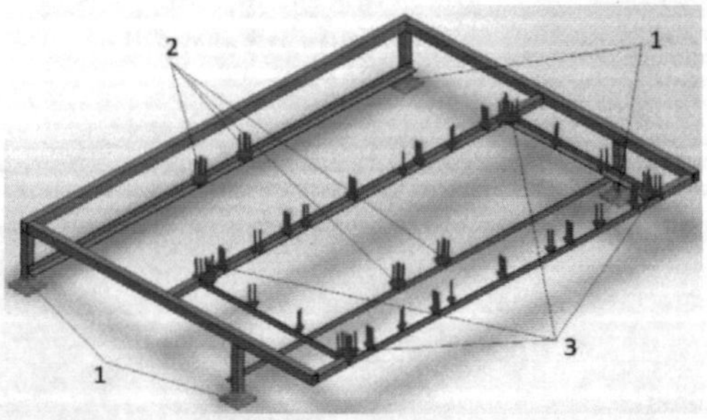

Figure 8. Design diagram of the metal structure: 1 - hard constraint; 2 - loads from the welded panel; 3 - loads from the carriage.

As the places for fixing the metal structure of the skid, hard constraints (rigid supports), located at the base of racks 1 (figure 8), are taken.

There are number of loads on the metal structure of the skid with the below values:

- load from the carriage to the running track - 153 N;

- load from the welded panel - 80 N.

Equal to the carriage, the calculation of the stress-strain state of the metal structure was carried done (figure 9).

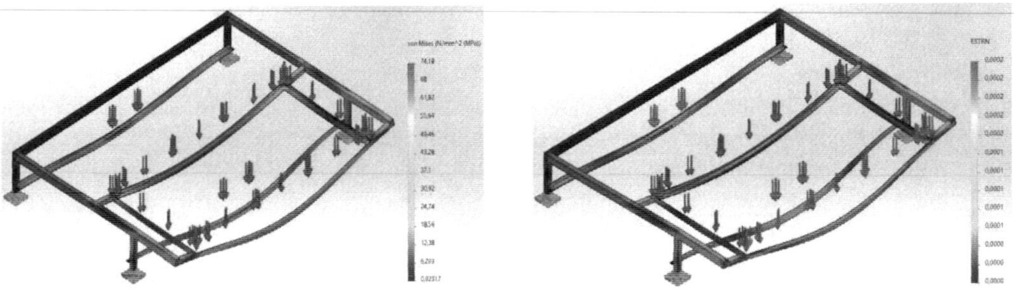

(a) (b)

Figure 9. The result of calculating the metal structure: (a) - stress distribution; (b) – strain distribution.

Assembling the skid based on the obtained results. The metal structure of the manual contact welding skid was made with the ability of a horizontal movement of the working body (figure 10, 11). It should be noted that before purchasing metal for the skid and starting work, it is necessary to study the physical and mechanical characteristics of the metal [20].

(a) (b)

Figure 10. Elements of the skid for manual contact welding: (a) - carriage; (b) - skid metal structure.

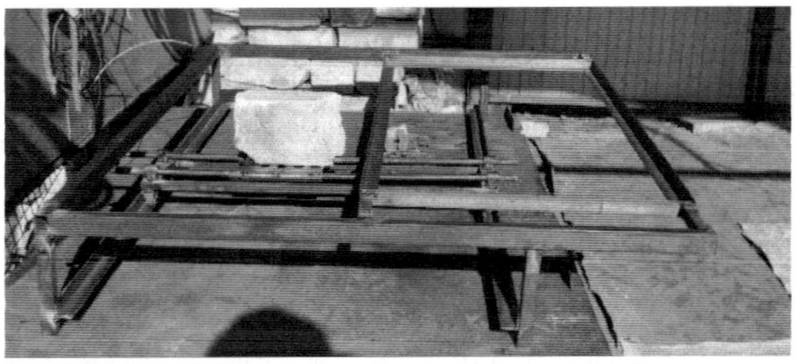

Figure 11. Skid of the manual contact welding machine.

3. Results and discussions

The obtained values of stresses and strains of the carriage and metal structure of the manual contact welding skid are as follows:

- maximum carriage stresses - 16.6 MPa;

- maximum carriage strain - 0.1 mm;
- maximum stresses of the metal structure - 74.2 MPa;
- maximum strain of the metal structure - 0.2 mm.

These values fit the conditions of strength and stiffness.

4. Conclusions

Based of the calculated and projected 3D model, a prototype of the metal structure of the manual contact welding skid was made. Calculations of the stress-strain state of the model revealed an excess safety margin of the carriage and an allowable safety margin of the metal structure. In this regard, in the future, it is advisable to carry out the optimal design of the carriage and to analyses the new values of the stress-strain state of the skid metal structure.

References

[1] Available at: www.tecna.ru (accessed date 09.10.2020)
[2] Kulinowski P 2014 Simulation method of designing and selecting tensioning systems for mining belt conveyors *Archives of Mining Sciences* chapter 59 pp 123-38
[3] Kumar D and Mandloi R K 2013 Analysis & prospects of modification in belt conveyors *Int. J. of Engineering Research and Applications* chapter 3 pp 581-7
[4] Lagerev A V and Dunaev V P 2009 Conveyors with suspension carrying belt - new type of continuous transport machines *Handbook. An Engineering J.* (Bryansk) chapter 10 pp 9-14
[5] Nosko A L, Safronov E V and Soloviev V A 2018 Study of friction and wear characteristics of the friction pair of centrifugal brake rollers *J. of Friction and Wear* (Moscow) chapter 39 pp 145-51
[6] Nosko A L, Safronov E V and Kirillov D V 2019 Selection of materials of friction lining centrifugal brake rollers *J. of Friction and Wear* (Moscow) chapter 40 pp 218-22
[7] Safronov E V and Nosko A L 2019 Method to determine allowable speed for a unit load in a pallet flow rack *Acta Mechanica et Automatica* (Moscow) chapter 13 pp 80-5
[8] Agasiev T and Karpenko A 2017 The program system for automated parameter tuning of optimization algorithms *Procedia Computer Science* chapter 103 pp 347-54
[9] Sakharov M and Karpenko A 2018 A new way of decomposing search domain in a global optimization problem *Advances in Intelligent System and Computing* chapter 679 pp 398-407
[10] Karpenko A and Sakharov M 2019 *New Adaptive Multi-Memetic Global Optimization Agorithm* (Moscow: Herald of the Bauman Moscow State Technical University, Experimental sciences series) chapter 2 pp 17 – 31
[11] Boslovyak P V and Emelyanova G A 2018 *Optimization Mathematical Modeling of the Weight of Metal Structure of Suspended Belt Conveyor Linear Section* (Moscow: IFAC-PapersOnLine) chapter 51 pp 616-9
[12] Boslovyak P V and Lagerev A V 2019 *Optimization of the conveyor transport cost* (Moscow: IFAC-PapersOnLine) chapter 52 pp 397-402
[13] Boslovyak P V and Tolkachev E N 2019 *Metal Construction Optimization of Drive Suspension of Special Belt Conveyor* (Moscow: IFAC-PapersOnLine) chapter 52 pp 198-202
[14] Alexandrov M P, Kolobov L N, Lobov N A, Nikolskaya T A and Polkovnikov V S 1986 *Hoisting machinery* (SPb: Mashinostroenie) p 400
[15] Kazak S A, Duse V E and Kuznetsov E 1989 *Course design of hoisting machines* (Moscow: Higher school) p 319
[16] Available at: www.solidworks.com.ru (accessed date 09.10.2020).
[17] Melnikov N P 1976 *Steel structures Constructor's reference 3rd ed* (Moscow: Stroyizdat) p 238
[18] Biryulev V V, Koshin I I, Krylov I I and Sylvestorov A V 1990 *Design of metal structures* (Moscow: Stroyizdat) p 432
[19] Available at: www.autocad-lessons.ru/solidworks/stati-solidworks (accessed date 09.10.2020).
[20] Zolotorevsky V S 1983 *Mechanical properties of metals* (Moscow: Metallurgy) p 352

IOP Conf. Series: Materials Science and Engineering 1129 (2021) 012068 doi:10.1088/1757-899X/1129/1/012068

Design of mechanisms of a robotic system for rehabilitation of the lower limbs

A A Voloshkin, L G Virabyan, V A Mamaev and L R Kholoshevskaya

Belgorod State Technological University named after V.G. Shukhov, 46 Kostyukova street, Belgorod, 308012, Russia

rl_bgtu@intbel.ru

Abstract. The article describes the main stages of developing a robotic system for the rehabilitation of the lower limbs based on the "tripteron" robot. The structure of a robotic system has been designed, consisting of a 3-PRRR parallel mechanism, which provides the angles of rotation of all joints of the patient's leg required for rehabilitation and a passive orthosis for supporting the limb. At the first stage, the positions of the active mechanism links are determined. The output link of only one kinematic chain, making a translational movement in the vertical direction, will experience a maximum load. A design diagram of this kinematic chain was built, and the reactions of the supports were determined at the second stage. At the third stage, kinematic dependences were obtained, and an engine was selected that would provide the required torque at the required speeds.

1. Introduction

There are currently many problems in practical health care, the optimal solution to which is the use of robotic tools. These tasks concern the treatment and rehabilitation of patients with disorders of the musculoskeletal system and the implementation of the functions of their self-service, social adaptation, and replenishment of the lost motor and communication functions. It should be noted that the need to create robotic aids for people with disabilities is steadily growing.

In any of the rehabilitation procedures, the treatment of patients with disabilities with impaired motor functions of the lower and upper limbs occurs in a critical sitting or lying position, since patients at this stage cannot control the movements of their limbs. Treating them using only the BWS body weight support system is difficult, as it requires a certain level of physical fitness. Continuous passive motion (CPM) is one of the most common therapies in the early stages of treatment when patients have low or uncontrollable limbs. Many CPM-based mechanisms intended for continuous or repeated therapeutic treatment cannot provide the required quality when performing continuous passive movement due to their complex design, high dynamic loads, and structural cumbersomeness. These mechanisms include the commercially available devices for rehabilitation Motion maker [1] (Switzerland), and the Lambda mechanisms [2], the professional trainer Supine [3], NEUROBike [4], Physiotherabot [5], and AIST-Tsukuba [6]. The relevance of stationary rehabilitation robotization is due to the low cost relative to solutions such as exoskeleton [7]. The low-price segment of rehabilitation equipment consists of systems that provide circumferential movement of the legs, due to the pedal drive, this method covers only a small part of the required operations in gait rehabilitation. On the other hand, the exoskeleton is the ultimate tool for rehabilitation, but the cost of such equipment is estimated at millions of rubles. The article proposes a system for rehabilitating the lower limbs based on a parallel manipulator [8-12]. The proposed solution will occupy a separate niche as a multifunctional, affordable device for rehabilitation.

2. Desine model

In figure 1 shows a block diagram of the proposed rehabilitation system. In the rehabilitation system structure, an active 3-PRRR parallel robot of the tripteron type proposed by Kong and Gosselin [13] and a passive orthosis were used. The active 3-PRRR parallel mechanism provides the required rehabilitation angles of all the patient's joints, the passive orthosis is used to support the patient's lower limb. The active mechanism consists of three kinematic chains $A_iB_iC_iD_i, i=3$ and has three degrees of freedom - translational movements along each of the axes. The rehabilitation platform's position, which is an equilateral triangle $D_1D_2D_3$ with a center at point P and a radius of the circumscribed circle R, is determined by linear displacements $q=(q_1q_2q_3)$. We introduce the following notation: a_i - the distance between points A_i and B_i, b_i – between B_i and C_i, c_i – between C_i and D_i, d_i – between B_i and D_i. A serial RRRR mechanism is used to secure the human leg. Its hinges match the patient's joints. The E hinge provides two movements of the hip joint: rotation in the sagittal plane with angle α and abduction of the leg with angle γ - between the projection of the EF link on the XOY plane and the OY axis. In the hinge F of the knee joint, the rotation of the FP link relative to EF by an angle θ is provided. In the process of rehabilitation, it is required to provide angles in the hip, knee and ankle joint when walking: $\alpha \in [-20°,10°]$, $\theta \in [120°,180°]$, $\gamma \in [0°,25°]$ [14].

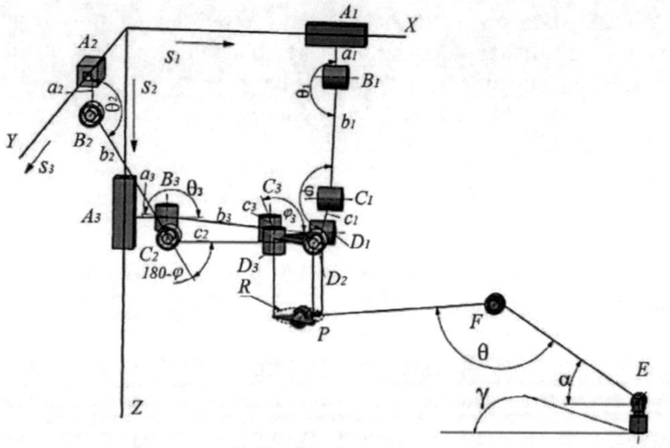

(a)

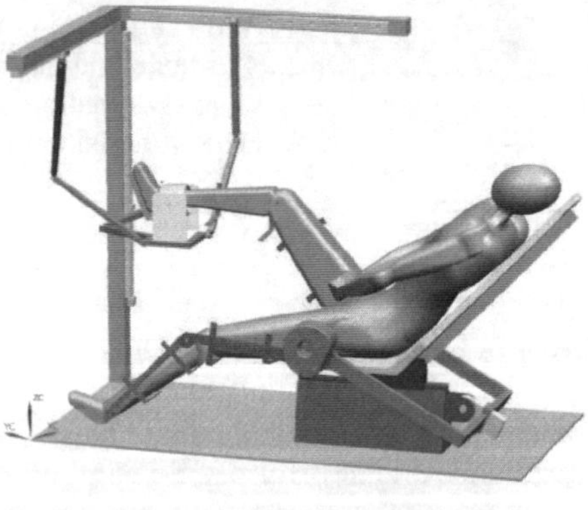

(b)

Figure 1. Lower limb rehabilitation system: (a) - block diagram of the system, (b) - 3D model of the system.

At the first stage of designing a rehabilitation system, it is important to determine the maximum useful work that the robot should provide. Based on the structural diagram (figure 1), we can conclude that with the links' unification, the maximum load falls on the link that moves along the Z axis. This is because, in one of the directions, this link will perform work aimed at overcoming gravity, as opposed to the remaining links, the direction of movement of which is perpendicular to the gravity vector. Thus, if we carry out the strength calculation of the kinematic chain 3, and the solution satisfies the conditions, it can be argued that the remaining links with a margin will be provided with the necessary engine operation. The robot's kinematic scheme is built so that the working area is 10-15% larger than the trajectory described by the human foot while walking, with a person's height up to 190 cm [14].

Consider the system in a static state. The link's peak load will be provided at the maximum extension of the kinematic chain 3 (figure 2).

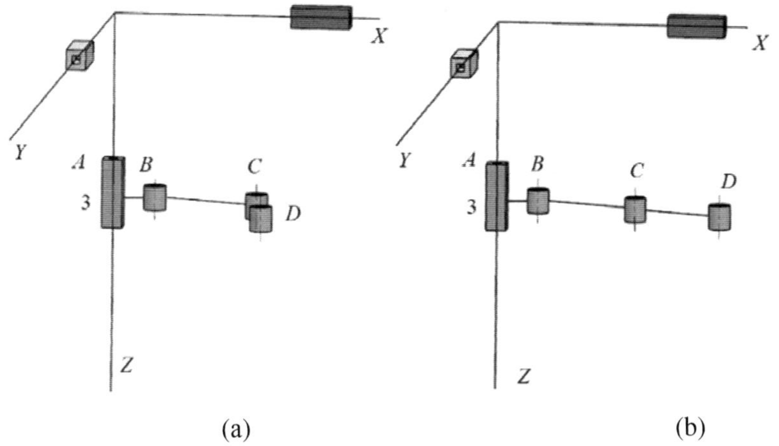

Figure 2. Position of the kinematic chain: (a) - arbitrary; (b) - at maximum load.

Let's build a design scheme to determine the reaction of the support of the input link. This scheme can be represented as a static beam with a rigid fixing. Thus we determine the static load on the structure, taking into account the metal consumption and the patient's weight (figure 3).

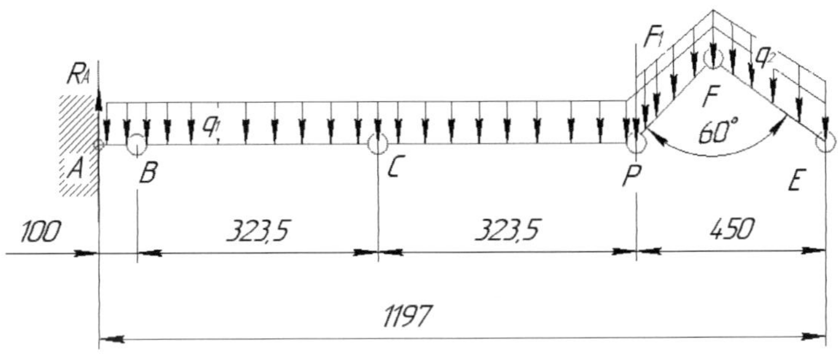

Figure 3. Design diagram of the kinematic chain 3 connected to a passive orthosis.

To calculate the structure, it is necessary to set some parameters of the proposed robot. The load is generated as follows:

q_1 – distributed load throughout the entire structure link, reflecting the own weight of the elements to which the hinge *B* belongs, the profile pipe 2 mm thick and about 323.5 mm long - the *BC* link, the *C* hinge, the profile pipe 2 mm thick, and about 323.5 mm long - *CP* link, orthosis *PE* - consists of a hinge and two profiled pipes with a thickness of 2 mm and a total length of about 900 mm. The calculation uses a 40x40 mm pipe profile. Thus, we obtain a structure mass of about 8.5 kg, which is approximately equal to 0.07 $kN \cdot m^{-1}$ distributed load on a section of 1.2 m;

F_1 – the force acting on the pivot *P*. This force replaces the weight of the brace attachment elements as well as the movement mechanism of the ankle, which will include the servo, pivot, support and restraint. The total weight of the elements should not exceed 5 kg;

F_2 – the force acting on the hinge *P*. It takes into account unexpected loads such as plaster and all kinds of tires;

R_A – support reaction at point A;

M_A – bending moment about point A;

q_2 – distributed load throughout the orthotic, including the mass of the rehabilitated limb. We find the mass of the patient's lower limb by the formula [15]:

$$m_x = B_1 + B_1 m + B_2 H, \tag{1}$$

Where m is the mass of the whole body, kg; H – body length, cm; $B_0 B_1 B_2$ – coefficients of the regression equation. Substituting the tabular values, the average leg mass for a person with a height of 190 cm is 15 kg, taking into account a margin of 20%. We take the effort required to bend the knee, 0.5 kN·m, on a plot of 0.45 m. As a result, we get the input data for the calculation: $L = 1.2$ m; $q_1 = 0.07$ kN·m^{-1}; $q_2 = 0.5$ kN·m^{-1}; $F_1 = 0.05$ kN; $F_2 = 0.02$ kN.

The design of the orthosis is unknown at this stage of design. Suppose the maximum flexion angle is no more than 60°. Consequently, the structural elements form an equilateral triangle with the center of mass located in the center. It follows from this that it is possible to simplify the design scheme for preliminary calculation. The links PF and FE are represented as a straight section of PE with a length z_2 compensating for the load q_2 (figure 4). A more detailed calculation can be made after an accurate determination of metal consumption and design optimization.

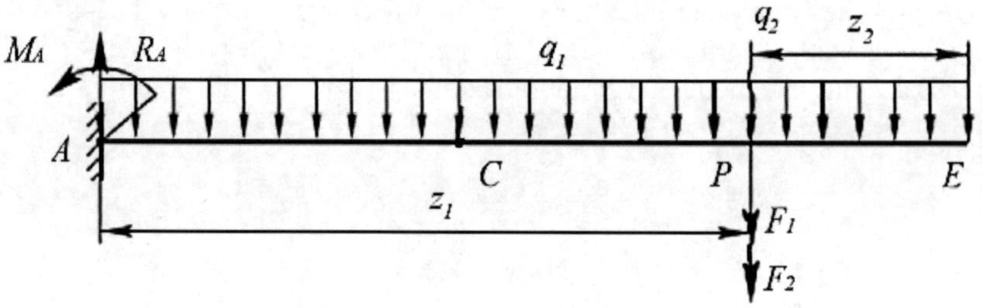

Figure 4. Design model for determining shear force and bending moment.

We compose the equations of static equilibrium [16]:

$$\sum F_y = -q_1 \cdot 1.2 - q_2 \cdot 0.45 - F_1 - F_2 + R_A = 0 \tag{2}$$

Plot №1 ($0 \le z_1 \le 0.747$ m)

$$Q_y = R_A - q_1 \cdot z_1,$$

$$M_x = R_A \cdot z_1 - M_A - q_1 \cdot \frac{z_1^2}{2}.$$

Plot №2 ($0 \le z_2 \le 0.45$ m)

$$Q_y = q_1 \cdot z_2 + q_2 \cdot z_2,$$

$$M_x = -q_1 \cdot \frac{z_2^2}{2} - q_2 \cdot \frac{z_2^2}{2},$$

where Q_y - shear force, kN; M_x - bending moment, kN·m.

As a result of solving the equations, we construct diagrams of transverse and bending stresses.

The main elements of this robot are the ball screw, as well as the guide and carriage. Since the calculation performed shows a static load, it is worth considering a margin of 20-30% when choosing components. Thus, based on the list of products, the most suitable is LLTHC 30 U Bottom Bracket.

After determining the link's load, it is necessary to determine the engine power to provide the required torque. In this robot, the translational movement of the link is provided by rotating the screw. Therefore, the required torque will depend on the gear ratio of the ball-screw transmission, and thus the pitch and diameter of the screw become the key variable. Since the screw's pitch will affect the linear speed of the slide, it is necessary to carry out an analytical calculation that will allow us to understand the required engine power at a given speed. The model SFU01605-4 was chosen as the ball screw (ball screw). Let's set the initial parameters: screw diameter $d_S = 0.016$ m; screw thread pitch $P_B = 0.01$ m; coefficient of friction $\mu = 0.2$; screw length $l_s = 0.7$ m; efficiency applied force $\eta = 0.88$, taking into account a margin of 30% N = 517 N; travel time $t = 3$ s.

The ball screw pitch P_B is the distance the nut moves when the shaft is turned one revolution, i.e. gear ratio of ball screws. Using the gear ratio, we calculate the engine torque M_l required to apply a given force N to the node:

$$M_1 = \frac{N \cdot P_B}{2\pi \cdot \eta}, \text{N·m} \tag{3}$$

To determine the number of engine revolutions, it is necessary to determine the slider's speed of linear movement. The angular velocity of link 2 will be equal $\omega = 0.7$ rad·s^{-1}[15].

We compose a design scheme for calculating the speed of the links (figure 5). Since points P and D have the same speed of movement, it is enough to find the speed of one point, therefore, we represent point P as a hinge fixed on the translational pair 3, which will allow us to determine the dependence of linear and translational speeds. The lengths of links 1 and 2 are 0.38 m.

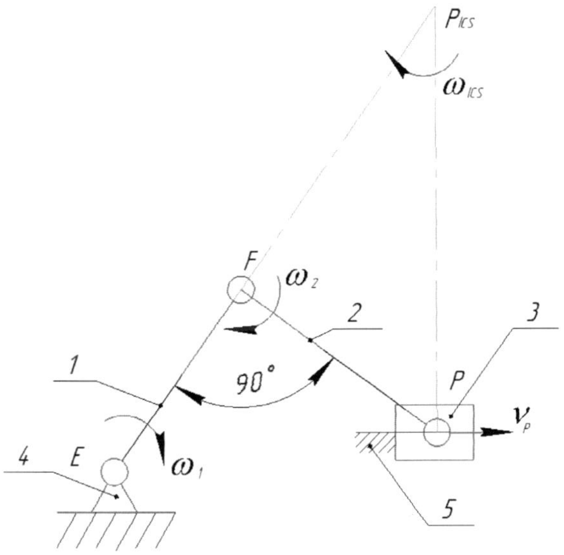

Figure 5. Design scheme of a passive orthosis.

Thus, the speed of linear movement of the slider 3 can be determined using the method of the instantaneous center of speeds. Point P_{ICS} - instantaneous center of speeds (ICS) for the FP link, we find the distance to the ICS:

$$P_{ICS}P = \sqrt{FP^2 + EF^2},\ \text{m} \tag{4}$$

$$\upsilon_P = \omega_{FP} \cdot P_{ICS}P,\ \text{m·s}^{-1} \tag{5}$$

Then the required number of engine revolutions n at a given speed of movement v_P is calculated by the formula:

$$n = \frac{\upsilon_P}{P_B},\ \text{rpm} \tag{6}$$

The gearbox is not used in the design, therefore,

$$n_l = n_m, \tag{7}$$

where n_l is the rotational speed of the screw, n_m is the speed of the motor shaft.

Calculate the load power:

$$P_0 = \frac{2\pi \cdot n_m \cdot M_l}{60},\ \text{W} \tag{8}$$

3. Results and discussion

On the constructed diagrams (figures 6, 7), the distribution of the load on the robot kinematic chain is clearly shown, it can be concluded that the required metal consumption of the chain in each of its sections.

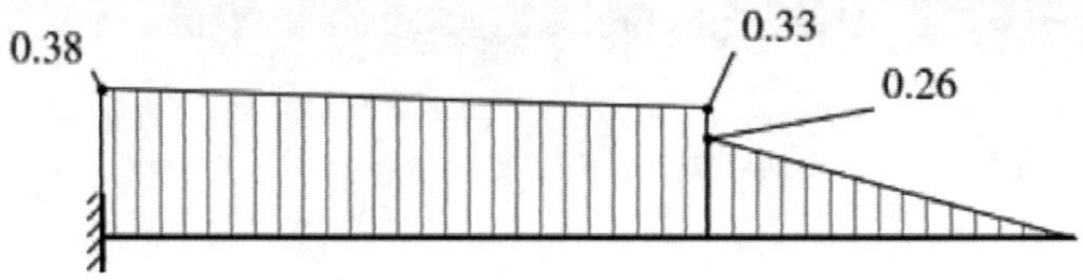

Figure 6. Diagram Q_y (shear force).

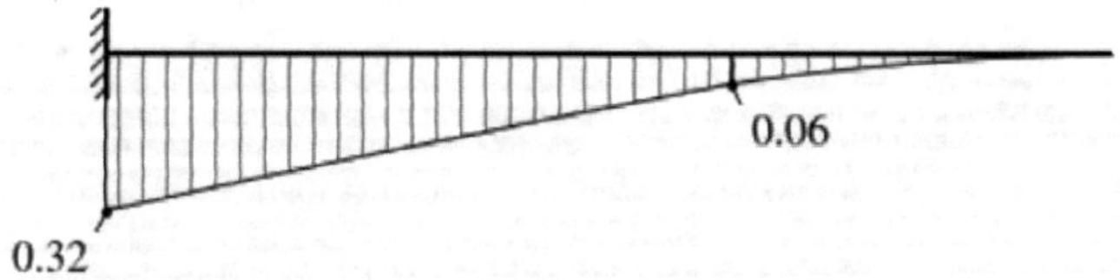

Figure 7. Diagram M_x (bending moment).

The maximum shear force is 0.38 kN. Given the input characteristics, we will select the engine to provide the required performance. Thus, 2250 rpm characterizes the speed of rotation of the engine while providing a force of 0.93 kN·m. These characteristics can be achieved with both a stepper and a servo motor. However, in the case of a stepper motor, maximum torque is obtained at low rpm, and the servo motor provides maximum torque at high rpm. Rehabilitation of patients is often carried out at a relatively low speed, but significant efforts are made to flex the patient's joints with cerebral palsy. Consequently, the priority in the selection of equipment is motors with a predominance of parameters, similar to stepper types.

We will choose the SME60S-0040-30AB model as the engine, the technical characteristics of which are shown in figure 8.

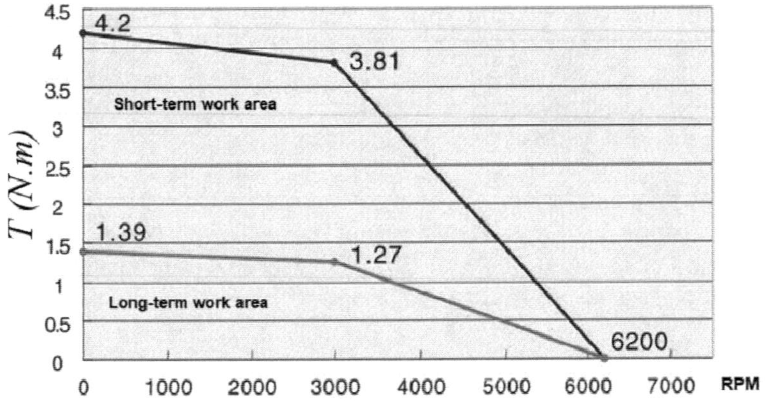

Figure 8. Dependence of the torque on the engine speed diagram.

4. Conclusion

As a result of the design of the robot, it was determined that the kinematic chain 3 was maximally loaded, while a peak stress is arised when the kinematic chains 1 and 2 reach their extreme positions other than zero. The maximum transverse stress falls on the input link of the kinematic chain 3 and is 0.38 kN, and the bending moment in this link was 0.32 kN·m. It was calculated that with a screw pitch of 10 mm, the engine must reach 2250 rpm so that the tripteron can provide the orthotic flexion rate equal to $\omega = 0.7$ rad·s^{-1}. Since the task of rehabilitation of the lower limbs requires the preservation of torque at low engine speeds and at an estimated 2250 rpm, the SME60S-0040-30AB engine was chosen, which would also provide a power reserve for short-term dynamic loads.

Acknowledgments

This work was supported by the Russian Science Foundation, Agreement No. 19-19-00692.

References

[1] Schmitt C, Metrailler P and Al-Khodairy A 2004 The Motion Maker: A rehabilitation system combining an orthosis with closed-loop electrical muscle stimulation *Proc. of the 8th Vienna Int. Workshop on Functional Electrical Stimulation* pp 117-20

[2] Bouri M, Le Gall B and Clavel R 2009 A new concept of parallel robot for rehabilitation and fitness: The Lambda *IEEE Int. Conf. on Robotics and Biomimetics* pp 2503-08

[3] Lim F M, Foong R and Yu H 2014 A supine gait training device for stroke rehabilitation *Proc. of the 15th Int. Conf. on Biomedical Engineering* **43** pp 512-15

[4] Monaco V, Galardi G, Coscia M, Martelli D and Micera S 2012 Design and evaluation of NEUROBike: a neurorehabilitative platform for bedridden post-stroke patients *IEEE Transactions on Neural Systems and Rehabilitation Engineering* **20** pp 845-52

[5] Akdoğan E and Adli M A 2011 The design and control of a therapeutic exercise robot for lower limb rehabilitation: Physiotherabot *Mechatronics* **21** pp 509-22

[6] Homma K, Fukuda O, Sugawara J, Nagata Y and Usuba M 2003 A wire-driven leg rehabilitation system: Development of a 4-dof experimental system *Proc. of the Int. Conf. on Advanced Intelligent Mechatronics* pp 908-13

[7] Vorobyev A A, Andryushchenko F A, Zasypkina O A, Solovyova I O, Krivonkin P S and Pozdnyakov A M 2015 Terminology and classification of exoskeletons *Bulletin of the Volgograd State Medical University* **55(3)** pp 71–78

[8] Mohan S, Mohanta J K, Behera L, Rybak L A and Malyshev D I 2019 Robust operational-space motion control of a sitting-type lower limb rehabilitation robot *Advances in Intelligent Systems and Computing. Advances in Artificial Systems for Medicine and Education III* **1126** pp 161-72

[9] Malyshev D I, Nozdracheva A V, Dubrovin G M, Rybak L A and Mohan S 2020 A numerical method for determining the workspace of a passive orthosis based on the RRRR mechanism in the lower limb rehabilitation system *New Trends in Mechanism and Machine Sci.* **89** pp 138-45

[10] Diveev A I, Dubrovin G M, Malyshev D I and Nozdracheva A V 2019 Geometric parameters and workspace optimization of sitting-type lower limb rehabilitation robot *Industrial and Robotic Systems* **86** pp 279-89

[11] Mohan S, Sunilkumar P, Rybak L, Malyshev D, Khalapyan S and Nozdracheva 2020 A conceptual design and control of a sitting-type lower-limb rehabilitation system established on a spatial 3-PRRR parallel manipulator *Advances in Service and Industrial Robotics* **84** pp 345-55

[12] Malyshev D, Rybak L, Behera L and Mohan S 2020 Workspace modelling of a parallel robot with relative manipulation mechanisms based on optimization methods *Robotics and Mechatronics* **78** pp 151-63

[13] Kong X and Gosselin C M 2002 Kinematics and singularity analysis of a novel type of 3-CRR 3 DOF translational parallel manipulator *The Int. J. of Robotics Research* **21(9)** pp 791–8

[14] Malyshev D I, Nozdracheva A V and Kholoshevskaya L R 2020 Identification of the parallel 3-PRRR manipulator parameters considering the workspace boundaries and the passive orthosis movements *J. of Physics: Conf. Series* **1582(1)** 012062

[15] Zatsiorsky V M, Aruin A S and Seluyanov V N 1981 *Biomechanics of the human motor apparatus* (Moscow: Physical culture and sport) p 143

[16] Feodosiev V I 1999 *Strength of materials* (Moscow: MSTU name after N.E. Bauman) p 590

Calculation and comparative analysis of bucket of the belt elevator

P V Boslovyak and V R Shagimardanov

Bauman Moscow State Technical University, ul. Baumanskaya 2-ya, 5/1, Moscow Postcode: 105005

boslovyak89@mail.ru

Abstract. The calculations of the bucket of a belt elevator for two working cases are presented - at the moment of scooping up the load by the bucket and at the moment of transporting the load into the vertical direction. A 3D model of the bucket was developed in the SolidWorks software package. The calculations of strength and stiffness of the bucket were performed. A comparative analysis of one size range of a bucket with different thicknesses of its walls is carried out.

1. Introduction

Currently, there is a continuous improvement and development of engineering directions, involving the robotic industry [1], various transport systems [2-4], including friction and wear in the nodes of conveyor transport [5, 6].

Continuous transport machines are one of the main means for the bulk cargo transportation in various industries. During the operations at mining enterprises, different kinds of this type of transport could be involved into the conjunction with the construction and road machines that perform various operations for handling, sorting and ragging the transported cargo.

In the process of the vertical handling of bulk cargo, it is advisable to use bucket elevators [7-9] of various types (belt and chain), with the metal buckets as working (carrying) bodies (figure 1). Depending on the pitch of their attachment to the traction elements (for belt elevators - a conveyor belt, for chain elevators - a chain or a pair of chains) and the height of the elevator itself, which on average reaches 40 meters [7-9], the number of buckets would differ.

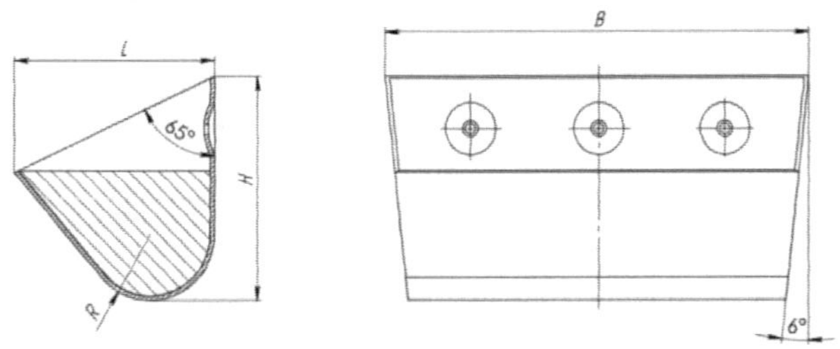

Figure 1. Design sketch of a deep type metal bucket [4].

Therefore, it is advisable to carry out the optimal design of the metal structure of the elevator [10-15], which consists in reducing the mass of the bucket while meeting the conditions of strength and rigidity of its structure, as since it is an urgent task at the design stage of this machine. Achievement of the set task would lead to the decrease in the load acting on the metal structure of the elevator and, as a consequence, to the decrease in its stress-strain state.

2. Design model

A deep type G bucket (figure 1) with the following characteristics was chosen as the load-carrying element of the belt elevator: width B = 500 mm; overhang L = 235 mm; height H = 255 mm; radius R = 75 mm; wall thickness t = 4 mm; material - carbon sheet steel; capacity - 12 litres. The transported cargo is oats with a density of $0.6\ t \cdot m^{-3}$.

During the operation of the belt elevator, at first the buckets scoop up the load, after that - they move it into the vertical direction. It follows from the above, that the design of bucket of the belt elevator must be carried out for two working cases.

When the bucket scoops up the load, specific resistance arises, acting evenly on the leading edge of the bucket

$$W_{sc} = K_{sc} \cdot q_l \qquad (1)$$

K_{sc} – load coefficient;
q_l – distributed load from the mass of the cargo.

In the process of vertical movement of the load, the pressure of the load is carried out on the bottom P_{bottom} [16, 18], side vertical walls P_{side} and on the inclined walls of the bucket P_{iw}.

$$P_{bottom} = \frac{m_c \cdot g}{A_{bottom}}, \qquad (2)$$

m_c - weight of cargo in the bucket;
g – gravitational acceleration;
A_{bottom} – bucket bottom area.

Pressure on the side vertical walls of the bucket P_{side} [17-19]

$$P_{side} = k \cdot P_{botton}, \qquad (3)$$

k – lateral ratio.

Pressure on the bucket inclined walls[17-19]

$$P_{iw} = [1 - (1 - k) \cdot \sin\beta] \cdot P_{bottom}, \qquad (4)$$

β – bucket sidewall angle.

Bucket gravity:

$$F_{buck} = m_{buck} \cdot g, \qquad (5)$$

m_{buck} – bucket weight.

After sequentially performing calculations on the dependences (1) - (5), the following results were obtained for further research: W_{sc}= 137 N, P_{bottom} = 858 N; P_{side} = 378 N; P_{iw} = 545 N; F_{buck} = 95 N.

A 3D model of a belt elevator bucket was created in the SolidWorks software package (figure 2). Later, it was transformed into a finite element model (figure 3).

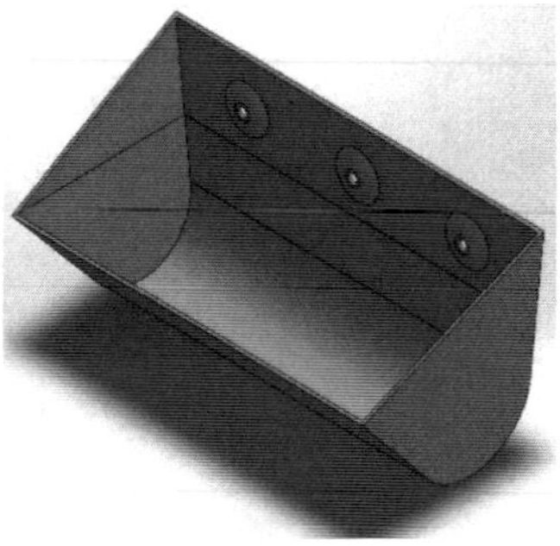

Figure 2. 3D-model of the belt elevator bucket.

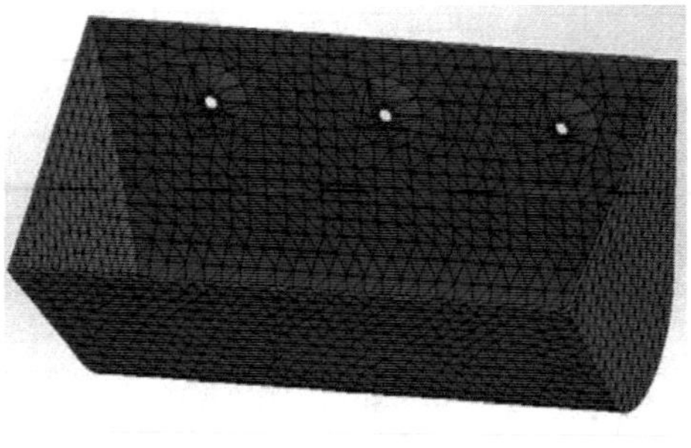

Figure 3. Finite element model of a belt elevator bucket.

On the basis of the obtained model, the design scheme of the bucket was formed for each of the two working cases. The number of rigid restrictions were set (places of attachment of the bucket to the belt) and loads (the gravity of the bucket and the external effect on the bucket from the load side) (figures 4, 5).

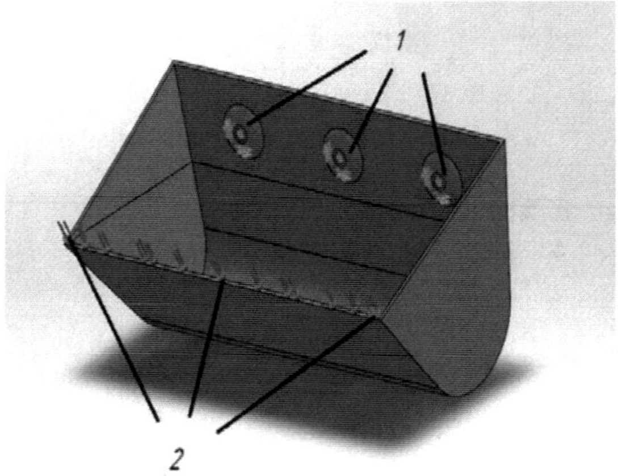

Figure 4. Design scheme for scooping up the load with a bucket: 1 - hard constraint; 2 - external impact on the leading edge of the bucket

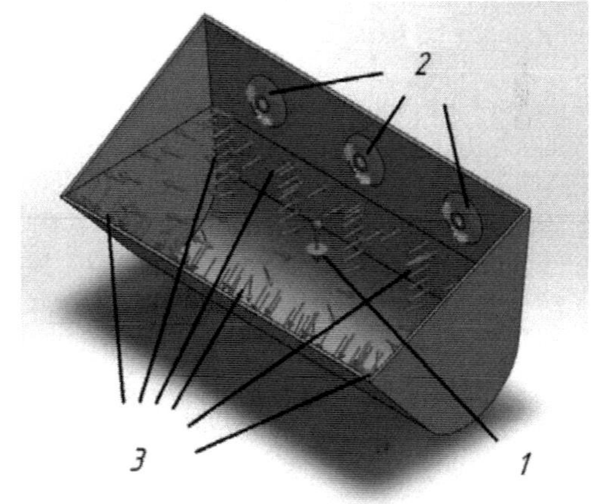

Figure 5. Calculation diagram of the vertical movement of the load by the bucket: 1 - the gravity of the bucket; 2 – hard constraint; 3 - pressure on the bottom and walls of the bucket

For two working cases, the strength and stiffness calculations of the bucket were carried out. The results of stresses and strains are shown in figures 6, 7.

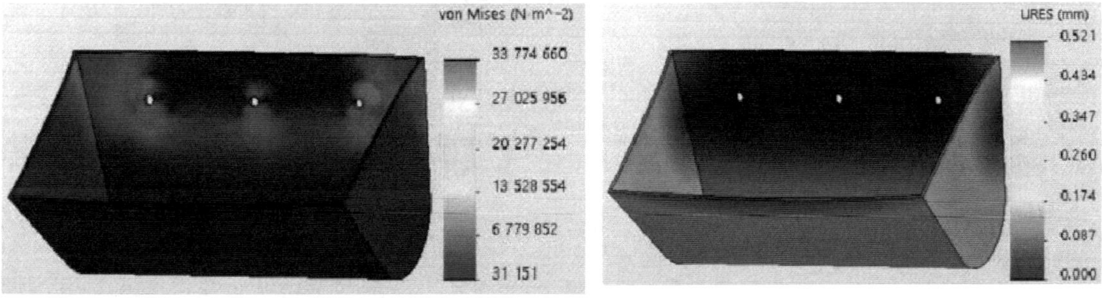

(a) (b)

Figure 6. Result of calculating the bucket during scooping up the load: a - stress distribution; b - strain distribution.

IOP Conf. Series: Materials Science and Engineering 1129 (2021) 012069 doi:10.1088/1757-899X/1129/1/012069

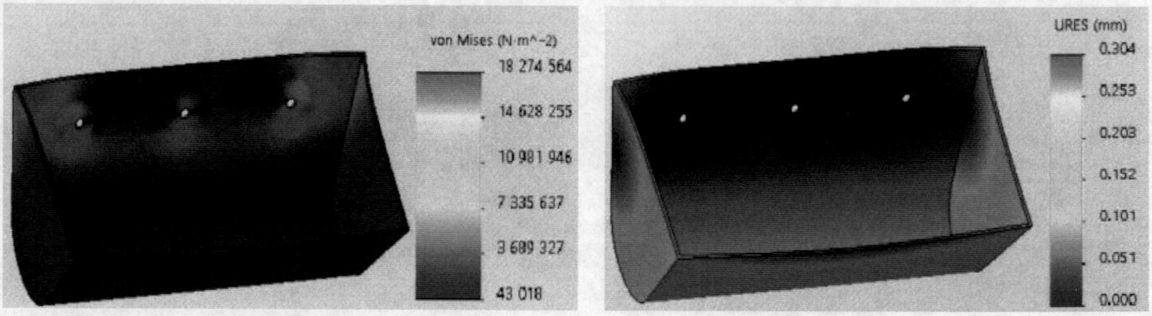

(a) (b)

Figure 7. Result of calculating the bucket during the vertical movement of the load: a - stress distribution; b - strain distribution

The calculations were done for a bucket with a wall thickness of 4 mm.

3. Results and discussion

As it could be seen from the results of the calculation of the bucket, the maximum stresses during the scooping up the load occur at the points of the bolted connection of the bucket with the conveyor belt. They are equal to 37.8 MPa. The maximum stresses during the vertical movement of the load also arise at the points of the bolted connection of the bucket with the conveyor belt. They are equal to 18.3 MPa.

By analogy, the structures of the bucket with wall thicknesses of 3 and 2 mm were calculated. The results of the stress-strain state are represented in table 1.

Table 1. Results of the stress-strain state of the bucket.

№	Wall thickness (mm)	Stress (MPa)		Deflections (mm)		Bucket weight (kg)
		during scooping up the load with a bucket	during the vertical movement of the cargo	during scooping up the load with a bucket	during the vertical movement of the cargo	
1	4	18.3	33.8	0.3	0.5	9.8
2	3	19.8	39.1	0.5	1.1	7.3
3	2	42.1	94.8	1.1	2.4	4.9

Comparative analysis of the results obtained for buckets (based on table 1):

1. The mass of a bucket with a wall thickness of 2 mm is 2 times less than the mass of a bucket with a wall thickness of 4 mm.
2. Maximum stresses in a bucket with a wall thickness of 4 mm, during scooping up a load, are 2.3 times less than the maximum stresses in a ladle with a wall thickness of 2 mm.
3. The maximum deflection occurs in a bucket with a wall thickness of 2 mm and is equal to 2.4 mm, which is 4.8 times higher than the maximum deflection in a bucket with a wall thickness of 4 mm.
4. Stresses and deflections of the bucket when moving the load in the vertical direction are greater than during scooping up the load with a bucket.

4. Conclusions

The below conclusions could be done, based on the results of the represented work:

1. Design diagrams were formed and calculation of two working cases of a belt elevator bucket was carried out. The obtained results stress-strain states do not exceed the maximum values.

2. A comparative analysis of the results of buckets with different wall thicknesses was done. For oats transportation with a belt bucket elevator, buckets with a smaller wall thickness than the minimum standard thickness of 4 mm are acceptable. The minimum permissible wall thickness for these conditions could be taken as a thickness of 2 mm. It is impractical to take less, as since the rapid abrasion of the bottom and walls of the bucket with a wall thickness of less than 2 mm on the transported load is possible during the operating process.

References

[1] Antonov A Yu, Vorotnikov S A, Vukolov A V, Saschenko D V and Shashurin G V 2017 Mathematical model of 3-P wheel-legged mobile robotic platform *J. Int. Review of Mechanical Engineering* **5(11)** pp 311-9

[2] Kulinowski P 2014 Simulation method of designing and selecting tensioning systems for mining belt conveyors *J. Archives of Mining Sciences* **59** pp 123-38

[3] Kumar D and Mandloi R K 2013 Analysis & prospects of modification in belt conveyors *Int. J. of Engineering Research and Applications (IJERA)* **3** pp 581-7

[4] Lagerev A V and Dunaev V P 2009 Conveyors with suspension carrying belt - new type of continuous transport machines *H. An Engineering J.* **10** pp 9-14

[5] Safronov E V and Nosko A L 2019 Method to determine allowable speed for a unit load in a pallet flow Rack *J. Acta Mechanica et Automatica* **13** DOI 10.2478/ama-2019-0011

[6] Nosko A L, Safronov E V and Kirillov D V 2019 Selection of materials of friction lining centrifugal brake rollers *J. of Friction and Wear* **40** DOI 10.3103/S1068366619030085

[7] Spivakovsky A O and Dyachkov V K 1983 *Conveying machines 3rd ed* (Moscow: Mashinostroenie) p 487

[8] Zenkov R L, Ivanov I I and Kolobov L N 1987 *Continuous transport machines 2nd ed* (Moscow: Mashinostroenie) p 432

[9] Romakin N E 2008 *Continuous transport machines* (Moscow: Academy) p 432

[10] Sakharov M and Karpenko A 2018 A new way of decomposing search domain in a global optimization problem *J. Advances in Intelligent System and Computing* **679** DOI 10.1007/978-3-319-68321-8_41

[11] Agasiev T and Karpenko A 2017 The program system for automated parameter tuning of optimization algorithms *J. Procedia Computer Science* **103** DOI10.1016/j.procs.2017.01.120

[12] Karpenko A and Sakharov M 2019 New adaptive multi-memetic global optimization agorithm *J. Herald of the Bauman Moscow State Technical University Series Experimental sciences* **2** DOI 10.18698/1812-3368-2019-2-17-31

[13] Boslovyak P V and Emelyanova G A 2018 Optimization mathematical modeling of the weight of metal structure of suspended belt conveyor linear section *J. IFAC-PapersOnLine* **51** DOI 10.1016/j.ifacol.2018.11.223

[14] Boslovyak P V and Lagerev A V 2019 Optimization of the conveyor transport cost *J. IFAC-PapersOnLine* **52** DOI 10.1016/j.ifacol.2019.12.569

[15] Boslovyak P V and Tolkachev E N 2019 Metal construction optimization of drive suspension of special belt conveyor *J. IFAC-PapersOnLine* **52** DOI 10.1016/j.ifacol.2019.12.472

[16] Alferov K V and Zenkov R L 1955 *Bunker installations. Design, calculation and operation.* (Vladimir: Printing house No. 1 of the State publishing house of literature on construction and architecture) p 305

[17] Gyachev L V 1986 *Basics of the theory of bunkers and silos* (Barnaul: Editorial and Publishing Department of Altai Polytechnic Institute) p 84

[18] Fedoseev V B, Gordeeva A B and Zatsarinnaya I A 2011 Theoretical calculation of pressure in containers filled with discrete medium *J. Bulletin of the Don State Technical University* vol 11 **2(53)** pp 163-8

[19] Fedoseev V B and Gordeeva A B 2010 Lateral coefficient and pressure in the wall embankment of bulk material *J. Proceedings of universities North Caucasian region Natural sciences* **2** issn 0321-3005

Design and strength estimation of composite multi-cavity pressure vessels

O Yu Sklemina[1,2], N A Tatus'[1] and A N Polilov[1]

[1] Mechanical Engineering Research Institute of the Russian Academy of Sciences, 4 Maly Kharitonyevsky Pereulok, Moscow, 101990, Russia
[2] Moscow Polytechnic University, 38 Bolshaya Semyonovskaya str., Moscow, 107023, Russia

sklemina97@yandex.ru

Abstract. The article discusses a layer-by-layer method for calculating of composite products, and the "thread analogy" is provided for pressure vessels calculating. Stress fields at the multi-cavity cylindrical and spherical pressure vessels are calculated using FEM. Gas injection system is designed.

1. Introduction

Composites are widely used in various industries, for example, in aircraft and automobile construction, in military and space technology, and in civil engineering. In particular, the problem of increasing the efficiency of using composites in multi-cavity pressure vessels is urgent.

Purpose of the study: to carry out the strength calculation of composite gas tanks and substantiate their advantages over steel counterparts.

The following tasks were set:

1. Conduct a design calculation of composite pressure vessels based on the simplest "thread analogy and using the layer-by-layer calculation method for monolayers and for pairs of layers."
2. Calculate using the FEM stress fields in a multi-cavity pressure vessel of cylindrical and spherical shape.
3. Design a system for pumping gas into a multi-cavity gas tank.

2. Thread analogy for calculating pressure vessels

A simpler scheme for calculating the winding gas tanks is to use three assumptions, the so-called 'thread analogy' [3]:

1. The fibers work only on tension condition and carry the entire load, and the matrix is not loaded at all.
2. All fibers are equally stressed.
3. Failure of the structure occurs as a result of reaching in all fibers (simultaneously - according to the second assumption) the ultimate stress σ_0.

Assumption 2 is very "strong", it limits the considered winding systems only to equally stressed ones: there are no extra fibers, everyone works at their limit.

IOP Conf. Series: Materials Science and Engineering 1129 (2021) 012070 doi:10.1088/1757-899X/1129/1/012070

As an example, consider the scheme of reinforcement of the cylindrical part of a gas tank by two families of fibers: 1 – with orientation $\pm\alpha_1$ and with layer thickness h_1 and 2 – with orientation $\pm\alpha_2$, with thickness h_2. The total wall thickness $h=h_1+h_2$.

The average axial σ_z and circumferential σ_θ stresses

$$\sigma_z = \frac{pR}{2h}, \; \sigma_\theta = \frac{pR}{h} \tag{1}$$

in a critical state can be expressed through the limiting stress along the fibers σ_0, summing up the forces created in the two families of layers,

$$\begin{aligned} \frac{A}{2} &= h_1 \cos^2\alpha_1 + h_2 \cos^2\alpha_2; \\ A &= h_1 \sin^2\alpha_1 + h_2 \sin^2\alpha_2, \end{aligned} \tag{2}$$

Two equations (2) contain four design parameters, two of which can be found after arbitrary specifying the other two. The question arises, is it really possible to set two winding angles arbitrarily, i.e. can any angle values be specified? Let us illustrate this with a few examples. Let's pretend that $h_2 = 0$. Two parameters remain, which are found from (2): $h = h_1 = \frac{3A}{2}$; $\alpha_1 = \alpha^* = 54^0 44'$; $(\mathrm{tg}^2\alpha^* = 2)$.

Choosing angles α_1, α_2 equal to (±30/90), (±45/90), (0/±60), we will make sure from the table above that the total wall thickness remains the same 3*A* /2 (table 1).

Table 1. Reinforcement angles.

h_0	h_{90}	α	h_α	h
0	0	54°	$\frac{3}{2}A$	$\frac{3}{2}A$
0	$\frac{5}{6}A$	30°	$\frac{2}{3}A$	$\frac{3}{2}A$
0	$\frac{1}{2}A$	45°	A	$\frac{3}{2}A$
$\frac{1}{2}A$	A	0°	0	$\frac{3}{2}A$
0	$-\frac{1}{2}A$	60°	$2A$	$\frac{3}{2}A$
$\frac{1}{6}A$	0	60°	$\frac{4}{3}A$	$\frac{3}{2}A$
$\frac{5}{2}A$	0	30°	$4A$	$\frac{3}{2}A$

The same will remain for the algebraic sum of the thicknesses and for a combination of angles (±60/90), (0/±45) or (0/±30), only one of the thicknesses will be negative, which means that it is impossible to provide equilibria in two stresses if two families of fibers have an orientation angle both

IOP Conf. Series: Materials Science and Engineering 1129 (2021) 012070 doi:10.1088/1757-899X/1129/1/012070

smaller or both larger than α^*. More precisely, it is not the thickness that turns out to be negative, but the stresses in one of the families must be compressive in order to ensure uniform stress - equal absolute values of stresses [2].

For determination average axial and circumferential stresses in cylindrical part of the cylinder were applied formulas (1). For example, with p = 5 MPa, average radius R = 50 mm and wall thickness h = 1.6 mm.

$$\sigma_\theta = 5 \cdot \frac{0.05}{0.0016} \approx 156 \text{ MPa.}$$

Accordingly,

$$\sigma_z = 5 \cdot \frac{0.05}{2 \cdot 0.0016} \approx 78 \text{ MPa.}$$

3. Layer-by-layer method for calculating composite products

Layer-by-layer method - one of the main methods of computer calculation stress states and limit loads for layered composite elements (laminates): plates and shells manufactured from unidirectional prepregs. The problem is set as follows: a design or a part consisting of a set unidirectional monolayers (prepregs) [3] (prepregs) is given.

As a first approximation, the previously popular "thread analogy", according to which only fibers bear the entire load, is considered. The requirement for the fibers to be evenly stressed makes it possible to design rational reinforcement structures and to estimate the possible weight of an "ideal" (optimal) balloon in which all families of fibers are equally loaded.

Calculation of a balloon reinforced with symmetrical pairs of layers.

The following technical material constants are specified:

$E_1 = 45$ GPa, $E_2 = E_3 = 10$ GPa,

$\mu_{12} = \mu_{13} = 0.3,\ \mu_{21} = 0.067,$

$G_{12} = G_{13} = 5$ GPa,

$G_{12} \approx 3.8$ GPa.

1. For the first reinforcement structure: $\alpha_1 = \pm 45°;\ \alpha_2 = 90°$.

$p = 100$ atm $= 10$ MPa,

$R = 150$ mm, $h = 3$ mm.

$\sigma_z = \dfrac{pR}{2h} = 250$ MPa, $\sigma_\theta = \dfrac{pR}{h} = 500$ MPa.

Number layers with angles orientations $\pm 45° - 5$;$90° - 10$.

2. For comparison was performed calculation stresses for angles of reinforcement $\pm 30° - 5$; $90° - 10$.

Table 2. Results of the layer-by-layer calculation method for pairs of layers.

	ε_z,%	ε_z,%	$\sigma_z(\alpha)$, MPa	$\sigma_\theta(\alpha)$, MPa	$\sigma_z(90°)$, MPa	$\sigma_\theta(90°)$, MPa
$(\pm 45° / 90°)$	1.1	1.4	395	432	194	907
$(\pm 30° / 90°)$	0.8	1.6	357	254	170	959

An example of a layer-by-layer calculation method for monolayers

Initial data:
α is the angle of reinforcement; h is the wall thickness;
$n = 29$ is the number of layers; R is the radius of the pipe.
$p = 2.5$ MPa; $R = 150$ mm; $h = 8$ mm.
$E(0) = 14$ GPa;
$E(90) = 0.9$ GPa;
$G_{12} = 0.7$ GPa;
$E_{12} = 2.5$ GPa.

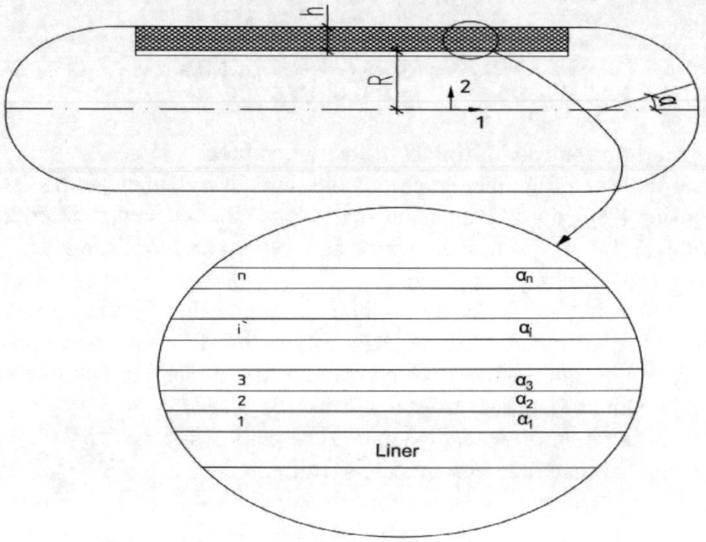

Figure 1. Cylindrical pressure vessel with designations for fiber placement angles and coordinate axes.

1. Calculation of average axial and circumferential stresses.

$$\sigma_z = \frac{pR}{2h} = \frac{2.5 \cdot 150}{16} \approx 23.5 \text{ MPa};$$
$$\sigma_\theta = \frac{pR}{h} = \frac{2.5 \cdot 150}{8} \approx 47 \text{ MPa}.$$

2. Calculation of the average Young's moduli for a package of monolayers.

$$\overline{E_z} = \frac{1}{n}\sum_{i=1}^{n} E(0)\cos^4\alpha + E(90)\sin^4\alpha + (2E_{12} + 4G_{12})s^2c^2 = 5.4 \text{ GPa};$$
$$\overline{E_\theta} = \frac{1}{n}\sum_{i=1}^{n} E(0)\sin^4\alpha + E(90)\cos^4\alpha + (2E_{12} + 4G_{12})s^2c^2 = 6.5 \text{ GPa}.$$

3. Calculation of the average axial and circumferential deformations.

$$\overline{\varepsilon_z} = \frac{\overline{\sigma_z}}{\overline{E_z}};\ \overline{\varepsilon_z} = \frac{23.5}{5381} = 0.00435;$$
$$\overline{\varepsilon_\theta} = \frac{\overline{\sigma_\theta}}{\overline{E_\theta}};\ \overline{\varepsilon_\theta} = \frac{46.9}{6490} = 0.00722.$$

4. Checking the calculation for the average axial and circumferential stresses.

$$\overline{\sigma_z} = \overline{E_z} \cdot \overline{\varepsilon_z} = 5381 \cdot 0.00435 \approx 23.5 \text{ MPa};$$

$$\overline{\sigma_\theta} = \overline{E_\theta} \cdot \overline{\varepsilon_\theta} = 6490 \cdot 0.00722 \approx 47 \text{ MPa}.$$

5. Application of strength criteria for composite pipes.

Applying one or another strength criterion, the load is determined at the first destruction of one of the layers. Sometimes this load is taken as critical for the entire package, but such a calculation for the first failure provides too much safety margin. A more accurate result is given by taking into account the sequential accumulation of destruction.

For this, the problem of the destruction of the layers is consistently solved. If the fibers break, the layer (element) is turned off from work, if the matrix is destroyed, then the shear and Young moduli across the fibers are zeroed in this element, and then the calculation is carried out again up to a complete loss of bearing capacity.

The calculated axial and circumferential stresses are shown as a point in figure 2. In the same figure 2, the limiting surface (ellipse, rectangle) is constructed for a given pair of layers. If the point lies inside the limiting surface, then there is no destruction, and it is possible to determine how many times the pressure must be increased so that this point, moving along the beam with the same stress ratio, reaches the limiting surface. This number of "times" will be the safety factor, which is different for different pairs of layers with different laying angles. The layer in which the safety factor is lower will be the first to collapse. And in the "optimal" structure, one must strive for the safety factor for all pairs of layers to be the same, as in the thread analogy.

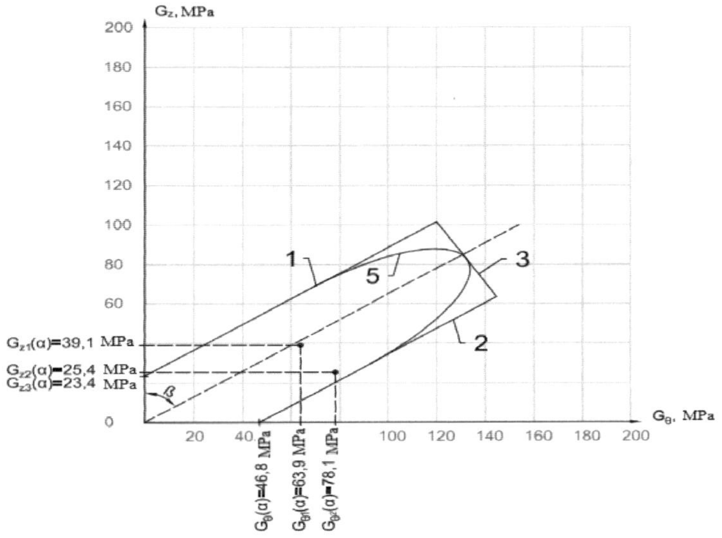

Figure 2. Limiting surfaces in stress space for a cylinder.

4. Design model

In the software complex ANSYS WB, a computational model of the cylindrical and spherical parts of the balloon was created. In figure 3 and figure 4 shows the calculation of axial and circumferential stresses in spherical and cylindrical shells [1].

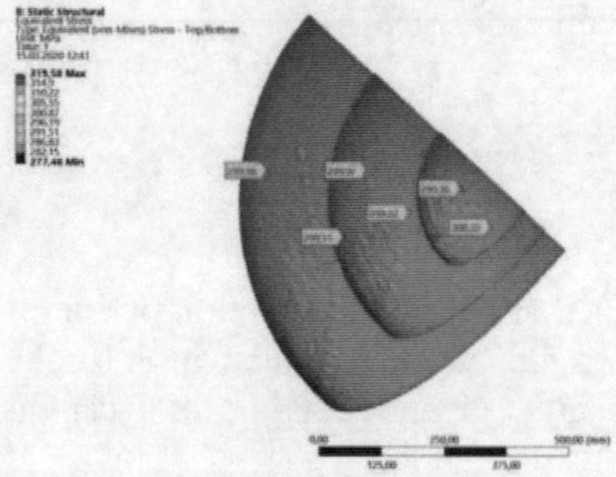

Figure 3. Results of calculating stresses in a spherical pressure vessel.

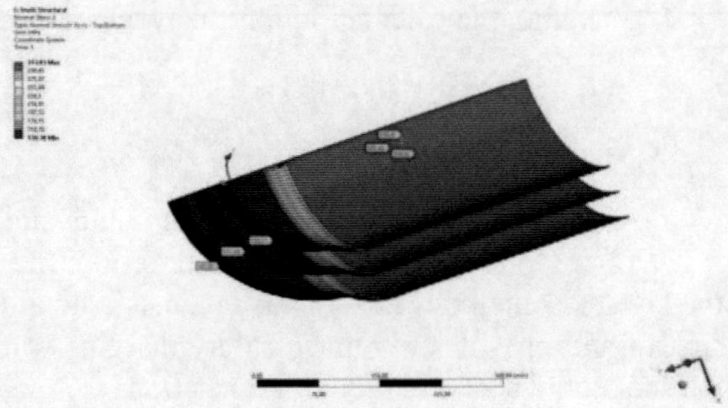

Figure 4. Results of calculating stresses in a cylindrical pressure vessel.

The performed analytical calculation of stresses in a multi-cavity gas tank was verified by the finite element method. In each shell, a pressure was set, which was calculated analytically. The circumferential stresses in each spherical shell turned out to be approximately equal to 300 MPa, which corresponded to the estimates by formulas (1).

5. FEM design cases

When calculating the cylindrical gas tank, internal pressure was applied and stresses, deformations, and safety factor were calculated. The pressure varied from 2.7 MPa to 9.9 MPa, since the working pressure of the cylinder is 2.7 MPa, and according to the current standards, the safety factor should be about three.

In figure 5 shows displacements in a cylindrical container, and figure 6 - safety factor.

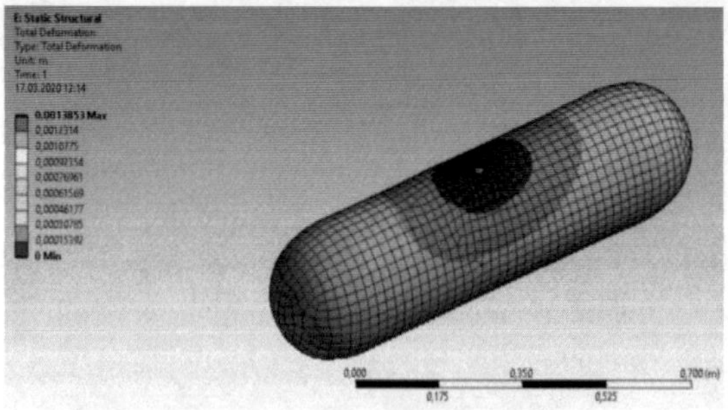

Figure 5. Deformations.

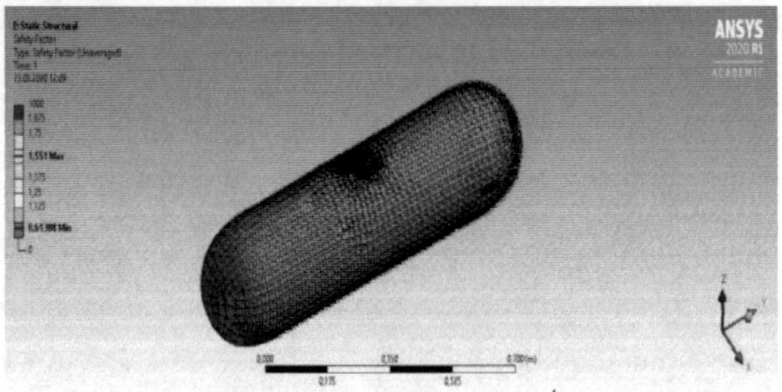

Figure 6. Safety factor.

In figure 7 and 8 show the results of calculating normal stresses and equivalent stresses according to von Mises.

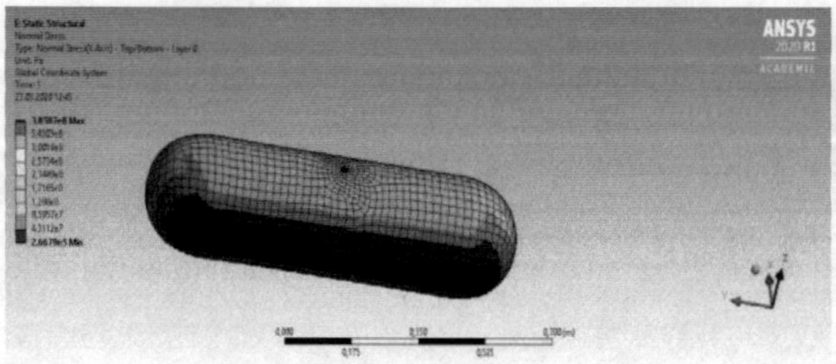

Figure 7. Normal stress.

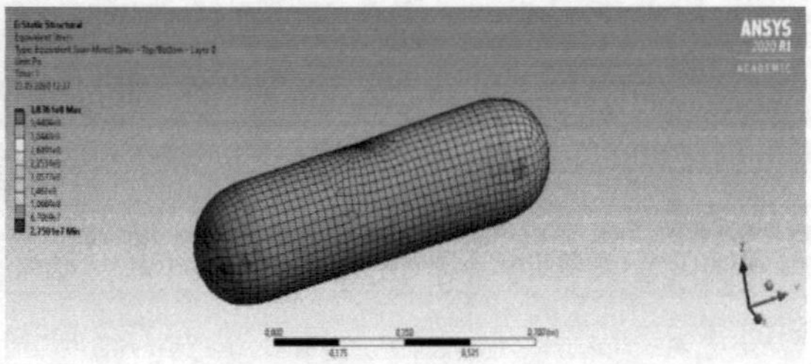

Figure 8. Equivalent von Mises stresses.

6. Designing a gas injection unit

A metal-composite multi-cavity gas tank (MCB) includes spherical shells mounted with a gap one into the other and forming cavities for gas placement. The shells are reinforced with pipelines with the possibility of gas flow. The cylinder is equipped with a filling device - a gas inlet/outlet connection that provides the required, different pressures in the cavities.

As a result of using a multi-cavity gas tank, a larger amount of gas can be pumped in the same external dimensions.

The advantages of composite gas tanks:

- weight efficiency;
- safety (MCB is destroyed without fragments);
- reliability and durability (not afraid of scratches and shocks during operation).

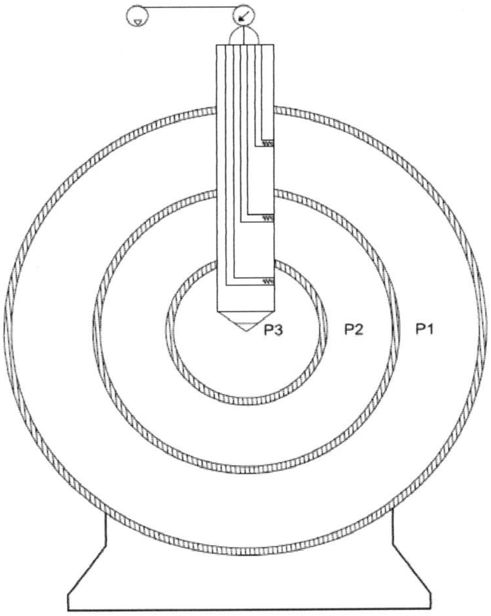

Figure 9. 2D model of MCB.

Conclusions

1. A variant of the design of the gas injection unit for a metal-composite multi-cavity spherical pressure vessel is proposed.
2. FEM-calculation of stress fields in the cylindrical and bottom parts of the balloon, and in the zone of the pole hole.
3. The applied layer-by-layer method calculation of composite products, taking into account the successive destruction of monolayers, makes it possible to estimate the critical pressure in the cylinder and the safety factor.

References

[1] Sklemina O Yu, Tatus N A and Polilov A N 2020 Analytical and finite element analysis of composite multi-cavity compressed gas cylinders *Materials of the Jubilee LXX Open Int. Student Scientific Conf. (SNK-2020) of the Moscow Poly* pp 700-4

[2] Polilov A N, Tatus N A and Sklemina O Yu 2019 Three etude problems on composite fuel gas tank *IOP Conf. Series: Materials Science and Engineering*

[3] Polilov A N 2016 *Experimental Mechanics of Composites 2nd ed* (Moscow: Publishing house of MSTU im. N.E. Bauman) p 376

IOP Conf. Series: Materials Science and Engineering 1129 (2021) 012071 doi:10.1088/1757-899X/1129/1/012071

Magnetic memory. Mechanical tests and proposed mechanisms

O Yu Sklemina[1,2], A A Skvortsov[2], D E Pshonkin[2] and L P Tsilenko[2]

[1] Mechanical Engineering Research Institute of the Russian Academy of Sciences (IMASH RAN), 4 Maly Kharitonyevsky Pereulok, Moscow, 101990, Russia
[2] Moscow Polytechnic University, 38 Bolshaya Semyonovskaya str., Moscow, 107023, Russia

sklemina97@yandex.ru

Abstract. The article considers the issue of an attempt to experimentally study the effect of constant magnetic fields on the processes of plastic deformation of aluminum alloys. An analysis of the influence of a magnetic field on the creep of an Al alloy with iron inclusions is presented. Estimates of the supposed mechanisms of the influence of the magnetic field on the creep processes are given.

1. Introduction

The effect of magnetic fields on the physical and mechanical properties of nonmagnetic materials has been studied as a phenomenon for several decades [1]. Investigators mark the effect of magnetic memory as the most interesting and relevant of its kind. Magnetic memory is a change in the physical and mechanical properties of a crystal after exposure to a magnetic field due to magnetic action on sensitive centers in the crystal structure of the material. Such changes in mechanical properties can persist for a certain time.

At the atomic level, there is no definitive understanding of the principle of operation of the magnetic memory effect, so the theoretical question remains open. An external magnetic field affects the processes of spin-dependent reactions in the dislocation-stopper complex. However, there is a whole chain of physical processes that is reliably recorded at the macroscopic level, such as: magnetically stimulated dynamics of dislocations and its effect on plastic deformation of the material. Various additives, such as ferromagnetic inclusions, can significantly affect the picture of plastic deformation of samples after exposure to a magnetic field. These data make it possible to recreate a qualitative picture of the effect of magnetic stimulation on the processes of plastic deformation of some non-magnetic alloys.

The main task of this study is to attempt to experimentally study the effect of magnetic memory of some aluminum alloys containing Fe-inclusions, as well as the processes of stimulating plastic deformation of aluminum alloys using an external magnetic field.

2. Materials and equipment

Creep tests were carried out on a German WP-600 Creep Testing Machine. Figure 1 shows the installation and a dial gauge that records the absolute elongation.

Figure 1. Lever Type WP-600 Creep Testing Machine.

The test piece was placed in special clamps, then the load was set. In the course of measurements, simultaneously with the application of the load, the elongation of the sample was recorded using a micrometer at each stage of loading. The test setup provided a constant load in the measurement process, as well as smooth loading and unloading.

The experiments were performed using aluminum-magnesium alloy 6063 with microscopic iron inclusions with a volume concentration of ~ 2.5 %. Previously, some of the samples were exposed in the field of a permanent magnet with an induction $B \leq 1$ T for 30 min. After that, the samples that have already been processed and the witness samples were tested for creep under uniaxial loading (figure 2).

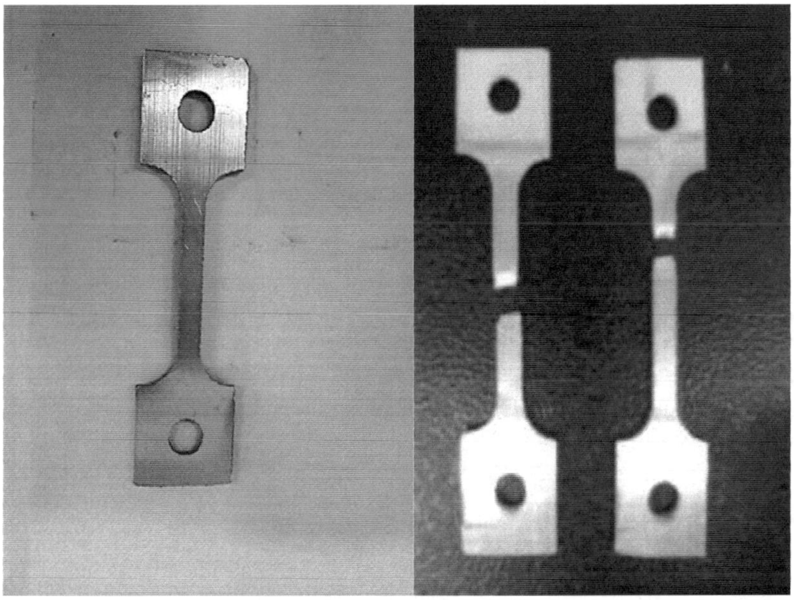

Figure 2. Samples before and after the creep test.

Test data for determining the magnetostriction of ferromagnetic inclusions, which provides local mechanical stresses and the generation of fresh dislocations in the vicinity of the inclusions, are the main reason for magnetoplasticity [2].

3. Results and discussion

The test results showed an increase in plastic deformation of the exposed samples by 25% compared to the reference samples.

A preliminary analysis of the chemical composition of the samples showed the presence of iron in the paramagnetic matrix of aluminum, the subsequent analysis of scanning electron and transmission microscopy confirmed the presence of solid-state precipitates of the second phase in the alloy (figure 3).

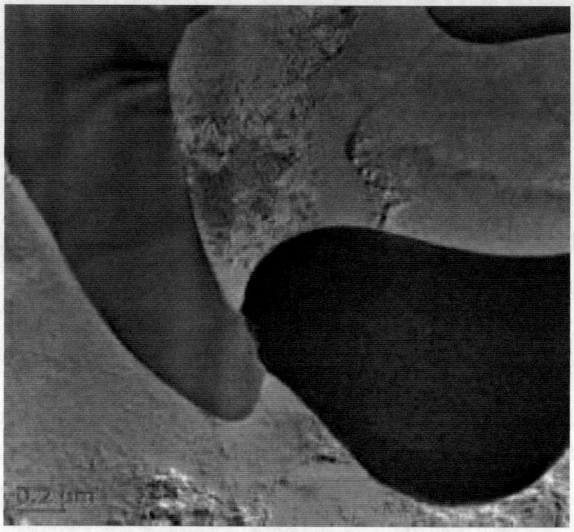

Figure 3. Fe-containing inclusion in the aluminum matrix of the sample.

The results of experiments for tensile loads in the range of 150 ... 180 N are shown in figure 4.

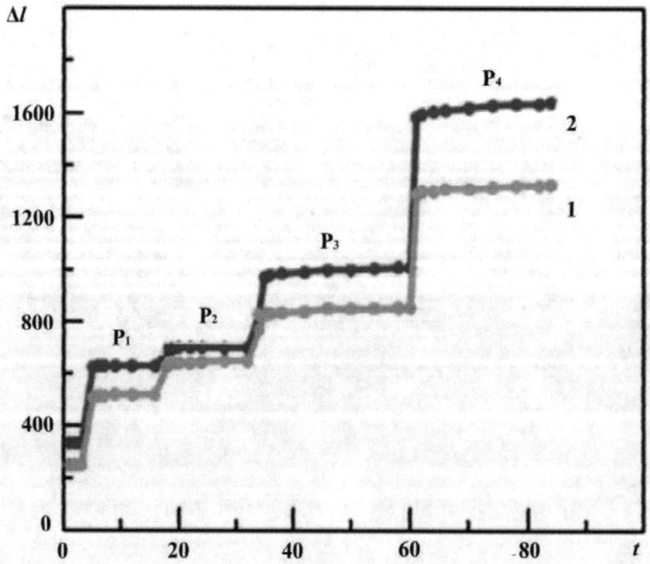

Figure 4. Creep curves for Al, which were recorded at step voltage in the range (150 ... 180 N) at room temperature. 1 - witness sample; 2 - exposure of the sample in a magnetic field.

Evaluation of the obtained experimental deformation curves made it possible to determine the assumed mechanisms of creep. The calculated values of the activation energy allow us to assume that the unsteady stage of creep is controlled by the processes of crossing or cutting of dislocations in orthogonal slip planes. From this assumption it follows that the process of intersection of dislocations is a thermally activated process with an activation energy at zero stress ΔE_i.

The activation energy is calculated by the formula:

$$\Delta E_n = \Delta E_i - \gamma(\sigma - \sigma_{VN}), \tag{1}$$

where γ is the activation volume;

σ_{VN} – average local internal stress caused by other dislocations in parallel slip planes.

The creep rate is determined as follows:

$$\dot{\varepsilon} = N \cdot A \cdot b \cdot v \cdot \exp\left\{\frac{-\Delta E_i + \gamma(\sigma - \sigma_{VN})}{RT}\right\} \tag{2}$$

where N is the total number of dislocation sources; A is the average area covered by the expanding loop; b is the Burgers vector; v is the natural frequency of vibrations of a single dislocation loop.

In the course of the experiments, the values of the activation energy of the creep process of the samples were obtained before and after exposure to a constant magnetic field (table 1).

Table 1. Calculated values of the magnitude ΔE_i before and after exposure to a magnetic field under normal conditions.

P, N	150	160	170	180
E_a, eV, $B=0$	0.124	0.124	0.124	0.124
E_a, eV, $B \neq 0$	0.125	0.125	0.125	0.125

From the data obtained, it can be concluded that the magnetic field has an insignificant effect on the activation energy of the creep process [3]. The magnetic field noticeably affects the dynamics of the activation volume of the γ process, which is associated with the density of dislocations in the material matrix near *Fe*-containing inclusions.

The magnetoplastic effect is the localization of mechanical stresses in the vicinity of inclusions in a magnetic field, which is sufficient to generate fresh dislocations and soften the material. The experimentally observed magnetoplastic effect can be explained by the fact that this alloy Fe_xAl_{1-x} has high magnetostriction constants at x = 86-87%.

The formula (3) calculates mechanical stresses in a magnetic field.

$$\sigma_m = \lambda_m \cdot B, \tag{3}$$

where $\lambda_m = 8 \cdot 10^8 N \cdot (T \cdot \mathrm{m}^2)^{-1}$ is the magnetostriction coefficient of the alloy Fe_xAl_{1-x}.

Normal stress $\sigma_m = 560$ MPa in a magnetic field $B = 0.7$ T. The results obtained exceed the yield point stress of $\sigma_Y = 120$ MPa, on this basis, an increase in creep can be observed. If r_0 is the radius of inversely proportional dependence σ_m on the distance to the center of the inclusion, then the radius of plastic deformation of the zone has the form $r = r_0 \cdot \frac{\sigma_m}{\sigma_Y}$: 6 μm. This is comparable to an average inter-inclusion spacing of 12 μm.

Based on the above, it can be concluded that MF can lead to magnetization of *Fe*-magnetic inclusions. Each particle forms local σ_m ones using the magnetostriction effect and increases the plastically deformed zone around the inclusion. These changes are saved even after the MP is turned off. These estimates are the most probable, but they do not exactly prove that the magnetostrictive mechanism is the main one.

4. Material destruction work

In the first approximation, the process of plastic deformation of a material can be considered from the point of view of the law of conservation of energy. If we neglect the heat losses due to the friction of the crystal lattice [4], then all the deformation energy expended will be spent on the work of destruction.

$$A_r = \int_0^{\varepsilon_{cr}} \sigma(\varepsilon) \cdot d\varepsilon, \quad (4)$$

where $\sigma(\varepsilon)$ is the acting stress under uniaxial tension, ε_{cr} is the maximum deformation of the material.

Since the acting stress depends on the plastic deformation, we assume that

$$\sigma(\varepsilon) = \sigma_p \cdot (1+\varepsilon), \quad (5)$$

where σ_p is the ultimate tensile stress. Then (4) after transformations will look as follows

$$A_p = \frac{\varepsilon_{cr}^2}{2} + \sigma_p \cdot \varepsilon_{cr} : \sigma_p \cdot \varepsilon_{cr} \quad (6)$$

On the other hand, this same fracture work can be shown graphically in a tensile diagram, in which the total work performed will correspond to the area under the deformation curve (figure 5).

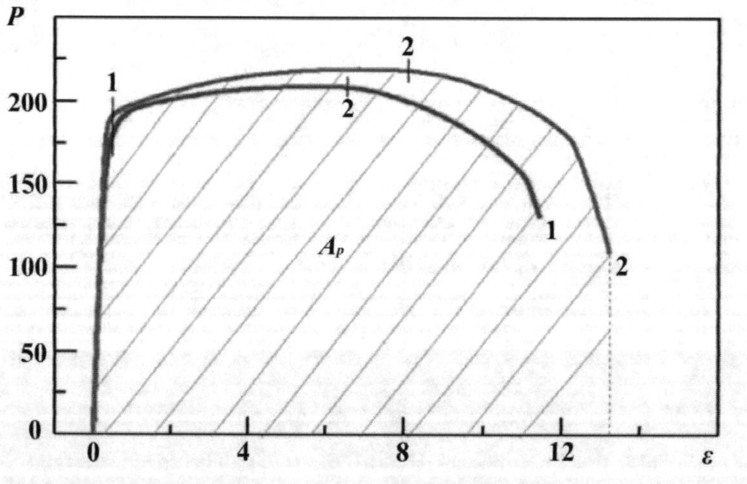

Figure 5. Diagram of stretching of samples before and after exposure to a magnetic field. 1 - witness sample; 2- sample after exposure to a constant magnetic field with an induction of 0.7 T. Zone 1 is the beginning of macroplastic deformation. Zone 2 - the beginning of the dynamic return stage.

As can be seen from the graph, changes in the nature of deformation are observed from the beginning of the stage of strain hardening. Estimating A_p according to equation (6), based on the experimental tensile curves, we obtain an increase in the work of plastic deformation of the material by 20% compared to the witness sample.

Such changes are provoked by the magnetostrictive effects of inclusions in the material matrix, which significantly affect the dislocation dynamics and, as a result, plastic deformation in general.

5. Conclusions

1. The analysis of the influence of MF on the creep of an *Al* alloy with iron inclusions has been carried out.
2. It was revealed that the preliminary exposure to the MF ($B \leq 0.7$ T) promotes an increase in the deformation of the alloy (up to 30%), the value of the activation volume of the alloy γ. The changes caused by the MF are confirmed by the dynamics of the initial creep rates at the unsteady stage of creep.
3. Estimates are given of the supposed mechanisms of the influence of the magnetic field on the creep processes, which are associated with the dynamics of dislocations and the magnetostriction of *Fe*-inclusions.
4. Numerical calculations of $\sigma_m = 560$ MPa are obtained, which contribute to an increase in the plastically deformed zone around the inclusion.

References

[1] Alshits V I, Darinskaya E V and Perekalina T M 1987 *On the motion of dislocations in NaCl crystals under the influence of a constant magnetic field* **2** pp 467-71

[2] Skvortsov A A 2019 Softening of the Al-Mg-Si-Fe alloy under magnetostriction of FeAl microinclusions *J. of Applied Physics* **125** p 10

[3] Skvortsov A A and Karizin A V 2012 *Magnetoplasticity and diffusion in silicon single crystals* **1** pp 96-100

[4] Alshits V I, Darinskaya E V, Koldaeva M V and Petrzhik E A 2013 *Anisotropy of resonant magnetoplasticity of NaCl crystals in the earth's magnetic field* pp 318-25

The influence of 3D printing layer height on frictional properties of ABS plastic

V E Slavkina, A O Shitov, Yu A Lopatina

Federal State Budgetary Scientific Institution "Federal Scientific Agricultural Engineering Center VIM", 1st Institutsky proezd 5, Moscow, Russia

slavkina.vim@gmail.com

Abstract. This work is devoted to the study of tribotechanical properties of 3D printed samples from ABS plastic. The results of determining the coefficient of friction and mass wear of the samples obtained by 3D-printing with different printing heights during the friction testing on machine II 5018 are presented. It is shown that in order to reduce the coefficient of friction and mass wear of the ABS plastic specimen, it is recommended to produce specimens with a higher print height. 3D printed parts with 0.3 mm layer height have shown the best results because of greater layer adhesion.

1. Introduction

A polymer is one of the best structural material in mechanical engineering. Polymers application expediency is primarily determined by the possibility of making the end product cheaper. This change can improve important technical and economic machine parameters: lower total weight, increase product reliability, etc. [1-4]. Poorly studied physical and mechanical properties and lack of information about behavior during friction processes are the main problems preventing polymer materials from becoming widely used in modern production and additive technology. Any finished sample has a ribbed surface due to the nature of 3D printing technology: the model is formed by sequentially applied layers of plastic. Ribs intensity on the surface is determined by the layer height which is set before printing (figure 1). The layer surface state determines the operational and tribotechnical properties and wear intensity of the workpiece [5-8]. Thus, the main goal of this article is to find the influence of 3D printing layer height on the tribotechnical characteristics of the ABS plastic parts. ABS is one of the most common materials used for 3D printing due to its high hardness, tensile strength, high impact toughness, low price and it is easy to print with. Products based on this material keep dimensions stable and have high chemical resistance [9-10].

(a) (b)

Figure 1. The surface of a 3D-printed part with different layer heights: (a) – 0.1 mm, (b) – 0.2 mm.

2. Materials and methods

The ABS plastic pads samples (30 x 25 x 5 mm in size) for this study (figure 2) were printed with Picaso Designer XPro. The printing parameters for these samples are in table 1.

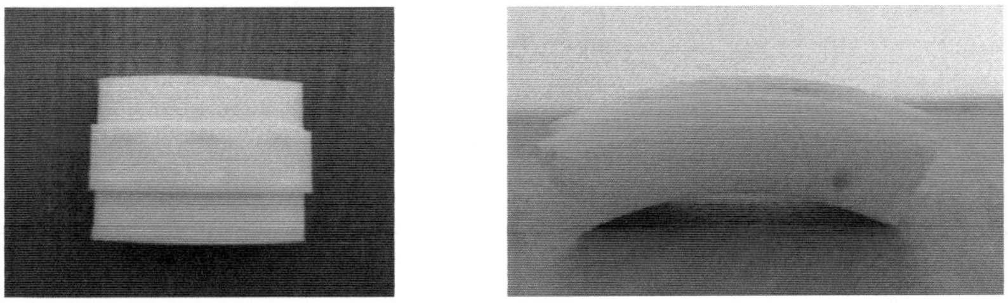

(a) (b)

Figure 2. ABS plastic samples made on a 3D printer for friction tests.

Table 1. 3D-printing parameters.

Plastics	3D printing layer height, mm	$T_{printing}$, °C	Fill percentage, %	k_{flow}
ABS	0.1 0.2 0.3	230	100	0.94

To take into account the thermal expansion of the material and its elastic properties the flow coefficient (k_{flow}) is locked to 0.94. This coefficient links theoretical polymer volume extrusion with the actual one. For tribotechnical sample properties research the friction machine II 5018 (figure 3) was used. Experiments were carried out with a disk-pad friction scheme (figure 4). A steel roller 50 mm in diameter and 12 mm in width acted as an opposite friction body.

Figure 3. Friction machine II 5018: 1 - specimen holder; 2 - steel disc (opposite friction body); 3 – the spring-loaded mechanism for creating pressure; 4 - handle for pressure adjusting.

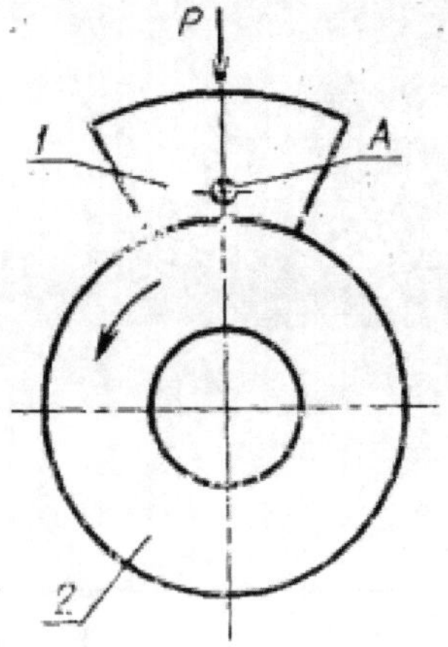

Figure 4. Disk-pad friction scheme for II 5018 machine: 1 - stationary specimen (pad); 2 - steel disk (opposite friction body); P – pressure.

The operation principle of the machine is to conduct scuffing friction of pair of bodies pressed against each other with a given load. During the experiment, the values of friction force moment, friction coefficient, pressing force, and friction distance are recorded and displayed on the screen. To carry out tests, a polymer specimen was fixed (figure 3) in holder 1 over a steel counter body 2. The specimens are loaded with the help of a spring mechanism 3. The load value is tuned by knob 4.

The tests were conducted in the load range of 80 - 117 N and at a disk rotation speed of 150 rpm, the friction distance was set to 2000 m. The load values measured during the test were converted into contact pressure values using equation 1. The measured friction torque values were converted into friction coefficient values using equation 2.

$$P = \frac{N}{S_k}, \tag{1}$$

In equation 1: P - pressure, MPa; N - load, N; S_k - contact surface area of the pad (40 mm^2).

$$f_{fr} = \frac{F_{fr}}{N} = \frac{M}{rN}, \tag{2}$$

In equation 2: f_{fr} - friction coefficient, F_{fr} - friction force, N; M - friction torque, Nm; r - radius of the roller, mm; N – load, N.

3. Results and discussion

The obtained data from friction tests for ABS specimens with different print layer heights are presented in table 2.

Table 2. Result data gained with friction machine II 5018.

Plastics	Sample	3D printing layer height, mm	Average force, N	Average torque, N·m	Contact pressure	Friction coefficient	Average friction coefficient
	1		104.7	2.50	2.62	0.48	
	2	0.1	81.8	2.34	2.05	0.57	0.54
	3		79.5	2.25	1.99	0.57	
	1		100.3	2.38	2.51	0.47	
ABS	2	0.2	97.3	1.87	2.43	0.38	0.45
	3		104.7	2.50	2.62	0.48	
	1		103.9	2.26	2.60	0.44	
	2	0.3	98.9	2.15	2.47	0.43	0.42
	3		105.5	2.11	2.64	0.40	

Samples with a print height of 0.3 mm have the lowest friction coefficient. The difference between the maximum (for the sample with a print height of 0.1 mm) and the minimum (for the sample with a print height of 0.3 mm) obtained values of the friction coefficient is about 30%. The average values of the friction coefficient for each of the series of samples with different print heights are presented as a histogram in figure 5.

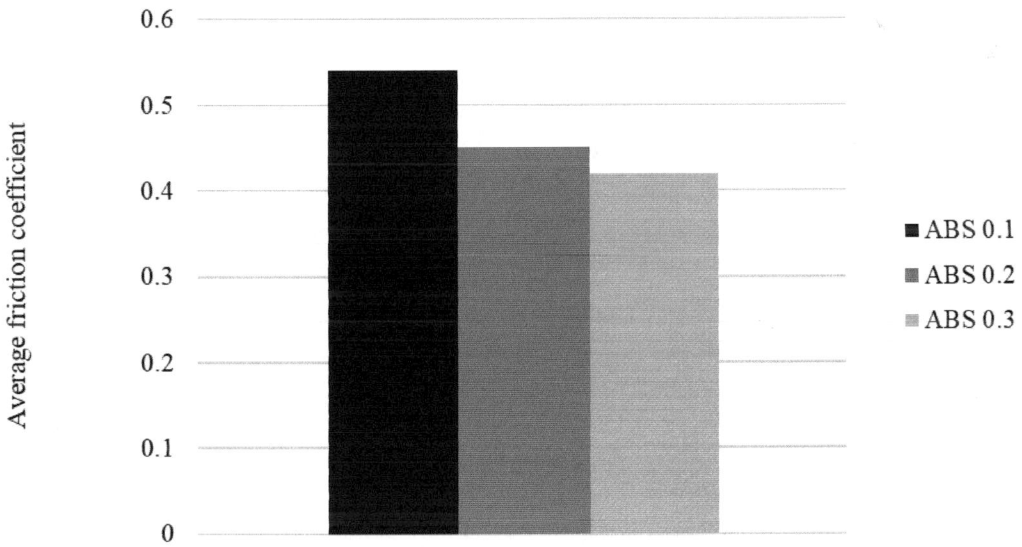

Figure 5. Average values of the friction coefficient of ABS plastic samples with different print heights.

The results of mass wear measurement are presented in table 3, the average mass wear of samples with different print heights are in figure 6.

Table 3. Result data gained with friction machine II 5018.

Sample	Sample mass, g		Mass wear, g	Average mass wear, g
	Before	After		
	5.74	5.26	0.48	
ABS 0.1	5.71	5.41	0.3	0.41
	5.75	5.31	0.44	
	5.63	5.13	0.5	
ABS 0.2	5.64	5.63	0.01	0.31
	5.63	5.21	0.42	
	5.84	5.67	0.17	
ABS 0.3	5.57	5.27	0.3	0.23
	5.55	5.33	0.22	

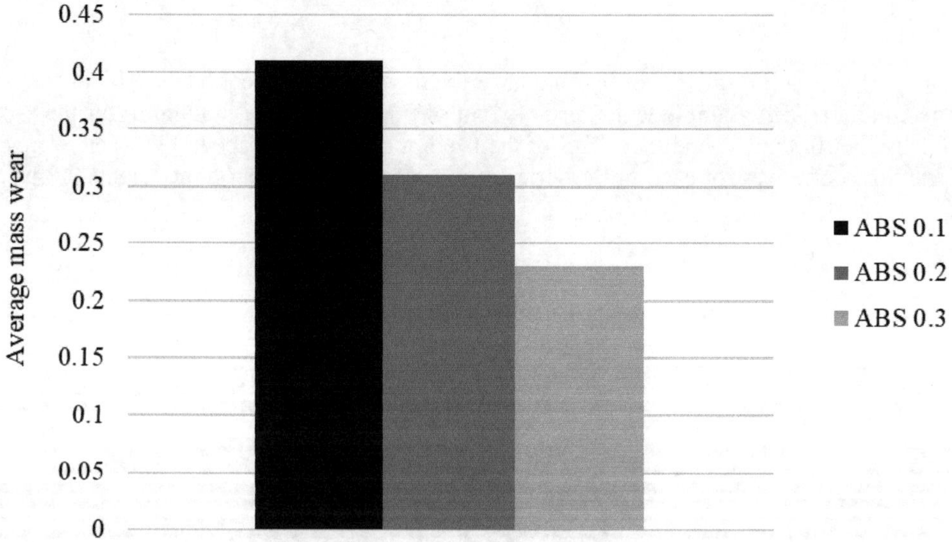

Figure 6. Average values of ABS plastic samples mass wear with different print heights.

The value of mass wear is in inverse dependence on the height of the print layer: the higher the printing layer – the less wear occurs. In all shown cases, the mass loss of the sample is less than 7% of the total mass. The change of wear with layer height is the same as the change of friction coefficient with layer height. Which proves that received influence data is reliable.

4. Conclusions

Experiments have shown that samples with a print height of 0.3 mm have a 30% lower coefficient of friction than samples with a print height of 0.1 mm. Samples with a print height of 0.3 mm have the least mass wear, the mass loss of these samples is about 40% less than 0.1 mm ones. The reason for this pattern probably is better adhesion between layers when printing at 0.3 mm height. This leads to better surface friction properties. The higher the temperature of the melted polymer and the slower it cools down, the better the adhesion between the layers. More polymer is going of the printer nozzle per second with a higher printing layer height, so a larger volume of material can retain heat better and cool down slower. It should also be noted that when the layer height is increased, the printing time decreases drastically. Required time to print samples with a print height of 0.3 mm is 2-3 times less than the required time to print samples with a height of 0.1 mm.

Thus, 3D printing with a 0.3 mm layer height is the best option for the production of friction parts, as it succeeds in manufacturing speed and can deliver better friction properties of the finished workpiece.

References

[1] Tertyshnaya Yu V, Levina N S, Popov A A, Moskovskii M N and Izmailov A Y 2019 Hydrolytic destruction of agrofiber made of natural polymers *FIB. CHEM.* **51** pp 117-20

[2] Sviridov A, Lopatina Yu and Kurganova Yu 2019 3D-printed polyether ether ketone samples mechanical properties estimation *IOP Conf. Series: Materials Science and Engineering* **589** Number of articles 012021 DOI:10.1088/1757-899X/589/1/012021

[3] Lopatina Yu A and Slavkina V E 2020 Development of a high strength polymeric composite material using 3D-printing and vacuum impregnation technology *IOP Conference Series: Materials Science and Engineering* **934** №012019 DOI:10.1088/1757-899X/934/1/012019

[4] Mirzaev M A, Smirnov I G, Dyshekov A I, Slavkina V E, Shereujev M A and Aliev S A 2019 Design of the solar panel frame of the spacecraft by 3D printing with composite materials *IOP Conference Series: Materials Science and Engineering* **747** №012009 DOI:10.1088/1757-899X/747/1/012009

[5] Lavrov N A, Kiemov S N and Kryzhanovskii V K 2019 Tribotechnical properties of composite materials based on epoxy polymers *Polymer Science: Series D* **12** **(2)** pp 182-5

[6] Kobernic N V, Mikheev R S, Kalashnokov I E, Kobeleva L I and Bolotova L K 2017 Tribological properties of babbitt alloy coating modified with carbon nanotubes *Inorganic Materials: Applied Research* **8** **(3)** pp 428-33

[7] Fernandes C M, Martins R C and Seabra J H 2016 Coefficient of friction equation for gears based on a modified hersey parameter *Tribology International* **101** pp 204-17

[8] Dorokhov A S, Khamaev V G and Lepeshkin K 2018 Modernization of grinding machines of grain cleaning machines *MATEC Web of Conference* p 05009

[9] Sharon O 2016 Plating on acrylonitrile-butadiene-styrene (ABS) plastic: a review *J. Mater. Sci.* pp 3657-74

[10] Kun K 2016 Reconstruction and development of a 3D printer using FDM technology *Procedia Engineering* pp 203-11

AUTHOR INDEX